CAMBRIDGE LIBRARY COLLECTION

Books of enduring scholarly value

Physical Sciences

From ancient times, humans have tried to understand the workings of the world around them. The roots of modern physical science go back to the very earliest mechanical devices such as levers and rollers, the mixing of paints and dyes, and the importance of the heavenly bodies in early religious observance and navigation. The physical sciences as we know them today began to emerge as independent academic subjects during the early modern period, in the work of Newton and other 'natural philosophers', and numerous sub-disciplines developed during the centuries that followed. This part of the Cambridge Library Collection is devoted to landmark publications in this area which will be of interest to historians of science concerned with individual scientists, particular discoveries, and advances in scientific method, or with the establishment and development of scientific institutions around the world.

Mathematical and Physical Papers

William Thomson, first Baron Kelvin (1824–1907), is best known for devising the Kelvin scale of absolute temperature and for his work on the first and second laws of thermodynamics, though throughout his 53-year career as a mathematical physicist and engineer at the University of Glasgow he investigated a wide range of scientific questions in areas ranging from geology to transatlantic telegraph cables. The extent of his work is revealed in the six volumes of his *Mathematical and Physical Papers*, published from 1882 until 1911, consisting of articles that appeared in scientific periodicals from 1841 onwards. Volume 5, published in 1911, includes articles from the period 1847–1908. Topics covered include thermodynamic and electrodynamic research, as well as some works on issues of geological physics such as the possible age of the sun's heat.

Cambridge University Press has long been a pioneer in the reissuing of out-of-print titles from its own backlist, producing digital reprints of books that are still sought after by scholars and students but could not be reprinted economically using traditional technology. The Cambridge Library Collection extends this activity to a wider range of books which are still of importance to researchers and professionals, either for the source material they contain, or as landmarks in the history of their academic discipline.

Drawing from the world-renowned collections in the Cambridge University Library, and guided by the advice of experts in each subject area, Cambridge University Press is using state-of-the-art scanning machines in its own Printing House to capture the content of each book selected for inclusion. The files are processed to give a consistently clear, crisp image, and the books finished to the high quality standard for which the Press is recognised around the world. The latest print-on-demand technology ensures that the books will remain available indefinitely, and that orders for single or multiple copies can quickly be supplied.

The Cambridge Library Collection will bring back to life books of enduring scholarly value (including out-of-copyright works originally issued by other publishers) across a wide range of disciplines in the humanities and social sciences and in science and technology.

Mathematical and Physical Papers

VOLUME 5

LORD KELVIN
EDITED BY JOSEPH LARMOR

CAMBRIDGE
UNIVERSITY PRESS

CAMBRIDGE UNIVERSITY PRESS

Cambridge, New York, Melbourne, Madrid, Cape Town,
Singapore, São Paolo, Delhi, Tokyo, Mexico City

Published in the United States of America by Cambridge University Press, New York

www.cambridge.org
Information on this title: www.cambridge.org/9781108029025

© in this compilation Cambridge University Press 2011

This edition first published 1911
This digitally printed version 2011

ISBN 978-1-108-02902-5 Paperback

MATHEMATICAL

AND

PHYSICAL PAPERS

CAMBRIDGE UNIVERSITY PRESS
London: FETTER LANE, E.C.
C. F. CLAY, Manager

Edinburgh: 100, PRINCES STREET
Berlin: A. ASHER AND CO.
Leipzig: F. A. BROCKHAUS
New York: G. P. PUTNAM'S SONS
Bombay and Calcutta: MACMILLAN AND CO., Ltd.

MATHEMATICAL

AND

PHYSICAL PAPERS

VOLUME V

THERMODYNAMICS
COSMICAL AND GEOLOGICAL PHYSICS
MOLECULAR AND CRYSTALLINE THEORY
ELECTRODYNAMICS

BY THE RIGHT HONOURABLE

SIR WILLIAM THOMSON, BARON KELVIN

O.M., P.C., G.C.V.O., LL.D., D.C.L., SC.D., M.D., ...
PAST PRES. R.S., FOR. ASSOC. INSTITUTE OF FRANCE,
GRAND OFFICER OF THE LEGION OF HONOUR, KT PRUSSIAN ORDER *POUR LE MÉRITE*,
CHANCELLOR OF THE UNIVERSITY OF GLASGOW
FELLOW OF ST PETER'S COLLEGE, CAMBRIDGE

ARRANGED AND REVISED WITH BRIEF ANNOTATIONS BY

SIR JOSEPH LARMOR, D.Sc., LL.D., Sec. R.S.

LUCASIAN PROFESSOR OF MATHEMATICS IN THE UNIVERSITY OF CAMBRIDGE
AND FELLOW OF ST JOHN'S COLLEGE
REPRESENTATIVE OF THE UNIVERSITY IN PARLIAMENT

CAMBRIDGE:

AT THE UNIVERSITY PRESS

1911

𝕮ambridge:

PRINTED BY JOHN CLAY, M.A.

AT THE UNIVERSITY PRESS

PREFACE

IT had been intended that the present volume should complete the collected edition of Lord Kelvin's Scientific Papers. It was soon found, however, that extensive omissions would be necessary in order to compress the remaining papers into this space; and eventually it was decided to run on into a sixth volume, on the ground that material of inferior direct importance would be valuable to the historical student in obtaining a view of the progress of knowledge. Thus it is mainly on this ground that the sub-section headed "Electrical Contributions to Engineering Societies," pp. 541–595, consisting largely of excerpts from the reported discussions at the "Society of Telegraph Engineers," now the "Institution of Electrical Engineers," has been included in the reprint. There remain over for the sixth volume the later papers on Electrionic Theory and Radio-activity, including cognate papers of the same period on Voltaic Phenomena, and also some supplementary matter.

The present volume begins with a section on Thermodynamics in the wider sense of that term. It has been thought desirable to reprint some material, in part controversial (cf. pp. 1–10, and 38–46), which Lord Kelvin had himself omitted in its own earlier connexion. Looking back, we can now recognize that there is ample room for all the illustrious names which are associated with the historical development of this subject, and that their independent efforts, when placed in full light of comparison, give a richer texture and a more human interest to the great fundamental advance in our outlook upon nature with which they were concerned.

The next sub-group of papers is concerned with the principle
of Dissipation of Energy and its relation to Dynamics. The first
of the series, pp. 11–20, was read in Feb. 1874, not long after
Maxwell's exposition of the limitation of that principle to
phenomena involving statistical aggregates (*Theory of Heat*, 1872,
near the end), which he enforced by the invention of his ideal
"demons" such as would be competent to reverse the downward
course of nature without requiring any compensation elsewhere.
In this paper the probability of the occurrence of a specified state
of a molecular system appears in science, probably for the first
time, as a subject of calculation; it is illustrated by the working
out of the problem of the relative probabilities of the various
degrees of incomplete mixture that may occur in two gases pervading the same space. Three years later, in 1877, Boltzmann
took the further step (in the *Wiener Berichte*, see *Wissenschaftliche
Abhandlungen*, vol. II. pp. 117, 165) of identifying the continual
increase of the entropy with the continual progress of the system
into more probable states, thus initiating a point of view which
plays a prominent part in recent thermodynamic developments.
Lord Kelvin returned to this subject in 1879 in a popular discourse
on "The Sorting Demon of Maxwell," pp. 21–23.

In the next paper, of date 1898, the conception of the
"Motivity" of a system—the energy "Available" or undissipated
at each constant temperature,—after having been employed
as early as 1855 to establish in a few words the whole physical
theory of systems undergoing transformation at constant temperature, is now at length formally developed as the appropriate basis
of Thermodynamics, in its wider province which is concerned with
dynamical phenomena involving variation of temperature. Meantime the same subject had been developed by Willard Gibbs with
masterly completeness from the side of Clausius' idea of entropy.
A note which analyses the interaction of electric and thermal
phenomena in voltaic contacts forms an application of this
principle.

Some very striking fragments (pp. 47–56) on the physics of
the structure of natural ice, and of liquid films, then follow,

now reprinted from the *Baltimore Lectures* into a more suitable context. In regard to the early advance of thermodynamic ideas, as developed in view of their connexion with special problems such as the behaviour of ice and crystals under stress, interesting historical material will appear in correspondence between Lord Kelvin and his brother, Prof. James Thomson, well-known as the pioneer in that subject, which is now prepared for publication in the collected *Papers on Physics and Engineering* of the latter author.

The remaining part of the section on Thermodynamics (pp. 64–133) is occupied by papers of more special type, the nature of which will be seen from the Table of Contents.

The following section (pp. 134–283) on "Cosmical and Geological Physics" collects together the outstanding material on topics such as the age and rigidity of the Earth, its thermal history, the duration of the Sun's heat, and the formation of stars and nebulae. The section is made complete by the usual references to the papers on the same subjects which have been already published in the earlier volumes.

A main original aim of these cosmical papers, in which Lord Kelvin was followed up by Helmholtz, was to point out the limitations that physical principles must impose on the extreme uniformitarianism which was then in vogue in Geology. But later the complaint had become widely prevalent that in the hands of Lord Kelvin and the physicists the evolution of the terrestrial strata, and of the life of which they contain the evidence, had been hurried up too much. It has been thought well to reprint, in an Appendix, the heads of a discussion on this subject, which was initiated by Prof. Perry.

The following section (pp. 284–358), under the general heading "Molecular and Crystalline Theory," begins with the papers in which Lord Kelvin drew concordant conclusions, from a wealth of very various physical phenomena, as to the degree of coarseness of structure in matter apparently continuous,—in fact as to the size of molecules. The section goes on into a series of papers in which he developed the properties of homogeneous arrangements in space, with a view to the theory of crystals and their physical properties, including their electric response to change of pressure or temperature.

A general section on Electrodynamics aims at collecting what remains over of his electrical papers. It begins with some early notes now of historical interest, and with the papers on methods of measurement of resistance and on early forms of standard batteries. As the stage of electrical engineering succeeded that of submarine telegraphy, the subject changes to the design of dynamos and the conditions for economy in electrical transmission. A group of papers follows on the screening of a region from electric or magnetic disturbances by metallic gratings; and, as already mentioned, the volume ends with contributions to discussions at the Society of Electrical Engineers.

As in the previous volume, the care and experience of Dr George Green have been available in the correction of the proofs; while the admirable work and the methodical arrangements of the Cambridge University Press have much facilitated the handling of the fragmentary material of which this volume largely consists.

J. L.

Cambridge,
Feb. 24, 1911.

CONTENTS

THERMODYNAMICS.

Year	No.		Page
1864	104	On the Protection of Vegetation from Destructive Cold every Night	58
1865	105	On the Dynamical Theory of Heat [Thermal Dissipation of Energy of Vibration of Solids]	59
1866	106	On the Dissipation of Energy	61
1870	107	Dr Balfour Stewart's Meteorological Blockade	62
1871	108	On the Ultramundane Corpuscles of Le Sage	64
1880	109	On Steam-Pressure Thermometers of Sulphurous Acid, Water, and Mercury	77
1880	110	On a Sulphurous Acid Cryophorus	88
1880	111	On a Realised Sulphurous Acid Steam-Pressure Thermometer, and on a Sulphurous Acid Steam-Pressure Differential Thermometer; also a Note on Steam-Pressure Thermometers	90
1880	112	On a Differential Thermoscope founded on Change of Viscosity of Water with Change of Temperature	96
1880	113	On a Thermomagnetic Thermoscope	97
1880	114	On a Constant Pressure Gas Thermometer	99
1880	115	On the Elimination of Air from Water	106
1880	116	On a method of determining the critical temperature for any liquid and its vapour without mechanism	107
1881	117	On the Sources of Energy in Nature available to Man for the Production of Mechanical Effect	109
1881	118	Accélération Thermodynamique du Mouvement de Rotation de la Terre	109
1884	119	On the Efficiency of Clothing for Maintaining Temperature	109
1897	120	On Osmotic Pressure against an Ideal Semi-permeable Membrane	110
1897	121	On a Differential Method for Measuring Differences of Vapour Pressures of Liquids at one Temperature and at Different Temperatures	113
1903	122	Animal Thermostat	118
1853	123	The Power required for the Thermodynamic Heating of Buildings	124

COSMICAL AND GEOLOGICAL PHYSICS.

1859	124	Recent Investigations of M. Le Verrier on the Motion of Mercury	134
1860	125	On the Variation of the Periodic Times of the Earth and Inferior Planets, produced by Matter falling into the Sun	138
1861	126	Physical Considerations regarding the possible Age of the Sun's Heat	141

MOLECULAR AND CRYSTALLINE THEORY.

ELECTRODYNAMICS.

CONTENTS   XV

CORRIGENDUM

Page 109, No. 119, *for* $k/2e$ *read* $2k/e$. (Increase of external radius diminishes the emission of heat only when this condition is satisfied.)

THERMODYNAMICS

85. ON THE DISSIPATION OF ENERGY. By Prof. P. G. TAIT.

[From *Philosophical Magazine*, Vol. VII. May 1879, pp. 344—346.]

To Sir W. Thomson, F.R.S.

MY DEAR THOMSON,

I address you as one of the Editors of the *Philosophical Magazine*, but also specially as the first propounder of the doctrine of the Dissipation of Energy. I do so because Prof. Clausius, in the second part of the new edition of his work on Thermodynamics, has challenged your claim to the well-known expression for the amount of heat dissipated in a non-reversible cycle. I think that the time has come for you to speak out on the subject, so as, if possible, to prevent further unnecessary discussions.

I shall endeavour, so far as I can, to keep to matters of scientific importance; but I must introduce the subject by a reference to the comments made by Prof. Clausius upon a somewhat slipshod passage (§ 178) of my little work on *Thermodynamics*. That passage refers to the integral

$$\int \frac{dq}{t},$$

to which I believe Rankine first called attention, but which is essentially connected with your doctrine.

I cannot altogether complain of Prof. Clausius's comments, because I cannot account for my having called the above integral (in the way in which I have employed it) a *positive* quantity, except by supposing that in the revision of the first proof of my

1

book I had thoughtlessly changed the word "negative" to "positive." This might easily happen from my having used a novel term, "practical value," in a somewhat ambiguous manner, at one place confounding it with "realized value." That the whole section was meant to bear the construction forced on it by Prof. Clausius is, I think, sufficiently disproved by its opening sentence, not to speak of the fact that no one in this country has so interpreted it.

But there is a graver matter involved than any such mere slips of the pen; for Prof. Clausius asserts that the method I employ (and which I certainly obtained from your paper of 1852) is inapplicable to any but reversible cycles. This, I think, is equivalent to denying altogether your claims in the matter. I therefore quote the whole passage, correcting, however, the above-mentioned slip, and slightly extending the latter part to make my meaning perfectly clear.

"§ 178. The real dynamical value of a quantity, dq, of heat is Jdq, whatever be the temperature of the body which contains it. But the extreme practical value is only

$$J \frac{t - t_0}{t} dq,$$

where t is the temperature of the body, and t_0 the lowest available temperature. This value may be written in the form

$$Jdq - Jt_0 \frac{dq}{t}.$$

Hence, in any cyclical process whatever, if q_1 be the whole heat taken in, and q_0 that given out, the practical value is

$$J(q_1 - q_0) - Jt_0 \int \frac{dq}{t} *.$$

Now the realized value is

$$J(q_1 - q_0)$$

by the first law; and if the cycle be *reversible*, this must be equal to the extreme practical value. Hence, in this particular case,

$$\int \frac{dq}{t} = 0.$$

* On this formula Prof. Clausius remarks, " Die Unrichtigkeit dieses Resultates lässt sich leicht aus dem blossen Anblicke der Formel erkennen "!

But in general this integral has a finite negative value, because in non-reversible cycles the realized value of the heat is always less than

$$J(q_1 - q_0) - Jt_0 \int \frac{dq}{t},$$

which is the extreme practical value.

"Hence the amount of heat lost needlessly, *i.e.* rejected in excess of what is necessarily rejected to the refrigerator for producing work, is

$$- t_0 \int \frac{dq}{t}.$$

This is Thomson's expression for the amount of heat *dissipated* during the cycle (*Phil. Mag.* and *Proc. R. S. E.* 1852, "On a Universal Tendency in Nature to Dissipation of Energy"). It is, of course, an immediate consequence of his important formula for the work of a perfect engine.

"[It is very desirable to have a word to express the *availability* for work of the heat in a given magazine; a term for that possession, the waste of which is called dissipation.]"

As I based the greater part of the last chapter of my work on your papers, mainly because they appeared to me to be greatly superior to all others on the subject in the three very important qualities of simplicity, conciseness, and freedom from hypothesis, I am anxious to know whether the above passage meets with your approval.

From Prof. Clausius's comments it appears, as I have already said, that he considers the method I have adopted from you to be one which cannot be applied except to *reversible* cycles, and which, therefore, it is absurd to employ in any argument connected with dissipation of energy.

Prof. Clausius also disputes the correctness of my reference to your paper in the *Philosophical Magazine*, as containing the above expression for the heat dissipated. You ought to be a competent authority on such a question as this.

I do not now reply to the many other remarks of Prof. Clausius, simply because they refer to myself, my motives, and my book, and not to the principles or the history of science. As the matter affects you, however, I may mention that Professor Clausius attributes to me the real authorship of the paper on "Energy" which we jointly wrote for *Good Words,* and which has been often referred to in the *Philosophical Magazine.*

But the passage in brackets in the extract above indicates a want of proper nomenclature, which would, I think, be well met by the publication of the paper on Thermodynamic Motivity, read by you some years ago to the Royal Society of Edinburgh.

Yours truly,

P. G. TAIT.

38 George Square, Edinburgh,
    March 17*th,* 1879.

NOTE BY SIR W. THOMSON ON THE PRECEDING LETTER*.

[From *Phil. Mag.* Vol. VII. May 1879, pp. 346—348 ; *Journ. de Phys.* Vol. VIII. 1879, pp. 236, 237.]

The passage quoted, with amendments, by Professor Tait from his *Thermodynamics,* seems to me perfectly clear and accurate. Taken in connexion with the sections which preceded it in the original, its meaning was unmistakable ; and a careful reader could have found little or no difficulty in making for himself the necessary corrections with which Professor Tait now presents it. It is certainly not confined to reversible cycles ; but, on the contrary, it gives an explicit expression for the amount of energy dissipated, or, as I put it, "absolutely and irrecoverably wasted," in operations of an irreversible character. My original article " On a Universal Tendency in Nature to the Dissipation of Mechanical Energy," communicated to the Royal Society of Edinburgh in April 1852, and published in the *Proceedings* of the Society for that date, and republished in the *Philosophical Magazine* for 1852, second half-year†, is a sufficient answer to the challenge referred to in the opening sentence of Professor Tait's letter.

I think Professor Tait quite right in referring also to that paper for the formula $t_0 \int dq/t$. The whole matter is contained in

---

* [Cf. the historical account of Lord Kelvin's thermodynamic activity, drawn up without knowledge of this Note, contained in the Obituary Notice, *Proc. Roy. Soc.* Vol. LXXXI. (1908), pp. xxix—xliv.]

† [*Math. and Phys. Papers,* Vol I. pp. 511—514.]

the formula $we^{-\frac{1}{J}\int_{T}^{S}\mu dt}$, which is given explicitly in that paper.
At the top of the next page in the *Philosophical Magazine* reprint
the following passage occurs*:—"If the system of thermometry
adopted be such that $\mu = J/(t + a)$, that is, if we agree to call
$J/\mu - a$ the *temperature* of a body, for which μ is the *value* of
Carnot's function (a and J being constants), &c."; and on the
word "adopted" the following footnote is given: "According to
Mayer's 'hypothesis' this system coincides with that in which
equal differences of temperature are defined as those with which
the same mass of air under constant pressure has equal differences
of volume, provided J be the mechanical equivalent of the thermal
unit, and $1/a$ the coefficient of expansion of air." Here the true
foundation of the absolute thermodynamic scale now universally
adopted was, I believe, for the first time given. I had previously,
in Part III. of my "Dynamical Theory of Heat," published in the
Transactions of the Royal Society of Edinburgh, and in the *Philosophical Magazine* for 1852, second half-year †, taking advantage of
a suggestion made to me by Joule, in a letter of date December 9,
1848, shown that the assumption $\mu = J/(a + t)$ reduces the formula
$we^{-\frac{1}{J}\int_{T}^{S}\mu dt}$ to $w(a + T)/(a + S)$: and I used this transformation in
the concluding formulas of the article referred to by Professor
Tait (corrected in the *errata* of *Phil. Mag.* 1853, first half-year).
It was not, however, until the experiments by Joule and myself,
made in the course of the years 1852, 1853, and the early part of
1854, on the thermal effects of forcing air and other gases through
porous plugs, had proved that my proposed thermodynamic scale
agreed as nearly with the scale of an air-thermometer as different
air-thermometers agreed with one another, that I definitively
adopted it in fundamental formulas of thermodynamics. Thus,
for example, in Part VI. ("Thermo-electric Currents") of my
"Dynamical Theory of Heat," published in the *Transactions of
the Royal Society of Edinburgh* for May 1854, and in the *Philosophical Magazine* for 1855, first half-year‡, the formula

$$\frac{H_t}{t} + \frac{H_{t'}}{t'} + \cdots + \frac{H_{t^{(n-1)}}}{t^{(n-1)}} + \frac{H_{t^{(n)}}}{t^{(n)}} = 0$$

is given as an equivalent for

$$\Sigma a_t = \Sigma a_t \left(1 - \epsilon^{-\frac{1}{J}\int_{T}^{t}\mu dt}\right),$$

* [*Loc. cit.* p. 513.]   † [*Loc. cit.* p. 199.]   ‡ [*Loc. cit.* p. 237.]

which was first published in the *Proceedings of the Royal Society of Edinburgh* for 1851, and *Phil. Mag.* 1852, first half-year*. Tait had actually quoted the formula from my 1854 paper in § 176 of his book, and so left absolutely no foundation for Professor Clausius' objection to his saying "This is Thomson's expression &c.," quoted in his letter above.

As to the *Good Words* article on "Energy" which appeared under our joint names, Professor Tait and I are equally responsible for its contents. I claim my full share of the "scientific patriotism" commended in that article, and cannot assent to Professor Clausius' giving all the credit of it to Professor Tait.

In compliance with the concluding sentence of Professor Tait's letter, I hope in the course of a few days to write out, and send to the *Philosophical Magazine* for publication, a short statement of the communication on "Thermodynamic Motivity" which I made *vivâ voce* to the Royal Society of Edinburgh on April 3rd, 1876.

ON THERMODYNAMIC MOTIVITY.

[From *Philosophical Magazine*, Vol. VII. May 1879, pp. 348—352.; repeated from *Math. and Phys. Papers*, Vol. I. pp. 456—459.]

After having for some years felt with Professor Tait the want of a word "to express the Availability for work of the heat in a given magazine, a term for that possession the waste of which is called Dissipation"†, I suggested three years ago the word *Motivity* to supply this want, and made a verbal communication to the Royal Society of Edinburgh defining and illustrating the application of the word; but as the communication was not given in writing, only the title of the paper, "Thermodynamic Motivity," was published. In consequence of Professor Tait's letter to me, published in the present Number of the *Philosophical Magazine*, I now offer, for publication in the *Proceedings of the Royal Society*

---

* [*Loc. cit.* p. 317.]

† Tait's *Thermodynamics*, first edition (1868), § 178: quoted also in Professor Tait's letter [*supra*].

of Edinburgh and in the *Philosophical Magazine*, the following short abstract of the substance of that communication.

In my paper "On the Restoration of Energy from an Un-equally Heated Space," published in the *Philosophical Magazine* in January 1853*, I gave the following expression for the amount of "mechanical energy" derivable from a body, *B*, given with its different parts at different temperatures, by the equalization of the temperature throughout to one common temperature† *T*, by means of perfect thermodynamic engines,

$$W = J \iiint dx\, dy\, dz \int_T^t c\, dt \left(1 - \epsilon^{-\frac{1}{J}\int_T^t \mu\, dt}\right) \quad \ldots\ldots\ldots(1),$$

where *t* denotes the temperature of any point *x, y, z* of the body, *c* the thermal capacity of the body's substance at that point and that temperature, *J* Joule's equivalent, and *μ* Carnot's function of the temperature *t*.

Further on in the same paper a simplification is introduced thus :—

"Let the temperature of the body be measured according to an absolute scale, founded on the values of Carnot's function, and expressed by the following equation,

$$t = \frac{J}{\mu} - \alpha,$$

---

* [*Loc. cit.* pp. 554—558.]

† In the present article I suppose this temperature to be the given temperature of the medium in which *B* is placed; and thermodynamic engines to work with their recipient and rejectant organs respectively in connexion with some part of *B* at temperature *t*, and the endless surrounding matter at temperature *T*. In the original paper this supposition is introduced subordinately at the conclusion. The chief purpose of the paper was the solution of a more difficult problem, that of finding the value of *T*—a kind of average temperature of *B* to fulfil the condition that the quantities of heat rejected and taken in by organs of the thermodynamic engines at temperature *T* are equal. The burden of the problem was the evaluation of this thermodynamic average; and I failed to remark that when the value which the solution gave for *T* is substituted in the formula of the text it reduces to $J \iiint dx\, dy\, dz \int_T^t c\, dt$, which was not instantly obvious from the analytical form of my solution, but which we immediately see must be the case by thinking of the physical meaning of the result; for the sum of the excesses of the heats taken in above those rejected by all the engines must, by the first law of thermodynamics, be equal to the work gained by the supposed process. This important simplification was first given by Professor Tait in his *Thermodynamics* (first and second editions). It does not, however, affect the subordinate problem of the original paper, which is the main problem of this one.

where α is a constant which may have any value, but ought to have for its value the reciprocal of the expansibility of air, in order that the system of measuring temperature here adopted may agree approximately with that of the air-thermometer. Then we have

$$\epsilon^{-\frac{1}{J}\int_0^t \mu\, dt} = \frac{\alpha}{t+\alpha} \quad\ldots\ldots\ldots\ldots\ldots\ldots(2)."$$

It was only to obtain agreement with the zero of the ordinary Centigrade scale of the air-thermometer that the α was needed; and in the joint paper by Joule and myself, published in the *Transactions of the Royal Society* (London) for June 1854, we agreed to drop it, and to define temperature simply as the reciprocal of Carnot's function, with a constant coefficient proper to the unit or degree of temperature adopted. Thus definitively, in equation (6) of § 5 of that paper, we took $t = J/\mu$, and have used this expression ever since as the expression for temperature on the arbitrarily assumed thermodynamic scale. With it we have

$$\epsilon^{-\frac{1}{J}\int_T^t \mu\, dt} = \frac{T}{t} \quad\ldots\ldots\ldots\ldots\ldots\ldots(3);$$

and by substitution (1) becomes

$$W = J \iiint dx\,dy\,dz \int_T^t c\,dt \left(1 - \frac{T}{t}\right) \quad\ldots\ldots\ldots\ldots(4).$$

Suppose now B to be surrounded by other matter all at a common temperature T. The work obtainable from the given distribution of temperature in B by means of perfect thermodynamic engines is expressed by the formula (4). If, then, there be no circumstances connected with the gravity, or elasticity, or capillary attraction, or electricity, or magnetism of B in virtue of which work can be obtained, that expressed by (4) is what I propose to call the whole Motivity of B in its actual circumstances. If, on the other hand, work is obtainable from B in virtue of some of these other causes, and if V denote its whole amount, then

$$\mathfrak{M} = V + W \ldots\ldots\ldots\ldots\ldots\ldots\ldots(5)$$

is what I call the whole Motivity of B in its actual circumstances according to this more comprehensive supposition.

We may imagine the whole Motivity of B developed in an infinite variety of ways. The one which is obvious from the formula (5) is first to keep every part of B unmoved, and to take all the work producible by perfect thermodynamic engines equalizing its temperature to T; and then keeping it rigorously at this temperature, to take all the work that can be got from it elastically, cohesively, electrically, magnetically, and gravitationally, by letting it come to rest unstressed, diselectrified, demagnetized, and in the lowest position to which it can descend. But instead of proceeding in this one definite way, any order of procedure whatever leading to the same final condition may be followed; and, provided nothing is done which cannot be undone (that is to say, in the technical language of thermodynamics, provided all the operations be reversible), the same whole quantity of work will be obtained in passing from the same initial condition to the same final condition, whatever may have been the order of procedure. Hence the Motivity is a function of the temperature, volume, figure, and proper independent variables for expressing the cohesive, the electric, and the magnetic condition of B, with the gravitational potential of B simply added (which, when the force of gravity is sensibly constant and in parallel lines, will be simply the product of the gravity of B into the height of its centre of gravity above its lowest position). So also is the *Energy* of a body B (as I first pointed out, for the case of B a fluid, in Part V. of my "Dynamical Theory of Heat," in the *Transactions of the Royal Society of Edinburgh* for December 15, 1851, entitled, "On the Quantities of Mechanical Energy contained in a Fluid in Different States as to Temperature and Density"). Consideration of the *Energy* and the *Motivity*, as two functions of all the independent variables specifying the condition of B completely in respect to temperature, elasticity, capillary attraction, electricity, and magnetism, leads in the simplest and most direct way to demonstrations of the theorems regarding the thermodynamic properties of matter which I gave in Part III. of the "Dynamical Theory of Heat" (March 1851); in Part VI. of "Dynamical Theory, Thermoelectric Currents" (May 1, 1854); in a paper in the *Proceedings* for 1858 of the Royal Society of London, entitled, "On the Thermal Effect of Drawing out a Film of Liquid"; and in a communication to the Royal Society of Edinburgh (*Proc. R. S. E.* 1869—70), "On the Equilibrium of Vapour at the Curved Surface of a Liquid"; and in my article "On the Thermoelastic and Thermomagnetic Properties of Matter," in the first number of the *Quarterly Journal of Mathematics* (April 1855); and in short articles in *Nichol's Cyclopædia,* under the titles " Thermomagnetism, Thermoelectricity, and Pyroelectricity," put together and republished with additions in the *Philosophical Magazine* for January 1878, under the title "On the Thermo-elastic, Thermomagnetic, and Pyroelectric Properties of Matter."

It would be beyond the scope of the present article to enter in detail into these applications, which were merely indicated in my communication to the Royal Society of Edinburgh of three years ago, as a very short and simple analytical method of setting forth the whole non-molecular theory of Thermodynamics.

86. THE KINETIC THEORY OF THE DISSIPATION OF ENERGY.

[From *Edinb. Roy. Soc. Proc.* Vol. VIII. 1875, pp. 325—334 [read Feb. 16, 1874]; *Nature*, Vol. IX. April 9, 1874, pp. 441—444; *Phil. Mag.* Vol. XXXIII. March 1892, pp. 291—299.]

IN abstract dynamics the instantaneous reversal of the motion of every moving particle of a system causes the system to move backwards, each particle of it along its old path, and at the same speed as before, when again in the same position. That is to say, in mathematical language, any solution remains a solution when t is changed into $-t$. In physical dynamics this simple and perfect reversibility fails, on account of forces depending on friction of solids; imperfect fluidity of fluids; imperfect elasticity of solids; inequalities of temperature, and consequent conduction of heat produced by stresses in solids and fluids; imperfect magnetic retentiveness; residual electric polarization of dielectrics; generation of heat by electric currents induced by motion; diffusion of fluids, solution of solids in fluids, and other chemical changes; and absorption of radiant heat and light. Consideration of these agencies in connexion with the all-pervading law of the conservation of energy proved for them by Joule, led me twenty-three years ago to the theory of the dissipation of energy, which I communicated first to the Royal Society of Edinburgh in 1852, in a paper entitled "On a Universal Tendency in Nature to the Dissipation of Mechanical Energy."

The essence of Joule's discovery is the subjection of physical phenomena to dynamical law. If, then, the motion of every particle of matter in the universe were precisely reversed at any instant, the course of nature would be simply reversed for ever after. The bursting bubble of foam at the foot of a waterfall would reunite and descend into the water; the thermal motions would reconcentrate their energy, and throw the mass up the fall in drops re-forming into a close column of ascending water. Heat which had been generated by the friction of solids and dissipated by conduction, and radiation with absorption, would come again to the place of contact, and throw the moving body back against the force to which it had previously yielded. Boulders would recover from the mud the materials required to rebuild them into their previous jagged forms, and would become reunited to the mountain peak from which they had formerly broken away. And if also the materialistic hypothesis of life were true, living creatures would grow backwards, with conscious knowledge of the future, but no memory of the past, and would become again unborn. But the real phenomena of life infinitely transcend human science; and speculation regarding consequences of their imagined reversal is utterly unprofitable. Far otherwise, however, is it in respect to the reversal of the motions of matter uninfluenced by life, a very elementary consideration of which leads to the full explanation of the theory of dissipation of energy.

To take one of the simplest cases of the dissipation of energy, the conduction of heat through a solid—consider a bar of metal warmer at one end than the other, and left to itself. To avoid all needless complication of taking loss or gain of heat into account, imagine the bar to be varnished with a substance impermeable to heat. For the sake of definiteness, imagine the bar to be first given with one-half of it at one uniform temperature, and the other half of it at another uniform temperature. Instantly a diffusion of heat commences, and the distribution of temperature becomes continuously less and less unequal, tending to perfect uniformity, but never in any finite time attaining perfectly to this ultimate condition. This process of diffusion could be perfectly prevented by an army of Maxwell's "intelligent demons*," stationed at the surface, or interface as we may call it with Professor James Thomson, separating the hot from the cold part of the bar. To see precisely how this is to be done, consider rather a gas than a solid, because we have much knowledge regarding the molecular motions of a gas, and little or no knowledge of the molecular motions of a solid. Take a jar with the lower half occupied by cold air or gas, and the upper half occupied

---

* The definition of a demon, according to the use of this word by Maxwell, is an intelligent being endowed with free-will and fine enough tactile and perceptive organization to give him the faculty of observing and influencing individual molecules of matter.

with air or gas of the same kind, but at a higher temperature; and let the mouth of the jar be closed by an air-tight lid. If the containing vessel were perfectly impermeable to heat, the diffusion of heat would follow the same law in the gas as in the solid, though in the gas the diffusion of heat takes place chiefly by the diffusion of molecules, each taking its energy with it, and only to a small proportion of its whole amount by the interchange of energy between molecule and molecule; whereas in the solid there is little or no diffusion of substance, and the diffusion of heat takes place entirely, or almost entirely, through the communication of energy from one molecule to another. Fourier's exquisite mathematical analysis expresses perfectly the statistics of the process of diffusion in each case, whether it be "conduction of heat," as Fourier and his followers have called it, or the diffusion of substance in fluid masses (gaseous or liquid), which Fick showed to be subject to Fourier's formulas. Now, suppose the weapon of the ideal army to be a club, or, as it were, a molecular cricket bat; and suppose, for convenience, the mass of each demon with his weapon to be several times greater than that of a molecule. Every time he strikes a molecule he is to send it away with the same energy as it had immediately before. Each demon is to keep as nearly as possible to a certain station, making only such excursions from it as the execution of his orders requires. He is to experience no forces except such as result from collisions with molecules, and mutual forces between parts of his own mass, including his weapon. Thus his voluntary movements cannot influence the position of his centre of gravity, otherwise than by producing collision with molecules.

The whole interface between hot and cold is to be divided into small areas, each allotted to a single demon. The duty of each demon is to guard his allotment, turning molecules back, or allowing them to pass through from either side, according to certain definite orders. First, let the orders be to allow no molecules to pass from either side. The effect will be the same as if the interface were stopped by a barrier impermeable to matter and to heat. The pressure of the gas being by hypothesis equal in the hot and cold parts, the resultant momentum taken by each demon from any considerable number of molecules will be zero; and therefore he may so time his strokes that he shall never move to any considerable distance from his station. Now, instead of stopping and turning all the molecules from crossing his allotted area, let each demon permit a hundred molecules chosen arbitrarily to cross it from the hot side; and the same number of molecules, chosen so as to have the same entire amount of energy and the same resultant momentum, to cross the other way from the cold side. Let this be done over and over again within certain small equal consecutive intervals of time, with care that if the specified balance of energy and momentum is not exactly fulfilled in respect to each successive hundred molecules crossing each way, the error will be carried forward, and as nearly as may be corrected, in respect to the next hundred. Thus, a certain perfectly regular diffusion of the gas both ways across the interface goes on, while the original different temperatures on the two sides of the interface are maintained without change.

Suppose, now, that in the original condition the temperature and pressure of the gas are each equal throughout the vessel, and let it be required to disequalize the temperature, but to leave the pressure the same in any two portions A and B of the whole space. Station the army on the interface as previously described. Let the orders now be that each demon is to stop all molecules from crossing his area in either direction except 100 coming from A, arbitrarily chosen to be let pass into B, and a greater number, having among them less energy but equal momentum, to cross from B to A. Let this be repeated over and over again. The temperature in A will be continually diminished and the number of molecules in it continually increased, until there are not in B enough of molecules with small enough velocities to fulfil the condition with reference to permission to pass from B to A. If after that no molecule be allowed to pass the interface in either direction, the final condition will be very great condensation and very low temperature in A; rarefaction and very high temperature in B; and equal pressures in A and B. The process of disequalization of temperature and density might be stopped at any time by changing the orders to those previously specified, and so permitting a certain degree of diffusion each way across the interface while maintaining a certain uniform difference of temperatures with equality of pressure on the two sides.

If no selective influence, such as that of the ideal "demon," guides individual molecules, the average result of their free motions and collisions must be to equalize the distribution of energy among them in the gross; and after a sufficiently long time, from the supposed initial arrangement, the difference of energy in any two equal volumes, each containing a very great number of molecules, must bear a very small proportion to the whole amount in either; or, more strictly speaking, the probability of the difference of energy exceeding any stated finite proportion of the whole energy in either is very small. Suppose now the temperature to have become thus very approximately equalized at a certain time from the beginning, and let the motion of every particle become instantaneously reversed. Each molecule will retrace its former path, and at the end of a second interval of time, equal to the former, every molecule will be in the same position, and moving with the same velocity, as at the beginning; so that the given initial unequal distribution of temperature will again be found, with only the difference that each particle is moving in the direction reverse to that of its initial motion. This difference will not prevent an instantaneous subsequent commencement of equalization, which, with entirely different paths for the individual molecules, will go on in the average according to the same law as that which took place immediately after the system was first left to itself.

By merely looking on crowds of molecules, and reckoning their energy in the gross, we could not discover that in the very special case we have just considered the progress was towards a succession of states, in which the distribution of energy deviates more and more from uniformity up to a certain time. The number of molecules being finite, it is clear that small finite deviations from absolute precision in the reversal we have supposed would not obviate the resulting disequalization of the distribution of energy. But the greater the number of molecules, the shorter will be the time during which the disequalizing will continue; and it is only when we regard the number of molecules as practically infinite that we can regard spontaneous disequalization as practically impossible. And, in point of fact, if any finite number of perfectly elastic molecules, however great, be given in motion in the interior of a perfectly rigid vessel, and be left for a sufficiently long time undisturbed except by mutual impact and collisions against the sides of the containing vessel, it must happen over and over again that (for example) something more than $\frac{9}{10}$ths of the whole energy shall be in one-half of the vessel, and less than $\frac{1}{10}$th of the whole energy in the other half. But if the number of molecules be very great, this will happen enormously less frequently than that something more than $\frac{6}{10}$ths shall be in one-half, and something less than $\frac{4}{10}$ths in the other. Taking as unit of time the average interval of free motion between consecutive collisions, it is easily seen that the probability of there being something more than any stated percentage of excess above the half of the energy in one-half of the vessel during the unit of time from a stated instant, is smaller the greater the dimensions of the vessel and the greater the stated percentage. It is a strange but nevertheless a true conception of the old well-known law of the conduction of heat, to say that it is very improbable that in the course of 1000 years one-half of the bar of iron shall of itself become warmer by a degree than the other half; and that the probability of this happening before 1,000,000 years pass is 1000 times as great as that it will happen in the course of 1000 years, and that it certainly will happen in the course of some very long time. But let it be remembered that we have supposed the bar to be covered with an impermeable varnish. Do away with this impossible ideal, and believe the number of molecules in the universe to be infinite; then we may say one-half of the bar will never become warmer than the other, except by the agency of external sources of heat or cold. This one instance suffices to explain the philosophy of the foundation on which the theory of the dissipation of energy rests.

Take, however, another case, in which the probability may be readily calculated. Let an hermetically sealed glass jar of air contain 2,000,000,000,000 molecules of oxygen, and 8,000,000,000,000 molecules of nitrogen. If examined any time in the infinitely distant future, what is the number of chances against one that all the molecules of oxygen and none of nitrogen shall be found in one stated part of the vessel equal in volume to $\frac{1}{5}$th of the whole? The number expressing the answer in the Arabic notation has about 2,173,220,000,000 of places of whole numbers. On the other hand, the chance against there being exactly $\frac{2}{10}$ths of the whole number of particles of nitrogen, and at the same time exactly $\frac{2}{10}$ths of the whole number of particles of oxygen in the first specified part of the vessel, is only 4021×10^9 to 1.

APPENDIX.

Calculation of probability respecting Diffusion of Gases.

For simplicity, I suppose the sphere of action of each molecule to be infinitely small in comparison with its average distance from its nearest neighbour; thus, the sum of the volumes of the spheres of action of all the molecules will be infinitely small in proportion to the whole volume of the containing vessel. For brevity, space external to the sphere of action of every molecule will be called free space: and a molecule will be said to be in free space at any time when its sphere of action is wholly in free space; that is to say, when its sphere of action does not overlap the sphere of action of any other molecule. Let A, B denote any two particular portions of the whole containing vessel, and let a, b be the volumes of those portions. The chance that at any instant one individual molecule of whichever gas shall be in A is $a/(a+b)$, however many or few other molecules there may be in A at the same time; because its chances of being in any specified portions of free space are proportional to their volumes; and, according to our supposition, even if all the other molecules were in A, the volume of free space in it would not be sensibly diminished by their presence. The chance that of n molecules in the whole space there shall be i stated individuals in A, and that the other $n-i$ molecules shall be at the same time in B, is

$$\left(\frac{a}{a+b}\right)^i \left(\frac{b}{a+b}\right)^{n-i}, \text{ or } \frac{a^i b^{n-i}}{(a+b)^n}.$$

Hence the probability of the number of molecules in A being exactly i, and in B exactly $n-i$, irrespectively of individuals, is a fraction having for denominator $(a+b)^n$, and for numerator the term involving $a^i b^{n-i}$ in the expansion of this binomial; that is to say, it is

$$\frac{n(n-1)\ldots(n-i+1)}{1.2\,\ldots\ldots\, i}\left(\frac{a}{a+b}\right)^i \left(\frac{b}{a+b}\right)^{n-i}$$

If we call this T_i, we have

$$T_{i+1} = \frac{n-i}{i+1}\frac{a}{b} T_i.$$

Hence T_i is the greatest term, if i is the smallest integer which makes

$$\frac{n-i}{i+1} < \frac{b}{a};$$

this is to say, if i is the smallest integer which exceeds

$$n\frac{a}{a+b} - \frac{b}{a+b}.$$

Hence if a and b are commensurable, the greatest term is that for which

$$i = n\frac{a}{a+b}.$$

To apply these results to the cases considered in the preceding article, put in the first place

$$n = 2 \times 10^{12},$$

this being the number of particles of oxygen; and let $i = n$. Thus, for the probability that all the particles of oxygen shall be in A, we find

$$\left(\frac{a}{a+b}\right)^{2\times10^{12}}$$

Similarly, for the probability that all the particles of nitrogen are in the space B, we find

$$\left(\frac{b}{a+b}\right)^{8\times10^{12}}$$

Hence the probability that all the oxygen is in A and all the nitrogen in B is

$$\left(\frac{a}{a+b}\right)^{2\times10^{12}} \times \left(\frac{b}{a+b}\right)^{8\times10^{12}}$$

Now by hypothesis

$$\frac{a}{a+b} = \frac{2}{10},$$

and therefore

$$\frac{b}{a+b} = \frac{8}{10};$$

hence the required probability is

$$\frac{2^{26\times10^{12}}}{10^{10^{13}}}$$

Call this $1/N$, and let log denote common logarithm. We have

$$\log N = 10^{13} - 26 \times 10^{12} \times \log 2 = (10 - 26 \log 2)$$
$$\times 10^{12} = 2173220 \times 10^{6}.$$

This is equivalent to the result stated in the text above. The logarithm of so great a number, unless given to more than thirteen significant places, cannot indicate more than the number of places of whole numbers in the answer to the proposed question, expressed according to the Arabic notation.

The calculation of T_i, when i and $n - i$ are very large numbers, is practicable by Stirling's theorem, according to which we have approximately

$$1.2 \ldots i = i^{i + \frac{1}{2}} \epsilon^{-i} \sqrt{2\pi},$$

and therefore

$$\frac{n(n-1)\ldots(n-i+1)}{1.2 \ \ldots\ldots \ i} = \frac{n^{n+\frac{1}{2}}}{\sqrt{2\pi}\, i^{i+\frac{1}{2}}(n-i)^{n-i+\frac{1}{2}}}$$

Hence for the case

$$i = n\frac{a}{a+b},$$

which, according to the preceding formulæ, gives T_i its greatest value, we have

$$T_i = \frac{1}{\sqrt{2\pi n e f}},$$

where
$$e = \frac{a}{a+b} \text{ and } f = \frac{b}{a+b}.$$

Thus, for example, let $n = 2 \times 10^{12}$;

$$e = \cdot 2, \qquad f = \cdot 8,$$

we have
$$T_i = \frac{1}{800000\sqrt{\pi}} = \frac{1}{1418000}.$$

This expresses the chance of there being 4×10^{11} molecules of oxygen in A, and 16×10^{11} in B. Just half this fraction expresses the probability that the molecules of nitrogen are distributed in exactly the same proportion between A and B, because the number of molecules of nitrogen is four times greater than of oxygen.

If n denote the number of the molecules of one gas, and n' that of the molecules of another, the probability that each shall be distributed between A and B in the exact proportion of the volume is

$$\frac{1}{2\pi e f \sqrt{nn'}}.$$

The value for the supposed case of oxygen and nitrogen is

$$\frac{1}{2\pi \times \cdot 16 \times 4 \times 10^{12}} = \frac{1}{4021 \times 10^{9}},$$

which is the result stated at the conclusion of the text above.

87. THE SORTING DEMON OF MAXWELL.

[Abstract of Royal Institution Lecture, Feb. 28, 1879, *Roy. Institution Proc.*
Vol. IX. p. 113. Reprinted in *Popular Lectures and Addresses*, Vol. I.
pp. 137—141.]

THE word "demon," which originally in Greek meant a super-
natural being, has never been properly used to signify a real or
ideal personification of malignity.

Clerk Maxwell's "demon" is a creature of imagination having
certain perfectly well defined powers of action, purely mechanical
in their character, invented to help us to understand the "Dissipation of Energy" in nature.

He is a being with no preternatural qualities, and differs from
real living animals only in extreme smallness and agility. He
can at pleasure stop, or strike, or push, or pull any single atom
of matter, and so moderate its natural course of motion. Endowed
ideally with arms and hands and fingers—two hands and ten
fingers suffice—he can do as much for atoms as a pianoforte
player can do for the keys of the piano—just a little more, he can
push or pull each atom *in any direction.*

He cannot create or annul energy; but just as a living animal
does, he can store up limited quantities of energy, and reproduce
them at will. By operating selectively on individual atoms he
can reverse the natural dissipation of energy, can cause one-half
of a closed jar of air, or of a bar of iron, to become glowingly hot
and the other ice cold; can direct the energy of the moving
molecules of a basin of water to throw the water up to a height
and leave it there proportionately cooled (1 deg. Fahrenheit for
772 ft. of ascent); can "sort" the molecules in a solution of salt
or in a mixture of two gases, so as to reverse the natural process
of diffusion, and produce concentration of the solution in one

portion of the water, leaving pure water in the remainder of the space occupied; or, in the other case, separate the gases into different parts of the containing vessel.

"Dissipation of Energy" follows in nature from the fortuitous concourse of atoms. The lost motivity is essentially not restorable otherwise than by an agency dealing with individual atoms; and the mode of dealing with the atoms to restore motivity is essentially a process of assortment, sending this way all of one kind or class, that way all of another kind or class.

The classification, according to which the ideal demon is to sort them, may be according to the essential character of the atom; for instance, all atoms of hydrogen to be let go to the left, or stopped from crossing to the right, across an ideal boundary; or it may be according to the velocity each atom chances to have when it approaches the boundary: if greater than a certain stated amount, it is to go to the right; if less, to the left. This latter rule of assortment, carried into execution by the demon, disequalises temperature, and undoes the natural diffusion of heat; the former undoes the natural diffusion of matter.

By a combination of the two processes, the demon can decompose water or carbonic acid, first raising a portion of the compound to dissociational temperature (that is, temperature so high that collisions shatter the compound molecules to atoms) and then sending the oxygen atoms this way, and the hydrogen or carbon atoms that way; or he may effect decomposition against chemical affinity otherwise, thus:—Let him take in a small store of energy by resisting the mutual approach of two compound molecules, letting them press as it were on his two hands, and store up energy as in a bent spring, then let him apply the two hands between the oxygen and the double hydrogen constituents of a compound molecule of vapour of water, and tear them asunder. He may repeat this process until a considerable proportion of the whole number of compound molecules in a given quantity of vapour of water, given in a fixed closed vessel, are separated into oxygen and hydrogen at the expense of energy taken from translational motions. The motivity (or energy for motive power) in the explosive mixture of oxygen and hydrogen of the one case, and the separated mutual combustibles, carbon and oxygen, of the other case, thus obtained, is a transformation of the energy found in the substance in the form of kinetic energy of the thermal motions of the compound molecules. Essentially different is the decomposition of carbonic acid and water in the natural growth of plants, the resulting motivity of which is taken from the undulations of light or radiant heat, emanating from the intensely hot matter of the sun.

The conception of the "sorting demon" is purely mechanical, and is of great value in purely physical science. It was not invented to help us to deal with questions regarding the influence of life and of mind on the motions of matter, questions essentially beyond the range of mere dynamics.

The discourse was illustrated by a series of experiments.

88. On Thermodynamics founded on Motivity and Energy.

[From *Proc. Royal Soc. Edinburgh*, March 21, 1898, Vol. XXII. pp. 126—130.]

1. In a verbal communication to the Royal Society of Edinburgh in 1876, under the title "Thermodynamic Motivity," I suggested the name *motivity* to express energy, whether thermal or of any other kind, available to generate velocity in molar matter, or to move molar matter against resisting force. By molar matter I mean matter as we know it, consisting, as we believe, of vast numbers of atoms or molecules. Only the title of this communication was published in the *Proceedings of the Royal Society*: a short report of it was published in the *Philosophical Magazine* for May 1879 *, in which it was pointed out that a "very short and simple analytical method of setting forth the whole non-molecular theory of thermodynamics" might be founded on the consideration of *Motivity* and *Energy* as two functions of all the independent variables specifying a body, or a system of bodies, or some definite apparatus under consideration:—apparatus, as I shall call it for brevity, to include every case, even such as a single crystal. The object of the present communication is to carry out this proposal.

2. Let the apparatus be given all at one temperature t. Denote by g_1, g_2, g_3, etc., the other variables by which its condition is specified. These will in many cases be geometrical, specifying elements or co-ordinates, such as strain-components, expressing change of bulk or shape of a piece of crystal or other elastic solid under stress; or positions of pistons in a pneumatic apparatus; or area, or curvature, of a free liquid surface in an application to theory of capillary attraction; or positions of electrified bodies, or electrostatic capacities, in an electrostatic system. Or, considered as generalised co-ordinates, the independent variables may be physical qualities, such as proportion of vapour to liquid in an enclosure, with or without a piston; or quantities of electricity on

---

* [*Supra*, p. 6. The motivity (energy available at constant temperature) is identical with the free energy of Helmholtz, and the thermodynamic potential of Willard Gibbs.]

particular insulated pieces of metal in an electrical system; or proportion of salt to solvent in an osmotic application.

3. Let $m_T(t, g_1, g_2, g_3, \ldots)$ and $e(t, g_1, g_2, g_3, \ldots)$ denote the motivity and the energy of the system, the latter being absolute, the former being relative to a temperature T, the lowest available for carrying off heat. These expressions written in full denote that m_T and e are functions of the independent variables t, g_1, g_2, etc.; but generally, except when it is convenient to be reminded that they are such functions, we shall for brevity denote their values simply by m and e.

4. First suppose the temperature of the apparatus kept constant at T, and let the other independent variables be augmented from $g_1 - \frac{1}{2}dg_1$, $g_2 - \frac{1}{2}dg_2$, etc., to $g_1 + \frac{1}{2}dg_1$, $g_2 + \frac{1}{2}dg_2$, etc., through infinitesimal ranges dg_1, dg_2, etc. Let

$$M_1(T) . dg_1 + M_2(T) . dg_2 + \text{etc.} \ldots\ldots\ldots\ldots(1)$$

denote the quantity of heat (positive or negative) which must be taken in from without to keep the temperature constant at T, and let

$$P_1(T) . dg_1 + P_2(T) . dg_2 + \text{etc.} \ldots\ldots\ldots\ldots(2)$$

denote the mechanical work required to produce the change. This work simply contributes its own amount to the motivity of the apparatus, because, as in Carnot's theory, we have an infinite river or ocean at temperature T, always ready to give or take freely any heat to be taken in by, or rejected from, the apparatus to keep it at constant temperature T. Hence

$$dm_T(T, g_1, g_2, \ldots) = P_1(T) . dg_1 + P_2(T) . dg_2 + \text{etc.} \ldots(3).$$

On the other hand, the energy is augmented, not only by the mechanical work done on it, but, in addition to this, by the dynamical equivalent of the quantity (positive or negative) of heat taken in. Hence, if J denote the dynamical equivalent of the thermal unit, we have

$$d\left[e(T, g_1, g_2, \ldots) - m_T(T, g_1, g_2, \ldots)\right]$$
$$= JM_1(T) . dg_1 + JM_2(T) . dg_2 + \text{etc.} \ldots(4).$$

The second members of (3) and (4) are complete differentials of functions g_1, g_2, etc., on the supposition that T is constant. Hence

$$\frac{d}{dg_1} P_2(T) = \frac{d}{dg_2} P_1(T); \quad \frac{d}{dg_2} P_3(T) = \frac{d}{dg_3} P_2(T); \quad \text{etc.} \ldots(5)$$

and

$$\frac{d}{dg_1} M_2(T) = \frac{d}{dg_2} M_1(T); \quad \frac{d}{dg_2} M_3(T) = \frac{d}{dg_3} M_2(T); \text{ etc....(6).}$$

5. Passing now from the supposition that the temperature is kept constant at T and going back to § 3, remark that whatever heat is taken in at temperature t imparts motivity to the apparatus to an amount equal to the proportion $(t-T)/t$ of its dynamical equivalent. Hence, if N denote the thermal capacity of the apparatus with g_1, g_2, etc., each constant; and if M_1, M_2, etc., and P_1, P_2, etc., denote the coefficients in (1) and (2) for any variable temperature t instead of the constant temperature T, we now have, instead of (3),

$$dm = J\frac{t-T}{t} Ndt + \left(P_1 + J\frac{t-T}{t} M_1\right) dg_1$$
$$+ \left(P_2 + J\frac{t-T}{t} M_2\right) dg_2 + \text{etc. ...(7);}$$

and for the energy we have simply

$$de = JNdt + (P_1 + JM_1)dg_1 + (P_2 + JM_2)dg_2 + \text{etc. ...(8).}$$

From this and (7) we have

$$d(e-m) = \frac{JT}{t}(Ndt + M_1 dg_1 + M_2 dg_2 + \text{etc.}) \quad ......(9).$$

6. From the conditions that the second members of (7) and (9) are complete differentials of all the independent variables, we now find from (9), as formerly from (4), in respect to g_1, g_2, etc.,

$$\frac{dM_2}{dg_1} = \frac{dM_1}{dg_2}; \quad \frac{dM_3}{dg_2} = \frac{dM_2}{dg_3}; \text{ etc. } \quad ....(10)$$

and from these in conjunction with (7),

$$\frac{dP_2}{dg_1} = \frac{dP_1}{dg_2}; \quad \frac{dP_3}{dg_2} = \frac{dP_2}{dg_3}; \text{ etc. } \quad ....(11).$$

Lastly, in respect to t, g_1; t, g_2; etc., we find from (9) and (8)

$$\frac{dN}{dg_1} = \frac{dM_1}{dt} - \frac{M_1}{t}; \quad \frac{dN}{dg_2} = \frac{dM_2}{dt} - \frac{M_2}{t}; \text{ etc. } \quad ......(12)$$

and $\quad J\dfrac{dN}{dg_1} = J\dfrac{dM_1}{dt} + \dfrac{dP_1}{dt}; \quad J\dfrac{dN}{dg_2} = J\dfrac{dM_2}{dt} + \dfrac{dP_2}{dt}; \text{ etc. (13).}$

From these, by elimination of N, we find

$$JM_1 = -t\frac{dP_1}{dt}; \quad JM_2 = -t\frac{dP_2}{dt}; \text{ etc.........(14).}$$

These equations (11), (12), (13), and (14) express all the knowledge regarding properties of matter which can be derived, according to suggestions of Carnot and Clapeyron, from the combined Carnot and Joule thermodynamic theory.

7. For some applications the following condensation of notation and corresponding modification of formulas will be found convenient. Let w denote the mechanical work performed in altering the apparatus from any one configuration $(t, g_1', g_2', ...)$ to any other configuration $(t, g_1, g_2, ...)$, both at the same temperature t, and let H be the heat absorbed during the process. Equations (11) and (10) demonstrate that w and H are independent of the particular succession of configurations through which the apparatus is brought from the initial to the final configuration, *provided heat is given to it, or taken from it, by external agency, so as to keep it at one unchanged temperature t throughout the process.* With this notation (11) and (10) are equivalent to the following:—

$$w = \chi(t, g_1, g_2, ...) - \chi(t, g_1', g_2', ...) \quad ......(15),$$

$$H = \psi(t, g_1, g_2, ...) - \psi(t, g_1', g_2', ...) \quad ......(16),$$

where χ and ψ denote two functions of the variables; and we have

$$P_1 = \frac{dw}{dg_1}; \quad P_2 = \frac{dw}{dg_2}; \text{ etc.} ......(17),$$

$$M_1 = \frac{dH}{dg_1}; \quad M_2 = \frac{dH}{dg_2}; \text{ etc.} ......(18).$$

In terms of this condensed notation we find as an equivalent for (14) the following single equation:—

$$JH = -t\frac{dw}{dt} \quad ......(19),$$

and by integration of (8), (9), and (7) with reference to g_1, g_2, etc.

$$e = w + JH + e(t, g_1', g_2', ...) ......(20),$$

$$e - m = \frac{JTH}{t} + e(t, g_1', g_2', ...) - m_T(t, g_1', g_2', ...)...(21),$$

$$m = w + J\frac{t-T}{t}H + m_T(t, g_1', g_2', ...) ......(22).$$

And, eliminating H by (19), we have finally e and m in terms of w as follows:—

$$e = w - t\,\frac{dw}{dt} + e\,(t,\,g_1{}',\,g_2{}',\,\ldots) \ldots\ldots\ldots(23),$$

$$m = w - (t - T)\,\frac{dw}{dt} + m_T\,(t,\,g_1{}',\,g_2{}',\,\ldots) \ldots\ldots(24).$$

8. For the particular case $t = T$ we fall back on the formulas of § 4, and we see that there is in that case no distinction between motivity and mechanical work done with the apparatus kept constantly at temperature T.

9. Our present notation, w, H, and e, is exactly that which I used forty-three years ago in my paper "On the Thermo-elastic and Thermo-magnetic Properties of Matter," published in the first number of the *Quarterly Journal of Mathematics* (April 1855), and republished with additions in the *Philosophical Magazine* for January 1878, and in Vol. I. of my *Mathematical and Physical Papers* (Art. XLVIII. Part vii.). Equations (6) and (8) of that article, found originally without the very convenient aid to thought given by the idea of motivity, are now reproduced as equations (19) and (23) above. The application to the thermo-elastic properties of fluids, of non-crystalline elastic solids, and of crystals, and to Thermo-magnetism, and to Pyro-electricity or the Thermo-electricity of non-conducting crystals, which that article contains, and my paper* on "Thermodynamics of Volta-Contact Electricity," read before the Royal Society of Edinburgh at its recent meeting of February 21, may be referred to as sufficiently illustrating the system of generalised co-ordinates and thermodynamic formulas of the present communication.

* [The paper next following here.]

89. ON THE THERMODYNAMICS OF VOLTA-CONTACT ELECTRICITY.

[From *Proc. Royal Soc. Edinburgh*, Vol. XXII. Feb. 21, 1898, pp 118—125.]

1. LET X and Y be two metals, of which X is electrically positive to Y in the Volta-contact series for dry metals, and make an incomplete circuit of them as indicated in the diagram, with X and Y metallically connected at an interface J, and free surfaces II, KCC exposed, with ether or air, or any gas or insulating fluid, between them. Let CC be a movable slab of the Y-metal, resting frictionlessly on a fixed surface KK of the same metal. If left to itself the movable slab would, in virtue of electric attraction, as we shall see, oscillate on the two sides of the middle position in which the whole of its upper surface is opposite to II. We shall suppose it held by applied force F in any position, or moved, or allowed to move, from any one position to any other, at our pleasure.

2. Suppose now our apparatus to be given with no excess of either electricity above the other, and to be insulated in air or ether at a distance from all other bodies great in comparison with its own dimensions, and with no electrified body near enough to it to produce sensible electrification through influence. Every part of the X-surface will be found positively and every part of the Y-surface negatively electrified, provided each surface is of perfectly uniform Volta-quality throughout its extent; but the electric surface-densities of the opposite electrifications will be everywhere exceedingly small, except in and near the portions of II and CC closely opposed to one another. Hence, if the slab is drawn outwards, an electric current will flow from II through the X-metal, and will cross the junction J from X to Y and flow through Y to compensate negative electricity on the portions of CC passing from close opposition to II. Our thermodynamic operations, of which we will arrange a Carnot cycle, are drawing in and out the slab CC, sometimes with the whole metal coated with an ideal varnish

impermeable to heat, and sometimes with the whole surface kept
at one temperature by giving heat to it, or taking heat from it,
where required for the fulfilment of this condition.

3. Suppose the relative thermo-electric quality of the two
metals is such that, when a complete metallic circuit is made of

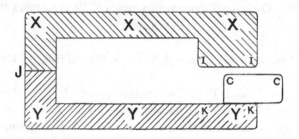

them, and the two junctions are kept at different temperatures,
the thermo-electric current is (according to the old French rule for
bismuth and antimony) "against the alphabet through hot"—that
is, from Y to X through the hot junction and from X to Y through
the cold junction. It is in this direction for some pairs X and Y
(as, for example, X zinc, Y gold), of which X is Volta-positive to
Y, and is in the opposite direction for others; but, by allowing
negative values to some of the quantities, this latter case is
included in our supposition which we make as a preliminary to
avoid circumlocutions.

4. Consider now the Peltier thermal effect of the current
produced as in § 2 by drawing the movable slab of Y-metal outwards. The current crosses the junction J in the same direction
as the natural thermo-electric current in a closed circuit with J
the cold junction: and the thermal effect is *therefore* production
of heat at J; according to a thermodynamic hypothesis which I
adopted as fulfilled so far as the *sign* of the Peltier effect was
concerned, in Peltier's splendid original discovery for bismuth and
antimony, and verified* by myself experimentally for copper and
iron below 280° C. (*Trans. Roy. Soc. Edin.*, May 1854, "Dynamical
Theory of Heat," part vi. §§ 105, 106; *Proc. Roy. Soc. Lond.*

---

* Verified by Le Roux and Jahn for several other pairs of metals for which they
also measured the value of the Peltier effect. Their results verify the thermodynamic hypothesis absolutely in respect to the sign, and tend to confirm it in
respect to the magnitude, of the Peltier effect. Le Roux (1867), *Ann. de Chim. et
de Phys.* (4), Vol. x. p. 201; Jahn (1888), *Wied. Ann.* Vol. xxxiv. p. 767.

May 1854, "Experimental Researches in Thermo-electricity," § II.
Both republished in Arts. XLVIII. and LI., Vol. I., *Mathematical
and Physical Papers.* See pp. 239—241 and 464—465).

But we must now consider, also, a quasi-Peltier effect produced
by electricity crossing the border between air or ether at the
surface of either metal, and the homogeneous metal inside. We
have absolutely no thermodynamic or molecular-hypothetic guide
to even guess the sign or magnitude of this effect at the surface of
either metal. It is conceivable that it may have opposite signs
for different metals, or that it is essentially of the same sign in all;
but it seems to me exceedingly improbable that it is non-existent,
when I consider Pellat's and Murray's discoveries of change of
Volta-surface-potential produced by scratching and by burnishing,
without any change of chemical constitution of the surface layer
of a metal.

5. Let Q denote the total quantity of heat produced by the
Peltier effect at J and the quasi-Peltier effect at the surfaces II
and CC, per unit quantity of electricity flowing from II through J
to CC, in virtue of motion of CC outwards. The part of Q which
is produced at J is, as we have seen, positive when X and Y are
in the thermo-electric order stated in § 3; but the total amount
of Q may be either positive or negative.

6. Our Carnot cycle will consist of the following four operations :—

I. ("Adiabatic"—according to Rankine's nomenclature.) The
whole apparatus being ideally coated with varnish impermeable to heat, draw out CC so slowly that the
temperature of the whole apparatus remains uniform
throughout, while rising from t to t', on absolute thermodynamic scale.

II. (Isothermal.) The whole apparatus being kept by proper
surface appliances at constant temperature t', let CC
move inwards very slowly, until a certain quantity of
heat, H', has been taken into the apparatus from
without.

III. (Adiabatic.) Let the slab move further inwards very
slowly till the temperature of the whole apparatus,
ideally coated with impermeable varnish, sinks to t.

IV. (Isothermal.) The whole apparatus being kept by proper surface appliances at constant temperature t, draw the slab outwards to its primitive position. Let H be the quantity of heat which must be removed to prevent lowering of temperature.

7. Remark that if Q of § 5 is positive, H' and H are both positive and $H' > H$; but if Q is negative, H' and H are both negative and $-H > -H'$. In either case we have essentially by my definition of absolute temperature (*Math. and Phys. Papers*, Vol. III. Art. XCII. part ii. §§ 34, 35)

$$H'/H = t'/t;$$

in the former case $t' > t$, in the latter $t' < t$.

By working out analytically all the details of this cycle of operations, and taking into account Joule's law of equivalence between heat and work, we arrive at the result

$$JQ = dV/d(\log t)$$

given as our final result in equation (7) below.

But we arrive at it more easily, and in some respects more conveniently, by founding on the doctrine of motivity, as follows:—

8. Let V be the Volta-difference of potential in air or ether between the opposed X and Y metallic surfaces.

Let ξ, ξ' be the electrostatic capacities of the variable condenser CC, II, for two positions of the movable slab CC, which for brevity we call position ξ and position ξ'.

The work required to pull out the slab from position ξ to position ξ' will be

$$\tfrac{1}{2}V^2(\xi - \xi').$$

The quantity of electricity passing from II across J to KK during this operation will be

$$V(\xi - \xi');$$

and therefore the quantity of heat which would have to be removed from the junction J and the surfaces II, CC, to prevent rise of temperature (§ 5 above), is

$$QV(\xi - \xi'), \text{ or } -QV(\xi' - \xi).$$

Hence if $e(\xi, t)$ denote the energy of the apparatus in the condition (ξ, t), we have

$$\frac{de(\xi, t)}{d\xi} = -\left(\tfrac{1}{2}V^2 - JQV\right) \dots\dots\dots\dots\dots(1).$$

Considering now change of temperature with ξ constant, we have

$$\frac{de(\xi, t)}{dt} = JN(\xi, t) \dots\dots\dots\dots\dots(2),$$

where $N(\xi, t)$ denotes the thermal capacity of the apparatus with ξ constant. When we consider the possible or probable quasi-Peltier productions of heat at II and CC, and the probability that these are different for the two metals, and the Peltier effect at J, we must regard N as probably varying with ξ. We must therefore regard it as a function of ξ and t.

9. Consider now the motivity* of our apparatus, which we shall denote by $m(\xi, t)$, being, as is the energy, a function of ξ and t.

When the slab CC is drawn out so as to diminish the capacity from ξ to ξ', the heat $QV(\xi - \xi')$, which, to prevent temperature from rising, must be given to external matter at temperature t, contributes to the recipient an amount of motivity equal to

$$\frac{t-T}{t}JQV(\xi - \xi'),$$

where T denotes the lowest temperature of neighbouring matter, available for receiving heat. Hence we have

$$\frac{dm(\xi, t)}{d\xi} = -\left(\tfrac{1}{2}V^2 - \frac{t-T}{t}JQV\right) \dots\dots\dots(3).$$

And if we raise the temperature of the whole apparatus infinitesimally from $t - \tfrac{1}{2}dt$ to $t + \tfrac{1}{2}dt$, we add to its motivity an amount

$$\frac{t-T}{t}JN(\xi, t)\,dt.$$

Hence $\qquad \dfrac{dm(\xi, t)}{dt} = \dfrac{t-T}{t}JN(\xi, t) \dots\dots\dots\dots(4).$

From (1) and (2) with the equation

$$\frac{d}{d\xi}\frac{de}{dt} = \frac{d}{dt}\frac{de}{d\xi};$$

---

* *Proc. Roy. Soc. Edin.* 1876; *Phil. Mag.* May 1879 ; *Math. and Phys. Papers*, Vol. I. Art. i. ; *Proc. Roy. Soc. Edin.* March 21, 1898 [*supra*].

and from (3) and (4) with the equation

$$\frac{d}{d\xi}\frac{dm}{dt} = \frac{d}{dt}\frac{dm}{d\xi};$$

we find
$$J\frac{dN}{d\xi} = -\left\{V\frac{dV}{dt} - J\frac{d(QV)}{dt}\right\} \quad\ldots\ldots\ldots(5)$$

and

$$\frac{t-T}{t}J\frac{dN}{d\xi} = -\left\{V\frac{dV}{dt} - \frac{t-T}{t}J\frac{d(QV)}{dt} - \frac{T}{t^2}JQV\right\} \quad\ldots\ldots(6).$$

Eliminating $\frac{dN}{d\xi}$ between (5) and (6) we find finally

$$\frac{dV}{dt} = \frac{J}{t}Q: \quad \text{or,} \quad JQ = \frac{dV}{d(\log t)} \quad\ldots\ldots\ldots(7).$$

10. The quantity of heat absorbed or produced in virtue of change of electric density on the surfaces II, CC must in all probability in every case be too small to be detected by direct observation. But the difference between the quantities produced at the two surfaces of the two different metals, per unit of electric quantity added to one surface or taken from the other, which is $Q - \Pi$ if Π be the Peltier effect at the junction J, can by aid of equation (7) be readily and surely determined for any two metals for which the Peltier effect is known. In fact, it is easy to arrange apparatus for measuring V through a considerable range of temperature, say from 0° to 100° C., by the now well-known compensation method introduced independently by Pellat and myself, and thus to measure dV/dt and so find Q by equation (7). I am at present commencing experiments for this purpose with air, and with carbonic acid gas, between the two opposed metal surfaces, so that we may judge what precautions, if any, will be necessary to eliminate disturbances due to different condensations of gas on the metals at different temperatures. Very interesting and important experiments by Pellat and by Erskine Murray have shown large temperature effects on the Volta-electric force between two plates, whether of the same or of different metal, when one of them is heated and the other kept at the ordinary atmospheric temperature; but this does not supply what we now want from experiment, which is, the variation of the Volta-electric force between two metals at one and the same temperature, when this temperature is varied.

Addition, of date 26th March 1898.

A number of preliminary experiments, carried on with the assistance of Mr W Craig Henderson, have shown large but largely irregular temperature variations of the Volta E.M.F. between copper and zinc, with all parts of the Volta circuit at the same temperature. The substitution of carbonic acid gas for air, and re-introduction of air for carbonic acid, seemed to make but little difference on the results. The irregularities seemed to have been chiefly due to permanent or sub-permanent changes in the copper plate. At all events we found somewhat more nearly regular results with zinc and gold. The zinc plate used had never, so far as I know, been polished or much disturbed by touching its surface since experimented on by Mr Erskine Murray three years ago. The "gold" was a brass plate gilded for me about 1859. It was one of the two "standard gold" plates in Erskine Murray's experiments, and, so far as I know, it has never been rubbed or polished since 1861.

We found in a range from 16° C. to 50° C. an augmentation of Volta-contact difference at rough average rate of about ·002 of a volt, or ·2 × 10⁶ C.G.S. units, per degree Centigrade. This is 800 times the thermo-electric difference of zinc and gold given as 250 C.G.S. units per degree Centigrade in Jenkins' *Electricity and Magnetism*, p. 176, and Everett's *Physical Units*, 1886, p. 173. And (§ 3 above) gold, zinc are as Y, X in respect to orders in the Volta series and in the thermo-electric series of metals. Hence, according to the secure thermo-dynamic formula (7) above, and the old probable thermo-dynamic hypothesis for thermo-electricity (§ 4 above), the quasi-Peltier effect at the interfaces gold-air and air-zinc in the Volta circuit would be 799 times the Peltier effect at the zinc-gold interface, if the preceding figure 800 were exact.

The *sign* of the quasi-Peltier effect for gold-zinc is such that on the whole heat is produced by vitreous electricity travelling from gold to gold-air frontier, and an equal quantity of vitreous electricity from air-zinc frontier to zinc; and, on the whole, cold is produced by equal electric motions in the opposite direction.

On Electric Equilibrium between Uranium and an Insulated Metal in its Neighbourhood. By Lord Kelvin,
J. Carruthers Beattie, and M. Smoluchowski de Smolan.

[*Roy. Soc. Edin.* (read March 1, 1897); accidentally omitted from *Proc.* for
Session 1896—1897, p. 417, but published in *Nature* for March 11, 1897.]

THE wonderful fact that uranium held in the neighbourhood
of an electrified body diselectrifies it was first discovered by
H. Becquerel. Through the kindness of Prof. Moissan we have
had a disc of this metal, about 5 cm. diameter and ½ cm. thickness,
placed at our disposal.

We made a few preliminary observations on its diselectrifying
property. We observed first the rate of discharge when a body
was charged to different potentials. We found that the quantity
lost per half-minute was very far from increasing in simple proportion to the voltage, from 5 volts up to 2100 volts; the electrified
body being at a distance of about 2 cm. from the uranium disc.
[*Added* March 9, 1897.—We have to-day seen Prof. Becquerel's
paper in *Comptes Rendus* for March 1. It gives us great pleasure
to find that the results we have obtained on discharge by uranium
at different voltages have been obtained in another way by the
discoverer of the effect. A very interesting account will be found
in Prof. Becquerel's paper, which was read to the French Academy
of Sciences on the same evening, curiously enough, as ours was
read before the Royal Society of Edinburgh.]

These first experiments were made with no screen placed
between the uranium and the charged body. We afterwards
found that there was also a discharging effect, though much slower,
when the uranium was wrapped in tinfoil. The effect was still
observable when an aluminium screen was placed between the
uranium, wrapped in tinfoil, and the charged body.

To make experiments on the electric equilibrium between
uranium and a metal in its neighbourhood, we connected an
insulated horizontal metal disc to the insulated pair of quadrants
of an electrometer. We placed the uranium opposite this disc
and connected it and the other pair of quadrants of the
electrometer to the sheath. The surface of the uranium was

parallel to that of the insulated metal disc, and at a distance of
about 1 cm. from it. It was so arranged as to allow of its easy
removal.

With a polished aluminium disc as the insulated metal, and
with a similar piece of aluminium placed opposite it in place
of the uranium, no deviation from the metallic zero was found
when the pairs of quadrants were insulated from one another.
With the uranium opposite the insulated polished aluminium, a
deviation of − 84 scale-divisions from the metallic zero was found
in about half a minute. [Sensibility of electrometer 140 scale-
divisions per volt.] After that, the electrometer-reading remained
steady at this point, which we may call the uranium-rays-zero for
the two metals separated by air which was traversed by uranium
rays. If, instead of having the uranium opposite to the aluminium,
with only air between them, the uranium was wrapped in a piece
taken from the same aluminium sheet, and then placed opposite
to the insulated polished aluminium disc, no deviation was produced. Thus in this case the rays-zero agreed with the metallic
zero.

With polished copper as the insulated metal, and the uranium
separated only by air from this copper, there was a deviation of
about + 10 scale-divisions. With the uranium wrapped in thin
sheet aluminium, and placed in position opposite the insulated
copper disc, a deviation from the metallic zero of + 43 scale-
divisions was produced in two minutes, and at the end of that
time a steady state had not been reached.

With oxidised copper as the insulated metal, opposed to the
uranium with only air between them, a deviation from the metallic
zero of about + 25 scale-divisions was produced.

When the uranium, instead of being placed at a distance of
1 cm. from the insulated metal disc, was placed at a distance of
2 or 3 mm., the deviation from the metallic zero was the same.

These experiments lead us to infer that two polished metallic
surfaces connected to the sheath and the insulated electrode of an
electrometer give, when the air between them is influenced by the
uranium rays, a deflexion from the metallic zero the same in
direction, and of about the same amount, as when the two metals
are connected by a drop of water. The uranium itself may be one
of the two metals.

90. NOTICE OF STIRLING'S AIR-ENGINE.

[From *Glasg. Phil. Soc. Proc.* Vol. II. pp. 169—170 ; read April 21, 1847.]

ATTENTION was called to the circumstance that, in accordance with Carnot's theory*, of which an explanation had been given by Professor Gordon at a previous meeting of the Society, the mechanical effect to be obtained by an Air Engine, from the transmission of a given quantity of heat, depends on the difference between the temperatures of the air in the cold space above and the heated space below the plunger; as this difference is considerably greater than that which exists between the boiler and the condenser in the best condensing Steam Engines, it appears that, if the practical difficulty in the construction of an efficient Air Engine can ever be removed to nearly the same extent as already has been done in the case of the Steam Engine, a much greater amount of mechanical effect would be obtained by the consumption of a given quantity of fuel.

Some illustrations, afforded by the Air Engine, of general physical principles, were also noticed. If the Air Engine be turned *forwards*, by the application of power, and if no heat be applied, the space below the plunger will become colder than the surrounding atmosphere, and the space above hotter. Expenditure of *work* will be necessary to turn the Engine, after this difference of temperatures, contrary to that which is necessary to *cause* the Engine to turn forwards, has been established. If, however, we prevent the temperature in one part from rising, and in the other from sinking, the Engine may be turned without the expenditure of any work (except what is necessary in an actual machine for overcoming friction, &c.). One obvious way of retaining the two parts at the same temperature, is to keep the machine immersed in a stream of water; but there is another way in which this may be done, if we can find a solid body which melts at the temperature at which it is required to retain the Engine. For instance, let this temperature be 32°; let a stream of water at 32° be made to run across the upper part of the Engine, and let the lower part of the vessel containing the plunger, which is protected from the stream, be held in a bason of

* An account of this theory is given in a paper by Clapeyron " On the Motive Power of Heat," of which a translation is published in Taylor's *Scientific Memoirs*, Vol. I.

water at 32°. When the Engine is turned forwards, heat will be taken from the space below the plunger and deposited in the space above. Now, this heat must be supplied by the water in the bason, which will, therefore, be gradually converted into ice at 32°. Hence we see that water at 32° may be converted into ice at 32°, without the expenditure of any work. This may also be very easily proved in the following manner :

Let a syringe be constructed of perfectly non-conducting materials, except the lower end of the cylinder, which is to be stopped by a solid plate, a perfect conductor. The syringe being at first full of air, at atmospheric pressure, and at the temperature of 32°, let the lower end be dipped in a stream of water at 32°, and the piston be pushed down. Let the syringe be then placed with its lower end in a bason of water at 32°, and the piston be allowed to rise. The mechanical effect given out in this part of the operation will be equal to the work spent in the former, and a portion of the water in the bason will be turned into ice.

NOTE.—To avoid perplexity, in the account which was given, it was supposed that the temperature of the air is always the same as that of the vessel in which it is contained, which will only be strictly true, even were the action of the plunger perfect in altering the temperature of the air, when the motion is very slow.

91. NOTE ON THE EFFECT OF FLUID FRICTION IN DRYING STEAM WHICH ISSUES FROM A HIGH-PRESSURE BOILER THROUGH A SMALL ORIFICE INTO THE OPEN AIR.

[From *Philosophical Magazine*, Vol. I. June 1851, p. 474 ; Vol. II. October 1851, pp. 273, 274.]

IN a letter to Mr Joule written last October, and since published in the *Philosophical Magazine**, I pointed out that the remarkable discovery (made independently by Messrs Rankine and Clausius), that steam allowed to expand requires heat to be added to it to prevent any part of it from becoming liquefied, can only be reconciled with the known fact of the dryness of steam issuing into the open air from a high-pressure boiler through a small aperture, by taking into account the heat developed by the fluid friction in the neighbourhood of the aperture. I may add,

* Vol. XXXVII. p. 387 (December 1850). [*Math. and Phys. Papers*, Vol. I. pp. 170—173.]

that the immediate mechanical effect of the work done by the
steam in pressing out, is the generation of *vis viva* in the fluid;
and all of this *vis viva*, except the very small proportion of it
retained by the steam after leaving the *rapids*, is lost in friction
before the fluid reaches the locality where its pressure is equal
sensibly to that of the atmosphere; that is, is converted into
thermal *vis viva* or heat. M. Clausius*, in an investigation of the
circumstances of this case of expanding steam, points out distinctly
that work is done, but overlooks the mechanical effect produced;
and in consequence arrives at the conclusion that my proposition,
that the steam, when it has expanded till its pressure is equal to
that of the atmosphere, could not be dry without the heat, or some
of the heat, it gets from fluid friction, is false. These remarks
are, I trust, sufficient to show that M. Clausius's objections to my
reasoning are groundless.

Second Note on the Effect of Fluid Friction in drying Steam which issues from a High-Pressure Boiler into the Open Air.

In the August Number of this Magazine, M. Clausius has
replied to a Note, published in the June Number, in which I
endeavoured to show that the objections he had made to my
reasoning regarding the condition of steam issuing from a high-
pressure boiler, were groundless. I cannot perceive that this
reply at all invalidates any of the statements made in my two
former communications†, to which I refer the reader who desires
to ascertain what my views are, and to judge as to the correctness
of the reasoning by which they are supported. An analytical
investigation, according to the principles discovered by Mr Joule,
of the thermo-dynamical circumstances of the rushing of any fluid
through a small orifice, is given in a paper communicated last
April to the Royal Society of Edinburgh, and since published in
the *Transactions* (Vol. xx. Part ii.) under the title "On a Method
of discovering Experimentally the Relation between the Mechanical
Work spent and the Heat produced by the compression of a gaseous
fluid§."

* In an article recently published in Poggendorff's *Annalen*, and republished in
the last Number (May) of the *Philosophical Magazine*.

† *Phil Mag.* Vol. xxxvii. p. 387 (Nov. 1850), and Vol. i. 4th Ser., p. 474 (June 1851).

§ [*Math. and Phys. Papers*, Vol. i. pp. 210—222].

I take the present opportunity of correcting a mistaken expression in my first communication regarding steam issuing from a high-pressure boiler, by which I gave a false, or an inadequate, representation of the connexion of that application of Mr Joule's general principles which I was bringing forward, with one which he had himself made in one of his published papers. The following is the passage of my communication (addressed as a letter to Mr Joule), which requires correction :—

" The pretended explanation of a corresponding circumstance connected with the rushing of air from one vessel to another in Gay-Lussac's experiment, on which you have commented, is certainly not applicable in this case, since, instead of receiving heat from without, the steam must lose a little in passing through the stop-cock or steam-pipe by external radiation and convection * "

I wrote this under the impression that Mr Joule had, in his paper " On the Changes of Temperature produced by the Condensation and Rarefaction of Air†," pointed out the incorrectness of an explanation often given of Gay-Lussac's experiment‡, and shown that the phenomenon could be truly explained only by taking into account the heat developed in the air by friction in its passage from one vessel to the other through the stop-cock. I find, however, on looking to the paper, which I had not by me when I wrote, that it contains no reference to Gay-Lussac's experiment, but the following passage, referring to Mr Joule's own experiments on the heat developed by the compression of air, and the heat absorbed by air allowed to expand from a vessel into which it has been compressed, through a small orifice, into the atmosphere, from which I obtained the idea of considering the heat developed by the friction of steam issuing from a high-pressure boiler.

" It is quite evident that the reason why the cold in the experiments of Table IV. was so much inferior in quantity to the heat evolved in those of Table I., is, that all the force of the air, over and above that employed in lifting the atmosphere, was applied in overcoming the resistance of the stop-cock, and was there converted back again into its equivalent of heat§."

* *Phil. Mag.* Ser. 3, Vol. xxxvii. p. 388. [*Math. and Phys. Papers*, Vol. i. p. 171.]
† *Phil. Mag.* Ser. 3, Vol. xxvi. p. 369 (May 1845).
‡ See Lamé, *Cours de Physique*, Vol. i. § 352.
§ *Phil. Mag.* Ser. 3, Vol. xxvi. p. 381.

92. On the Mechanical Action of Radiant Heat or Light.
     On the Power of Animated Creatures over Matter.
     On the Sources available to Man for the Production
     of Mechanical Effect.

[From *Edin. Roy. Soc. Proc.* Vol. III. 1857, pp. 108—113 [read Feb. 2, 1852];
    Phil. Mag. Vol. IV. Oct. 1852, pp. 256—260; *Math. and Phys. Papers,*
    Vol. I. Art. lviii. pp. 505—510.]

93. On a Universal Tendency in Nature to the Dissi-
     pation of Mechanical Energy.

[From *Edin. Roy. Soc. Proc.* Vol. III. [read April 19, 1852], pp. 139—142;
    Phil. Mag. Vol. IV. Oct. 1852, pp. 304—306; *Math. and Phys. Papers,*
    Vol. I. Art. lix. pp. 511—514.]

94. On the Restoration of Mechanical Energy from an
     unequally heated Space.

[From *Phil. Mag.* Vol. V. Feb. 1853, pp. 102—105; *Math. and Phys. Papers,*
    Vol. I. Art. lxiii. pp. 554—558.]

95. On Thermo-elastic and Thermo-magnetic Properties
     of Matter, Part I.

[From *Quart. Journal of Math.* Vol. I. 1857 [dated from Glasgow College,
    March 10, 1855], pp. 57—77; *Phil. Mag.* Vol. V. 1878 [with additions],
    pp. 4—27; *Math. and Phys. Papers,* Vol. I. Art. xlviii. pp. 291—316.]

96. On the Origin and Transformations of Motive
     Power.

[From *Roy. Institution Proc.* Vol. II. 1854—1858 [Feb. 29, 1856], pp. 199—204;
    Chemist, Vol. III. 1856, pp. 607—612; *Math. and Phys. Papers,* Vol. II.
    Art. xc. pp. 182—188; *Popular Lectures and Addresses,* Vol. II. pp. 418—
    432.]

97. ON THE DISCOVERY OF THE TRUE FORM OF CARNOT'S FUNCTION.

[From *Phil. Mag.* Vol. XI. June 1856, pp. 447, 448.]

To the Editors of the Philosophical Magazine and Journal.

GENTLEMEN,

In a paper communicated to the Royal Society of Edinburgh in 1851*, Prof. W. Thomson ascribed to Mr J. P. Joule the discovery of the theorem, that Carnot's function, which Clapeyron expressed by C, and Thomson by the fraction $1/\mu$, " is nothing more than the absolute temperature multiplied by the equivalent of heat for the unit of work." I have hitherto avoided mentioning this point in my papers; principally because I have so high an esteem for the labours of the physicist for whom Prof. Thomson claims priority, that I was anxious to avoid even the appearance of wishing to lessen his deserts. But as Prof. Thomson has since then frequently repeated that assertion, —among other places in the paper in the March Number of this Journal, where, in page 215, he calls that theorem " Mr Joule's conjecture,"—I think it necessary to say a few words on the subject.

Holtzmann established the same formula for the function C in a paper which appeared as early as 1845†; and Helmholtz, in his pamphlet published in 1847, " On the Conservation of Force," citing Holtzmann's paper, calculated several values obtained by this formula, and compared them with those arrived at by Clapeyron in a different manner. But the views upon which Holtzmann founded his speculations do not agree with the mechanical theory of heat as at present received; so that after this had been recognized, the correctness of the formula found by

---

* *Edin. Trans.* Vol. XX.; and *Phil. Mag.* 4th Ser. Vol. IX.

† *On the Heat and Elasticity of Gases and Vapours.* By C. Holtzmann. Mannheim, 1845.

him was, naturally, again rendered doubtful. On this account, in a paper communicated to the Berlin Academy in February 1850*, "On the Moving Force of Heat," in which I brought Carnot's proposition in agreement with the mechanical theory of heat, I again endeavoured to determine his function more accurately. Therein I arrived at the same formula as Holtzmann, and I believe that I then, for the first time, correctly explained the principles upon which this formula is based.

In presence of these facts, Prof. Thomson, to justify his statement, says†, "It was suggested to me by Mr Joule, in a letter dated December 9, 1848, that the true value of μ might be inversely as the temperature from zero." Against this I must beg to urge,—*First*, that, as far as I am aware, it is usual, in determining questions of priority in scientific matters, only to admit such statements as have been *published*. And I believe that this custom ought to be conscientiously adhered to, especially in theoretical investigations; for it usually requires continued and laborious research in order to give to a thought, after it has been first entertained, and perhaps casually communicated to a friend, that degree of certainty which is necessary before venturing upon its publication. *Secondly*, that since Thomson does not say that Mr Joule had proved the theorem, but only that he had offered it as an opinion, I do not see why this opinion should have the priority over that which Holtzmann had arrived at three years before.

In conclusion allow me to make one remark. In a more recent paper, "On a Modified Form of the Second principal Theorem in the Mechanical Theory of Heat‡," I have introduced, instead of Carnot's function C, another function of the temperature, which I have designated by T, and by which all developments are very much simplified. This function has a determinate relation to that of Carnot's, which I have expressed by the equation

$$\frac{dT}{dt} \bigg/ T = \frac{A}{C},$$

in which t represents the temperature, and A the equivalent of heat for unit of work. It is easily recognized, that, according

---

* Poggendorff's *Annalen*, Vol. LXXIX.; and *Phil. Mag.* 4th Ser. Vol. II.

† *Edin. Trans.* Vol. XX. p. 279.

‡ Poggendorff's *Annalen*, Vol. XCIII.

to this equation, the functions C and T are in general to be considered different; but that for the special case, in which C is proportional to the absolute temperature, T must be also proportional to it. And in fact I have shown, from the same principles which before led me to the determination of C, that in all probability T is simply the absolute temperature itself.

I remain, Gentlemen,

With great respect, yours &c.,

R. CLAUSIUS.

Zurich, *March* 20, 1856.

To the Editors of the Philosophical Magazine and Journal.

Oakfield, Moss Side, Manchester,
May 12, 1856.

GENTLEMEN,

M. Clausius, in a letter of date March 20, 1856, addressed to yourselves, and published this month in your Magazine, objects to a statement he supposes me to have made in 1851, and to have frequently repeated since that time, that Mr Joule had discovered "the theorem, that Carnot's function (C or $1/\mu$) 'is nothing more than the absolute temperature multiplied by the equivalent of heat for the unit of work.'" He attributes the discovery of the true form of Carnot's function to Holtzmann, who gave the formula referred to in a paper which appeared as early as 1845; but he believes that in his own paper "On the Moving Force of Heat," communicated to the Berlin Academy in 1850, the principles upon which that formula is based were first correctly explained.

Allow me to answer the charge he makes against me by quoting what I said with reference to the "discovery" for which M. Clausius claims priority.

"This formula was suggested to me by Mr Joule, in a letter dated December 9, 1848, as probably a true expression for μ, being required to reconcile the expression derived from Carnot's theory (which I had communicated to him) for the heat evolved in terms of the work spent in the compression of a gas, with the hypothesis that the latter of these is exactly the mechanical equivalent of the former, which he had adopted in consequence of its being, at least approximately, verified by his own experiments.

This, which will be called Mayer's hypothesis, from its having been first assumed by Mayer, is also assumed by Clausius without any reason from experiment; and an expression for μ, the same as the preceding, is consequently adopted by him as the foundation of his mathematical deductions from elementary reasoning regarding the motive power of heat*."

This passage is the sequel to the extract quoted by M. Clausius in his letter to you, and appeared in the same Part of the *Transactions*, and in the same volume of the *Philosophical Magazine*. When it is read, I think it will be admitted that I did not do injustice to his claims in writing the following sentence two years later, of which the words "Mr Joule's conjecture" have called forth his reclamation :—

"A more convenient assumption has since been pointed to by Mr Joule's conjecture, that Carnot's function is equal to the mechanical equivalent of the thermal unit divided by the temperature by the air thermometer from its zero of expansion; an assumption which experiments on the thermal effects of air escaping through a porous plug, undertaken by him in conjunction with myself for the purpose of testing it (*Phil. Mag.* October 1852), have shown to be not rigorously, but very approximately true."

I remain, Gentlemen,

Yours very faithfully,

WILLIAM THOMSON.

98. DISCUSSION OF J. P. JOULE'S PAPER ON "A SURFACE CONDENSER."

[From *Inst. Mech. Engrs. Proc.* 1856, pp. 191—192, 194. (Glasgow Meeting, Sept. 17, 18, 1856.)]

[Mainly remarks on the practical working of the steam-condenser in which he had assisted Joule, especially on the importance of a good vacuum. There is a later memoir by Joule, "On the Surface Condensation of Steam," *Phil. Trans.* 1860, Joule's *Scientific Papers*, Vol. I. pp. 502—531.]

* "On a Method of discovering experimentally the Relation between the Mechanical Work spent, and the Heat produced by the Compression of a Gaseous Fluid." (*Trans. Roy. Soc. Edinb.* April 17, 1851; or *Phil. Mag.* December 1852.) [*Math. and Phys. Papers*, Vol. I. pp. 210—222.]

99. Remarks on the Interior Melting of Ice.

In a letter to Prof. Stokes, Sec. R.S.

[From *Roy. Soc. Proc.* Vol. IX. 1857—1859, pp. 141—143 (read Jan. 25, 1858); *Phil. Mag.* Vol. XVI. Oct. 1858, pp. 303, 304; reprinted in *Baltimore Lectures*, Appendix F, pp. 577—579.]

In the Number of the *Proceedings* just published, which I received yesterday, I see some very interesting experiments described in a communication by Dr Tyndall, "On some Physical Properties of Ice." I write to you to point out that they afford direct ocular evidence of my brother's theory of the plasticity of ice, published in the *Proceedings* of the 7th of May last; and to add, on my own part, a physical explanation of the blue veins in glaciers, and of the lamellar structure which Dr Tyndall has shown to be induced in ice by pressure, as described in the sixth section of his paper.

Thus, my brother, in his paper of last May, says, "If we commence with the consideration of a mass of ice perfectly free from porosity, and free from liquid particles diffused through its substance, and if we suppose it to be kept in an atmosphere at or above 0° Cent., then, as soon as pressure is applied to it, pores occupied by liquid water must instantly be formed in the compressed parts, in accordance with the fundamental principle of the explanation I have propounded—the lowering, namely, of the freezing-point or melting-point, by pressure, and the fact that ice cannot exist at 0° Cent. under a pressure exceeding that of the atmosphere." Dr Tyndall finds that when a cylinder of ice is placed between two slabs of box-wood, and subjected to gradually increasing pressure, a dim cloudy appearance is observed, which he finds is due to the melting of small portions of the ice in the interior of the mass. The permeation into portions of the ice, for a time clear, "by the water squeezed against it from such parts as may be directly subjected to the pressure," theoretically demonstrated by my brother, is beautifully illustrated by Dr Tyndall's statement, that "the hazy surfaces produced by the compression of the mass were observed to be in a state of intense commotion, which followed closely upon the edge of the surface as it advanced through the solid. It is finally shown that these surfaces are due to the liquefaction of the ice in planes perpendicular to the pressure."

There can be no doubt but that the "oscillations" in the melting-point of ice, and the distinction between strong and weak pieces in this respect, described by Dr Tyndall in the second section of his paper, are consequences of the varying pressures which different portions of a mass of ice must experience when portions within it become liquefied.

The elevation of the melting temperature which my brother's theory shows must be produced by diminishing the pressure of ice below the atmospheric pressure, and to which I alluded as a subject for experimental illustration, in the article describing my experimental demonstration of the lowering effect of pressure (*Proceedings, Roy. Soc. Edin.* Feb. 1850), demonstrates that a vesicle of water cannot form in the interior of a solid of ice except at a temperature higher than 0° Cent. This is a conclusion which Dr Tyndall expresses as a result of mechanical considerations: thus, "Regarding heat as a mode of motion," "liberty of liquidity is attained by the molecules at the surface of a mass of ice before the molecules at the centre of the mass can attain this liberty."

The physical theory shows that a removal of the atmospheric pressure would raise the melting-point of ice by $\frac{3}{400}$ths of a degree Centigrade. Hence it is certain that the interior of a solid of ice, heated by the condensation of solar rays by a lens, will rise to at least that excess of temperature above the superficial parts. It appears very nearly certain that cohesion will prevent the evolution of a bubble of vapour of water in a vesicle of water forming by this process in the interior of a mass of ice, until a high "negative pressure" has been reached, that is to say, until cohesion has been called largely into operation, especially if the water and ice contain little or no air by absorption (just as water freed from air may be raised considerably above its boiling-point under any non-evanescent hydrostatic pressure). Hence it appears nearly certain that the interior of a block of ice originally clear, and made to possess vesicles of water by the concentration of radiant heat, as in the beautiful experiments described by Dr Tyndall in the commencement of his paper, will rise very considerably in temperature, while the vesicles enlarge under the continued influence of the heat received by radiation through the cooler enveloping ice and through the fluid medium (air and a watery film, or water) touching it all round, which is necessarily at 0° Cent. where it touches the solid.

I find I have not time to execute my intention of sending you to-day a physical explanation of the blue veins of glaciers which occurred to me last May, but I hope to be able to send it in a short time.

100. On the Stratification of Vesicular Ice by Pressure.

In a letter to Prof. Stokes, Sec. R.S.

[From *Roy. Soc. Proc.* Vol. ix. 1857—1859, pp. 209—213 [read Apr. 22, 1858]; *Phil. Mag.* Vol. xvi. Dec. 1858, pp. 463—466. Reprinted in *Baltimore Lectures*, pp. 579—583.]

In my last letter to you I pointed out that my brother's theory of the effect of pressure in lowering the freezing-point of water, affords a perfect explanation of various remarkable phenomena involving the internal melting of ice, described by Professor Tyndall in the number of the *Proceedings* which has just been published. I wish now to show that the stratification of vesicular ice by pressure observed on a large scale in glaciers, and the lamination of clear ice described by Dr Tyndall as produced in hand specimens by a Bramah's press, are also demonstrable as conclusions from the same theory.

Conceive a continuous mass of ice, with vesicles containing either air or water distributed through it; and let this mass be pressed together by opposing forces on two opposite sides of it. The vesicles will gradually become arranged in strata perpendicular to the lines of pressure, *because of the melting of ice in the localities of greatest pressure and the regelation of the water in the localities of least pressure, in the neighbourhood of groups of these cavities.* For, any two vesicles nearly in the direction of the condensation will afford to the ice between them a relief from pressure, and will occasion an aggravated pressure in the ice round each of them in the places farthest out from the line joining their centres; while the pressure in the ice on the far sides of the two vesicles will be somewhat diminished from what it would be were their cavities filled up with the solid, although not nearly as much diminished as it is in the ice between the two. Hence, as demonstrated by my brother's theory and my own experiment, the melting temperature of the ice round each vesicle will be highest on its side nearest to the other vesicle, and lowest in the localities on the whole farthest from the line joining the centres. Therefore, ice will melt from these last-mentioned localities, and, if each vesicle have water in it, the partition between the two will thicken by freezing on each side of it. Any two vesicles, on the other hand, which are nearly in a line perpendicular to the direction of pressure will agree in leaving an aggravated pressure to be borne by the solid between them, and will each direct away some of the pressure from the portions of the solid next itself on the two sides farthest from the plane through the centres, perpendicular to the line of pressure. This will give rise to an increase of pressure on the whole in the solid all round the two cavities, and nearly in the plane perpendicular to the pressure, although nowhere else so much as in the part between them. Hence these two vesicles will gradually extend towards one another by the melting of the intervening ice, and each will become flattened in towards the plane through the centres perpendicular to the direction of pressure, by the freezing of water on the parts of the bounding surface farthest from this plane. It may be similarly shown that two vesicles in a line oblique to that of condensation will give rise to such variations of pressure in the solid in their neighbourhood, as to make them, by melting and freezing, to extend, each obliquely *towards* the other and *from* the parts of its boundary most remote from a plane midway between them, perpendicular to the direction of pressure.

The general tendency clearly is for the vesicles to become flattened and arranged in layers, in planes perpendicular to the direction of the pressure from without.

It is clear that the same general tendency must be experienced even when there are bubbles of air in the vesicles, although no doubt the resultant effect would be to some extent influenced by the running down of water to the lowest part of each cavity.

I believe it will be found that these principles afford a satisfactory physical explanation of the origin of that beautiful veined structure which Professor Forbes has shown to be an essential organic property of glaciers. Thus the first effect of pressure not equal in all directions, on a mass of snow, ought to be, according to the theory, to convert it into a stratified mass of layers of alternately clear and vesicular ice, perpendicular to the direction of maximum pressure. In his remarks "On the Conversion of the Névé into Ice *," Professor Forbes says, *"that the conversion into ice is simultaneous"* (and in a particular case referred to *"identical"*) *" with the formation of the blue bands* ;...and that these bands are formed where the pressure is most intense, and where the differential motion of the parts is a maximum, that is, near the walls of a glacier." He farther states, that, after long doubt, he feels satisfied that the conversion of snow into ice is due to the effects of pressure on the loose and porous structure of the former; and he formally abandons the notion that the blue veins are due to the freezing of infiltrated water, or to any other cause than the kneading action of pressure. All the observations he describes seem to be in most complete accordance with the theory indicated above. Thus, in the thirteenth letter, he says, " the blue veins are formed where the pressure is most intense and the differential motion of the parts a maximum."

Now the theory not only requires pressure, but requires difference of pressure in different directions to explain the stratification of the vesicles. Difference of pressure in different directions produces the " differential motion " referred to by Professor Forbes. Further, the difference of pressure in different directions must be continued until a very considerable amount of this differential motion, or distortion, has taken place, to produce any sensible degree of stratification in the vesicles. The absolute amount of distortion experienced by any portion of the viscous mass is therefore an index of the persistence of the differential pressure, by the continued action of which the blue veins are induced. Hence also we see why blue veins are not formed in any mass, ever so deep, of snow *resting* in a hollow or corner.

As to the direction in which the blue veins appear to lie, they must, according to the theory, be something intermediate between the surfaces perpendicular to the greatest pressure, and the surfaces of sliding; since they will commence being formed exactly perpendicular to the direction of greatest pressure, and will, by the differential motion accompanying their formation, become gradually laid out more and more nearly parallel to the sides of

* Thirteenth Letter on Glaciers, section (2), dated Dec. 1846.

the channel through which the glacier is forced. This circumstance, along with the comparatively weak mechanical condition of the white strata (vesicular layers between the blue 'strata), must, I think, make these white strata become ultimately, in reality, the surfaces of "sliding" or of "tearing," or of chief differential motion, as according to Professor Forbes's observations they seem to be. His first statement on the subject, made as early as 1842, that "the blue veins seem to be perpendicular to the lines of maximum pressure," is, however, more in accordance with their mechanical origin, according to the theory I now suggest, than the supposition that they are *caused* by the tearing action which is found to take place along them when formed. It appears to me, therefore, that Dr Tyndall's conclusion, that the vesicular stratification is produced by pressure in surfaces perpendicular to the directions of maximum pressure, is correct as regards the mechanical origin of the veined structure; while there seems every reason, both from observation and from mechanical theory, to accept the view given by Professor Forbes of their function in glacial motion.

The mechanical theory I have indicated as the explanation of the veined structure of glacial ice is especially applicable to account for the stratification of the vesicles observed in ice originally clear, and subjected to differential pressure, by Dr Tyndall; the formation of the vesicles themselves being, as remarked in my last letter*, anticipated by my brother's theory, published in the *Proceedings* for May, 1857.

I believe the theory I have given above contains the true explanation of one remarkable fact observed by Dr Tyndall in connexion with the beautiful set of phenomena which he discovered to be produced by radiant heat, concentrated on an internal portion of a mass of clear ice by a lens; the fact, namely, that the planes in which the vesicles extend are generally parallel to the sides when the mass of ice operated on is a flat slab; for the solid will yield to the "negative" internal pressure due to the contractility of the melting ice, most easily in the direction perpendicular to the sides. The so-called negative pressure is therefore least, or which is the same thing, the positive pressure is greatest in this direction. Hence the vesicles of melted ice, or of vapour caused by the contraction of melted ice, must, as I have shown, tend to place themselves parallel to the sides of the slab.

The division of the vesicular layers into leaves like six-petaled flowers is a phenomenon which does not seem to me as yet so easily explained; but I cannot see that any of the phenomena described by Dr Tyndall can be considered as having been proved to be due to ice having mechanical properties of a uniaxal crystal.

[It now seems to me most probable Tyndall was right in attributing the six-rayed structure to the molecular mechanics of a uniaxal crystal. K., *Dec.* 13, 1903.]

101. On the Thermal Effect of drawing out a Film of Liquid.

[From *Roy. Soc. Proc.* Vol. IX. 1857—1859, pp. 255, 256 [read June 10, 1858]; *Phil. Mag.* Vol. XVII. Jan. 1859, pp. 61, 62.]

Being an extract of two letters to J. P. Joule, F.R.S., dated Feb. 2 and 3, 1858.

A very novel application of Carnot's cycle has just occurred to me in consequence of looking this morning into Waterston's paper on Capillary Attraction, in the January Number of the *Philosophical Magazine*. Let T be the contractile force of the surface (by which in Dr Thomas Young's theory the resultant effect of cohesion on a liquid mass of varying form is represented), so that, if Π be the atmospheric pressure, the pressure of air within a bubble of the liquid of radius r shall be $4T/r + \Pi$. Then if a bubble be blown from the end of a tube (as in blowing soap-bubbles), the work spent, per unit of augmentation of the area of one side of the film, will be equal to $2T$.

Now since liquids stand to different heights in capillary tubes at different temperatures, and generally to less heights at the higher temperatures, T must vary, and in general decrease, as the temperature rises, for one and the same liquid. If T and T' denote the values of the capillary tension at temperatures t and t' of our absolute scale, we shall have $2(T - T')$ of mechanical work gained, in allowing a bubble on the end of a tube to collapse so as to lose a unit of area at the temperature t and blowing it up again to its original dimensions after having raised its temperature to t'. If $t' - t$ be infinitely small, and be denoted by $\mathfrak{T}$, the gain of the work may be expressed by

$$- \frac{2dT}{dt} \times \mathfrak{T} ;$$

and by using Carnot's principle as modified for the Dynamical Theory, in the usual manner, we find that there must be an absorption of heat at the high temperature, and an evolution of heat at the low temperature; amounting to quantities differing from one another by

$$\frac{1}{J} \times \frac{-2dT}{dt} \times \mathcal{C},$$

and each infinitely nearly equal to the mechanical equivalent of this difference, divided by Carnot's function, which is J/t, if the temperature is measured on our absolute scale. Hence if a film such as a soap-bubble be enlarged, its area being augmented in the ratio of 1 to m*, it experiences a cooling effect, to an amount calculable by finding the lowering of temperature produced by removing a quantity of heat equal to

$$m \frac{t}{J} \times \frac{-dT}{dt},$$

from an equal mass of liquid unchanged in form.

For water $T = 2 \cdot 96$ gr. per lineal inch.

Work per square inch spent in drawing out a film $= 5 \cdot 92$, say 6 inch-grains; $dT/dt = \frac{1}{550}T$, or thereabouts.

Suppose $t'/J = 300/(1390 \times 12)$, then the quantity of heat to be removed, to produce the cooling effect, per square inch of surface of augmentation of film, will be $\frac{1}{6100}$. Suppose, then, 1 grain of water to be drawn out to a film of 16 square inches, the cooling effect will be $\frac{16}{6100}$ of a degree Centigrade, or about $\frac{1}{320}°$. The work spent in drawing it out is $16 \times 6 = 96$ inch-grains, and is equivalent to a heating effect of $96/(12 \times 1390) = \frac{1}{174}°$. Hence the total energy (reckoned in heat) of the matter is increased $\frac{1}{174} + \frac{1}{320}$ of a degree Centigrade, when it is drawn out to 16 square inches.

* [This should read 'augmented by an area $\frac{1}{2}m$ on each side.']

102. On the Importance of making Observations on Thermal Radiation during the coming Eclipse of the Sun.

[From *Roy. Astron. Soc. Monthly Notices*, Vol. xx. 1860, pp. 317, 318.]

Abstract of Letter to the Editors dated June 5, 1860.

Prof. Grant thinks I ought to call the attention of the Astronomical Society to the importance of making observations on thermal radiation during the eclipse of the sun, by writing to you on the subject.

All that I have to suggest is, that differential observations on the effects experienced by three or four thermometers with their bulbs placed, one in the centre of the image of the moon, in the focus of a lens or speculum, and the others in different positions round her disk, during the time of total obscuration of the sun, would probably give decided results as to the difference of radiation from the moon, and from the portion of space round the sun. The "photosphere," which I suppose is always seen round the moon's disk during a total eclipse of the sun, and which must be due to her light reflected from atmosphere or dust round the sun itself, must radiate heat, and would probably show a sensible effect, as compared with radiation from an equal angular area of the moon.

The thermometers should have their bulbs blackened with soot or lamp-black, and be all of the same size. The bulb of each should be considerably smaller than the sun's image. This, if the focal length of the lens or mirror is 57 inches, would be a circle half an inch diameter, so that very ordinary thermometers would answer. I would recommend spirits-of-wine thermometers for superior sensitiveness; but, perhaps, mercury, with a rather fine bore, would be sensitive enough, and surer.

A mirror would be better than a lens, because the hot dust round the sun must produce radiant heat of such colour as that of a hot stone or metal not at a bright red heat; and which, therefore, would be much absorbed by glass.

There would need good arrangements to preserve the thermometers from disturbing influences. Until the totality, the radiation from the sun would need to be strictly excluded. At the instant of complete obscuration the cover would be removed, and the thermometers all observed. One observer could note them in succession. The sketch represents the positions such as I would have the thermometers in, the circle denoting the moon's unseen disk.

There seems no doubt but that the thing is practicable, and would lead to results; but it is to be considered whether arrangements can be made, and an observer or observers secured, to execute it. Will you consult members of the Astronomical Society on the occasion of its meeting on Friday, if you find any who are disposed to consider the subject?

From what Prof. Grant says, it appears that radiation experiments, as hitherto made or contemplated, have not attempted any definite concentration by means of a lens or otherwise. This appears to be as discriminating a way of investigating the phenomenon as it would be to attempt to record visible appearances by a photograph made by diffused light, with no lens to make a picture.

103. ON THE CONVECTIVE EQUILIBRIUM OF TEMPERATURE IN THE ATMOSPHERE.

[From *Manchester Phil. Soc. Proc.* Vol. II. 1860—1862, pp. 170—176 [Jan. 21, 1862]; *Manchester Phil. Soc. Mem.* Vol. II. 1865, pp. 125—131. Reprinted in *Math. and Phys. Papers*, Vol. III. Art. xcii. pp. 255—260.]

104. ON THE PROTECTION OF VEGETATION FROM DESTRUCTIVE COLD EVERY NIGHT.

[From *Edinb. Roy. Soc. Proc.* Vol. V. 1866, pp. 203, 204 [read April 4, 1864]. Reprinted in *Popular Lectures and Addresses*, Vol. II. pp. 1—5.]

105. On the Dynamical Theory of Heat [Thermal
    Dissipation of Energy of Vibration of Solids].

[From *Edinb. Roy. Soc. Proc.* Vol. v. 1866, pp. 510—512; read Dec. 18, 1865.]

THIS paper commences with a condensed re-statement of the
fundamental principles and formulæ of the Dynamical Theory of
Heat, from the first six parts of the author's treatment of the subject previously communicated to the Royal Society of Edinburgh,
and his articles "On the Thermo-elastic Properties of Matter,"
in the *Quarterly Mathematical Journal* (April, 1855), and on
"Thermo-magnetism," and "Thermo-electricity," in Nichol's
Cyclopedia (Edinburgh, 1860).

The chief object of the paper is the deduction of numerical
values in absolute measure for the thermo-electric effects which
form the subject of Part VI. of this series (*Transactions of the
Royal Society of Edinburgh*, 1854; and *Phil. Mag.* 1854, second
half year, and 1855, first half year) especially for differences of
temperature produced by electric convection of heat, and for the
changes of temperature due to strain in elastic solids, investigated
in the article on thermo-elastic properties of matter above referred
to. The very valuable results, recently published, of the experiments of Forbes and Ångström for determining in absolute
measure the thermal conductivities of iron and copper, supply a
very important element, previously wanting, for definite estimates
of those changes of temperature, and are taken advantage of in
the present paper. Thus, the author has been enabled to give
that practical character to some of his former conclusions, of
which, when they were first published, he pointed out the want.
In particular, with reference to elastic solids, the *apparent* value
of Young's modulus* when the stress is applied and removed, or

---

* The amount of the force divided by the elongation produced by it, when any
force within practical limits of elasticity is applied to elongate a bar rod or wire, of
the substance, one square centimetre in section.

reversed so rapidly that the loss of thermal effect by conduction
and radiation is insensible, is proved to be given by the following
formula :—

$$M' = M\left(1 + \frac{e^2 t M}{Js\rho}\right),$$

where M denotes the Young's modulus of the substance for
constant temperature, s its specific heat (per unit mass, as usual),
e its longitudinal (linear) expansion per degree of elevation of
temperature, ρ its density or specific gravity*, and t its actual
temperature from absolute zero (*Dynamical Theory of Heat*,
Part VI., § 100), temperature centigrade with 274 added. Of
course, if M is reckoned (Thomson and Tait's *Natural Philosophy*,
§§ 220, 221, 238), in gravitation measure (weight of one gramme,
the unit of mass), J must be reckoned in gravitation measure
(grammes weight working through one centimetre), in which
case its numerical value is 42,400, being Joule's number (1390),
reduced from feet to centimetres. Values of surface resistance to
gain or loss of heat in absolute measure, derived from experiments
by the author, are used to estimate the effect of radiation and
convection in dissipating energy in virtue of the thermo-dynamic
change of temperature in a rod executing longitudinal vibrations.
The velocity of propagation of longitudinal vibrations (as in the
transmission of sound along a bar) being equal to the velocity
acquired by a body in falling through a height equal to half the
"length of the modulus†," is, of course, half as much affected as
the modulus, by changes of temperature. In iron, for instance,
the effect of change of temperature, when there is no dissipation,
is an increase of about one-third per cent. on the Young's modulus,
and of about one-sixth per cent. on the velocity of sound along a
bar. The effect of the conduction of heat in diminishing the
differences of temperature in a rectangular bar executing flexural
vibrations, is investigated from the solution invented by Fourier
for expressing periodical variations of underground temperature.
Its absolute amount for bars of iron or copper, of stated dimensions,
vibrating in stated periods, is determined from Forbes' and
Ångström's conductivities. It is proved that the loss of energy

---

* Which, when the French system (unit bulk of water being of mass unity) is
followed, mean the same thing.

† The "length of the modulus" is $M \div \rho$, if M be the modulus in grammes
weight per square centimetre.—Thomson and Tait's *Natural Philosophy*, § 689.

due to this effect at its maximum is not by any means insensible, though it is not sufficient to account for the whole loss of energy which the author has found in experiments on flexural vibrations of metal springs, which therefore prove imperfectness in the elasticity of flexure, such as he had previously proved for the elasticity of torsion*

* *Proceedings of the Royal Society of London*, May, 1865. W. Thomson, " On the Elasticity and Viscosity of Metals."

106. On the Dissipation of Energy.

The Rede Lecture, Cambridge, May 23, 1866.

Extract, ' On the Observations and Calculations required to find the Tidal Retardation of the Earth's Rotation,' published in *Popular Lectures and Addresses*, Vol. II. 1894, pp. 65—72.

107. Dr Balfour Stewart's Meteorological Blockade.

[From *Nature*, Vol. i. Jan. 20, 1870, p. 306.]

Imagine a line drawn round any district, and consider all the air that passes over that line outwards and inwards in any time. Let the whole quantity of vapour of water carried across the boundary line by this air be determined. If during any particular interval of time there is just as much vapour carried outwards as inwards, there must be in that interval either no rainfall on, and no evaporation from the district, or there must be just as much rainfall as evaporation. If more vapour is carried out than in, there must be more evaporation than rainfall. Or, if more vapour enters than leaves, the difference falls in excess of rain above evaporation.

Dr Stewart proposes to establish a *cordon* of meteorological stations, and to arrange a reduction of observations taken at them, so as to keep, as far as possible, an exact account of the quantity of vapour entering and leaving the space over the surrounded district. This appears to me a most valuable proposal, which, if well carried out, must have a very important influence, tending to raise meteorology from its present empirical condition to the rank of a science. The object of the present notice is to suggest that the same system of account-keeping ought to be applied to electricity.

Whatever we may think as to the nature of electricity, it certainly has, in common with true matter, the property of being invariable in quantity. This property is conveniently enough expressed, as it were, mechanically, by the one-fluid hypothesis, which asserts that positive or negative electrification of any piece of matter consists in the presence of more or less than a certain quantum of the electric fluid; that quantum being the amount possessed when the matter in question exercises no attractive or repulsive force, varying with artificial variations of the electric condition of the testing body presented to it. On this hypothesis, "quantities of electricity," positive and negative, are excesses of the quantity of the hypothetical fluid above or below the "quantum" corresponding to zero of the electric tests.

The ordinary fair-weather condition in our latitudes presents us with negatively electrified air in the lowest stratum, extending at least as high as our ordinary houses above the surface of the earth ; and positive electricity of greater amount, on the whole, in the higher regions. The atmospheric electrometer indicates in absolute electrostatic units the total quantity of electricity in the atmosphere, over a certain area of the place of observation ; being the excess of the amount of positive above the amount of negative electricity in the whole column. This excess in fair weather is generally positive. The fact that it is not the electricity in the lower regions alone, but an effect depending on the whole electricity of the atmosphere from lowest to highest, that is the thing observed in the ordinary observation of atmospheric electricity, renders this subject more suitable even than moisture for the application of Dr Stewart's blockade. Thus, the hygrometrical blockade is complete only if both moisture and the effective component of wind are known at all heights above the surface ; the electrical blockade is complete when, besides the electrometer measurements at the observatory, the effective component of the wind at all heights is known. But among the many unknown quantities involved, the two departments of the blockade combined will give means for eliminating some and estimating others for which the hygrometrical blockade alone, or the electrical one alone, would be insufficient.

108. On the Ultramundane Corpuscles of Le Sage.

[From *Edinb. Roy. Soc. Proc.* Vol. VII. 1872, pp. 577—589 [read Dec. 18, 1871]; *Phil. Mag.* Vol. XLV. May, 1873, pp. 321—332.]

Le Sage, born at Geneva in 1724, devoted the last sixty-three years of a life of eighty to the investigation of a mechanical theory of gravitation. The probable existence of a gravific mechanism is admitted, and the importance of the object to which Le Sage devoted his life pointed out, by Newton and Rumford* in the following statements:—

"It is inconceivable that inanimate brute matter should, without the mediation of something else, which is not material, operate upon, and affect other matter without mutual contact; as it must do, if gravitation, in the sense of *Epicurus*, be essential and inherent in it. And this is the reason why I desired you would not ascribe innate gravity to me. That gravity should be innate, inherent, and essential to matter, so that one body may act upon another at a distance through a *vacuum*, without the mediation of any thing else, by and through which their action and force may be conveyed from one to another, is to me so great an absurdity, that I believe no man who has in philosophical

---

* On the other hand, by the middle of last century the mathematical naturalists of the Continent, after half a century of resistance to the Newtonian principles (which, both by them and by the English followers of Newton, were commonly supposed to mean the recognition of gravity as a force acting simply at a distance without mediation of intervening matter), had begun to become more "Newtonian" than Newton himself. On the 4th February, 1744, Daniel Bernoulli wrote as follows to Euler: "Uebrigens glaube ich, dass der Aether sowohl *gravis versus solem*, als die Luft versus terram sey, und kann Ihnen nicht bergen, dass ich über diese Puncte ein völliger Newtonianer bin, und verwundere ich mich, dass Sie den Principiis Cartesianis so lang adhäriren; es möchte wohl einige Passion vielleicht mit unterlaufen. Hat Gott können eine *animam*, deren Natur uns unbegreiflich ist, erschaffen, so hat er auch können eine attractionem universalem materiæ imprimiren, wenn gleich solche attractio *supra captum* ist, da hingegen die Principia Cartesiana allzeit *contra captum* etwas involviren."

matters a competent faculty of thinking, can ever fall into it. Gravity must be caused by an agent acting constantly according to certain laws; but whether this agent be material or immaterial, I have left to the consideration of my readers."— NEWTON'S *Third Letter to Bentley*, February 25th, 1692–3.

"Nobody surely, in his sober senses, has ever pretended to understand the mechanism of gravitation; and yet what sublime discoveries was our immortal Newton enabled to make, merely by the investigation of the laws of its action*."

Le Sage expounds his theory of gravitation, so far as he had advanced it up to the year 1782, in a paper published in the *Transactions* of the Royal Berlin Academy for that year, under the title "Lucrèce Newtonien." His opening paragraph, entitled "But de ce mémoire," is as follows :—

"Je me propose de faire voir : que si les premiers Epicuriens avoient eu ; sur la Cosmographie des idées aussi saines seulement, que plusieurs de leurs contemporains, qu'ils négligeoient d'écouter†; et sur la Géométrie, une partie des connoissances qui étoient déjà communes alors : ils auroient, très probablement, découvert sans effort; les Loix de la Gravité universelle, et sa Cause mécanique. *Loix*; dont l'invention et la démonstration, font la plus grande gloire du plus puissant génie qui ait jamais existé : et *Cause*, qui après avoir fait pendant longtems, l'ambition des plus grands Physiciens ; fait à présent, le désespoir de leurs successeurs. De sorte que, par exemple, les fameuses Règles de *Kepler*; trouvées il y a moins de deux siècles, en partie sur des conjectures gratuites, et en partie après d'immenses tâtonnemens ; n'auroient été que des corollaires particuliers et inévitables, des lumières générales que ces anciens Philosophes pouvoient puiser (comme en se jouant) dans le mécanisme proprement dit de la Nature. Conclusion ; qu'on peut appliquer exactement aussi, aux Loix de *Galilée* sur la chûte des Graves sublunaires ; dont la découverte a été plus tardive encore, et plus contestée : joint à ce que, les expériences sur lesquelles cette découverte étoit établie ; laissoient dans leurs résultats (nécessairement grossiers),

---

* "An Inquiry concerning the Source of the Heat which is excited by Friction." By Count Rumford, *Philosophical Transactions*, 1798.

† "Vobis (Epicureis) minùs notum est, quemadmodum quidque dicatur. Vestra enim solùm legitis, vestra amatis; cæteros, causâ incognità, condemnatis." Cicéron, *De natura Deorum* II. 29.

une latitude, qui les rendoit également compatibles avec plusieurs autres hypothèses; qu'aussi, l'on ne manqua pas de lui opposer: au lieu que, les conséquences du choc des Atoms; auroient été absolument univoques en faveur du seul principe véritable (des Accélérations égales en Tempuscules égaux)."

If Le Sage had but excepted Kepler's third law, it must be admitted that his case, as stated above, would have been thoroughly established by the arguments of his "mémoire"; for the epicurean assumption of parallelism adopted to suit the false idea of the earth being flat, prevented the discovery of the law of the inverse square of the distance, which the mathematicians of that day were quite competent to make, if the hypothesis of atoms moving in all directions through space, and rarely coming into collision with one another, had been set before them, with the problem of determining the force with which the impacts would press together two spherical bodies, such as the earth and moon were held to be by some of the contemporary philosophers to whom the epicureans "would not listen." But nothing less than direct observation, proving Kepler's third law—Galileo's experiment on bodies falling from the tower of Pisa, Boyle's guinea and feather experiment, and Newton's experiment of the vibrations of pendulums composed of different kinds of substance—could give either the idea that gravity is proportional to mass, or prove that it is so to a high degree of accuracy for large bodies and small bodies, and for bodies of different kinds of substance. Le Sage sums up his theory in an appendix to the "Lucrèce Newtonien," part of which translated (literally, except a few sentences which I have paraphrased) is as follows:—

Constitution of Heavy Bodies.

1. Their indivisible particles are cages; for example, empty cubes or octahedrons vacant of matter except along the twelve edges.

2. The diameters of the bars of these cages, supposed increased each by an amount equal to the diameter of one of the gravific corpuscles, are so small relatively to the mutual distance of the parallel bars of each cage, that the terrestrial globe does not intercept even so much as a ten-thousandth part of the corpuscles which offer to traverse it.

3.   These diameters are all equal, or if they are unequal, their inequalities sensibly compensate one another [in averages].

Constitution of Gravific Corpuscules.

1.   Conformably to the second of the preceding suppositions, their diameters added to that of the bars is so small relatively to the mutual distance of parallel bars of one of the cages, that the weights of the celestial bodies do not differ sensibly from being in proportion to their masses.

2.   They are isolated. So that their progressive movements are necessarily rectilinear.

3.   They are so sparsely distributed, that is to say, their diameters are so small relatively to their mean mutual distances, that not more than one out of every hundred of them meets another corpuscule during several thousands of years. So that the uniformity of their movements is scarcely ever troubled sensibly.

4.   They move along several hundred thousand millions of different directions; in counting for one same direction all those which are [within a definite very small angle of being] parallel to one straight line.   The distribution of these straight lines is to be conceived by imagining as many points as one wishes to consider of different directions, scattered over a globe as uniformly as possible, and therefore separated from one another by at least a second of angle; and then imagining a radius of the globe drawn to each of those points.

5.   Parallel, then, to each of those directions, let a current or torrent of corpuscules move; but, not to give the stream a greater breadth than is necessary, consider the transverse section of this current to have the same boundary as the orthogonal projection of the visible world on the plane of the section.

6.   The different parts of one such current are sensibly equi-dense; whether we compare, among one another, collateral portions of sensible transverse dimensions, or successive portions of such lengths that their times of passage across a given surface are sensible. And the same is to be said of the different currents compared with one another.

7.   The mean velocities, defined in the same manner as I have just defined the densities, are also sensibly equal.

8. The ratios of these velocities to those of the planets are several million times greater than the ratios of the gravities of the planets, towards the sun, to the greatest resistance which secular observations allow us to suppose they experience. For example, [these velocities must be] some hundredfold a greater number of times the velocity of the earth, than the ratio of 190,000* times the gravity of the earth towards the sun, to the greatest resistance which secular observations of the length of the year permit us to suppose that the earth experiences from the celestial masses.

CONCEPTION, *which facilitates the Application of Mathematics to determine the mutual Influence of these Heavy Bodies and these Corpuscles.*

1. Decompose all heavy bodies into molecules of equal mass, so small that they may be treated as attractive points in respect to theories in which gravity is considered without reference to its cause; that is to say, each must be so small that inequalities of distance and differences of direction between its particles and those of another molecule, conceived as attracting it and being attracted by it, may be neglected. For example, suppose the diameter of the molecule considered to be a hundred thousand times smaller than the distance between two bodies of which the mutual gravitation is examined, which would make its apparent semi-diameter, as seen from the other body, about one second of angle.

2. For the surfaces of such a molecule, accessible but impermeable to the gravific fluid, substitute one single spherical surface equal to their sum.

3. Divide those surfaces into facets small enough to allow them to be treated as planes, without sensible error [&c., &c.].

Remarks.

1. It is not necessary to be very skilful to deduce from these suppositions all the laws of gravity, both sublunary and universal (and consequently also those of Kepler, &c.), with all the accuracy with which observed phenomena have proved those laws. Those

---

* To render the sentence more easily read, I have substituted this number in place of the following words:—" le nombre de fois que le firmament contient le disque apparent du soleil."

laws, therefore, are inevitable consequences of the supposed constitutions.

2.   Although I here present these constitutions crudely and without proof, as if they were gratuitous hypotheses and hazarded fictions, equitable readers will understand that on my own part I have at least some presumptions in their favour (independent of their perfect agreement with so many phenomena), but that the development of my reasons would be too long to find a place in the present statement, which may be regarded as a publication of theorems without their demonstrations.

3.   ......There are details upon which I have wished to enter on account of the novelty of the doctrine, and which will readily be supplied by those who study it in a favourable and attentive spirit.   If the authors who write on hydro-dynamics, aerostatics, or optics, had to deal with captious readers, doubting the very existence of water, or air, or light, and therefore not adapting themselves to any tacit supposition regarding equivalencies or compensations not expressly mentioned in their treatises, they would be obliged to load their definitions with a vast number of specifications which instructed or indulgent readers do not require of them.   One understands "*à demi-mot*" and "*sano sensu*" only familiar propositions towards which one is already favourably inclined.

Some of the details referred to in this concluding sentence of the appendix to his " Lucrèce Newtonien," Le Sage discusses fully in his *Traité de Physique Mécanique*, edited by Pierre Prévost, and published in 1818 (Geneva and Paris).

This treatise is divided into four books.

I.   " Exposition sommaire du système des corpuscules ultra-mondains."

II.   " Discussion des objections qui peuvent s'élever contre le système des corpuscules ultramondains."

III.   " Des fluides élastiques ou expansifs."

IV.   "Application des théories précédentes à certaines affinités."

It is in the first two books that gravity is explained by the impulse of ultramundane corpuscles, and I have no remarks at present to make on the third and fourth books.

From Le Sage's fundamental assumptions, given above as nearly as may be in his own words, it is, as he says himself, easy to deduce the law of the inverse square of the distance, and the law of proportionality of gravity to mass. The object of the present note is not to give an exposition of Le Sage's theory, which is sufficiently set forth in the preceding extracts, and discussed in detail in the first two books of his posthumous treatise. I may merely say that inasmuch as the law of the inverse square of the distance, for every distance, however great, would be a perfectly obvious consequence of the assumptions, were the gravific corpuscules infinitely small, and therefore incapable of coming into collision with one another, it may be extended to as great distances as we please, by giving small enough dimensions to the corpuscules relatively to the mean distance of each from its nearest neighbour. The law of masses may be extended to as great masses as those for which observation proves it (for example the mass of Jupiter), by making the diameters of the bars of the supposed cage-atoms, constituting heavy bodies, small enough. Thus, for example, there is nothing to prevent us from supposing that not more than one straight line of a million drawn at random towards Jupiter and continued through it, should touch one of the bars. Lastly, as Le Sage proves, the resistance of his gravific fluid to the motion of one of the planets through it, is proportional to the product of the velocity of the planet into the average velocity of the gravific corpuscules; and hence by making the velocities of the corpuscules great enough, and giving them suitably small masses, they may produce the actual forces of gravitation, and not more than the amount of resistance which observation allows us to suppose that the planets experience. It will be a very interesting subject to examine minutely Le Sage's details on these points, and to judge whether or not the additional knowledge gained by observation since his time requires any modification to be made in the estimate which he has given of the possible degrees of permeability of the sun and planets, of the possible proportions of diameters of corpuscules to interstices between them in the "gravific fluid," and of the possible velocities of its component corpuscules. This much is certain, that if hard indivisible atoms are granted at all, his principles are unassailable; and nothing can be said against the probability of his assumptions. The only imperfection of his theory is that which is inherent to every supposition of hard, indivisible atoms.

They must be perfectly elastic or imperfectly elastic, or perfectly inelastic. Even Newton seems to have admitted as a probable reality hard, indivisible, unalterable atoms, each perfectly inelastic.

Nicolas Fatio is quoted by Le Sage and Prévost, as a friend of Newton, who in 1689 or 1690 had invented a theory of gravity perfectly similar to that of Le Sage, except certain essential points; had described it in a Latin poem not yet printed; and had written, on the 30th March 1694, a letter regarding it, which is to be found in the third volume of the works of Leibnitz, having been communicated for publication to the editor of those works by Le Sage. Redeker, a German physician, is quoted by Le Sage as having expounded a theory of gravity of the same general character, in a Latin dissertation. published in 1736, referring to which Prévost says, " Où l'on trouve l'exposé d'un système fort semblable à celui de Le Sage dans ses traits principaux, mais dépourvu de cette analyse exacte des phénomènes qui fait le principal mérite de toute espèce de théorie." Fatio supposed the corpuscules to be elastic, and seems to have shown no reason why their return velocities after collision with mundane matter should be less than their previous velocities, and therefore not to have explained gravity at all. Redeker, we are told by Prévost, had very limited ideas of the permeabilities of great bodies, and therefore failed to explain the law of the proportionality of gravity to mass; " he enunciated this law very correctly in section 15 of his dissertation; but the manner in,which he explains it shows that he had but little reflected upon it. Notwithstanding these imperfections, one cannot but recognise in this work an ingenious conception which ought to have provoked examination on the part of naturalists, of whom many at that time occupied themselves with the same investigation. Indeed, there exists a dissertation by Segner on this subject*. But science took another course, and works of this nature gradually lost appreciation. Le Sage has never failed on any occasion to call attention to the system of Redeker, as also to that of Fatio†."

Le Sage shows that to produce gravitation those of the ultramundane corpuscules which strike the cage-bars of heavy bodies must either stick there or go away with diminished velocities. He supposed the corpuscules to be inelastic (*durs*), and points

* De Causa gravitatis Redekeriana.

† Le Sage was remarkably scrupulous in giving full information regarding all who preceded him in the development of any part of his theory.

out that we ought not to suppose them to be permanently lodged in the heavy body (*entassés*), that we must rather suppose them to slip off; but that being inelastic, their average velocities after collision must be less than that which they had before collision*.

That these suppositions imply a gradual diminution of gravity from age to age was carefully pointed out by Le Sage, and referred to as an objection to his theory. Thus he says, "...Donc, la durée de la gravité seroit *finie* aussi, et par conséquent la durée du monde.

"*Réponse*. Concedo; mais pourvu que cet obstacle ne contribue pas à faire finir le monde plus promptement qu'il n'auroit fini sans lui, il doit être considéré comme nul†."

Two suppositions may be made on the general basis of Le Sage's doctrine:—

1. (Which seems to have been Le Sage's belief.) Suppose the whole of mundane matter to be contained within a finite space, and the infinite space round it to be traversed by ultramundane corpuscules; and a small proportion of the corpuscules coming from ultramundane space to suffer collisions with mundane matter, and get away with diminished gravific energy to ultramundane space again. They would never return to the world were it not for collision among themselves and other corpuscules. Le Sage held that such collisions are extremely rare; that each collision, even between the ultramundane corpuscules themselves, destroys some energy‡; that at a not infinitely remote past time they were set in motion for the purpose of keeping gravitation throughout the world in action for a limited period of time; and that both by their mutual collisions, and by collisions with mundane atoms, the whole stock of gravific energy is being gradually reduced, and therefore the intensity of gravity gradually diminishing from age to age.

2. Or, suppose mundane matter to be spread through all space,

---

* Le Sage estimated the velocity after collision to be two-thirds of the velocity before collision.

† Posthumous. *Traité de Physique Mécanique*, edited by Pierre Prévost. Geneva and Paris, 1818.

‡ Newton (*Optics*, Query 30, edn. 1721, p. 373) held that two equal and similar atoms, moving with equal velocities in contrary directions, come to rest when they strike one another. Le Sage held the same; and it seems that writers of last century understood this without qualification when they called atoms hard.

but to be much denser within each of an infinitely great number of finite volumes (such as the volume of the earth) than elsewhere. On this supposition, even were there no collisions between the corpuscules themselves, there would be a gradual diminution in their gravific energy through the repeated collisions with mundane matter which each one must in the course of time suffer. The secular diminution of gravity would be more rapid according to this supposition than according to the former, but still might be made as slow as we please by pushing far enough the fundamental assumptions of very small diameters for the cage-bars of the mundane atoms, very great density for their substance, and very small volume and mass, and very great velocity for the ultra-mundane corpuscules.

The object of the present note is to remark that (even although we were to admit a gradual fading away of gravity, if slow enough), we are forbidden by the modern physical theory of the conservation of energy to assume inelasticity, or anything short of perfect elas-ticity, in the ultimate molecules, whether of ultramundane or of mundane matter; and, at the same time, to point out that the assumption of diminished exit velocity of ultramundane corpuscules, essential to Le Sage's theory, may be explained for perfectly elastic atoms, consistently both with modern thermodynamics, and with perennial gravity.

If the gravific corpuscules leave the earth or Jupiter with less energy than they had before collision, their effect must be to con-tinually elevate the temperature throughout the whole mass. The energy which must be attributed to the gravific corpuscules is so enormously great, that this elevation of temperature would be sufficient to melt and evaporate any solid, great or small, in a fraction of a second of time. Hence, though outward-bound cor-puscules must travel with less velocity, they must carry away the same energy with them as they brought. Suppose, now, the whole energy of the corpuscules approaching a planet to consist of trans-latory motion; a portion of the energy of each corpuscule which has suffered collision must be supposed to be converted by the collision into vibrations, or vibrations and rotations. To simplify ideas, suppose for a moment the particles to be perfectly smooth elastic globules. Then collision could not generate any rotatory motion; but if the cage-atoms constituting mundane matter be each of them, as we must suppose it to be, of enormously great mass in comparison with one of the ultramundane globules, and if the substance of the latter, though perfectly elastic, be much less rigid than that of the former, each globule that strikes one of the cage-bars must (Thomson & Tait's *Natural Philosophy*, § 301), come away with diminished velocity of translation, but with the corresponding deficiency of energy altogether converted into vibration of its own mass. Thus the condition required by Le Sage's theory is fulfilled without violating modern thermo-dynamics; and, according to Le Sage, we might be satisfied not to inquire what becomes of those ultramundane corpuscules which have been in collision either with the cage-bars of mundane matter or with one another; for at present, and during ages to come, these would be merely an inconsiderable minority, the great majority being still fresh with original gravific energy unimpaired by collision. Without entering on the purely metaphysical question,—Is any such supposition satisfactory? I wish to point out how gravific energy may be naturally restored to corpuscules in which it has been impaired by collision.

Clausius has introduced into the kinetic theory of gases the very important consideration of vibrational and rotational energy. He has shown that a multitude of elastic corpuscules moving through void, and occasionally striking one another, must, on the average, have a constant proportion of their whole energy in the form of vibrations and rotations, the other part being purely translational. Even for the simplest case,—that, namely, of smooth elastic globes,—no one has yet calculated by abstract dynamics the ultimate average ratio of the vibrational and rotational, to the translational energy. But Clausius has shown how to deduce it for the corpuscules of any particular gas from the experimental determination of the ratio of its specific heat, pressure constant, to its specific heat, volume constant[*]. He found that

$$\beta = \frac{2}{3} \frac{1}{\gamma - 1},$$

if γ be the ratio of the specific heats, and β the ratio of the whole energy to the translational part of it. For air, the value of γ found by experiment, is $1 \cdot 408$, which makes $\beta = 1 \cdot 634$. For steam, Maxwell says, on the authority of Rankine, β "may be as much

---

[*] Maxwell's *Elementary Treatise on Heat*, Chapter XXII. Longmans, 1871.

as 2·19, but this is very uncertain." If the molecules of gases are
admitted to be elastic corpuscules, the validity of Clausius' principle is undeniable; and it is obvious that the value of the ratio β
must depend upon the shape of each molecule, and on the distribution of elastic rigidity through it, if its substance is not
homogeneous. Farther, it is clear that the value of β for a set of
equal and similar corpuscules will not be the same after collision
with molecules different from them in form or in elastic rigidity,
as after collision with molecules only of their own kind. All that
is necessary to complete Le Sage's theory of gravity in accordance
with modern science, is to assume that the ratio of the whole
energy of the corpuscules to the translational part of their energy
is greater, on the average, after collisions with mundane matter
than after inter-collisions of only ultramundane corpuscules*.
This supposition is neither more nor less questionable than that of
Clausius for gases, which is now admitted as one of the generally
recognised truths of science. The corpuscular theory of gravity is
no more difficult in allowance of its fundamental assumptions than
the kinetic theory of gases as at present received; and it is more
complete, inasmuch as, from fundamental assumptions of an
extremely simple character, it explains all the known phenomena
of its subject, which cannot be said of the kinetic theory of gases
so far as it has hitherto advanced.

Postscript, April 1872.

In the preceding statement I inadvertently omitted to remark
that if the constituent atoms are aeolotropic in respect to permeability, crystals would generally have different permeabilities in

---

* [It has been generally considered to be a consequence of the Maxwell-Boltzmann
doctrine of equipartition of energy among molecular freedoms that there can be
only one ultimate distribution of energy in the molecule however it is produced,
and that there remains only the question of the time it takes for the distribution
to become established. Thus Maxwell has objected that on the Le Sage hypothesis
the thermal energy of the universe would become transferred to his corpuscles.
Lord Kelvin here in effect puts the question whether the trend to uniform distribution would be disturbed, if the corpuscles collided only with the atoms and
not with one another. For his objections to the general law of equipartition
cf. Vol. IV. pp. 484 *seq*.: for subsequent literature cf. Larmor, Bakerian Lecture,
Proc. Roy. Soc. 83 A. (1909) pp. 82—95.

The interest of the present paper, mainly historical, is somewhat enhanced by
the obvious incidental analogies to the behaviour of free electrons that are presented
by Le Sage's particles.]

different directions, and would therefore have different weights according to the direction of their axes relatively to the direction of gravity. No such difference has been discovered, and it is certain that if there is any it is extremely small. Hence, the constituent atoms, if aeolotropic as to permeability, must be so, but to an exceedingly small degree. Le Sage's second fundamental assumption given above, under the title " *Constitution of Heavy Bodies*," implies sensibly equal permeability in all directions, even in an aeolotropic structure, unless much greater than Jupiter, provided that the atoms are isotropic as to permeability.

A body having different permeabilities in different directions would, if of manageable dimensions, give us a means for drawing energy from the inexhaustible store laid up in the ultramundane corpuscules, thus:—First, turn the body into a position of minimum weight; Secondly, lift it through any height; Thirdly, turn it into a position of maximum weight; Fourthly, let it down to its primitive level. It is easily seen that the first and third of those operations are performed without the expenditure of work ; and, on the whole, work is done by gravity in operations 2 and 4. In the corresponding set of operations performed upon a moveable body in the neighbourhood of a fixed magnet, as much work is required for operations 1 and 3 as is gained in operations 2 and 4; the magnetisation of the moveable body being either intrinsic or inductive, or partly intrinsic and partly inductive, and the part of its aeolotropy (if any), which depends on inductive magnetisation, being due either to magne-crystallic quality of its substance, or to its shape*.

---

* " Theory of magnetic induction in crystalline and non-crystalline substances," *Phil. Mag.*, March 1851 ; " Forces experienced by inductively magnetized ferromagnetic and dia-magnetic non-crystalline substances," *Phil. Mag.*, Oct. 1850 ; " Reciprocal action of dia-magnetic particles," *Phil. Mag.*, Dec. 1855 ; all to be found in a collection of reprinted and newly written papers on electrostatics and magnetism, nearly ready for publication (Macmillan, 1872).

109. On Steam-Pressure Thermometers of Sulphurous Acid, Water, and Mercury.

[From *Edinb. Roy. Soc. Proc.* Vol. x. pp. 432—441; read March 1, 1880. Reprinted in *Ency. Brit.* Ed. 9, article 'Heat,' and *Math. and Phys. Papers*, Vol. iii. pp. 156—165.]

THE first annexed diagram represents a thermometer constructed to show absolute temperature realised for the case of water and vapour of water as thermometric substance. The containing vessel consists of a tube with cylindric bulb like an ordinary thermometer; but, unlike an ordinary thermometer, the tube is bent in the manner shown in the drawing. The tube may be of from 1 to 2 or 3 millims. bore, and the cylindrical part of the bulb of about ten times as much. The length of the cylindrical part of the bulb may be rather more than $\frac{1}{100}$ of the length of the straight part of the tube. The contents, water and vapour of water, are to be put in and the glass hermetically sealed to enclose them, with the utmost precautions to obtain pure water as thoroughly freed from air as possible, after better than the best manner of instrument makers in making cryophoruses and water hammers. The quantity of water left in at the sealing must be enough to fill the cylindrical part of the bulb and the horizontal branch of the tube. When in use the straight part of the tube must be vertical with its closed

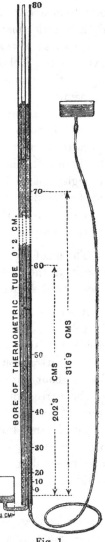

Fig. 1.

end up, and the part of it occupied by the manometric water-column must be kept at a nearly enough definite temperature by a surrounding glass jacket-tube of iced-water. This glass jacket-tube is wide enough to allow little lumps of ice to be dropped into it from its upper end, which is open. By aid of an india-rubber tube connected with its lower end, and a little movable cistern, as shown in the drawing, the level of the water in the jacket is kept from a few inches above to a quarter of an inch below that of the interior manometric column. Thus, by dropping in lumps of ice so as always to keep some unmelted ice floating in the water of the jacket, it is easy to keep the temperature of the top of the manometric water-column exactly at the freezing temperature. As we shall see presently, the manometric water below its free surface may be at any temperature from freezing to 10° C. above freezing without more than $\frac{1}{40}$ per cent. of hydrostatic error. The temperature in the vapour-space above the liquid column may be either freezing or anything higher. It ought not to be lower than freezing, because, if it were so, vapour would condense as hoar frost on the glass, and evaporation from the top of the liquid column would either cryophoruswise freeze the liquid there, or cool it below the freezing point.

The chief object of keeping the top of the manometric column exactly at the freezing-point is to render perfectly definite and constant the steam-pressure in the space above it.

A second object of considerable importance when the bore of the tube is so small as one millimetre, is to give constancy to the capillary tension of the surface of the water. The elevation by capillary attraction of ice-cold water in a tube of one millimetre bore is about 7 millims. The constancy of temperature provided by the surrounding iced water will be more than sufficient to prevent any perceptible error due to inequality of this effect. To avoid error from capillary attraction the bore of the tube ought to be very uniform, if it is so small as one millimetre. If it be three millimetres or more, a very rough approach to uniformity would suffice.

A third object of the iced-water jacket, and one of much more importance than the second, is to give accuracy to the hydrostatic measurement by keeping the density of the water throughout the long vertical branch definite and constant. But the density of water at the freezing point is only $\frac{1}{40}$ per cent. less than the maximum density, and is the same as the density at 8° C.; and therefore when $\frac{1}{40}$ per cent. is an admissible error on our thermometric pressure, the density will be nearly enough constant with any temperature from 0° to 10° C. throughout the column. But on account of the first object mentioned above, the very top of the water-column must be kept with exceeding exactness at the freezing temperature.

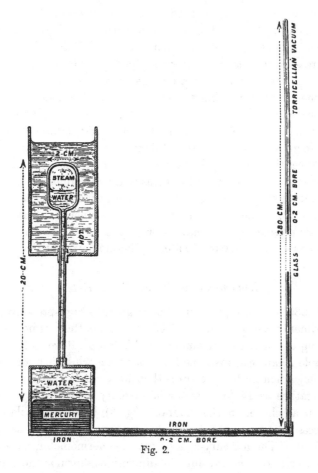

Fig. 2.

In this instrument the "thermometric substance" is the water and vapour of water in the bulb, or more properly speaking the portions of water and vapour of water infinitely near their separating interface. The rest of the water is merely a means of measuring hydrostatically the fluid pressure at the interface. When the temperature is so high as to make the pressure too great to be conveniently measured by a water column, the hydrostatic measurement may be done, as shown in the second annexed drawing (fig. 2), by a mercury column in a glass tube, surrounded by a glass water jacket not shown in the drawing, to keep it very accurately at some definite temperature so that the density of the mercury may be accurately known.

The simple form of steam thermometer represented with figured dimensions in the first diagram will be very convenient for practical use for temperatures from freezing to 60° C. Through this range the pressure of vapour of water, reckoned in terms of the balancing column of water of maximum density, increases from 6·25 to 202·3 centimetres; and for this, therefore, a tube of a little more than 2 metres will suffice. From 60° to 140° C. the pressure of steam now reckoned in terms of the length of a balancing column of mercury at 0° increases from 14·88 to 271·8 centimetres; and for this a tube of 280 centimetres may be provided. For higher temperatures a longer column, or several columns, as in the multiple manometer, or an accurate air pressure-gauge, or some other means, such as a very accurate instrument constructed on the principle of Bourdon's metallic pressure-gauge, may be employed, so as to allow us still to use water and vapour of water as thermometric substance.

High-pressure Steam Thermometer.

At 230° C., the superior limit of Regnault's high-pressure steam experiments, the pressure is 27·53 atmos, but there is no need for limiting our steam thermometer to this temperature and pressure. Suitable means can easily be found for measuring with all needful accuracy much higher pressures than 27 atmos. But at so high a temperature as 140° C., vapour of mercury measured by a water column, as shown in the diagram (fig. 3), becomes available for purposes for which one millimetre to the degree is a sufficient sensibility. The mercury-steam-pressure thermometer, with pressure measured by water-column, of dimensions shown in the drawing, serves from 140° to 280° C., and will have very ample sensibility through the upper half of its scale. At 280° C. its sensibility will be about 4¾ centimetres to the degree! For temperatures above 280° C. sufficient sensibility for most purposes is obtained by

substituting mercury for water in that simplest form of steam
thermometer shown in fig. 1, in which the pressure of the steam
is measured by a column of the liquid itself kept at a definite
temperature. When the liquid is mercury there is no virtue in
the particular temperature 0° C., and a stream of water as nearly
as may be of atmospheric temperature will be the easiest as well
as the most accurate way of keeping the mercury at a definite
temperature. As the pressure of mercury-steam is at all ordinary

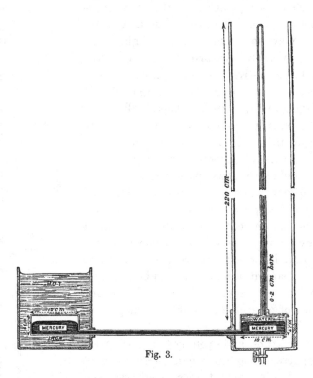

Fig. 3.

atmospheric temperatures quite imperceptible to the hydrostatic
test when mercury itself is the balancing liquid, that which was
the chief reason for fixing the temperature at the interface between
liquid and vapour at the top of the pressure-measuring column
when the balancing liquid was water, has no weight in the present
case; but, on the other hand, a much more precise definiteness
than the ten degrees latitude allowed in the former case for the
temperature of the main length of the manometric column is now
necessary. In fact, a change of temperature of 2·2° C. in mercury

at any atmospheric temperature produces about the same proportionate change of density as is produced in water by a change of temperature from 0° to 10° C., that is to say, about $\frac{1}{20}$ per cent.; but there is no difficulty in keeping, by means of a water jacket, the mercury column constant to some definite temperature within a vastly smaller margin of error than 2·2° C., especially if we choose for the definite temperature something near the atmospheric temperature at the time, or the temperature of whatever abundant water supply may be available. If the glass tube for the pressure-measuring mercury column be 830 centimetres long, the simple mercury-steam thermometer may be used up to 520° C., the highest temperature reached by Regnault in his experiments on mercury-steam. By using an iron bulb and tube for the part of the thermometer exposed to the high temperature, and for the lower part of the measuring column to within a few metres of its top, with glass for the upper part to allow the mercury to be seen, a mercury-steam-pressure thermometer can with great ease be made which shall be applicable for temperatures giving pressures up to as many atmospheres as can be measured by the vertical height available. The apparatus may of course be simplified by dispensing with the Torricellian vacuum at the upper end of the tube, and opening the tube to the atmosphere, when the steam-pressure to be measured is so great that a rough and easy barometer observation gives with sufficient accuracy the air-pressure at the top of the measuring column. The easiest, and not necessarily in practice the least accurate, way

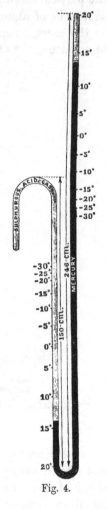

Fig. 4.

of measuring very high pressures of mercury-steam will be by enclosing some air above the cool, pressure-measuring column of mercury, and so making it into a compressed-air pressure-gauge, it being understood that the law of compression of the air under the pressures for which it is to be used in the gauge is known by accurate independent experiments such as those of Regnault on the compressibility of air and other gases.

The water-steam thermometer may be used, but somewhat precariously, for temperatures below the freezing-point, because water, especially when enclosed and protected as the portion of it in the bulb of our thermometer is, may be cooled many degrees below its freezing-point without becoming frozen; but, not to speak of the uncertainty or instability of this peculiar condition of water, the instrument would be unsatisfactory on account of insufficient thermometric sensibility for temperatures more than two or three degrees below the freezing-point. Hence, to make a steam thermometer for such temperatures some other substance than water should be taken, and none seems better adapted for the purpose than sulphurous acid, which, in the apparatus represented with figured dimensions in the accompanying diagram (fig. 4), makes an admirably convenient and sensitive thermometer for temperatures from $+20°$ to something far below $-30°$ C., as we see from the results of Regnault's measurements.

To sum up, we have, in the preceding description and drawings, a complete series of steam-pressure thermometers, of sulphurous acid, of water, and of mercury, adapted to give absolutely definite and highly sensitive thermometric indications throughout the wide range from something much below $-30°$ to considerably above $520°$ of the centigrade scale. The graduation of the scales of these thermometers to show absolute temperature is to be made by calculation from the thermodynamic formula

$$\log_e \frac{t}{t_0} = \int_{p_0}^{p} \frac{(1-\sigma)\,dp}{J\rho\kappa},$$

where t denotes the absolute temperature corresponding to steam-pressure p; t_0 the absolute temperature corresponding to steam-pressure p_0; κ the latent heat of the steam per unit mass; ρ the density of the steam; and σ the ratio of the density of the steam to the density of the liquid in contact with it. When the requisite experimental data, that is to say, the values of σ and $\rho\kappa$ for different values of p throughout the range for which each substance is to be used as thermometric fluid are available, the graduation of the scales of these thermometers to show absolute temperature can be performed in practice by calculation from the formula. Hitherto these requisites have not been given by direct experiment for any one of the three sub-stances with sufficient accuracy for our thermometric purpose through any range whatever. Water, naturally, is the one for which the nearest approach to the requisite information has been obtained. For it Regnault's experiments have given, no doubt with great accuracy, the values of p and of κ for all temperatures reckoned by his normal air thermometer, which we now regard merely as an arbitrary scale of temperature, through the range from $-30°$ to $+230°$. If he, or any other experimenter, had given us with similar accuracy through the same range the values of ρ and σ for temperatures reckoned on the same arbitrary scale, we should have all the data from experiment required for the graduation of our water-steam thermometer to absolute thermo-dynamic scale. For it is to be remarked that all reckoning of temperature is eliminated from the second member of the formula, and that, in our use of it, Regnault's normal thermometer has merely been referred to for the values of $\rho\kappa$ and of $1-\sigma$, which correspond to stated values of p. The arbitrary constant of integration, t_0, is truly arbitrary. It will be convenient to give it such a value that the difference of values of t between the freezing-point of water and the temperature for which p is equal to one atmo shall be 100, as this makes it agree with the centi-grade scale in respect to the difference between the numbers measuring the temperatures which on the centigrade scale are marked $0°$ and $100°$ C. Indirectly, by means of experiments on hydrogen gas, this assignation of the arbitrary constant of inte-gration would give 273 for the absolute temperature $0°$ C., and 373 for that of $100°$ C., as is proved in p. 56 of the article on "Heat," in the *Encyclopædia Britannica.* Meantime, as said above, we have not the complete data from direct experiments even on water-steam for graduating the water-steam thermo-meter; but, on the other hand we have, from experiments on air and on hydrogen and other gases, data which allow us to graduate indirectly any continuous intrinsic thermoscope according to the absolute scale. By thus indirectly graduating the water-steam thermometer, we learn the density of steam at different tempera-tures with more probable accuracy than it has hitherto been made known by any direct experiments on water-steam itself.

Merely viewed as a continuous intrinsic thermoscope, the steam-thermometer, in one or other of the forms described above

to suit different parts of the entire range from the lowest temperatures to temperatures somewhat above 520° C., is no doubt
superior in the conditions requisite for accuracy to every other
thermoscope of any of the different kinds hitherto in use; and it
may be trusted more surely for accuracy than any other as a
thermometric standard when once it has been graduated according
to the absolute scale, whether by practical experiments on steam,
or indirectly by experiments on air or other gases. In fact, the
use of steam-pressure measured in definite units of pressure, as a
thermoscopic effect, in the steam thermometer is simply a continuous extension to every temperature, of the principle already
practically adopted for fixing the temperature which is called
100° on the centigrade scale; and it stands on precisely the same
theoretical footing as an air thermometer, or a mercury-in-glass
thermometer, or an alcohol thermometer, or a methyl-butyrate
thermometer, in respect to the graduation of its scale according
to absolute temperature. Any one intrinsic thermoscope may be
so graduated ideally by thermodynamic experiments on the
substance itself without the aid of any other thermometer or any
other thermometric substance; but the steam-pressure thermometer has the great practical advantage over all others, except
the air thermometer, that these experiments are easily realisable
with great accuracy instead of being, though ideally possible,
hardly to be considered possible as a practical means of attaining
to thermodynamic thermometry. In fact, for water-steam it is
only the most easily obtained of experimental data, the measurement of the density of the steam at different pressures, that
has not already been actually obtained by direct experiment.
Whether or not when this lacuna has been filled up by direct
experiments, the data from water-steam alone may yield more
accurate thermodynamic thermometry than we have at present
from the hydrogen or nitrogen gas thermometer,—to be described
in a subsequent communication to the Royal Society * (*Proceedings*,
April 19, 1880)—we are unable at present to judge. But when
once we have the means, directly from itself, or indirectly from
comparison with hydrogen or nitrogen or air thermometers, of
graduating once for all a sulphurous acid steam thermometer, a
water-steam thermometer, or a mercury-steam thermometer, that is
to say, when once we have a table of the absolute thermodynamic

* [No. 114 *infra*, pp. 99—105.]

temperatures corresponding to the different steam-pressures of the substances sulphurous acid, water, and mercury, we have a much more accurate and more easily reproducible standard than either the air or gas thermometer of any form, or the mercury thermometer, or any liquid thermometer, can give. In fact, the series of steam thermometers for the whole range from the lowest temperatures can be reproduced with the greatest ease in any part of the world by a person commencing with no other material than a piece of sulphur and air to burn it in*, some pure water, some pure mercury, and with no other apparatus than can be made by a moderately skilled glass-blower, and with no other standard of physical measurement of any kind than an accurate linear measure. He may assume the force of gravity to be that calculated for his latitude with the ordinary rough allowance for his elevation above the sea, and his omission to measure with higher accuracy the actual force of gravity in his locality can lead him into no thermometric error which is not incomparably less than the inevitable errors in the reproduction and use of the air thermometer, or of mercury or other liquid thermometers. In temperatures above the highest for which mercury-steam pressure is not too great to be practically available, nothing hitherto invented but Deville's air thermometer with hard porcelain bulb suited to resist the high temperature is available for accurate thermometry.

The following statement is in the *Encyclopædia Britannica* article "Heat," appended to the description of steam-pressure thermometers which it contains:—"We have given the steam thermometer as our first example of thermodynamic thermometry because intelligence in thermodynamics has been hitherto much retarded, and the student unnecessarily perplexed, and a mere quicksand has been given as a foundation for thermometry, by building from the beginning on an ideal substance called perfect gas, with none of its properties realised rigorously by any real substance, and with some of them unknown, and utterly unassignable, even by guess. But after having been moved by this reason to give the steam-pressure thermometer as our first theoretical example, we have been led into the preceding carefully detailed examination of its practical qualities, and we have thus

---

* Practically, the best ordinary chemical means of preparing sulphurous acid, as from sulphuric acid, by heating with copper, might be adopted in preference to burning sulphur.

become convinced that though hitherto used in scientific investigations only for fixing the "boiling-point," and (through an
inevitable natural selection) by practical engineers for knowing
the temperatures of their boilers by the pressures indicated by
the Bourdon gauge, it is destined to be of great service both in
the strictest scientific thermometry, and as a practical thermometer for a great variety of useful applications."

110. On a Sulphurous Acid Cryophorus.

[From *Edinb. Roy. Soc. Proc.* Vol. x. pp. 442, 443 ; read March 1, 1880
(Abstract).]

THE instrument exhibited to the Royal Society consisted of a
U-shaped glass tube stopped at both ends, containing sulphurous
acid liquid and steam. The process by which the sulphurous
acid is freed from air, which was partially exhibited to the Royal
Society, is as follows :—

Begin with a glass U tube open at both ends, and attach to
each a small convenient, very fine, and perfectly gas-tight, stop-
cock. Placing it with the bend down in a freezing mixture,
condense pure well-dried sulphurous acid gas direct into it from
the generator till it is full nearly to the tops of the two branches.
Then close the stop-cock, detach from the generator, and remove
from the freezing mixture. Holding it still with the bend down,
apply gentle heat to the bend, by a warm hand or by aid of a
spirit-lamp, so as to produce boiling, the bubbles rising up in
either one or the other of the two branches. After doing this
for some time let the bend cool, and apply gentle heat to the
surface of the liquid in that one of the branches into which the
bubbles passed. With great care now open very slightly the
stop-cock at the top of this branch, until the liquid is up to very
near the top of the tube, and close the stop-cock before it begins
to blow out. Repeat the process several times, causing the bubbles
sometimes to rise up one branch, and sometimes up the other.
After this has been done two or three dozen times, it is quite
certain that only a very infinitesimal amount of air can have
remained in the apparatus. When satisfied that this is the case,
sink the bend once more into a freezing mixture, and with a
convenient blow-pipe and flame melt the glass tube below each
stop-cock so as to hermetically seal the two ends of the U tube,

and detach them from the stop-cocks. This completes the construction of the sulphurous acid cryophorus.

The instrument, if turned with the bend up and the two sealed
ends down, may be used as a cryophorus presenting interesting
peculiarities.

The most interesting qualities are those which it presents
when held with the bend down. In this position it constitutes a
differential thermometer of exceedingly high sensibility, founded
on the difference of sulphurous acid steam-pressure due to
difference of pressure in the two branches. One very remarkable
and interesting feature is the exceeding sluggishness with which
the liquid finds its level in the two branches when the external
temperature is absolutely uniform all round. In this respect it
presents a most remarkable contrast with a U tube, in other
respects similar, but occupied by water and water-steam instead
of sulphurous acid and sulphurous-acid steam. If the U tube of
water be suddenly inclined 10 or 20 degrees to the vertical in the
plane of the two branches, the water oscillates before it settles
with the free surfaces in the two branches at the same level.
When the same is done to the U tube of sulphurous acid, it
seems to take no notice of gravity; but in the course of several
minutes it is seen that the liquid is sinking slowly in one branch
and rising in the other towards identity of level. The reason is
obvious.

111. On a Realised Sulphurous Acid Steam-Pressure Thermometer, and on a Sulphurous Acid Steam-Pressure Differential Thermometer; also a Note on Steam-Pressure Thermometers.

[From *Edin. Roy. Soc. Proc.* Vol. x. pp. 532—536; read April 19, 1880.]

A SULPHUROUS acid steam-pressure thermometer, on the plan described in my communication on the subject to the Royal Society of March 1, has been actually constructed, with range up to 25° C., but not yet in a permanent form. The slight trials I have been able to make with it give promise that, in respect to sensibility and convenience for practical use, it will most satisfactorily fulfil all expectations, and have given some experience in respect to the overcoming of difficulties of construction, from which the following instructions are suggested as likely to be useful to any one who may desire to make such an instrument:—

(1) The sulphurous acid steam thermometer might more properly be called a cryometer than a thermometer, because it is not very convenient, except for measuring temperatures lower than the atmospheric temperature at the place and time of observation; for, it must be remarked, that the thermometric substance, that is to say, the infinitesimal layer of liquid and steam of sulphurous acid at the interface between the two in the bulb in the annexed drawing (fig. 1), must be at a lower temperature than any other part of the space of bulb and tube between it and the mercury surface in the shorter vertical column. It is satisfactory, however, that the instrument is really not needed for temperatures above + 10° C., because for such the water steam-pressure thermometer, represented in the first of the three diagrams of my former communication, has ample sensibility for most practical purposes. Hence, instead of the range up to 25° C. in the instrument already realised, and the great length

of tube (295 centimetres for the long vertical branch) which it requires, I propose in future to let $+10°$ C. be the superior limit of the temperatures to be measured by an ordinary sulphurous acid steam thermometer. For this, the long vertical branch need not be more than 175 centimetres; thus the instrument is much more easily made, and when made, is much less cumbrous.

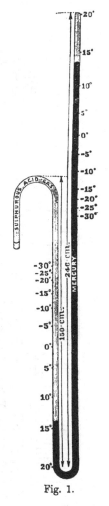

(2) The upper end of the long branch, being open to begin with, is to be securely cemented to a small and very perfectly air-tight iron stop-cock L, communicating with an iron pipe, bent at right angles, as shown in the drawing (fig. 2). This iron pipe is, in the first place, to be put into communication temporarily by an india-rubber junction with the generator, and with an air-pump, by means of a metal branch tube, with two stop-cocks R and S, as shown in the drawing.

(3) To begin, close R and open S and L; and exhaust moderately (down to half an inch of mercury will suffice). Warm the whole length of the bent tube moderately by a spirit lamp, or spirit lamps, to dry the inner surface sufficiently. Then, still maintaining the exhaustion by the air-pump, apply a freezing mixture to the bulb and shorter vertical tube, and all of the long vertical tube except a convenient length of a foot or two next its upper end, as shown in the drawing. Before joining the generator to Q, let enough of sulphurous acid gas be passed out through P to clear out fairly well the air from the generator and the purifying sulphuric acid wash-bottles and pumice tubes, &c. Then make the junction

Fig. 1.

at Q, close the stop-cock S and open R very gently taking care not to let air be sucked in by the safety tube of the generating apparatus *. Continue until liquid sulphurous acid is seen in the

* One of the sulphuric acid wash-bottles must be provided with a safety tube with overflow bulb. An ordinary pipette with its stem fitted into the india-rubber stopper of the bottle will serve for the purpose.

long vertical branch, about the top of the freezing mixture, then close the iron stop-cock L, disconnect at E, and remove from the freezing mixture (but beware of Professor Guthrie's cryohydrates).

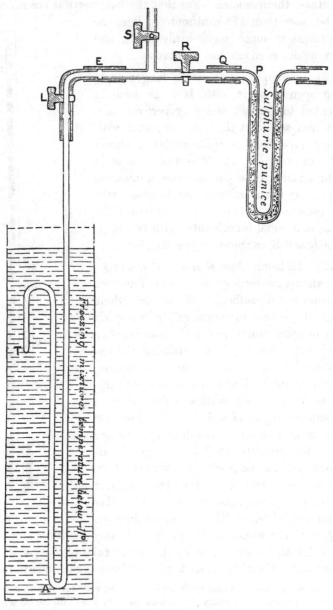

Fig. 2.

(4) Now, place the instrument with the long straight parts of the tube nearly horizontal, but sloping slightly upwards towards L, and by a hand or spirit lamp properly applied, boil the liquid, the stop-cock L being still closed. After boiling for some time cause the liquid to occupy the whole space from T to its free surface, then very carefully open the stop-cock L slightly, being ready to close it quickly and prevent the escape of any of the liquid in case of sudden ebullition. Repeat this process over and over again two or three dozen times, so as thoroughly to remove air or other gases more volatile than sulphurous acid from the liquid. To as much as possible remove water or any fluid less volatile than sulphurous acid, proceed as follows:—Apply heat at T and in the bend next T until the liquid leaves that part of the enclosure and stands nearly at a level in the short and long vertical branch, the instrument being held with A down. Apply a freezing mixture to T, taking care not to cool it to quite as low a temperature as $-11°$ C.; so that the pressure of the sulphurous acid liquid and steam may remain something above the external atmospheric pressure. Occasionally open the stop-cock L very slightly to prevent the liquid from being drawn up the short vertical branch through preponderance of temperature in the long vertical branch. Continue this until about a centimetre of liquid has been distilled over into the bulb T. Then open the stop-cock L very carefully until all the liquid in the two vertical branches is blown out, leaving that which has been distilled over into the bulb T, and then close L again.

(5) Then dip the end E under pure mercury, and by opening L very gently and warming the free surface of the liquid sulphurous acid, let gas escape bubbling up through the mercury. Close L again before or when the quantity of liquid in the bulb at T begins to be perceptibly diminished. Then apply a freezing mixture to T until mercury is drawn in. Incline the instrument with A up and L down, and watch until the mercury is drawn up to A, then incline with A down and let a little more mercury come in. Then close L. Lastly, keeping T still in the freezing mixture, melt the glass below L till it collapses and blows the mercury down, leaving Torricellian vacuum at the sealed end. The instrument is now complete and ready for use.

Sulphurous Acid Steam-Pressure Differential Thermometer.

This consists of a U tube, with its ends bent down, as shown in the drawing, containing mercury in the main bend and in the lower parts of the straight vertical branches, and sulphurous acid gas, steam, and liquid in the rest of the enclosure. Every other part of the enclosure must be kept somewhat warmer than the warmer of the two ends, T, T'.

The infinitesimal quantities of matter in the transitional layers, between liquid and steam, at T and T', constitute the thermometric substance. The gas between these and the manometric mercury, and the mercury serve merely the purpose of transmitting the steam-pressures, and measuring the difference between them.

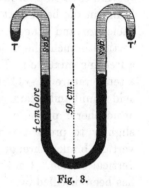

Fig. 3.

At $12\frac{1}{2}°$ C., the sensibility of the instrument, as we see by Regnault's tables of sulphurous acid steam-pressures, quoted in the article "Heat," of the eleventh volume of the *Encyclopædia Britannica*, is 7 centimetres difference of mercury levels, to 1° difference of temperature (that is temperatures 12°, 13°) at T, T' respectively.

At $22\frac{1}{2}°$ C. the sensibility similarly reckoned, is 12·2 cms. to 1° C.

Note on Steam-Pressure Thermometers.

If the bore of the vertical tube is less than three or four millimetres, there ought to be an enlargement at its upper end, or else there should not be quite enough of the liquid to fill the tall manometric tube; otherwise, if in the use of the instrument the liquid is pressed up to the top of the vertical tube, it is impossible to get it down again except by the tedious operation of distilling the whole liquid from the tube into the bulb, by applying heat by means of a spirit-lamp or a large vessel of hot water to the manometric tube, which, to facilitate this operation, may be held inclined with the closed end down. An instrument like that shown in fig. 1 of my former communication (March 1,

1880), with vertical tube of the diameter (2 or 3 millimetres) there stated is subject to this inconvenience, although, in my first attempts to realise the instrument, imperfect removal of air from the water and steam in the enclosed space prevented me from experiencing it. The difficulty, of course, might have been foreseen; but I did not think it would have been so great as I now find it to be with an instrument constructed exactly according to fig. 1 of my former communication, with air very perfectly removed from the enclosure by a proper process of boiling before sealing the instrument.

112. On a Differential Thermoscope founded on Change of Viscosity of Water with Change of Temperature.

[From *Edinb. Roy. Soc. Proc.* Vol. x. p. 537 ; read April 19, 1880.]

WATER flows from a little cistern or reservoir R through a wide vertical tube S, about two metres long; thence through a horizontal capillary tube C', 50 or 100 centimetres long; thence through a wide horizontal metal tube T, 20 or 30 centimetres long; thence through a second horizontal capillary C; and lastly, out by a little constant-level overflow cup L. A vertical glass manometric tube M, a metre and a half long, standing up above the end of T next C to measure the pressure in T by a water column; and a means of giving any uniform temperature to the outsides of S and C', and any other uniform temperature to the outsides of T and C'; complete the instrument.

Denote the heights of the levels of the water in R and M, above L, by h and $\frac{1}{2}h - x$. If C and C' are equal and similar, or otherwise so proportioned as to be equal in their resistances to the flow of the water at equal temperatures through them, we find from the formula by which Poiseuille expressed the results of his experiments on the flow of water through capillary tubes

$$x = \tfrac{1}{2}h \frac{\cdot 03368 \cdot \frac{1}{2}(t - t') + \cdot 000221 \cdot \frac{1}{2}(t^2 - t'^2)}{1 + \cdot 03368 \cdot \frac{1}{2}(t + t') + \cdot 000221 \cdot \frac{1}{2}(t^2 + t'^2)},$$

where t' and t denote the temperatures of the water as it flows through C' and C. By the arrangements described it is secured that t is very nearly the same as the temperature of the outsides of B and C. Thus, if $h = 200$ cms., $t' = 0°$, and $t = 1°$, we have $x = 3\cdot3$. Thus the sensibility is 33 mms. per 1° C.; and 1/30 of a degree would therefore be very perceptible.

Even with its high sensibility this instrument may not be frequently found convenient for thermal researches, and its chief use may be for illustration of Poiseuille's important discovery.

113. On a Thermomagnetic Thermoscope.

[From *Edinb. Roy. Soc. Proc.* Vol. x. [read April 19, 1880], pp. 538, 539.]

THIS thermoscope is founded on the change produced in the magnetic moment of a steel magnet by change of temperature. Several different forms suggest themselves: the one which seems best adapted to give good results is to be made as follows:—

(1) Prepare an approximately astatic system of two thin, hardened steel wires, $r\,b,\,r'\,b'$, each 1 cm. long, one of them, $r\,b$, hung by a single silk fibre, and the other hung bifilarly from it, by fibres about 3 cms. long, so attached that the projections of the two, on a horizontal plane, shall be inclined at an angle of about 01 of a radian (or ·57°) to one another.

(2) Hang a very small light mirror bifilarly from the lower of the two wires.

(3) Magnetise the two wires to very exactly equal magnetic moments in the dissimilar directions. This is easily done by a few successive trials, to make them rest as nearly as possible perpendicular to the magnetic meridian.

(4) Take two pieces of equal and similar straight steel wire, well hardened, each 2 cms. long, and about 04 cm. diameter; magnetise them equally and similarly; and mount them on a suitable frame to fulfil conditions (5) and (6). Call them $R\,B$ and $R'\,B'$, B and B' denoting the ends containing true north polarity (ordinarily marked B), and $R\,R'$ true south (ordinarily marked red). The small letters r, b, r', b' mark, on the same plan, the polarities of $r\,b$ and $r'\,b'$.

(5) The magnets $R\,B, R'\,B'$, are to be relatively fixed in line on their frame, with similar poles next one another, at a distance of about 2 cms. asunder; as thus $R\,B \dots B'\,R'$, with $B\,B' = 2$ cms.

(6)   This frame is to be mounted on a geometrical slide upon the case within which the astatic pair $r\ b,\ r'\ b'$ is hung, in such a manner that the line of $R\ B,\ B'\ R'$ bisects $r\ b$, approximately at right angles, and that $R\ B\ B'\ R'$ may be moved by a micrometer screw through about a millimetre on each side of its central position, the line of motion being the line of $R\ B,\ B'\ R'$, and the "central position" being that in which B and B' are equidistant from the centre of $r\ b$.

(7)   A lamp and scale, with proper focussing lens if the mirror is not concave, are applied to show and measure small deflections as in my mirror galvanometers and electrometer.

Use of the Thermoscope.

(8)   Place the instrument with the needles approximately perpendicular to the magnetic meridian, turning it so as to bring b and b' to the south side of the vertical plane bisecting the small angle between the projections of $r\ b$, $r'\ b$, and r and r' to the north side of it.

(9)   By aid of the micrometer screw bring the luminous image to its middle position on the scale.

(10)   Cause $R\ B,\ B'\ R'$ to have different temperatures.   The luminous image is seen to move in such a direction as is due to r approaching the cooler, and receding from the warmer of the two deflectors $B\ R,\ B'\ R'$.

114. ON A CONSTANT PRESSURE GAS THERMOMETER.

[From *Edinb. Roy. Soc. Proc.* Vol. x. [read April 19, 1880], pp. 539—545 ;
Math. and Phys. Papers, Vol. III, pp. 182—188.]

IN the article on "Heat" published in the eleventh volume
of the *Encyclopædia Britannica*, referred to in my previous communications to the Royal Society on Steam Pressure Thermometers,
it is shown that the Constant Pressure Air Thermometer is the
proper form of expansional thermometer to give temperature
on the absolute thermodynamic scale, with no other data as to
physical properties of the fluid than the thermal effect which it
experiences in being forced through a porous plug, as in the
experiment of Joule and myself on this subject*; and the thermal
capacity of the fluid under constant pressure.   These data for air,
hydrogen, and nitrogen have all been obtained with considerable
accuracy, and therefore it becomes an important object towards
promoting accurate thermometry, to make a practical working
thermometer directly adapted to show temperature on the absolute
thermodynamic scale through the whole range of temperature,
from the lowest attainable by any means, to the highest for which
glass remains solid.   This, I believe, may be done by avoiding the
objectionable expedient adopted by Pouillet and Regnault, of
allowing a portion (when high temperatures are to be measured
the greater portion) of the whole gas to be pressed into a cool
volumetric chamber, out of the thermometric chamber proper, by
the expansion of the portion which remains in; and instead fulfilling the condition, stated, but pronounced practically impossible,
by Regnault (*Expériences*, Vol. I. pp. 168, 169), that the thermometric gas "shall like the mercury of a mercury thermometer be
allowed to expand freely at constant pressure in a calibrated
reservoir maintained throughout at one temperature."   I have
accordingly designed a constant pressure gas thermometer to

---

* "Thermal Effects of Fluids in Motion," *Trans. Roy. Soc. Lond.*, June 1853,
June 1854, June 1860, and June 1862.

fulfil this condition. It is represented in the accompanying
drawing, and described in the following extract from the article
referred to:—

The vessel containing the thermometric fluid, which in this
case is to be either hydrogen or nitrogen*, consists in the main
of a glass bulb and tube placed vertically with bulb up and
mouth down; but there is to be a secondary tube of much finer
bore opening into the bulb or into the main tube near its top,
as may be found most convenient in any particular case. The
main tube which, to distinguish it from the secondary tube, will
be called the volumetric tube, is to be of large bore, not less
than 2 or 3 centimetres, and is to be ground internally to a truly
cylindric form. To allow this to be done it must be made of
thick, well-annealed glass like that of the French glass-barrelled
air-pumps. The secondary tube, which will be called the manometric capillary, is to be of round bore, not very fine, say from
half a millimetre to a millimetre diameter. Its lower end is to
be connected with a mercury manometer to show if the pressure
of the thermometric air is either greater or less than the definite
pressure to which it is to be brought every time a thermometric
measurement is made by the instrument. The change of volume
required to do this for every change of temperature is made and
measured by means of a micrometer screw† lifting or lowering

---

* Common air is inadmissible, because even at ordinary temperatures its
oxygen attacks mercury. The film of oxide thus formed would be very inconvenient at the surface of the mercury caulking, round the base of the piston, and
on the inner surface of the glass tube to which it would adhere. Besides, sooner
or later the whole quantity of oxygen in the air must be diminished to a sensible
degree by the loss of the part of it which combines with the mercury. So far as
we know, Regnault did not complain of this evil in his use of common air in his
normal air thermometer nor in his experiments on the expansion of air (*Expériences*, Vol. I.), though probably it has vitiated his results to some sensible degree.
But he found it to produce such great irregularities when, instead of common air,
he experimented on pure oxygen, that from the results he could draw no conclusion
as to the expansion of this gas (*Expériences*, Vol. I. p. 77). Another reason for the
avoidance of air or other gas containing free oxygen is to save the oil or other
liquid which is interposed between it and the mercury of the manometer from
being thickened or otherwise altered by oxidation.

† This screw is to be so well fitted in the iron sole-plate as to be sufficiently
mercury-tight without the aid of any soft material, under such moderate pressure
as the greatest it will experience when the pressure chosen for the thermometric
gas is not more than a few centimetres above the external atmospheric pressure.
When the same plan of apparatus i used for investigation of the expansion of

a long solid glass piston, fitting easily in the glass tube, and caulked air-tight by mercury between its lower end and an iron sole-plate by which the mouth of the volumetric tube is closed. To perform this mercury caulking, when the piston is raised and lowered, mercury is allowed to flow in and out through a hole in the iron sole-plate by an iron pipe, connected with two mercury cisterns at two different levels by branches each provided with a stopcock. When the piston is being raised the stopcock of the branch leading to the lower cistern is closed, and the other is opened enough to allow the mercury to flow up after the piston and press gently on its lower side, without entering more than infinitesimally into the space between it and the surrounding glass tube (the condition of the upper bounding surface of the mercury in this respect being easily seen by the observer looking at it through the glass tube). When the piston is being lowered, the stopcock in the branch leading from the upper cistern is closed, and the one in the branch leading to the lower cistern is opened enough to let the mercury go down before the piston, instead of being forced to any sensible distance into the space between it and the surrounding tube, but not enough to allow it to part company with the lower surface of the piston. The manometer is simply a mercury barometer of the form commonly called a siphon barometer, with

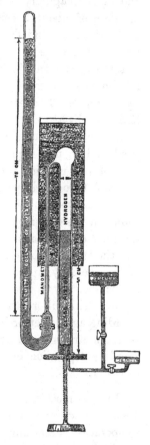

gases under high pressures, a greased leather washer may be used on the upper side of the screw-hole in the sole-plate, to prevent mercury from escaping round the screw. It is to be remarked that in no case will a little oozing out of the mercury round the screw while it is being turned introduce any error at all into the thermometric result; because the correctness of the measurement of the volume of the gas depends simply on the mercury being brought up into contact with the bottom of the piston, and not more than just perceptibly up between the piston and volumetric tube surrounding it.

its lower end not open to the air but connected to the lower end of the manometric capillary. This connection is made below the level of the mercury in the following manner. The lower end of the capillary widens into a small glass bell or stout tube of glass of about 2 centimetres bore and 2 centimetres depth, with its lip ground flat like the receiver of an air-pump. The lip or upper edge of the open cistern of the barometer (that is to say, the cistern which would be open to the atmosphere were it used as an ordinary barometer) is also ground flat, and the two lips are pressed together with a greased leather washer between them to obviate risk of breaking the glass, and to facilitate the making of the joint mercury tight. To keep this joint perennially good, and to make quite sure that no air shall ever leak in, in case of the interior pressure being at any time less than the external barometric pressure or being arranged to be so always, it is preserved and caulked by an external mercury jacket not shown in the drawing. The mercury in the thus constituted lower reservoir of the manometer is above the level of the leather joint, and the space in the upper part of the reservoir over the surface of the mercury, up to a little distance into the capillary above, is occupied by a fixed oil or some other practically vapourless liquid. This oil or other liquid is introduced for the purpose of guarding against error in the reckoning of the whole bulk of the thermometric gas, on account of slight irregular changes in the capillary depression of the border of the mercury surface in the reservoir.

In the most accurate use of the instrument, the glass and mercury and oil of the manometer are all kept at one definite temperature, according to some convenient and perfectly trustworthy intrinsic thermoscope, by means of thermal appliances not represented in the drawing but easily imagined. This condition being fulfilled, the one desired pressure of the thermometric gas is attained with exceedingly minute accuracy by working the micrometer screw up or down until the oil is brought precisely to a mark upon the manometric capillary.

In fact, if the glass and mercury and oil are all kept rigorously at one constant temperature, the only access for error is through irregular variations in the capillary depressions in the borders of the mercury surfaces. With so large a diameter as the 2 centimetres chosen in the figured dimensions of the drawing, the error from this cause can hardly amount to $\frac{1}{100}$ per cent. of the whole pressure, supposing this to be one atmo or thereabouts.

For ordinary uses of this constant-pressure gas thermometer, where the most minute accuracy is not needed, the rule will still be to bring the oil to a fixed mark on the manometric capillary; and no precaution in respect to temperature will be necessary except to secure that it is approximately uniform throughout the mercury and containing glass, from lower to higher level of the mercury. The quantity of oil is so small that, whatever its temperature may be, the bringing of its free surface to a fixed mark on the capillary secures that the mercury surface below the oil in the lower reservoir is very nearly at one constant point relatively to the glass, much more nearly so than it could be made by direct observation of the mercury surface, at all events without optical magnifying power. Now if the mercury surface be at a constant point of the glass, it is easily proved that the difference of pressures between the two mercury surfaces will be constant, notwithstanding considerable variations of the common temperature of the mercury and glass, provided a certain easy condition is fulfilled, through which the effect of the expansion of the glass is compensated by the expansion of the mercury. This condition is, that the whole volume of the mercury shall bear to the volume in the cylindric vertical tube from the upper surface to the level of the lower surface the ratio of $(\lambda - \frac{1}{3}\sigma)$ to $(\lambda - \sigma)$, where λ denotes the cubic expansion of the mercury and σ the cubic expansion of the solid for the same elevation of temperature, it being supposed for simplicity of statement that the tube is truly cylindric from the upper surface to the level of the lower surface, and that the sectional area of the tube is the same at the two mercury surfaces. The cubic expansion of mercury is approximately seven times the cubic expansion of glass. Hence

$$(\lambda - \tfrac{1}{3}\sigma)/(\lambda - \sigma) = (7 - \tfrac{1}{3})/6 = 1{\cdot}111.$$

Hence the whole volume of the mercury is to be about 1·111 times the volume from its upper surface to the level of the lower surface; that is to say, the volume from the lower surface in the bend to the same level in the vertical branch is to be $\frac{1}{9}$ of the volume in the vertical tube above this surface. A special experiment on each tube is easily made to find the quantity of mercury that must be put in to cause the pressure to be absolutely constant when the surface in the lower reservoir is kept at a fixed point relatively to the glass, and when the temperature is varied through such moderate differences of temperature as are to be found in the use of the instrument at different times and seasons.

A sheet-iron can containing water or oil or fusible metal, with external thermal appliances of gas or charcoal furnace, or low-pressure or high-pressure steam heater, and with proper internal stirrer or stirrers, is fitted round the bulb and manometric tube to produce uniformly throughout the mass of the thermometric gas the temperature to be measured. This part of the apparatus, which will be called for brevity the heater, must not extend so far down the manometric tube that when raised to its highest temperature it can warm the caulking mercury to as high a temperature as 40° C., because at somewhat higher temperatures than this the pressure of vapour of mercury begins to be perceptible, and would vitiate the thermometric use of the pure hydrogen or nitrogen of our thermometer. To secure sufficient coolness of the mercury it will probably be advisable to have an open glass jacket of cold water (not shown in the drawing) round the volumetric tube, 2 or 3 centimetres below the bottom of the heater, and reaching to about half a centimetre above the highest position of the bottom of the piston.

It seems probable that the constant-pressure hydrogen or nitrogen gas thermometer which we have now described may give even more accurate thermometry than Regnault's constant-volume air thermometers, and it seems certain that it will be much more easily used in practice.

We have only to remark here further that, if Boyle's law were rigorously fulfilled, thermometry by the two methods would be identical, provided the scale in each case is graduated or calculated so as to make the numerical reckoning of the temperature agree at two points,—for example, 0° C. and 100° C. The very close agreement which Regnault found among his different gas thermometers and his air thermometers with air of different densities, and the close approach to rigorous fulfilment of Boyle's law which he and other experimenters have ascertained to be presented by air and other gases used in his thermometers, through the ranges of density, pressure, and temperature at which they were used in these thermometers, renders it certain that in reality the difference between Regnault's normal air thermometry and thermometry by our hydrogen gas constant-pressure thermometer must be exceedingly small. It is therefore satisfactory to know that for all practical purposes absolute temperature is to be obtained with very great accuracy from Regnault's thermometric system by simply adding 273 to his numbers for temperature on the centigrade scale. It is probable that at the temperatures of 250° or 300° C. (or 523 or 573 absolute) the greatest deviation of temperature thus reckoned, from correct absolute temperature, is not more than half a degree.

115. On the Elimination of Air from Water.

[Read to Phys. Soc., *Proc.* Vol. IV. [May 8, 1880, title only], p. 4; reprinted from *The Telegraphic Journal.*]

Sir W. Thomson made a communication on the elimination of air from a water steam pressure thermometer, and on the construction of a water steam pressure thermometer. He said it was a mistake to suppose that air was expelled by boiling water because the water dissolved less air when warm than when cold. The fact was due to the relations between the density of air in water, and the density of air in water vapour. There was 50 times more air in the water vapour over water in a sealed tube than in the water below. If this air could be suddenly expelled only $\frac{1}{50}$ part of air would remain, and of this only $\frac{1}{2500}$ in the water, the rest being in the vapour. This suggested a means of eliminating air from water which he had employed with success. It consisted in boiling the water in a tube, and by means of a fluid mercury valve allowing a puff of the vapour to escape at intervals. Sir W. Thomson also described his proposed new water steam thermometer, now being made by Mr Casella. It is based on the relations of temperature and pressure in water steam, as furnished by Regnault's or other tables, and will consist of a glass tube with two terminal bulbs like a cryophorus, part containing water, part water steam, and the stem enclosed in a jacket of ice-cold water. Similar vapour thermometers will be formed in which sulphurous acid and mercury will be used in place of water or in conjunction with it. For low or ordinary temperatures they will be more accurate than ordinary thermometers.

116. On a method of determining the critical temperature
for any liquid and its vapour without mechanism.

[From *Nature*, Vol. XXIII. Nov. 25, 1880, pp. 87, 88.]

A PIECE of straight glass tube—60 centimetres is a convenient length—is to be filled with the substance in a state
of the greatest purity possible. It is to contain such a quantity
of the substance that, at ordinary atmospheric temperatures,
about 3 or 4 centimetres of the tube are occupied by steam of
the substance, and the remainder liquid. Fix the tube in an
upright position, with convenient appliances for warming the
upper 10 centimetres of the length to the critical temperature, or
to whatever higher or lower temperature may be desired; and
for warming a length of 40 centimetres from the bottom to some
lower temperature, and varying its temperature conveniently at
pleasure.

Commence by warming the upper part until the surface of
separation of liquid and steam sinks below 5 centimetres from
the top. Then warm the lowest part until the surface rises
again to a convenient position. Operate thus, keeping the
surface of separation of liquid and solid at as nearly as possible
a constant position of 3 centimetres below the top of the tube,
until the surface of separation disappears.

The temperature of the tube at the place where the surface
of separation was seen immediately before disappearance is the
critical temperature.

It may be remarked that the changes of bulk produced by the
screw and mercury in Andrews' apparatus are, in the method
now described, produced by elevations and depressions of temperature in the lower thermal vessel. By proper arrangements
these elevations and depressions of temperature may be made as
easily, and in some cases as rapidly, as by the turning of a screw.
The dispensing with all mechanism and joints, and the simplicity

afforded by using the substance to be experimented upon, and no other substance in contact with it, in a hermetically sealed glass vessel, are advantages in the method now described. It is also interesting to remark that in this method we have continuity through the fluid itself all at one equal pressure exceeding the critical pressure, but at different temperatures in different parts, varying continuously from something above the critical temperature at the top of the tube to a temperature below the critical temperature in the lower part of the tube.

The pressure may actually be measured by a proper appliance on the outside of the lower part of the tube to measure its augmentation of volume under applied pressure. If this is to be done, the lower thermal vessel must be applied, not round the bottom of the tube, but round the middle portion of it, leaving, as already described, 10 to 20 cms. above for observation of the surface of separation between liquid and vapour, and leaving at the bottom of the tube 20 or 30 cms. for the pressure-measuring appliance.

This appliance would be on the same general principle as that adopted by Prof. Tait in his tests of the *Challenger* thermometers under great pressure (*Proceedings Royal Soc. Edin.* 1880); a principle which I have myself used in a form of depth-gauge for deep-sea soundings; in which the pressure is measured, not by the compression of air, but by the flexure or other strain produced in brass or glass or other elastic solid.

117. On the Sources of Energy in Nature available to Man for the Production of Mechanical Effect.

[Presidential Address to the Mathematical and Physical Science Section of the British Association at York, Sept. 1. From *British Association Report*, 1881, pp. 513—518 ; *Nature*, Vol. XXIV. Sept. 8, 1881, pp. 433—436 ; *Franklin Inst. Journ.* Vol. LXXXII. Nov. 1881, pp. 376—385. Reprinted in *Popular Lectures and Addresses*, Vol. II. pp. 433—450.]

118. Accélération Thermodynamique du Mouvement de Rotation de la Terre.

[From *Paris Soc. Phys. Séances*, 1881 [Sept. 23, 1881], pp. 200—210; *Edinb. Roy. Soc. Proc.* Vol. XI. [read Jan. 16, 1882], pp. 396—405; *Journ. de Phys.* Vol. I. 1882, pp. 61—70; *Nuovo Cimento*, Vol. XI. 1882, pp. 240—243 ; *L'Astronomie*, Vol. IV. June 1885, pp. 230, 231. Reprinted in *Math. and Phys. Papers*, Vol. III. pp. 341—350.]

119. On the Efficiency of Clothing for Maintaining Temperature.

[From *Edinb. Roy. Soc. Proc.* Vol. XII. [March 3, 1884, title only], p. 563 ; *Nature*, Vol. XXIX. April 10, 1884, p. 567.]

Sir W. Thomson showed that if a body be below a certain size, the effect of clothing will be to cool it. In a globular body the temperature will only be kept up if the radius be greater than $k/2e$, where k is the conductivity of the substance and e its emissivity.

120. On Osmotic Pressure against an Ideal Semi-permeable Membrane.

[From *Edinb. Roy. Soc. Proc.* Vol. XXI. [read Jan. 18, 1897], pp. 323—325; *Nature*, Vol. LV. Jan. 21, 1897, pp. 272, 273.]

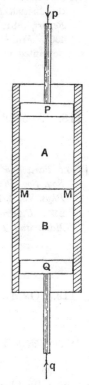

To approach the subject of osmotic pressure against an ideal impermeable membrane, consider first a vessel filled with any particular fluid divided into two parts, A and B, by an ideal surface, MM. Let a certain number of individual molecules of the fluid in A, any one of which we shall call D (the dissolved substance), be endowed with the property that they cannot cross the surface MM (the semi-permeable membrane); but let them continue to be in other respects exactly similar to every other molecule of the fluid in A, and to all the molecules of the fluid in B, any one of which we shall call S (the solvent), each of which can freely cross the membrane. Suppose now the containing vessel and the dividing membrane all perfectly rigid*. Let the apparatus be left to itself for so long time that no further change is perceptible in the progress towards final equilibrium of temperature and pressure. The pressures in A and B will be exactly the same as they would be with the same densities of the fluid if MM were perfectly impermeable, and all the molecules of the fluid were homogeneous in all qualities; and MM will be pressed on one side only, the side next A, with a force equal to the excess of the pressure in A above the pressure in B, and due solely to the impacts of D molecules striking it and rebounding from it.

If now, for a moment, we suppose the fluid to be "perfect gas," we should find the pressure on MM to be equal to that which

---

* In the drawing, the vessel is represented by a cylinder closed at each end by a piston to facilitate the consideration of what will happen if, instead of supposing it rigid, any arbitrary condition as to the pressures on the two sides of the membrane be imposed.

would be produced by the D molecules if they were alone in the space A; and this is, in fact, approximately what the osmotic pressure would be with two ordinary gases at moderate pressures, one of which is confined to the space A by a membrane freely permeable by the other. On this supposition the number of the S molecules per unit bulk would be the same on the two sides of the membrane. If, for example, there are 1000 S molecules to one D molecule in the space A, the pressure on the piston P would be 1001 times the osmotic pressure, and on Q 1000 times the osmotic pressure. But if the fluid be "liquid" on both sides of the membrane, we may annul the pressure on Q and reduce the pressure on P to equality with the osmotic pressure, by placing the apparatus under the receiver of an air-pump, or by pulling Q outwards with a force equal and opposite to the atmospheric pressure on it. When we do this, the annulment of the integral pressure of the liquid on the piston Q is effected through balancing by attraction, of pressure due to impacts, between the molecules of the liquid S and the molecules of the solid piston Q. We are left absolutely without theoretical guide as to the resultant force due to the impacts of S molecules and D molecules striking the other piston, P, and rebounding from it, and their attractions upon its molecules; and as to the numbers per unit volume of the S molecules on the two sides of MM, except that they are not generally equal. [*Addition, of date June* 30, 1897.—In an interesting article published in *Nature* of March 18, Prof. Willard Gibbs has shown that in the present ideal case the difference of pressures on the two sides of the ideal semi-permeable membrane fulfils van't Hoff's law. But this is only because of the identity of character of the S and D molecules in all qualities except in respect to action on the ideal semi-permeable membrane: and the demonstration essentially fails when the law of variation of pressure and density, according to height, differs in two vertical tubes, one of them containing S molecules alone, and the other containing a mixture of S and D molecules.]

No molecular theory can, for sugar or common salt or alcohol, dissolved in water, tell us what is the true osmotic pressure against a membrane permeable to water only, without taking into account laws quite unknown to us at present regarding the three sets of mutual attractions or repulsions: (1) between the molecules of the dissolved substance; (2) between the molecules of water;

(3) between the molecules of the dissolved substance and the molecules of water. Hence van't Hoff's well-known statement, applying to solutions Avogadro's law of gases, has manifestly no theoretical foundation at present; even though for some solutions other than mineral salts dissolved in water, it may be found somewhat approximately true; while for mineral salts dissolved in water it is wildly far from the truth. The subject is full of interest, which is increased, not diminished, by eliminating from it fallacious theoretical views. Careful consideration of how much we can really learn with certainty from theory (of which one example is the relation between osmotic pressure and vapour pressure at any one temperature) is exceedingly valuable in guiding and assisting experimental efforts for the increase of knowledge. All chemists and physicists who occupy themselves with the "theory of solutions," may well take to heart warnings, and leading views, and principles, admirably put before them by FitzGerald in his "Helmholtz Memorial Lecture" (*Trans. Chem. Soc.*, 1896) of January 1896 (pp. 898—909).

[From *Nature*, Vol. LV. Jan. 21, 1897, p. 272.]

IN last week's *Nature*, Lord Rayleigh gave*, *for an involatile liquid*, a rigorous and clear proof of "the Central Theorem" of osmotics. But this theorem, though highly interesting in itself, is not, so far as I can see, useful as a guide for experiment. Consider for example the typical cases of sugar, and of common salt, dissolved in water.

If water were absolutely non-volatile, the osmotic pressure of each solution against an ideal semi-permeable membrane separating it from pure water, would, according to the theorem, be equal to the calculable pressure of the ideal gas of the dissolved substance supposed alone in the space occupied by the solution. *This would be true whatever be the molecular grouping of the sugar or of the salt in the solution.* It is believed that experiment has verified the theorem, extended to volatile solvents, as approximately true for sugar and several other substances of organic origin, and of highly complex atomic structure; but has proved it to vastly under-estimate the osmotic pressure for common salt and many other substances of similarly simple composition.

BELFAST, *Jan.* 19.

* [*Scientific Papers*, Vol. IV. p. 267.]

121. On a Differential Method for Measuring Differences
of Vapour Pressures of Liquids at one Temperature
and at Different Temperatures.

[From *Edinb. Roy. Soc. Proc.* Vol. xxi. [read Jan. 18, 1897], pp. 429—432;
Nature, Vol. lv. Jan. 21, 1897, pp. 273, 274.]

1. Apparatus for realising the proposed method is represented in the accompanying diagram. Two Woulffe's bottles,
each having a vertical glass tube fitted air-tight into one of its
necks, contain the liquids the difference of whose vapour pressures
is to be measured. Second necks of the two bottles are connected
by a bent metal (or glass) pipe, with a vertical branch provided with three (metal or glass) stopcocks, as indicated in the
diagram. Each bottle has a third neck, projecting downwards
through its bottom, stopped by a glass stopcock which can be
opened for the purpose of introducing or withdrawing liquid.
The upper ends of the glass tubes are also connected (by short
india-rubber junctions or otherwise) with a bent metal pipe
carrying a vertical branch for connection with a Toepler* mercury
air-pump. This vertical branch is provided with a metal stop-
cock. The vertical branch of the pipe fitted into necks of the
two bottles is also connected to the air-pump as indicated in the
drawings.

2. To introduce the liquids, bring open vessels containing
them into such positions below the bottles that the necks project
downwards into them. Close the glass stopcocks of these lower
necks, open all the other six stopcocks, and produce a slight
exhaustion by a few strokes of the air-pump. Then, opening the

* An ordinary mechanical air-pump would not serve the purpose, because its
valves would not open properly to draw out the very small amount of air which
must be removed to avoid vitiating the observations by any sensible amount of
air-pressure added to the pressure of the vapour.

glass stopcocks very slightly, allow the desired quantities of the liquids to enter, and close them again. They will not be opened again unless there is occasion to remove the whole or some part of the liquid from either bottle; and, unless explicitly mentioned, will not be included among the stopcocks referred to in what follows. It will generally be convenient to make the quantities of the two liquids introduced such that they stand at as nearly as may be the same levels in the two bottles, as indicated in the drawing.

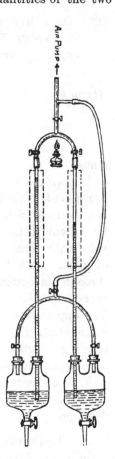

3. Operate now on one only of the liquids until it is got into equilibrium, with its upper level at some point in its glass tube, and nothing but its own vapour between this surface and the closed stopcock immediately above it. To do this proceed as follows :—Close and keep closed the two stopcocks of the liquid not operated on, and work the air-pump with the other four stopcocks all open until an exhaustion, not quite as perfect as is possible, of the air over the liquid operated on is produced.

4. Then close the lower air-pump stopcock, and go on working the pump until the liquid in the tube ceases to rise further above its level in the bottle. Close the two stopcocks of this liquid.

5. Operate similarly on the other liquid.

6. Close now the lower air pump stopcock, and equalise the pressures of air and vapour above the liquids in the two bottles by opening their neck stopcocks. If the levels of the liquids in the two columns are lower than convenient for observation, some air should be allowed very cautiously to run back from the airpump into the two bottles through the lower air-pump stopcock. After doing this, repeat the operation of § 4 for each liquid.

7. The operation of §§ 4, 5 must be continued long enough to distil out of the upper part of each liquid, in its glass tube, air or

any foreign* volatile substance sufficiently to prevent any sensible
pressure on the free surface other than that of the vapour of the
solvent.

8. By proper thermal appliances, indicated by the dotted lines
in the diagram, and the lamp under the upper bent metal tube
(inserted merely as an indication that somehow the metal tube is
to be always slightly warmer than the warmer of the two liquid
surfaces, in order that there may be no condensation of vapour in
it), bring the upper surfaces of the liquids to any one temperature,
or to two different temperatures. The difference of levels of the
liquids in the two tubes, with proper correction for the densities
of the two liquids at their actual temperatures in different parts
of their columns, gives the difference of vapour pressures for the
actual temperatures of the two liquids at their upper surfaces.

9. To facilitate and approximately determine the hydrostatic
correction for specific gravities at the actual temperatures of the
two liquids, open wide the stopcock above the top of one of the
two glass tubes, and let a little air run back from the air-pump, by
very cautiously and slightly opening our upper air-pump stopcock,
and closing it again before the liquid surface reaches the lower
end of its glass tube. Then open wide the stopcock over the top
of the other glass tube. After that, by cautiously opening and
closing our lower air-pump stopcock, let in a little air to the bottles
until the mean level of the liquids in the two columns rises to
nearly the same level as it had in the observed positions of § 8.
In the present circumstances, air in the upper bent metal tube
resists diffusion of vapour through it sufficiently to prevent any
important difference of temperatures from being produced by
evaporation and condensation at the two liquid surfaces, and
there is practically perfect hydrostatic equilibrium of equal liquid
pressures at the tops of the two columns.

10. The vapour pressure of water is accurately known through
a very wide range of temperature from Regnault's experiments;
hence, if pure water be taken for one of our two liquids, the mode
of experiment described above determines the vapour pressure of
the other liquid.

* Compare Ostwald, *Physico-Chemical Measurements*, translated by Walker
(Macmillan, 1892), last paragraph, p. 112.

11. The apparatus may be kept day after day with the same liquids in it (all the stopcocks to be closed, except when it is not in use for observations); and thus the observations for difference of vapour pressures may be repeated day after day; or a long series of observations may very easily be made to determine vapour pressures at different temperatures. Always before commencing observations the operation of § 7 must be repeated to remove air or other volatile impurity, if any has escaped from dissolution in either liquid into the vapour space above it, or if any air has leaked in by the upper stopcocks.

[From *Nature*, Vol. LV. 1896—97, Jan. 28, 1897, pp. 295, 296.]

THE distillation of vapour from one of the vertical tubes to the other, referred to at the end of Operation No. 4, in my communication published in last week's *Nature* (p. 273), may be wholly got quit of by the following simplified mode of procedure.

Operate first on one only of the liquids until it is got into equilibrium, with its upper level at any conveniently marked point in its glass tube, and nothing but its own vapour between this surface and the closed stopcock immediately above it; the upper-neck stopcock over the bottle for this liquid being also closed.

Operate similarly on the other liquid; and close both the air-pump stopcocks, so that now we have all the stopcocks closed.

Open now very gradually the upper-neck stopcocks of the two bottles. While doing so, prevent the liquid from rising in either tube above the marked point by working the air-pump and very slightly opening the lower air-pump stopcock. When both the upper-neck stopcocks are wide open, any adjustment that is considered desirable for the level of the liquid standing higher than the other in its glass tube, may be deliberately made by drawing out or letting in a little air through the lower air-pump stopcock.

Either or both liquids may be thoroughly stirred at any time to ensure homogeneousness by alternately exhausting and letting in air to the bottle or bottles by means of the air-pump and the lower air-pump stopcock, the upper three stopcocks being kept closed.

Operation No. 6 of my article on the subject, in last week's *Nature* (p. 274), must be performed as often as is found necessary. Every one of the stopcocks must be kept closed except when it is open for operation or observation.

The metal tube connecting the upper necks of the two bottles must be long enough, or of fine enough bore, to prevent diffusion of vapour to any sensible extent from either bottle to the other during the time of an observation. It ought to be kept at a temperature somewhat higher than that of the bottles, to prevent any liquid from condensing as dew on its inner surface.

122. ANIMAL THERMOSTAT.

[From *British Association Report*, 1902, pp. 543—546; *Nature*, Vol. LXVII. Feb. 26, 1903, pp. 401, 402; *Phil. Mag.* Vol. v. Feb. 1903, pp. 198—202.]

A THERMOSTAT is an apparatus, or instrument, for automatically maintaining a constant temperature in a space, or in a piece of solid or fluid matter with varying temperatures in the surrounding matter.

Where and of what character is the thermostat by which the temperature of the human body is kept at about 98°·4 Fahrenheit? It has long been known that the source of heat drawn upon by this thermostat is the combination of food with oxygen, when the surrounding temperature is below that of the body. The discovery worked out by Lavoisier, Laplace, and Magnus still holds good, that the place of the combination is chiefly in tissues surrounding minute tubes through which blood circulates through all parts of the body, and not mainly in the place where the furnace is stoked by the introduction of food, in the shape of chyle, into the circulation, nor in the lungs where oxygen is absorbed into the blood. It is possible, however, that the controlling mechanism by which the temperature is kept to 98°·4 may be in the central parts, about, or in, the pumping station (the heart); but it may seem more probable that it is directly effective in the tissues or small blood-vessels in which the combination of oxygen with food takes place.

But how does the thermostat act when the surrounding temperature is anything above 98°·4 and the atmosphere saturated with moisture so that perspiration could not evaporate from the surface? If the breath goes out at the temperature of the body and contains carbonic acid, what becomes of the heat of combustion of the carbon thus taken from the food? It seems as if a large surplus of heat must somehow be carried out by the breath: because heat is being conducted in from without across the skin all over the body; and the food and drink we may suppose to be at the surrounding temperature when taken into the body.

Much is wanted in the way of experiment and observation to test the average temperature of healthy persons living in a thoroughly moist atmosphere at temperatures considerably above 98°·4; and to find how much, if at all, it is above 98°·4. Experiments might also, safely I believe, be tried on healthy persons by keeping them for considerable times in baths at 106° Fahr. with surrounding atmosphere at the same temperature and thoroughly saturated with vapour of water. The temperature of the mouth (as ordinarily taken in medical practice) should be tested every two minutes or so. The temperature and quantity and moisture and carbonic acid of the breath should also be measured as accurately as possible.

P.S., December 5, 1902. Since the communication of this note my attention has been called to a most interesting paper by Dr Adair Crawford in the *Philosophical Transactions* for 1781 (Hutton's *Abridgments*, Vol. xv. p. 147), "Experiments on the Power that Animals, when placed in certain Circumstances, possess of producing Cold." Dr Crawford's title expresses perfectly the question to which I desired to call the attention of the British Association; and, as contributions towards answering it, he describes some very important discoveries by experiment in the following passage, which I quote from his paper:

· "The following experiments were made with a view to determine with greater certainty the causes of the refrigeration in the above instances*. To discover whether the cold produced by a living animal, placed in air hotter than its body, be not greater than what would be produced by an equal mass of inanimate matter, Dr Crawford took a living and a dead frog, equally moist, and of nearly the same bulk, the former of which was at 67°, the latter at 68°, and laid them on flannel in air which had been raised to 106°. In the course of twenty-five minutes the order of heating was as annexed†.

---

* Observations by Governor Ellis in 1758; teachings of Dr Cullen prior to 1765; very daring and important experiments by Dr Fordyce on himself in heated rooms, communicated to the Royal Society of London in 1774.

† In the two following experiments the thermometers were placed in contact with the skin of the animals under the axillæ.—*Orig.*

Min.	Air	Dead Frog	Living Frog
In 1	......	70½°	67½°
„ 2	102	72	68
„ 3	100	72½	69½
„ 4	100	73	70
„ 25	95	81¼	78¼

"The thermometer being introduced into the stomach, the internal heat of the animals was found to be the same with that at the surface. Hence it appears that the living frog acquired heat more slowly than the dead one. Its vital powers must therefore have been active in the generation of cold.

"To determine whether the cold produced in this instance depended solely on the evaporation from the surface, increased by the energy of the vital principle, a living and dead frog were taken at 75°, and were immersed in water at 93°, the living frog being placed in such a situation as not to interrupt respiration *.

Min.	Dead Frog	Living Frog
In 1	85°	81°
„ 2	88⅓	85
„ 3	90⅓	87
„ 5	91⅓	89
„ 6	91⅓	89
„ 8	91½	89

"These experiments prove, that living frogs have the faculty of resisting heat, or producing cold, when immersed in warm water; and the experiments of Dr Fordyce prove, that the human body has the same power in a moist as well as in a dry air: it is therefore highly probable, that this power does not depend solely on evaporation.

"It may not be improper here to observe, that healthy frogs, in an atmosphere above 70°, keep themselves at a lower temperature than the external air, but are warmer internally than at

* In the above experiment the water, by the cold frogs and by the agitation which it suffered during their immersion, was reduced nearly to 91½°.—*Orig.*

the surface of their bodies; for when the air was 77°, a frog was found to be 68°, the thermometer being placed in contact with the skin; but when the thermometer was introduced into the stomach, it rose to 70½°. It may also be proper to mention, that an animal of the same species placed in water at 61°, was found to be nearly 61¼° at the surface, and internally it was 66½°. These observations are meant to extend only to frogs living in air or water at the common temperature of the atmosphere in summer. They do not hold with respect to those animals, when plunged suddenly into a warm medium, as in the preceding experiments.

"To determine whether other animals also have the power of producing cold, when surrounded with water above the standard of their natural heat, a dog at 102° was immersed in water at 114°, the thermometer being closely applied to the skin under the axilla, and so much of his head being uncovered as to allow him a free respiration.

In 5 minutes the dog was 108°, water 112°.
,, 6 ,, ,, 109°, ,, 112°.
,, 11 ,, ,, 108°, ,, 112°, the respiration having become very rapid.
,, 13 ,, ,, 108°, ,, 112°, the respiration being still more rapid.
,, 30 ,, ,, 109°, ,, 112°, the animal then in a very languid state.

"Small quantities of blood being drawn from the femoral artery, and from a contiguous vein, the temperature did not seem to be much increased above the natural standard, and the sensible heat of the former appeared to be nearly the same with that of the latter.

"In this experiment a remarkable change was produced in the appearance of the venous blood; for it is well known, that in the natural state, the colour of the venous blood is a dark red, that of the arterial being light and florid; but after the animal, in the experiment in question, had been immersed in warm water for half an hour, the venous blood assumed very nearly the hue of the arterial, and resembled it so much in appearance, that it was difficult to distinguish between them. It is proper to observe, that the animal which was the subject of this experiment, had been previously weakened by losing a considerable quantity of blood a few days before. When the experiment was repeated with dogs which had not suffered a similar evacuation, the change in the colour of the venous blood was more gradual; but in every instance in which the trial was made, and it was repeated six times, the alteration was so remarkable, that the blood which was taken in the warm bath could readily be distinguished from that which had been taken from the same vein before immersion, by those who were unacquainted with the motives or circumstances of the experiment.

"To discover whether a similar change would be produced in the colour of the venous blood in hot air, a dog at 102° was placed in air at 134°. In ten minutes the temperature of the dog was $104\frac{1}{2}$°, that of the air being 130°. In fifteen minutes the dog was 106°, the air 130°. A small quantity of blood was then taken from the jugular vein, the colour of which was sensibly altered, being much lighter than in the natural state. The effect produced by external heat on the colour of the venous blood, seems to confirm the following opinion, which was first suggested by my worthy and ingenious friend Mr Wilson, of Glasgow. Admitting that the sensible heat of animals depends on the separation of absolute heat from the blood by means of its union with the phlogistic principle in the minute vessels, may there not be a certain temperature at which that fluid is no longer capable of combining with phlogiston, and at which it must of course cease to give off heat? It was partly with a view to investigate the truth of this opinion that Dr Crawford was led to make the experiments recited above."

These views of Dr Crawford and "his worthy and ingenious friend Mr Wilson*, of Glasgow," express, about as well as it was possible to express before the chemical discoveries of carbonic acid and oxygen, the now well-known truth that oxygen carried along with, but not chemically combined with, food in the arteries, combines with the carried food in the capillaries or surrounding tissues in the outlying regions, and yields carbonic acid to the returning venous blood: this carbonic acid giving the venous blood its darker colour, and being ultimately rejected from the blood and from the body through the lungs, and carried away in

---

* Who, no doubt, was Dr Alex. Wilson, first Professor of Astronomy in the University of Glasgow (1760—1784); best known now for his ingenious views regarding sun-spots.

the breath. Crawford's very important discovery that the venous blood of a dog which had been kept for some time in a hot-water bath at 112° Fahr. was almost undistinguishable from its arterial blood proves that it contained much less than the normal amount of carbonic acid, and that it may even have contained no carbonic acid at all. Chemical analysis of the breath in the circumstances would be most interesting; and it is to be hoped that this chemical experiment will be tried on men. It seems indeed, with our present want of experimental knowledge of animal thermodynamics, and with such knowledge as we have of physical thermodynamics, that the breath of an animal kept for a considerable time in a hot-water bath above the natural temperature of its body may be found to contain no carbonic acid at all. But even this would not explain the *generation of cold* which Dr Crawford so clearly and pertinaciously pointed out. Very careful experimenting ought to be performed to ascertain whether or not there is a surplus of oxygen in the breath; more oxygen breathed out than taken in. If this is found to be the case, the *animal cold* would be explained by deoxidation (unburning) of matter within the body. If this matter is wholly or partly water, free hydrogen might be found in the breath; or the hydrogen of water left by oxygen might be disposed of in the body, in less highly oxygenated compounds than those existing when animal heat is wanted for keeping up the temperature of the body, or when the body is dynamically doing work.

123. THE POWER REQUIRED FOR THE THERMODYNAMIC
HEATING OF BUILDINGS*.

[From *Cambridge and Dublin Mathematical Journal*, November, 1853.]

SOLUTION OF A PROBLEM.

(*a*) What horse-power would be required to supply a building with 1 lb. of air per second, heated mechanically from 50° to 80° Fahrenheit? Compare the fuel that an engine producing this effect as $\frac{1}{10}$ of the equivalent of the heat of combustion would consume, with that which would be required to heat directly the same quantity of air.

(*b*) Explain how this effect may be produced with perfect economy by operating on the air itself to change its temperature, and give dimensions &c. of an apparatus that may be convenient for the purpose.

(*c*) Show how the same apparatus may be adapted to give a supply of cooled air.

Ex. Let it be required to supply a building with 1 lb. of air per second, cooled from 80° to 50° Fahrenheit. Determine the horse-power wanted to work the apparatus in this case.

(*a*) Instead of heating the air directly, we can produce the required effect more economically by means of a perfect thermodynamic engine; and it is easy to show that this is the most economical way. We will consider the air heated pound by pound, and sent into the building at the end of the heating process. Generally, let T be the temperature of the unheated air, S the temperature to which we wish it heated. T, being the temperature of air, water, &c. external to the building, will be the temperature of our refrigerator; the pound of air to be heated will be our source (nominally), and by working the engine backwards instead of taking away, we will give heat to the source.

* [See "On the Economy of the Heating or Cooling of Buildings by means of Currents of Air," *Glasgow Phil. Soc. Proc.* Dec. 1852, reprinted, with an addition referring to Coleman's refrigerating process, in *Math. and Phys. Papers*, Vol. I. pp. 515—520.]

If a be the specific heat of air, adt units will be required to raise the temperature of the pound of air from t to $t + dt$, and the work which must be spent to supply this will be

$$Jadt \, \frac{t - T}{t + E^{-1}}*.$$

Let the whole work spent upon the pound of air be denoted by W; then we have

$$W = Ja \int_{T}^{S} \frac{t - T}{t + E^{-1}} \, dt \, ;$$

whence

$$W = Ja \left\{ (S - T) - \left(T + E^{-1} \right) \log \frac{S + E^{-1}}{T + E^{-1}} \right\}.$$

Ex. $S = 80°$, $T = 50°$, Fahrenheit.

As E is usually given with reference to units Centigrade, we prefer reducing to that scale.

$$S = 26° \, 66' \text{ and } T = 10° \text{ Cent.,} \quad E = ·00366.$$

$$E^{-1} = 273·2240437, \quad a = ·24, \quad J = 1390.$$

$$W = 1390 \times ·24 \left\{ 16·66' - 283·22404 \log \frac{299·89071}{283·22404} \right\}$$

$$= 1390 \times ·24 \left\{ 16·66' - 16·194713 \right\}$$

$$= 157·4438.$$

As one pound of air is heated per second, the H.P. of the engine will be got by dividing this by 550, so that

$$\text{H.P. of engine} = ·28626.$$

If an engine (probably a steam-engine) be employed to drive the heating machine, and economise only $\frac{1}{10}$th of the fuel, the fuel must have evolved $10 W/J$ units. To heat the pound directly $a (S - T)$ units must be supplied, and $\dfrac{10 W/J}{a (S - T)} \times 100$ gives the percentage. In the particular case we have been considering, we find that to heat the air by means of an engine economising $\frac{1}{10}$ would require $\dfrac{4·7195}{16·6'} \times 100$, or 28·317 per cent. of the fuel required for direct heating.

* For the formulas regarding the duty of a perfect engine, and the mechanical value of each of its cycles of operations, constantly to be employed in these solutions, we refer to a paper by Professor W. Thomson in the *Philosophical Transactions*, and which appears in the present number of this *Journal*.

(b) Conceive two double-stroke cylinders connected by tubes and valves in some convenient way, with a reservoir between them. Conceive the one to be made of perfectly conducting matter, so that there shall be no difference in temperature between internal and external air (practically this may be approximated to by immersion in running water); the other cylinder must, on the contrary, be perfectly non-conducting. The pressure in the reservoir being kept at an amount depending on the required heating effect, air is admitted (doing work as it enters) by the former, which we shall call the ingress cylinder, and is not allowed to cool below atmospheric temperature. It is pumped out by the latter, called the egress cylinder, and so heated by compression to the required temperature.

After these very general explanations, we proceed to mention more particularly the *details* of this process.

Let p', t' be respectively the atmospheric pressure and temperature, v' the volume of one pound of air under pressure p' and at temperature t', t the temperature to which we wish the air to be raised, p the pressure such that, if air under it be compressed to pressure p', the temperature will rise from t' to t, v the volume of one pound of air under pressure p and at temperature t', v_1 the volume of air under pressure p' and at temperature t. Now, by Poisson's formula and by the gaseous laws we have

$$\text{(A)} \quad \frac{E^{-1}+t'}{E^{-1}+t} = \left(\frac{p}{p'}\right)^{\frac{K-1}{K}}, \qquad p = p'\left\{\frac{E^{-1}+t'}{E^{-1}+t}\right\}^{\frac{K}{K-1}},$$

from which p may be determined. p must be, moreover, the pressure in the reservoir, as will afterwards appear.

Volume of Cylinder. As the apparatus has to supply one pound of air per second, it will be convenient to suppose the cylinders of such a size, as to contain one pound of air at pressure p and temperature t'.

Operations in Ingress Cylinder. Suppose the piston at the top of its stroke, and the lower part of the cylinder connected with the reservoir, and consequently filled with air at pressure p and temperature t'. Then external air admitted above the piston will push it down ($p' > p$). In this the first part of the stroke, admit so much air that, when secondly it is allowed to expand at constant temperature t', we will have reached the end of the stroke by the time that the pressure has fallen to p. The lower part of the cylinder having been connected with the reservoir, has given to the latter the pound of air it contained; and at the end of the down-stroke the upper part is filled with air ready to be sent in by the up-stroke.

In this operation it is plain that we obtain mechanical effect, and we will naturally spend it, in helping to pump the air out by the egress cylinder.

Operations in Egress Cylinder. Suppose, as before, the piston at the top of its stroke, and the cylinder filled with air at pressure p and temperature t'. During the whole stroke you allow air from the reservoir to enter above the piston. During the first part of the stroke you compress the air below the piston, until the pressure becomes p' and the temperature consequently t. Then expel this heated air into the building or whatever place you wish to heat.

Estimate of total work spent.

(1) In ingress cylinder:

mechanical effect obtained during the first part of the stroke

$$= (p' - p)\, v';$$

mechanical effect obtained during the second part

$$= p'v' \log \frac{v}{v'} - p\,(v - v').$$

(B) Whole gain in ingress cylinder

$$= p'v' \log \frac{v}{v'}.$$

(2) In egress cylinder:

mechanical effect spent during compression

$$= \frac{pv}{K-1} \left\{ \left(\frac{v}{v_1} \right)^{K-1} - 1 \right\} - p\,(v - v_1)$$

work spent during expulsion

$$= p'v_1 - pv_1 :$$

but

$$\frac{p'v_1}{pv} = \frac{E^{-1} + t}{E^{-1} + t'} = \left(\frac{v}{v_1} \right)^{K-1}$$

(C)   Whence whole work spent in egress cylinder

$$= \frac{pv}{K-1}\left\{\frac{E^{-1}+t}{E^{-1}+t'}-1\right\}+pv\left\{\frac{E^{-1}+t}{E^{-1}+t'}\right\}-pv,$$

$$= p'v'\frac{K}{K-1}\frac{t-t'}{E^{-1}+t'},$$

since $pv = p'v'$.

(D)   Amount of work spent in both

$$= p'v'\frac{K}{K-1}\frac{t-t'}{E^{-1}+t'}-p'v'\log\frac{v}{v'}.^*$$

Ratios of expansion :

in the first cylinder,

$$\frac{v'}{v}=\frac{p}{p'}=\left\{\frac{E^{-1}+t'}{E^{-1}+t}\right\}^{\frac{K}{K-1}},$$

in the second cylinder,

$$\frac{v_1}{v}=\left\{\frac{E^{-1}+t'}{E^{-1}+t}\right\}^{\frac{1}{K-1}}.$$

A slight consideration will shew, that the rates of the cylinders must be the same if we consider them as of the same size, and as each contains one pound of air at pressure p and temperature t', the rate will evidently be 30 double strokes per minute.

Let h be the height of the cylinder, and r the radius of the base ; then volume of cylinder $= v = \pi r^2 h$, and if V_0 be the volume of a pound of air under pressure p' and at 0° Cent.,

(E) $$v' = V_0(1 + Et'),$$

$$v = \frac{p'}{p}V_0(1+Et').$$

---

* Modifying this by means of the formulas

$$\frac{v}{v'}=\left\{\frac{E^{-1}+t}{E^{-1}+t'}\right\}^{\frac{K}{K-1}}\text{ and (E),}$$

we find the following as equivalent:

$$\frac{K}{K-1}Ep'V_0\left\{(t-t')-(t'+E^{-1})\log\frac{t+E^{-1}}{t'+E^{-1}}\right\};$$

which becomes identical with the expression given in division (a) when we substitute for $\frac{K}{K-1}Ep'V_0$ the value Ja, which it must have in consequence of the relation between the specific heats of air and the mechanical equivalent of the thermal unit established in another paper in this number of the *Journal*.

The preceding formulas give us the means of calculating readily the most useful results.

We will take as an example, to supply a building with one pound of air heated mechanically from 50° to 80° Fahr. (solved before, in question (a)). We have then $t' = 10°$ C., $t = 26.6'$ C., $E = .00366$, $1/E = 273.22404$, $p' = 2114$ pounds per square foot. $K = 1.41$. Then, by (A),

$$p = 2114 \left(\frac{283.22404}{299.89071} \right)^{\frac{141}{41}}$$

$$= 1736.6189 = 2114 \times .8214848$$

$$= 2114 \times \frac{1}{1.217308}.$$

All these forms are useful.

Volume of cylinder.

$$v' = V_0 (1 + 10E)$$

$$= 12.383 \times 1.0366 = 12.836218.$$

$$pv = p'v'. \quad v = \frac{p'}{p} v' = 1.217308 \times 12.836218.$$

Volume of cylinder

$$= v = 15.6256.$$

A practically useful height of cylinder might be 4 feet, the corresponding diameter to which is 2.2302 feet.

Again, we have, amount of work spent in heating in this way one pound of air (D)

$$= \frac{K}{K-1} p'v' \left\{ \frac{E^{-1} + t}{E^{-1} + t'} - 1 \right\} - p'v' \log \frac{v}{v'}$$

$$= \frac{141}{41} \times 2114 \times 12.836 \left\{ \frac{299.89071}{283.22404} - 1 \right\}$$

$$\qquad - 2114 \times 12.836 \log 1.217308$$

$$= 5491.54 - 5336.03$$

$$= 155.51 \text{ estimated in foot pounds.}$$

As one pound must be supplied per second, H.P. of engine required to drive the apparatus = .2827. This result ought, inasmuch as this apparatus possesses all the qualifications of a perfect engine, to be identical with the answer found in division (*a*) of this problem; we however find a difference of ·0034 between the two, due to the circumstance that the number ·24 which we employed as the value of the specific heat of air in the previous solution, also 1·41 for K in this, are only approximately true; but the true H.P. to two significant figures is 28.

Ratios of Expansion.   In first cylinder

$$v'/v = ·8214848,$$

so that 3·2859 feet of the stroke passed while air was being admitted at pressure 2114, and ·71406 feet in allowing this to expand to pressure of receiver.

In second cylinder

$$v_1/v = ·869825,$$

or 5207 feet of the stroke was spent in compressing the air from pressure p to pressure p' or 2114, the remaining 3·4793 feet in expelling it.

(*c*)   The first suggestion, we believe, of an apparatus for cooling buildings by compressing air, was to pump in air into a reservoir and allow it to cool to the temperature of the atmosphere, on the supposition that if then allowed to rush out by means of a stopcock, it would, in consequence of the expansion, fall in temperature.   Unfortunately however for this scheme, it has been found that there is only an almost imperceptible depression of temperature (after motion ceases in the air) due to a want of perfect rigour in Mayer's hypothesis.   The friction of the air in the orifice &c. almost entirely compensates for the cold of expansion.

The apparatus described in (*b*) can, however, be very simply applied.

Instead of allowing the air to rush out, and thus heat itself by friction, let it out slowly, and make it work a piston in a double-stroke cylinder, and we shall not only obtain the full benefit of the cold of expansion, but also gain so much work as to make the H.P. of the engine required to drive the apparatus a mere trifle.

The working of the apparatus, however, will not be so simple as in the last case, for as we are to use the same apparatus, we

cannot make the cylinders hold one pound of air, and cannot even
have the pistons moving at the same rate.

Let p' and t' be the atmospheric pressure and temperature
respectively, v' the volume of either cylinder, t the temperature
of the cooled air, p the pressure in the receiver, which will be
such that if air at pressure p and temperature t' be allowed to
expand to pressure p' the temperature will become t, v the
volume under pressure p of a quantity of air, which under
pressure p' would fill the cylinder, there being no change of
temperature, v_1 volume under pressure p and temperature t', of a
quantity of air which would fill the cylinder under pressure p'
and at temperature t.

$$\text{(A')} \qquad \frac{p}{p'} = \left\{ \frac{E^{-1}+t'}{E^{-1}+t} \right\}^{\frac{K}{K-1}}.$$

Operations in Ingress Cylinder. Suppose the piston at the
top of its stroke, the cylinder full of air at ordinary pressure.
Admitting external air above the piston, push the piston down
until the air below is compressed to pressure p, the temperature
being kept constant; and then send this compressed air into the
reservoir.

Operations in Egress Cylinder. In the first part of the
stroke allow so much air to enter the cylinder from the reservoir,
that if allowed to expand in the remaining part of the stroke,
the pressure and temperature at the end will be p' and t
respectively.

In Ingress Cylinder. Work spent in compressing air from p'
to p (temperature constant)

$$= p'v' \log \frac{v'}{v} - p'(v'-v).$$

Work spent in sending the air into the receiver

$$= pv - p'v.$$

(B') Total expenditure per stroke in first cylinder

$$= p'v' \log \frac{v'}{v}.$$

In Egress Cylinder. Mechanical effect gained in partially
filling the cylinder from reservoir

$$= pv_1 - p'v_1.$$

Mechanical effect gained during the rest of the stroke

$$= \frac{p'v'}{K-1}\left\{\left(\frac{v'}{v_1}\right)^{K-1} - 1\right\} - p'\left(v' - v_1\right).$$

(C') Total gain per single stroke

$$= \frac{p'v'}{K-1}\left\{\left(\frac{v'}{v_1}\right)^{K-1} - 1\right\} + pv_1 - p'v'$$

$$= \frac{p'v'}{K-1}\left\{\frac{E^{-1}+t'}{E^{-1}+t} - 1\right\} + p'v'\left\{\frac{E^{-1}+t'}{E^{-1}+t} - 1\right\}$$

$$= p'v'\frac{K}{K-1}\frac{t'-t}{E^{-1}+t}.$$

Ratios of Expansion. In ingress cylinder,

$$\frac{v}{v'} = \frac{p'}{p} = \left\{\frac{E^{-1}+t}{E^{-1}+t'}\right\}^{\frac{K}{K-1}}.$$

In egress cylinder,

$$\frac{v_1}{v'} = \left\{\frac{E^{-1}+t}{E^{-1}+t'}\right\}^{\frac{1}{K-1}}.$$

It would be a matter of no difficulty to give generally the number of pounds each cylinder would contain, and also the rate of each piston, but it would only tend to confusion of symbols, as we should have to take in the p's, t's, and v's of (b), as well as those of this case; we will give these details in the example, which is

$t' = 80°$ Fahrenheit $= 26\cdot6'$ Cent., $t = 50°$ Fahrenheit $= 10$ Cent., p' of course 2114, height of cylinder, as before, 4 feet, and consequently, diameter $2\cdot2302$ feet.

From (A') we have $p = 2114 \times 1\cdot217308$ pounds per square foot.

In (b) we found volume of cylinder $= 15\cdot6256 = v'$, according to our present notation.

From (B') we have total expenditure per single stroke in ingress cylinder

$$= p'v'\log\frac{v'}{v} = p'v'\log\frac{p}{p'}$$

$= 1\cdot217308 \times \{\text{the gain in ingress cylinder in } (b)\}$

$= 6495\cdot59.$

From (C′) we have total gain per single stroke in egress cylinder

$$= 1\cdot217308 \times \{\text{loss in egress cylinder in } (b)\}$$
$$= 6684\cdot9.$$

Ratios of Expansion. In ingress cylinder

$$v/v' = \cdot8214848,$$

and in egress

$$v_1/v' = \cdot869825,$$

as in the example attached to (b).

Let χ be the number of pounds the first cylinder contains at pressure p' (2114), and at temperature t' (26·6′ Cent.).

Volume of χ pounds $= 12\cdot383\chi(1 + E \times 26\cdot6) =$ volume of cylinder $= 12\cdot383(1 + E \times 10) \times 1\cdot217308$, as we found in (b); hence

$$\chi = \frac{283\cdot22404}{299\cdot89071} 1\cdot217308 = 1\cdot149655.$$

Number of single strokes per second

$$= \chi^{-1} = \cdot8698348.$$

Number of double strokes per minute

$$= 26\cdot095044.$$

In the second cylinder, let χ' be the number of pounds contained by it at p' (2114) and t (10° Cent.), we easily obtain

$$\chi' = 1\cdot217308.$$

Number of strokes per second

$$= \chi'^{-1} = \cdot8214848.$$

Number of double strokes per minute

$$= 24\cdot644544.$$

Whole work spent per second in ingress cylinder

$$= 6495\cdot59/\chi = 5650\cdot036.$$

Whole mechanical effect gained per second in egress cylinder

$$= 6684\cdot9/\chi' = 5491\cdot54.$$

Total motive power required

$$= 5650\cdot036 - 5491\cdot54 = 158\cdot496 \text{ ft. pounds per second.}$$

Hence, H.P. of engine required to drive the apparatus $= \cdot288$.

COSMICAL AND GEOLOGICAL PHYSICS.

124. RECENT INVESTIGATIONS OF M. LE VERRIER ON THE
MOTION OF MERCURY.

[From *Glasg. Phil. Soc. Proc.* Vol. IV. [read Dec. 14, 1859], pp. 263—266.]

PROFESSOR W. THOMSON stated that he had had his attention
called, a few days since, to recent investigations of M. Le Verrier
on the motion of Mercury, which, he believed, would interest the
Society, not only as constituting a new step of high importance
in the theory of the planetary motions, but as now affording that
kind of evidence of the existence of matter circulating round the
Sun within the earth's orbit, which, more than five years ago, in
publishing his theory of meteoric vortices * to account for the
Sun's heat and light, he had called for, from perturbations to be
observed in the motions of the known planets. M. Le Verrier,
in a letter to M. Faye, published in the *Comptes Rendus* for
September 12, 1859, which contains his account of the investigations referred to, writes as follows :—

"You have not perhaps forgotten how, in my studies regarding
the motions of our planetary system, I have encountered difficulties
in the way of establishing a complete agreement between theory
and observation. This agreement, said Bessel, thirty years ago,
is always affirmed, yet has never been hitherto verified in a
sufficiently serious manner.

"The study of the difficulties offered by the Sun has been
long and complicated. It has been necessary, in the first place,

* "On the Mechanical Energies of the Solar System," by Professor W.
Thomson—*Transactions of the Royal Society, Edinburgh*, and *Philosophical
Magazine*, 1854.

to revise the catalogue of the fundamental stars, so as to leave no systematic error there. I have next taken up the whole theory of the inequalities in the Earth's motion; in connection with which I have been led to discuss as many as 9,000 observations of the Sun, made in different observatories. This work has shown that the meridian observations may not always have had the precision which has been attributed to them, and that thus the discrepancies at first indicated as belonging to the theory must be rejected, because of uncertainty as to the observations.

"The theory of the Sun's apparent motion once put beyond question, it became possible to resume, with advantage, the study of the motion of Mercury. It is this work on which I wish now to engage your attention.

" While for the Sun we possess only meridian observations liable to great objections, we have, for the planet Mercury, a certain number of observations of an extremely accurate character, made in the course of a century and a-half. I mean the internal contacts of the disc of Mercury with the disc of the Sun, at the end of a transit of that planet. Provided that the place of observation is well known, and provided that the observer has had a passable telescope, and a clock showing time within a few seconds of perfect accuracy, his observation of the instant of the internal contact ought to afford an estimate of the distance between the centres of the two bodies, with no error exceeding a second of arc. From 1697 to 1848, we have twenty-one observations of this kind, which ought to be perfectly satisfied if the perturbations in the motions of the Earth and Mercury have been well calculated, and if correct values have been attributed to the disturbing masses.

" The results of my first studies on Mercury, published in 1842, did not represent with great accuracy the transit observations. Among other discrepancies was to be remarked a progressive error in the transits of the month of May, reaching as much as nine seconds of arc in the year 1753. Deviations such as this could not be attributed to errors of observation; but not having revised the theory of the Sun, I believe that I ought to abstain from drawing any conclusion from them.

" The use of the corrected tables of the Sun has not, however, in my new work, made these discrepancies appear;—systematic discrepancies for which the observations cannot be blamed, unless we suppose that such astronomers as Lalande, Cassini, Bouguer, &c., could have committed errors amounting to several minutes of time, and even varying progressively from one epoch to another!

"Now, what is remarkable is, that it suffices to augment by 38″ the centenary movement of the perihelion of Mercury's orbit to represent all the transit observations to within a second, and most of them even to less than half a second. This result, so precise and simple, which gives at once to all the comparisons an accuracy superior to that which has been hitherto attained in astronomical theories, shows clearly that the increase in the motion of the perihelion is necessary, and that when made to fulfil this condition, the tables of Mercury and the Sun possess all desirable accuracy."

M. Le Verrier proceeds in his letter to examine the different suppositions by which it may be attempted to explain the perturbation in Mercury's motion, which he has thus discovered. An increase of $\frac{1}{10}$th on the supposed mass of Venus would account for it; but the periodical disturbances in the Earth's motion, and the secular variation of the inclination of the Earth's axis to the plane of the ecliptic produced by Venus, do not allow any such change in our estimate of her mass. On the other hand, a planet revolving round the Sun, inside Mercury's orbit, might produce precisely the variation in the perihelion to be explained, without sensibly disturbing the motions of Venus, the Earth, or any of the superior planets. M. Le Verrier shows, for instance, that a planet equal in mass to Mercury, and revolving in a circular orbit in the same plane, at a little less than half the distance from the Sun, would fulfil these conditions; and, therefore, that a less mass in a larger orbit, or a greater mass in a smaller orbit, would do the same. But, considering that so large a mass could scarcely have escaped observation, either in its transits across the Sun's disc, or by its own brilliant illumination, which would render it visible to us during total eclipses of the Sun, even if, on account of its nearness to the Sun, not ordinarily seen as a morning and evening star alternately, M. Le Verrier thinks it most probable that the disturbing mass consists in reality of a series of corpuscules circulating between the Sun and Mercury.

Here, then, by a profound appreciation of purely astronomical data, the great French physical astronomer, leaving those remote regions where, independently of our own countryman, Adams, he tracked the unseen planet by its disturbing influence on the remotest body of the then known list, has been led to conclude that the innermost of the recognized planets is also disturbed by planetary matter, not previously reckoned among the influencing masses of our system.   Is this new planetary matter, like the other till discovered by its influence, unseen?   Surely, on the contrary, it is it that we see as the Zodiacal Light, long before conjectured to consist of a cloud of corpuscules circulating round the Sun, and, in the dynamical theory of the Sun's radiation, supposed to contain the reserve of force from which this Earth, as long as it continues a fit habitation for man as at present constituted, is to have its fresh supplies of heat and light.

125. On the Variation of the Periodic Times of the Earth and Inferior Planets, produced by Matter falling into the Sun.

[From *Glasg. Phil. Soc. Proc.* Vol. IV. [read Jan. 4, 1860], pp. 272—274.]

It may be remarked, in the first place, that the absolute effect on the periodic time of Mercury, producible by such a distribution of planetary matter as M. Le Verrier concludes must circulate between Mercury and the Sun, is not discoverable. The true mean distance of Mercury from the Sun is not, in fact, known with sufficient accuracy to allow us to judge whether or not the central force corresponding to its periodic time, when compared with the forces experienced by the other planets, deviates from the law of inverse squares of distances from the Sun's centre to such an extent as must be the case if M. Le Verrier's disturbing planetary matter is altogether inside Mercury's orbit. But it does not follow that the periodic time of Mercury, or even that of Venus, and the Earth, may not be sensibly influenced by the falling in of portions of that matter to the Sun. To discover the general character of this influence, and to estimate its amount, we may first consider the resultant force experienced by a planet under the joint influence of the Sun and a concentric circular ring in the plane of its orbit. It is easily seen that the Sun's force must be diminished by the attraction of the ring, if this lies outside the planet, but, on the contrary, increased, if inside. If the radius of the ring be very small in comparison with the distance of the planet, it is clear that to a first approximation the attraction of the ring may be calculated by supposing its mass collected at its centre. The full expression for the resultant attraction of the ring, being the following convergent series:—

$$\frac{m}{a^2}\left\{1 + 3\left(\frac{1}{2}\right)^2\left(\frac{r}{a}\right)^2 + 5\left(\frac{1.3}{2.4}\right)^2\left(\frac{r}{a}\right)^4 + 7\left(\frac{1.3.5}{2.4.6}\right)^2\left(\frac{r}{a}\right)^6 + \&c.\right\}$$

where m denotes the mass of the ring, r its radius, and a the distance of the planet from its centre, shows that the resultant force is in reality somewhat greater than it would be if the mass were collected at the centre. The planet's orbit being nearly circular, the attraction of such a ring as we have supposed will be represented to a second approximation by supposing a mass greater than its own in the ratio of 1 to $1 + \frac{3}{4}(r/a)^2$, collected at its centre, or, which would produce the same effect, distributed uniformly over the Sun's surface. Hence the gradual falling in of such a ring to the Sun will diminish the force experienced by the planet as much as would be done by a simple subtraction of $\frac{3}{4}(r/a)^2 m$, from the Sun's mass. The amount I have estimated as falling in annually to produce the solar heat and light is $\frac{1}{16200000}$ of the Sun's mass. The effect of this, if coming from a ring at distance r, would be to diminish the central force on a planet at distance a, in the ratio of

$$1 \text{ to } 1 + \frac{3}{4}\left(\frac{r}{a}\right)^2 \frac{1}{16{,}200{,}000}.$$

According to the investigation contained in addition No I. to my paper "On the Mechanical Energies of the Solar System," the effect of such a change in the Sun's mass would be to alter the angular motion of the planet in the inverse ratio of the square of the mass. The integral effect of this will, therefore, be a diminution of the planet's helio-centric longitude amounting to

$$\frac{3}{4}\left(\frac{r}{a}\right)^2 \frac{n^2}{16{,}200{,}000} \times 360°$$

in n revolutions. Merely for the sake of example, let us consider the effect on the Earth's motion, produced by matter falling in at the supposed rate, during a period of two thousand years, from a ring of double the Sun's diameter. In this case

$$\frac{r}{a} = \frac{1}{107{\cdot}5} \text{ and } n = 2{,}000.$$

Hence the disturbance in the Earth's longitude would be a diminution amounting to $360°/62400 = 20''{\cdot}8$ an amount of loss altogether undiscoverable over such a period of time.

To estimate the effect of the same transference of matter upon the motion of Mercury, we must take $r/a = 1/41{\cdot}7$. The period during which we have the most accurate knowledge of Mercury's motion is, as Le Verrier has remarked, from the year 1697 to

1848. This being about 624 of Mercury's revolutions round the Sun we may take $n = 624$; and we find by the preceding expression $13\frac{1}{2}''$ as the effect on the helio-centric longitude of the planet. This will amount to nearly $8\frac{1}{2}''$ of geo-centric arc—an error which could not possibly have escaped detection in the very thorough investigation which M. Le Verrier has applied to the motion of Mercury. It may be concluded that if matter has been really falling in at the rate supposed by my dynamical theory of the solar radiation, the place from which it has been falling must be either nearer the Sun or more diffused from the plane of Mercury's orbit than we have supposed in the preceding example. With a view to determining whether this theory is tenable or not, it will be necessary to consider whether the appearances presented by the zodiacal light, and the photo-sphere seen round the Sun in total eclipses, allow us to find a place for a sufficient future dynamic supply without supposing a denser distribution of meteors or meteoric vapours than is consistent with what we know of the motion of comets before and after passing very close to the Sun.

126. PHYSICAL CONSIDERATIONS REGARDING THE POSSIBLE AGE OF THE SUN'S HEAT.

[From *Brit. Assoc. Report*, 1861, pt. ii. pp. 27—28; *Phil. Mag.* Vol. XXIII. Feb. 1862, pp. 158—160. Incorporated in *Popular Lectures and Addresses*, Vol. I. pp. 349—368.]

THE author prefaced his remarks by drawing attention to some principles previously established. It is a principle of irreversible action in nature, that, "although mechanical energy is indestructible, there is a universal tendency to its dissipation, which produces gradual augmentation and diffusion of heat, cessation of motion, and exhaustion of potential energy, through the material universe." The result of this would be a state of universal rest and death, if the universe were finite and left to obey existing laws. But as no limit is known to the extent of matter, science points rather to an endless progress through an endless space, of action involving the transformation of potential energy through palpable motion into heat, than to a single finite mechanism, running down like a clock and stopping for ever. It is also impossible to conceive either the beginning or the con-tinuance of life without a creating and overruling power. The author's object was to lay before the Section an application of these general views to the discovery of probable limits to the periods of time *past* and *future*, during which the sun can be reckoned on as a source of heat and light. The subject was divided under two heads: 1, on the secular cooling of the sun; 2, on the origin and total amount of the sun's heat.

PART I. *On the Secular Cooling of the Sun.*

In the first part it is shown that the sun is probably an incan-descent liquid mass radiating away heat without any appreciable compensation by the influx of meteoric matter. The rate at which heat is radiated from the sun has been measured by Herschel and Pouillet independently; and, according to their results, the author estimates that if the mean specific heat of the sun were the same as that of liquid water, his temperature would be lowered by 1·4° Centigrade annually.   In considering what the sun's specific heat may actually be, the author first remarks that there are excellent reasons for believing that his substance is very much like the earth's.   For the last eight or nine years, Stokes's principles of solar and stellar chemistry have been taught in the public lectures on natural philosophy in the University of Glasgow; and it has been shown as a first result, that there *certainly is sodium in the sun's atmosphere.*   The recent application of these principles in the splendid researches of Bunsen and Kirchhoff (who made an independent discovery of Stokes's theory), has demonstrated with equal certainty that there are iron and manganese, and several of our other known metals in the sun.   The specific heat of each of these substances is less than the specific heat of water, which indeed exceeds that of every other known terrestrial solid or liquid.   It might therefore at first sight seem probable that the mean specific heat of the sun's whole substance is less, and very certain that it cannot be much greater, than that of water.   But thermodynamic reasons, explained in the paper, lead to a very different conclusion, and make it probable that, on account of the enormous pressure which the sun's interior bears, his specific heat is more than ten times, although not more than 10,000 times, that of liquid water.   Hence it is probable that the sun cools by as much as 14° C. in some time more than 100 years, but less than 100,000 years.

As to the sun's actual temperature at the present time, it is remarked that at his surface it cannot, as we have many reasons for believing, be incomparably higher than temperatures attainable artificially at the earth's surface.   Among other reasons, it may be mentioned that he radiates heat from every square foot of his surface at only about 7000 horse-power.   Coal burning at the rate of a little less than a pound per two seconds would generate the same amount; and it is estimated (Rankine, *Prime Movers*, p. 285, edit. 1859) than in the furnaces of locomotive engines, coal burns at from 1 lb. in 30 seconds to 1 lb. in 90 seconds per square foot of grate-bars.   Hence heat is radiated from the sun at a rate not more than from fifteen to forty-five times as high as that at which heat is generated on the grate-bars of a locomotive furnace, per equal areas.

The interior temperature of the sun is probably far higher than that at the surface, because conduction can play no sensible part in the transference of heat between the inner and outer portions of his mass, and there must be an approximate *convective* equilibrium of heat throughout the whole; that is to say, the temperatures at different distances from the centre must be approximately those which any portion of the substance, if carried from the centre to the surface, would acquire by expansion without loss or gain of heat.

PART II.   *On the Origin and Total Amount of the Sun's Heat.*

The sun being, for reasons referred to above, assumed to be an incandescent liquid now losing heat, the question naturally occurs, how did this heat originate? It is certain that it cannot have existed in the sun through an infinity of past time, because as long as it has so existed it must have been suffering dissipation; and the finiteness of the sun precludes the supposition of an infinite primitive store of heat in his body. The sun must therefore either have been created an active source of heat at some time of not immeasurable antiquity by an overruling decree; or the heat which he has already radiated away, and that which he still possesses, must have been acquired by some natural process following permanently established laws. Without pronouncing the former supposition to be essentially incredible, the author assumes that it may be safely said to be in the highest degree improbable, if, as he believes to be the case, we can show the latter to be not contradictory to known physical laws.

The author then reviews the meteoric theory of solar heat, and shows that, in the form in which it was advocated by Helmholtz[*], it is adequate, and it is the only theory consistent with natural laws which is adequate to account for the present condition of the sun, and for radiation continued at a very slowly decreasing rate during many millions of years past and future. *But neither this nor any other natural theory can account for solar radiation continuing at anything like the present rate for many hundred millions of years.* The paper concludes as follows:—" It seems therefore, on the whole, most probable that the sun has not illuminated the

---

[*] Popular Lecture delivered at Königsberg on the occasion of the Kant commemoration, February 1854.

earth for 100,000,000 years, and almost certain that he has not
done so for 500,000,000 years. As for the future, we may say
with equal certainty that inhabitants of the earth cannot continue
to enjoy the light and heat essential to their life for many million
years longer, unless new sources, now unknown to us, are prepared
in the great storehouse of Creation."

127. On the Convective Equilibrium of Temperature in the Atmosphere.

[From *Manchester Phil. Soc. Proc.* Vol. II. 1860—1862 [Jan. 21, 1862], pp.
170—176; reprinted in *Math. and Phys. Papers*, Vol. III. art. xcii.
pp. 255—260.]

128. On the Age of the Sun's Heat.

[From *Macmillan's Magazine*, Vol. v. March 1862, pp. 288—393; reprinted
in *Popular Lectures and Addresses*, Vol. I. pp. 349—368.]

129. On the Rigidity of the Earth.

[From *Glasg. Phil. Soc. Proc.* Vol. v. [read March 26, 1862], pp. 169—170;
Roy. Soc. Proc. Vol. XII. 1862—1863 [read May 15, 1862, abstract], pp.
103—104; *Phil. Trans.* Vol. CLIII. 1863, appendix added Jan. 2, 1864,
pp. 573—582; *Phil. Mag.* Vol. XXV. Feb. 1863, pp. 149—151; *Math. and
Phys. Papers*, Vol. III. art. xcv. pp. 312—336. Reprinted in Thomson
and Tait's *Natural Philosophy*, §§ 832—849.]

130. On the Secular Cooling of the Earth.

[From *Edinb. Roy. Soc. Proc.* Vol. IV. [read Apr. 28, 1862], pp. 610—611;
Edinb. Roy. Soc. Trans. Vol. XXIII. 1864, pp. 157—170; *Phil. Mag.*
Vol. XXV. Jan. 1863, pp. 1—14; *Edinb. New Phil. Journ.* Vol. XVI. 1862,
pp. 151—152. Reprinted in *Math. and Phys. Papers*, Vol. III. art. xciv.
pp. 295—311; Thomson and Tait's *Natural Philosophy*, Appendix D.]

131. Sur le Refroidissement séculaire du Soleil. De
la Température actuelle du Soleil. De l'Origine et
de la Somme totale de la Chaleur solaire.

[From *Les Mondes* III. 1863, pp. 473—480; translated from No. 128 *supra*,
omitting an introduction given in No. 126 *supra*, which is an earlier draft.]

I. *Sur le refroidissement séculaire du Soleil.*

De combien le Soleil s'est-il refroidi d'année en année, si tant
est qu'il se soit refroidi? Nous n'avons aucun moyen de le découvrir ou de l'évaluer, même approximativement. D'abord, nous
ne savons même pas s'il perd réellement sa chaleur, car il est
certain que *quelque chaleur* est engendrée dans son atmosphère,
par l'influence de la matière météorique, et il est possible que la
somme de chaleur ainsi engendrée d'année en année suffise à
compenser la perte causée par le rayonnement. Il est possible
aussi que le Soleil soit encore aujourd'hui une masse liquide
incandescente, rayonnant de la chaleur, primitivement créée ou
communiquée à sa substance, ou bien, ce qui semble beaucoup
plus probable, engendrée par la chute incessante des météores
pendant les âges écoulés.

Il a déjà été démontré que, si la première supposition était
fondée, les météores qui auraient produit la chaleur solaire durant
les vingt ou trente derniers siècles auraient dû, pendant tout ce
laps de temps, se trouver très en dedans de l'espace compris entre
la Terre et le Soleil et qu'ils ont dû s'approcher du corps central,
en décrivant des spirales de plus en plus resserrées, parce que si
la quantité de matière nécessaire à produire l'effet thermique dont
il s'agit était venue des régions situées au delà de l'orbite terrestre,
la longueur de l'année se serait trouvée très-sensiblement diminuée
par ces additions incessantes à la masse du Soleil. Dans cette
hypothèse, la quantité de matière absorbée annuellement a dû
être un quarante-septième de la masse de la Terre ou *un quinze-*

K. V.   10 *millionième* de la masse du Soleil. Il faudrait donc supposer que la masse de la lumière zodiacale s'élève à un cinq-millième au moins de la masse du Soleil, pour expliquer de la même manière l'approvisionnement de chaleur solaire pour les trois mille ans à venir. Lorsque ces conclusions furent publiées pour la première fois, on fit remarquer qu'il fallait étudier les perturbations du mouvement des planètes "visibles" comme pouvant fournir les moyens d'évaluer la somme probable de matière de la lumière zodiacale, et l'on crut, *a priori*, qu'elle serait loin de suffire à un approvisionnement de chaleur pour trente mille années, au taux actuel de la dépense. Ces *desiderata* ont été, jusqu'à un certain point, remplis par M. Le Verrier. En effet, ses grandes recherches sur le mouvement de Mercure ont récemment révélé l'existence d'une influence sensible qu'on ne peut attribuer qu'à la matière qui circule sous forme d'astéroïdes innombrables entre l'orbite de Mercure et le Soleil. Mais la somme de matière mise en évidence par cette théorie est très-petite; et, par conséquent, si la chute de matières météoriques est réellement la source d'une partie appréciable de la chaleur rayonnante, il faut admettre que cette matière circule autour du Soleil à de très-faibles distances de sa surface. Mais la densité de ce nuage météorique ne serait-elle pas tellement grande que les comètes n'auraient pas pu la traverser sans éprouver une résistance qui aurait modifié sensiblement leur marche quand leur distance au Soleil n'aurait plus été qu'un trentième du rayon solaire? Tout bien considéré, il semble peu probable que la perte de chaleur solaire par rayonnement soit compensée, d'une manière appréciable, par la chaleur provenant de la chute des météores, pour le moment du moins; et comme on ne peut pas trouver davantage cette compensation dans quelque action chimique, on est conduit à conclure que le plus probable est que le Soleil n'est aujourd'hui qu'une masse incandescente liquide en cours de refroidissement.

Il est donc très-important de connaître de combien il se refroidit d'année en année, mais nous sommes totalement incapables de résoudre aujourd'hui cette question. Il est vrai que nous avons des données sur lesquelles nous pourrions baser une évaluation probable, et desquelles nous pourrions déduire, avec un certain degré de confiance, des limites assez rapprochées, entre lesquelles doit se trouver la vraie loi du refroidissement du Soleil. Car nous savons, d'après les recherches séparées, mais concor-

dantes, d'Herschel et de Pouillet, que le Soleil rayonne chaque
année de toute sa surface environ 3×10^{30} fois (3 suivi de 30 zéros)
la chaleur suffisante pour élever de 1° C. la température d'un
kilogramme d'eau. Nous avons aussi de fort bonnes raisons de
croire que la substance du Soleil ressemble beaucoup à celle de
la Terre. Les principes de Stokes sur la constitution chimique
du Soleil et des étoiles ont été exposés depuis plusieurs années
dans l'université de Glasgow, et on y a enseigné, comme premier
résultat, que le sodium existe certainement dans l'atmosphère du
Soleil et dans celles de plusieurs étoiles, quoiqu'il y en ait dans
lesquelles on ne puisse pas le découvrir. La nouvelle application
de ces principes, dans les magnifiques recherches de Bunsen et de
Kirchhoff (qui ont fait une découverte indépendante de la théorie
de Stokes), a démontré, avec une égale certitude, qu'il existe dans
le Soleil du fer et du manganèse et plusieurs de nos autres
métaux. La chaleur spécifique de chacune de ces substances est
moindre que la chaleur spécifique de l'eau, qui, en effet, dépasse
celle de tout autre corps terrestre, solide ou liquide. Il pourrait
donc, à première vue, paraître probable que la chaleur spécifique
moyenne de toute la substance solaire est moindre que celle de
l'eau, et il est certain qu'elle ne saurait être beaucoup plus grande.
Si elle était égale à celle de l'eau, il n'y aurait qu'à diviser le
chiffre précédent (3×10^{30}), donné par Herschel et Pouillet, par le
nombre de kilogrammes de la masse du Soleil ($2 \cdot 2 \times 10^{30}$) pour
trouver $1° \cdot 4$ C. comme chiffre annuel du refroidissement actuel.
Il semblerait donc probable que le Soleil se refroidit de plus d'un
degré et quatre dixièmes (centigrade) par an, et il paraît presque
certain qu'il ne se refroidit pas moins. Mais si ce calcul est juste,
on serait également fondé à supposer que l'*expansibilité* du Soleil
par la chaleur ne doit pas beaucoup différer de celle de quelque
autre corps terrestre. Si, par exemple, la dilatation du Soleil
était égale à celle du verre solide, laquelle est d'un quarante-
millième du volume, et d'un cent-vingt-millième du diamètre, par
degré centigrade (et pour les liquides, surtout à des températures
élevées, ce chiffre est beaucoup plus considérable), et si la chaleur
spécifique était celle de l'eau à l'état liquide, il s'opérerait dans le
diamètre du Soleil, au bout de 860 ans, une contraction de un
pour cent qui n'aurait pas échappé aux observations astro-
nomiques. Mais il y a encore une plus forte raison de croire qu'il
ne peut s'être opéré une telle contraction, et, par conséquent, de

soupçonner que les circonstances physiques de la masse solaire
rendent la condition des substances qui le composent, relativement à la dilatation et à la chaleur spécifique, bien différente de
celle des mêmes substances lorsqu'on les traite dans nos laboratoires. La gravitation mutuelle entre les diverses parties de la
masse du Soleil, durant la contraction, doit produire une quantité
de travail que l'on ne saurait calculer avec certitude, parce que
l'on ne connaît pas la loi de la densité intérieure du Soleil. La
quantité de travail produit par une contraction d'un dixième pour
cent du diamètre, si la densité demeurait uniforme dans tout
l'intérieur, serait, ainsi que l'a démontré Helmholtz, égale à vingt
mille fois l'équivalent mécanique correspondant à la somme
de chaleur émanée du Soleil en un an, d'après les calculs de
M. Pouillet. Mais en réalité la densité du Soleil doit augmenter
considérablement vers son centre, et probablement dans des propositions très-variables, à mesure que la température s'abaisse et
que la masse se contracte. On ne peut donc pas dire si le travail
fait par la gravitation mutuelle durant une contraction d'un
dixième pour cent du diamètre serait plus ou moins grand que
l'équivalent correspondant à vingt mille fois la chaleur annuelle,
mais nous pouvons admettre comme très-probable qu'il ne serait
pas beaucoup de fois plus grand ou plus petit. Or, il est de toute
improbabilité que l'énergie mécanique puisse, en aucune façon,
augmenter dans un corps qui se contracte par le froid. Il est
certain qu'en réalité elle diminue notablement, d'après toutes les
expériences faites jusqu'ici. Il faut donc supposer que le Soleil
perd, par le rayonnement, une quantité de chaleur plus grande que
celle qui correspond à l'équivalent mécanique de Joule, déduit du
travail opéré dans la contraction de sa masse par la gravitation
mutuelle des parties. De là il faut conclure qu'en se contractant
d'un dixième pour cent de son diamètre, ou des trois dixièmes
pour cent de son volume, le Soleil doit rayonner un peu plus, ou
très-peu moins, de vingt mille fois sa chaleur annuelle. Ainsi,
même sans la preuve historique de l'invariabilité de son diamètre,
il paraît juste de conclure qu'aucune contraction, au delà de un
pour cent en 860 ans, n'a pu réellement avoir lieu. Il semble,
au contraire, probable que, d'après la somme actuelle de rayonnement, une contraction d'un dixième pour cent ne saurait avoir
lieu en beaucoup moins de 20,000 ans et qu'il serait à peine
possible qu'elle ait lieu en moins de 8600 ans. Si donc la chaleur

spécifique moyenne de la masse solaire, dans son état actuel, n'est pas plus de dix fois celle de l'eau, la dilatation en volume doit être inférieure à un quatre-millième par degré centigrade (c'est-à-dire moins d'un dixième de celle du verre solide), ce qui semble improbable. Mais quoique, d'après ces considérations, nous soyons amenés à penser comme probable que la chaleur spécifique du Soleil dépasse de beaucoup dix fois celle de l'eau (et que par conséquent sa masse perd beaucoup moins de 100 degrés en 700 ans, conclusion presque inévitable pour de simples raisons géologiques), les principes de physique sur lesquels nous nous basons ne nous fournissent aucun motif de supposer que la chaleur spécifique du Soleil soit plus de dix mille fois celle de l'eau, parce que nous ne pouvons pas dire que sa dilatation cubique soit probablement supérieure à un quatre-centième par degré centigrade. Et il y a encore, d'autre part, de puissantes raisons de croire que la chaleur spécifique est réellement bien inférieure à ces dix mille fois. Car il est presque certain que la température moyenne du Soleil est aujourd'hui de quatorze mille degrés centigrade, et la plus grande quantité de chaleur que nous puissions supposer, avec quelque probabilité, acquise naturellement par le Soleil (comme on le verra dans un troisième paragraphe), n'aurait pu élever sa masse à une telle température à aucune époque, à moins que sa chaleur spécifique n'eût été inférieure à dix mille fois celle de l'eau.

Nous pouvons donc admettre comme fort probable que la chaleur spécifique du Soleil est plus de dix fois et moins de dix mille fois celle de l'eau à l'état liquide. Il s'ensuivrait donc, avec certitude, que sa température s'abaisse de 100 degrés pendant une durée de 700 à 700,000 ans.

Que faut-il donc penser, par exemple, de l'évaluation géologique qui compte 300 millions d'années pour la dénudation du Weald ? Est-il plus probable que l'état physique de la matière solaire diffère mille fois plus de la matière que nous étudions, que la dynamique ne nous autorise à le supposer, et qu'une mer orageuse, aidée de marées d'une extrême violence, doit ronger une falaise calcaire mille fois plus rapidement que ne l'estime M. Darwin, qui trouve un pouce par siècle ?

II.   *De la température actuelle du Soleil.*

A sa surface, la température du Soleil ne saurait, pour plusieurs raisons, être incomparablement plus élevée que la température que nous pouvons obtenir dans nos laboratoires.

Entre autres raisons on peut donner celle-ci : que le Soleil rayonne de la chaleur de chaque pied carré de sa surface avec une force de sept mille chevaux seulement. La houille, brûlant sur le pied d'un peu moins d'un kilogramme en quatre secondes, donnerait la même force, et il a été trouvé (Rankine, *Premiers moteurs*, p. 285, Edimbourg, 1852) que dans les foyers des locomotives le charbon se consume depuis une livre par trente secondes jusqu'à une livre en quatre-vingt-dix secondes par pied carré de grille. De là on conclut que le Soleil rayonne de la chaleur avec une intensité de quinze à quarante-cinq fois la chaleur d'un foyer de locomotive à surfaces égales.

La température intérieure du Soleil est probablement beaucoup plus élevée que celle de la surface, parce que la conduction directe ne peut jouer aucun rôle appréciable dans la transmission de la chaleur entre les parties intérieures et extérieures de la masse, et il doit exister un équilibre de convection approximatif à travers toute la masse, si la masse est fluide. Ce qui revient à dire que les températures, à différentes distances du centre, doivent être approximativement celles que toute portion de la substance, portée du centre à la circonférence, acquerrait par l'expansion, sans perte ni gain de chaleur.

III.   *De l'origine et de la somme totale de la chaleur solaire.*

Le Soleil étant, par les raisons données ci-dessus, supposé un liquide incandescent perdant aujourd'hui sa chaleur, il se présente une question toute naturelle: quelle est l'origine de cette chaleur? Il est certain qu'elle ne peut pas avoir existé dans le Soleil depuis un temps infini, puisque tant qu'elle a existé elle a subi une déperdition, et le Soleil étant un corps fini, ce fait exclut la supposition qu'il y ait en lui une source primitive infinie de chaleur. Le Soleil doit donc avoir été créé comme source active de chaleur à quelque époque d'une antiquité calculable, par quelque décret tout-puissant; ou bien la chaleur qu'il a déjà rayonnée, et qu'il conserve encore, doit avoir été acquise par quelque moyen naturel, suivant une loi fixe. Sans affirmer que la première hypothèse soit absolument incroyable, on peut dire, avec assez de sécurité, qu'elle est improbable au plus haut degré, si l'on peut démontrer que la seconde n'est pas en opposition avec les lois physiques connues. Nous allons le démontrer et faire plus, en indiquant simplement certaines actions, s'opérant aujourd'hui sous nos yeux, et qui, si elles ont été suffisamment fréquentes à quelque époque passée, ont dû donner au Soleil une chaleur suffisante pour expliquer tout ce que nous savons de son rayonnement passé et de sa température actuelle.

Il n'est pas nécessaire, pour le moment, d'entrer dans de longs détails sur la théorie météorique, qui paraît avoir été énoncée sous une forme définie pour la première fois par Mayer et ensuite par Waterston; ni sur l'hypothèse modifiée des tourbillons météoriques, dont l'auteur de ce mémoire a démontré la nécessité, pour que la longueur de l'année, telle qu'elle est du moins depuis deux mille ans, n'ait pas été sensiblement altérée par les accroissements qu'a dû subir la masse du Soleil pendant cette période, si la chaleur rayonnée a toujours été compensée par celle due à l'influx météorique.

Par suite des raisons énoncées dans la première partie de cet article, on peut croire maintenant que toutes les théories de compensation météorique contemporaine, complète ou à peu près complète, doivent être rejetées, mais on peut cependant toujours supposer que: *l'action météorique…n'existe pas seulement comme une cause de la chaleur solaire, mais que c'est la seule de toutes les causes concevables dont nous connaissons l'existence par des preuves indépendantes.*

La théorie météorique qui paraît aujourd'hui la plus probable, et qui fut d'abord discutée selon les vrais principes de la thermodynamique par Helmholtz, consiste à supposer que le Soleil et sa chaleur tirent leur origine de l'agglomération de corpuscules tombant ensemble par l'effet d'une gravitation mutuelle, et développant, conformément à la grande loi démontrée par Joule, une chaleur équivalant exactement au mouvement perdu dans leur collision.

Qu'une théorie météorique de quelque nature soit certainement la vraie et complète explication de la chaleur solaire, c'est ce dont on ne saurait douter, en considérant les raisons suivantes:

1. Aucune autre explication naturelle, excepté l'action chimique, ne peut se concevoir;

2. La théorie d'une action chimique est tout à fait insuffisante, parce que l'action la plus énergique que nous connaissions, s'opérant entre des substances équivalant à toute la masse du Soleil, ne développerait que trois mille années de chaleur;

3. Au moyen de la théorie météorique on peut, sans difficulté, rendre raison de vingt millions d'années de chaleur.

On serait entraîné dans des calculs mathématiques trop longs si on voulait expliquer complètement sur quels principes est fondée la dernière évaluation. Qu'il suffise de dire que des corps, tous moindres que le Soleil, tombant ensemble d'un état de repos relatif, et à des distances mutuelles considérables eu égard à leurs diamètres, et formant un globe d'une densité uniforme, égale en masse et en diamètre au Soleil, développeraient une somme de chaleur qui, calculée d'après les principes et les expériences de Joule, s'est trouvée être justement vingt millions de fois la somme annuelle de rayonnement évaluée par Pouillet. La densité du Soleil doit, selon toute probabilité, s'accroître considérablement vers son centre, et par conséquent il faut supposer qu'une bien plus grande somme de chaleur a dû être engendrée si sa masse entière s'est formée par l'agglomération de corps comparativement petits. D'un autre côté, nous ne savons pas combien de chaleur a pu se perdre par la résistance et avant l'agglomération finale; mais il y a lieu de croire que même la réunion la plus rapide que nous puissions concevoir n'a pu donner au globe qui en est résulté qu'à peu près la moitié de la chaleur entière due à la somme d'énergie potentielle de la gravitation mutuelle qui s'est épuisée. Nous pouvons donc accepter comme le chiffre le plus bas de la chaleur initiale du Soleil dix millions de fois la chaleur d'une année actuelle, mais cinquante ou même cent millions de fois sont possibles, en conséquence de la plus grande densité du Soleil dans ses parties centrales.

Les considérations développées plus haut, concernant la chaleur spécifique possible du Soleil, la loi de son refroidissement et sa température superficielle, font supposer, avec beaucoup de probabilité, qu'il a dû être sensiblement plus chaud il y a un million d'années; par conséquent, que s'il a existé comme corps lumineux pendant dix ou vingt millions d'années, il doit avoir perdu, par le rayonnement, beaucoup plus qu'il n'en perd annuellement aujourd'hui.

Il semble donc, à tout prendre, fort probable que le Soleil n'a pas éclairé la Terre durant cent millions d'années, et il est presque certain qu'il ne l'a pas fait pendant cinq cents millions d'années. Pour ce qui est de l'avenir, on peut dire, avec une égale certitude, que les habitants de la Terre ne pourront continuer à jouir de la lumière et de la chaleur essentielles à leur existence pendant plusieurs millions d'années encore, à moins que des sources aujourd'hui inconnues ne soient préparées dans la grande réserve de la création.

132. THE NATURE OF THE SUN'S MAGNETIC ACTION ON THE EARTH.

[Note on two paragraphs of Mr C. Chambers' paper: from *Phil. Trans.* Vol. CLIII. 1863, pp. 515, 516.]

IF the sun were a magnet as intense on the average as the earth, the magnetic force it would exert at a distance equal to the earth's, or, let us suppose, 200 radii, would be only $\frac{1}{8000000}$ of the earth's surface magnetic force in the corresponding position relatively to its magnetic axis. Considering, therefore (according to the principles explained by Mr Chambers), the sun as a magnet having its magnetic axis nearly perpendicular to the ecliptic, we see that, with an average intensity of magnetization equal to the earth's, the effect of reversing the sun's magnetization would be to introduce, in a direction perpendicular to the ecliptic, a disturbing force equal to about $\frac{1}{8000000}$ of the earth's average polar force, and would therefore be absolutely insensible to the most delicate terrestrial magnetic observation yet practised. A disturbing force of this amount, acting perpendicularly to the direction of the terrestrial magnetic force about the equator, would produce a disturbance in declination of only half a second; and the sun's magnetization would therefore need to be 120 times as intense as the earth's to produce a disturbance of 1' in declination even by a *complete reversal* in the most favourable circumstances. These estimates appear to me to give strong evidence in support of the conclusion at which Mr Chambers has arrived by a careful examination of the disturbances actually observed, that no effect of the sun's action as a magnet is sensible at the earth.

The same estimates are applicable to the moon, her apparent diameter being the same as the sun's. It is of course most probable that the moon is a magnet; but she must be a magnet thousands or millions of times more intense than the earth to produce any sensible effect of the character of any of the observed terrestrial magnetic disturbances.

133. On the Elevation of the Earth's Surface Tempera-
    ture produced by Underground Heat.

[From *Edinb. Roy. Soc. Proc.* Vol. v. 1866 [read March 21, 1864],
pp. 200, 201.]

Peclet found, by his own experiments, that a body with any
common unpolished non-metallic surface, kept by heat from
within at 1° higher temperature than that of the air and other
objects round it, loses heat from each square metre of surface at
the rate of about nine kilogramme-water thermal units per hour,
or, which is the same, $\frac{1}{4000}$th of a gramme-water unit from each
square centimetre per second. The mean conductivity of the
three Edinburgh strata, in which Principal Forbes's underground
thermometers were placed, is $2\frac{3}{4}$ grain-water units per second per
square foot per 1° per foot rate of variation of temperature, as
I have shown previously*.

That of the Greenwich stratum is 2·6, in terms of the same
units, according to Professor Everett's recent reductions; and that
of certain strata (clay and sand) in Sweden is 1·64, according to
Ångström (Poggendorff's *Annalen*, last volume of 1861). The
mean of these three numbers is 2·33, which, reduced to the unit
of conductivity founded on the gramme-water-second-centimetre
units, is ·005. Taking this, therefore, as an average conductivity
in the earth's upper crust, we find that if the temperature increased downwards at the rate of 1° per 20 centimetres, the
quantity of heat lost by conduction outwards would be $\frac{1}{4000}$th;
and therefore, according to Peclet's result, this would keep the
surface just 1° warmer than it would be if there were no conduction
of heat from within. Hence, to warm the surface to 10° Fahr.
above what it would be if there were no conduction from within,

* *Trans. R. S. E.*, April 1860, " On the Reduction of Observations of Underground Temperature," § 42.

the rate of rise of temperature must be 1° Fahr. per 2 centimetres, or ·0656 of a foot (which would probably destroy the roots
of any large tree or plant), but at all events could not, as I have
shown*, be the real condition of the earth at any time later than
about 180 years after even a greater heating (7000° Fahr.) of the
whole globe than the greatest we can suppose it at all probable
the earth ever experienced. Hence it is certain that the climate
can never have been sensibly influenced from the earliest "geological" era by underground heat. This conclusion was stated in
§ 17 of the paper already referred to "On the Secular Cooling of
the Earth," as rendered certain by a rough general knowledge of
the circumstances, without any approach to an accurate estimate
of the absolute amount of radiation.

We now see, farther, that the present rate of underground
rise of temperature, estimated at 1° Fahr. per 50 feet, is only
$\frac{1}{75}$th of that which is required to warm the surface by 1°. Hence
the surface is only about $\frac{1}{75}$th of 1° Fahr. warmer at present than
it would be if there were no supply of heat from within.

* *Trans. R. S. E.*, April 1862, "On the Secular Cooling of the Earth," § 18.
[No. 130 *supra*.]

134. THE "DOCTRINE OF UNIFORMITY" IN GEOLOGY
BRIEFLY REFUTED.

[From *Edinb. Roy. Soc. Proc.* Vol. v. 1866 [read Dec. 18, 1865], pp. 512, 513;
reprinted in *Popular Lectures and Addresses*, Vol. II. pp. 6—9.]

135. On the Physical Cause of the Submergence and
Emergence of Land during the Glacial Epoch.

[Note on Mr Croll's paper on this subject: from *Phil. Mag.* Vol. xxxi. April
1866, pp. 305, 306; quoted in *Popular Lectures and Addresses*, Vol. ii.
pp. 320—323.]

Mr Croll's estimate of the influence of a cap of ice on the
sea-level is very remarkable in its relation to Laplace's celebrated
analysis, as being founded on that law of thickness which leads
to expressions involving only the first term of the series of
"Laplace's functions," or "spherical harmonics." The equation
of the level surface, as altered by any given transference of solid
matter, is expressed by equating the altered potential function
to a constant. This function, when expanded in the series of
spherical harmonics, has for its first term the potential due to
the whole mass supposed collected at its altered centre of gravity.
Hence a spherical surface round the altered centre of gravity is
the *first* approximation in Laplace's method of solution for the
altered level surface. Mr Croll has with admirable tact chosen,
of all the arbitrary suppositions that may be made foundations
for rough estimates of the change of sea-level due to variations
in the polar ice-crusts, *the* one which reduces to zero all terms
after the first in the harmonic series, and renders that first approximation (which always expresses the *essence* of the result)
the whole solution, undisturbed by terms irrelevant to the great
physical question.

Mr Croll, in the preceding paper, has alluded with remarkable clearness to the effect of the change in the distribution of
the water in increasing, by its own attraction, the deviation of
the level surface above that which is due to the *given* change in
the distribution of solid matter. The remark he makes, that it
is round the centre of gravity of the altered solid and altered
liquid that the altering liquid surface adjusts itself, expresses the

essence of Laplace's celebrated demonstration of the stability of
the ocean, and suggests the proper elementary solution of the
problem to find the true alteration of sea-level produced by a
given alteration of the solid. As an assumption leading to a
simple calculation, let us suppose the solid earth to rise out of
the water in a vast number of small flat-topped islands, each
bounded by a perpendicular cliff, and let the proportion of water-
area to the whole be equal in all quarters. Let all of these
islands in one hemisphere be covered with ice, of thickness
according to the law assumed by Mr Croll—that is, varying in
proportion to the [sine of the] latitude. Let this ice be removed
from the first hemisphere and similarly distributed over the islands
of the second. By working out according to Mr Croll's directions,
it is easily found that the change of sea-level which this will
produce will consist in a sinking in the first hemisphere and
rising in the second, through heights varying according to the
same law (that is, simple proportionality to sines of latitudes),
and amounting at each pole to

$$\frac{(1-\omega)it}{1-\omega w},$$

where t denotes the thickness of the ice-crust at the pole; i the
ratio of the density of ice, and w that of sea-water, to the earth's
mean density; and ω the ratio of the area of ocean to the whole
surface.

Thus, for instance, if we suppose $\omega = 2/3$, and $t = 6000$ feet,
and take $1/6$ and $1/5\frac{1}{2}$ as the densities of ice and water respectively, we find for the rise of sea-level at one pole, and depression
at the other,

$$\frac{1/3 \times 1/6 \times 6000}{1 - 2/3 \times 1/5\frac{1}{2}},$$

or approximately 380 feet.

It ought to be remarked that a transference of floating ice
goes for nothing in changing the sea-level, and that in estimating
the effect of grounded icebergs the excess of the mass of ice above
that of the water displaced by it is to be reckoned just as if so
much ice were laid on the top of an island.

136. ON THE OBSERVATIONS AND CALCULATIONS REQUIRED TO FIND THE TIDAL RETARDATION OF THE EARTH'S ROTATION.

[Part of the Rede Lecture at Cambridge, May 23, 1866; from *Phil. Mag.* Vol. XXXI. 1866, pp. 533—537 (Supplement). Reprinted in *Math. and Phys. Papers*, Vol. III. art. XCV. pp. 337—341.]

137. ON GEOLOGICAL TIME.

[From *Glasg. Geol. Soc. Trans.* Vol. III. 1871 [read Feb. 27, 1868], pp. 1—28; reprinted in *Popular Lectures and Addresses*, Vol. II. pp. 10—64, 320—323.]

138.   ON THE FRACTURE OF BRITTLE AND VISCOUS SOLIDS
BY "SHEARING."

[From *Roy. Soc. Proc.* Vol. XVII. 1869 [read Feb. 25, 1869], pp. 312, 313;
Phil. Mag. XXXVIII. July 1869, pp. 71—73.]

ON recently visiting Mr Kirkaldy's testing works, the Grove,
Southwark, I was much struck with the appearances presented
by some specimens of iron and steel round bars which had been
broken by torsion.   Some of them were broken right across, as
nearly as may be in a plane perpendicular to the axis of the bar.
On examining these I perceived that they had all yielded through
a great degree to distortion before having broken.   I therefore
looked for bars of hardened steel which had been tested similarly,
and found many beautiful specimens in Mr Kirkaldy's museum.
These, without exception, showed complicated surfaces of fracture,
which were such as to demonstrate, as part of the whole effect in
each case, a spiral fissure round the circumference of the cylinder
at an angle of about 45° to the length.   This is just what is to be
expected when we consider that if *ABDC* (fig. 1) represent an
infinitesimal square on the surface of a round bar with its sides
AC and *BD* parallel to the axis of the cylinder, before torsion,
and *ABD'C'* the figure into which this square becomes distorted

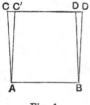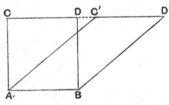

Fig. 1.   Fig. 2.

just before rupture, the diagonal *AD* has become elongated to
the length *AD'*, and the diagonal *BC* has become contracted to
the length *BC'*, and that therefore there must be maximum

tension everywhere, across the spiral of which BC' is an infinitely short portion. But the specimens are remarkable as showing in softer or more viscous solids a tendency to break parallel to the surfaces of "shearing" AB, CD, rather than in surfaces inclined to these at an angle of 45°. Through the kindness of Mr Kirkaldy, his specimens of both kinds are now exhibited to the Royal Society. On a smaller scale I have made experiments on round bars of brittle sealing-wax, hardened steel, similar steel tempered to various degrees of softness, brass, copper, lead.

Sealing-wax and hard steel bars exhibited the spiral fracture. All the other bars, without exception, broke as Mr Kirkaldy's soft steel bars, right across, in a plane perpendicular to the axis of the bar. These experiments were conducted by Mr Walter Deed and Mr Adam Logan in the Physical Laboratory of the University of Glasgow; and specimens of the bars exhibiting the two kinds of fracture are sent to the Royal Society along with this statement. I also send photographs exhibiting the spiral fracture of a hard steel cylinder, and the "shearing" fracture of a lead cylinder by torsion.

These experiments demonstrate that continued "shearing" parallel to one set of planes, of a viscous solid, developes in it a tendency to break more easily parallel to these planes than in other directions,—or that a viscous solid, at first isotropic, acquires "cleavage planes" parallel to the planes of shearing. Thus, if CD and AB (fig. 2) represent in section two sides of a cube of a viscous solid, and if, by "shearing" parallel to these planes, CD be brought to the position $C'D'$, relatively to AB supposed to remain at rest, and if this process be continued until the material breaks, it breaks parallel to AB and $C'D'$.

The appearances presented by the specimens in Mr Kirkaldy's museum attracted my attention by their bearing on an old controversy regarding Forbes's theory of glaciers. Forbes had maintained that the continued shearing motion, which his observations had proved in glaciers, must tend to tear them by fissures parallel to the surfaces of "shearing." The correctness of this view for a viscous solid mass, such as snow becoming kneaded into a glacier, or the substance of a formed glacier as it works its way down a valley, or a mass of débris of glacier ice, reforming as a glacier after disintegration by an obstacle, seems strongly confirmed by the experiments on the softer metals described above. Hopkins had argued against this view, that, according to the theory of elastic solids, as stated above, and represented by the first diagram, the fracture ought to be at an angle of 45° to the surfaces of "shearing." There can be no doubt of the truth of Hopkins's principle *for an isotropic elastic solid, so brittle as to break by shearing before it has become distorted through more than a very small angle*; and it is illustrated in the experiments on brittle sealing-wax and hardened steel which I have described. The various specimens of fractured elastic solids now exhibited to the Society may be looked upon with some interest, if only as illustrating the correctness of each of the two seemingly discrepant propositions of those two distinguished men.

139. NOTE ON THE METEORIC THEORY OF THE SUN'S HEAT.

[After Prof. Grant's paper on "The Physical Constitution of the Sun," *Glasg. Phil. Soc. Proc.* Vol. VII. [March 24, 1869], pp. 111, 112; *Glasg. Geol. Soc. Trans.* Vol. III. 1871, pp. 239, 240. Reprinted in *Popular Lectures and Addresses*, Vol. II. pp. 127—131.]

140. GEOLOGICAL DYNAMICS.

[From *Glasg. Geol. Soc. Trans.* Vol. III. 1871 [read April 5, 1869], pp. 215—240; *Geol. Mag.* Vol. VI. 1869, pp. 472—476. Reprinted in *Popular Lectures and Addresses*, Vol. II. pp. 73—127.]

141. ADDRESS TO THE BRITISH ASSOCIATION AT EDINBURGH.

[From *British Association Report*, Vol. XLI. 1871, pp. lxxxiv—cv; *Nature*, Vol. IV. Aug. 3, 1871, pp. 262—270; *Amer. Journ. Sci.* Vol. II. (3rd ser.), 1871, pp. 269—294; *Math. and Phys. Papers*, Vol. II. art. lxvi. pp. 25—27. Reprinted in *Popular Lectures and Addresses*, Vol. II. pp. 132—205.]

142. The Rigidity of the Earth.

[Letter to *Nature*, Vol. v. Jan. 18, 1872, pp. 223, 224. See however No. 144 *infra* as regards the rigidity which is produced by the inertia of fluid in rotation.]

I HAVE been urged from several quarters to defend my argument for the rigidity of the earth against attacks which are
supposed to have been made upon it. It has, in fact, never been
attacked to my knowledge, and I feel under no obligation to
defend it. There is, I believe, a general impression that grave
objections to it have been raised by M. Delaunay, and it seems
that even in this country some geological writers and teachers,
in their reluctance to abandon the hypothesis of a thin solid
crust, enclosing a wholly liquid mass, hastily concluded that all
dynamical arguments against it had been utterly overthrown by
Delaunay.

In point of fact Delaunay made no reference at all to the
tidal argument, and clearly was unaware that I had brought it
forward when he made his communication on the "Hypothesis of
the interior fluidity of the terrestrial globe*" to the French
Academy, three years and a half ago, objecting to Hopkins's
argument founded on precession and nutation, and merely quoting
me as having expressed acquiescence. On this subject I say
nothing at present, except that ten years ago, before I expressed
(in my first communication of the tidal argument to the Royal
Society) my assent to Hopkins's argument from precession and
nutation, I had thought of the objection to this argument since
brought forward by Delaunay, and had convinced myself of its
invalidity. But I hope to be able on some future occasion to

---

* *Comptes Rendus* for July 13, 1868.

return to the subject, and to prove that any degree of viscosity, acting in the manner and to the effect described by Delaunay, must in an extremely short time abolish the distinction between summer and winter.   My reason for writing to you at present is that I see in Mr Scrope's beautiful book on Volcanoes (just published as a second edition) a sentence ("Prefatory Remarks," page 24), written on the supposition that the tidal argument had been brought forward for the first time at the recent meeting of the British Association in Edinburgh.   I therefore take the liberty of suggesting to you that a reprint of the short abstract of my tidal argument, which appeared in the Proceedings of the Royal Society, for May 16, 1862 *, might not be inappropriate to your columns.   I ought, however, to inform you that the tidal argument was carefully re-stated in the first volume of the treatise on Natural Philosophy, by Prof. Tait and myself, published in 1867, but as the volume is at present out of print, you may not consider this objection fatal to my proposal.

[From *Nature*, Vol. v. Feb. 1, 1872, pp. 257—259.]

WE have been favoured with permission to reprint the following extract from a letter addressed by Sir Wm. Thomson to Mr G. Poulett Scrope:

THE UNIVERSITY, GLASGOW,
Jan. 15, 1872.

DEAR SIR,

I thank you very much for the copy of your beautiful book on Volcanoes, which you have been so kind as to send me through Professor Geikie.   It is full of matter most interesting to me, and I promise myself great pleasure in reading it.

I see with much satisfaction, in your prefatory remarks, that you "earnestly protest against the assertion of some writers, that the theory of the internal fluidity of the globe is or ought to be generally accepted by geologists on the evidence of its high internal temperature."   Your sentence upon the "attractive sensational idea that a molten interior to the globe underlies a thin superficial crust; its surface agitated by tidal waves and flowing freely towards any issue that may here and there be

* [Cf. *Math. and Phys. Papers*, Vol. III. pp. 312—336; Thomson and Tait's *Nat. Phil.* §§ 833—848.]

opened for its outward escape," in which you say that you "do not think it can be supported by reasoning, based on any ascertained facts or phenomena," is thoroughly in accordance with true dynamics. It will, I trust, have a great effect in showing that volcanic phenomena, far from being decisive, as many geologists imagine them to be, in favour of a thin crust enclosing a wholly liquid interior, tend rather, the more thoroughly they are investigated, to an opposite conclusion.

I must, however, take exception to your next sentence, that in which you say that "M. Delaunay has disposed of the well-known astronomical argument of Mr Hopkins and Sir W. Thomson, as to the entire or nearly entire solidity of the earth, derived from the nutation of its axis." Delaunay's deservedly high reputation as one of the first physical astronomers of the day, has naturally led many in this country to believe that his objection to the astronomical argument in favour of the earth's rigidity cannot but be valid. It has even been hastily assumed that the objection is founded on mathematical calculation, an error which the most cursory reading of Delaunay's paper corrects. His hypothesis of a viscous fluid breaks down utterly when tested by a simple calculation of the amount of tangential force required to give to any globular portion of the interior mass the precessional and nutational motions, which, with other physical astronomers, he attributes to the earth as a whole. Thus: taking the ratio of polar diameter to equatorial diameter as 299 to 300, and the density of the upper crust as half the mean density of the earth, I find (from the ordinary elementary formulæ) that when the moon's declination is $28\frac{1}{2}°$, the couple with which she tends to turn the plane of the earth's equator towards the plane of her own centre and the equinoctial line has for its moment a force of $\cdot285 \times 10^{18}$ times the gravity of one gramme at the earth's surface, or rather more than a quarter of a million million tons weight, on an arm equal to the earth's radius. A quadrant of the earth being ten thousand kilometres, the area is five hundred and nine million square kilometres, or 5·09 million million million square centimetres. Hence a force of $\cdot285 \times 10^{18}$ grammes weight distributed equally over two-thirds of the earth's area would give 084 of a gramme weight per square centimetre. This supposition is allowable (for reasons with which I need not trouble you) in estimating roughly the greatest amount of tangential force acting between the upper crust and a spherical interior mass in contact with it, from the preceding accurate calculation of the whole couple exerted by the moon on the upper crust. It is thus demonstrable that the earth's crust must, as a whole, down to depths of hundreds of kilometres, be capable of transmitting tangential stress amounting to nearly $\frac{1}{10}$ of a gramme weight per square centimetre. Under any such stress as this any plastic substance which could commonly be called a viscous fluid would be drawn out of shape with great rapidity. Stokes, who discovered the theory of fluid viscosity, and first made accurate investigations of its amount in absolute measure, found that a cubic centimetre of water, if exposed to tangential force of the millionth part of $\frac{1}{10}$ of a gramme weight on each of four sides, would, even under so small a distorting stress as this, become distorted so rapidly that at the end of a second of time its four corresponding right angles would become one pair of them acute and the other obtuse, by as much as a two-hundredth part of the angle whose arc is radius, that is to say by 29 of a degree. Not as much as a ten-million-millionth part of this distortion could be produced every second of time by the lunar influence in the material underlying the earth's crust without very sensibly affecting precession and nutation; for the effect of the maximum couple exerted on the upper crust by the moon is to turn the whole earth in a second of time through an angle of a one-hundred-million-millionth of ·57 of a degree, so as to give to it at the end of a second the position obtained by geometrically compounding this angular displacement with the angular displacement due simply to rotation. Hence the viscosity assumed by Delaunay, to produce the effect he attributes to it, must be more than ten million million million times the viscosity of water. How much more may be easily estimated with some degree of precision from Helmholtz's mathematical solution of the problem of finding the motion of a viscous fluid contained in a rigid spherical envelope urged by periodically varying couples*. The most interesting part of the application of this solution to the hypothetical problem regarding the earth, is to find how rapidly the obliquity of the ecliptic would be done away with by any assumed degree of viscosity in the interior; such an amount of

* Helmholtz and Piotrowski, "Ueber Reibung tropfbarer Flüssigkeiten," *Imp. Acad.*, Vienna, 1860. [Helmholtz' *Wiss. Abhandl.* Vol. i. pp. 172—222.]

viscosity, for example, as would render the excesses of precession
and nutation above their values for a perfectly rigid interior, not
greater than observation could admit.

The hypothesis of a continuous internal viscous fluid being
disposed of, the question occurs, what rigidity must the interior
mass have, even if enclosed in a perfectly rigid crust, to produce
the actual phenomena of precession and nutation ? The solutions
given by Lamé and myself of the problem of the vibrations of an
elastic solid globe, may be readily applied to determine the
influences on precession and on the several nutations, which
would be produced by elastic yielding with any assumed rigidity
short of infinite rigidity. This application I have no time at
present to make; but without attempting a rigorous investigation,
it is easy roughly to estimate an inferior limit to the admissible
rigidity. In the first place, suppose, with perfect elasticity, the
rigidity be so slight that distorting stress of $\frac{1}{10}$ of a gramme
weight would produce an angular distortion of a half degree or a
degree. The whole would not rotate as a rigid body round one
"instantaneous axis" at each instant, but the rotation would take
place internally, round axes deviating from the axis of external
figure, by angles to be measured in the plane through it and the
line perpendicular to the ecliptic, in the direction towards the
latter line. These angular deviations would be greater and
greater the more near we come to the earth's centre, and the
greatest angular deviation would be comparable with 1°. Hence
the moment of momentum round the solsticial line would be
sensibly less than if the whole mass rotated round the axis of
figure. Now suppose for a moment our measurement of force to
be founded on a year as the unit of time. We find the amount
of precession in a year by dividing the mean amount of the whole
couple due to the influence of moon and sun by the moment of
momentum of the earth's motion round the solsticial line. Hence
the amount of precession would be sensibly augmented by the
elastic yielding; for the motive couple is uninfluenced by the
elastic yielding, if we suppose the earth to be of uniform internal
density. An ordinary elastic jelly presents a specimen of the
degree of elasticity here supposed, as is easily seen when we
consider that the mass of a cubic centimetre of such material is a
gramme, and therefore that the weight of a cubic centimetre of
the substance is the "gramme weight" understood in the speci-

fication $\frac{1}{10}$ of a gramme weight per square centimetre. If then, the interior mass of the earth were no more rigid than an ordinary elastic jelly, and if the upper crust were rigid enough to resist any change of figure that could sensibly influence the result, the precession would be considerably more rapid than if the rigidity were infinite throughout. The lunar nineteen-yearly nutation proves a higher degree of elasticity than this; the solar semi-annual nutation still a higher degree; and still higher yet the imperceptibility of the lunar fortnightly nutation; provided always we suppose the interior mass to be of uniform density, and the upper crust rigid enough to permit no influential change of figure.

The motive of the nineteen-yearly precession may be mechanically represented by two circles of matter pivoted on diameters fixed in the plane of the earth's equator, bisecting one another perpendicularly at the earth's centre. These two circles must oscillate round their pivot-diameters, each through an angle of about 5° on one side and the other of the plane of the equator, in a period of about nineteen years, to produce the lunar nineteen-yearly nutation (more nearly eighteen years seven months). If the radius of each of the supposed material circles is equal to the moon's mean distance from the earth, the mass of each must be a little less than the moon's mass, and one of them a little less than the other*. The diameter on which the latter is pivoted is to be the equinoctial line. The latter alone causes the nutation in right ascension; the former the nutation in declination. The phases of maximum and of zero deflection, in the oscillations of the two circles, follow alternately at equal intervals of time, so that when either is in the plane of the earth's equator, the other is at its greatest inclination of 5° on either side. Taking one of the constituents of the nutational motive alone, we find, on the principles indicated above, $\frac{1}{100}$ of a gramme weight per square centimetre as a very rough estimate for the greatest tangential stress produced by it in the material underlying the earth's crust. Now it is clear that the central parts of the earth and the upper crust cannot, in the course of the nutatory oscillations, experience relative angular motions to any extent considerable in comparison

---

* The greater is equal to the moon's mass multiplied by the cosine of the obliquity of the ecliptic; the less is equal to the moon's mass multiplied by the cosine of twice the obliquity of the ecliptic.

with the nutation of the upper crust, without considerably affecting the whole amount of the nutation. The nutation in declination amounts to 9·25″ on each side of the mean position of the earth's poles, and therefore the tangential stress of $\frac{1}{100}$ of a gramme weight per square centimetre cannot produce an angular distortion considerable in comparison with 9″.

An angular distortion of 8″ is produced in a cube of glass by a distorting stress of about ten grammes weight per square centimetre. We may, therefore, safely conclude that the rigidity of the earth's interior substance could not be less than a millionth of the rigidity of glass without very sensibly augmenting the lunar nineteen-yearly nutation. The lunar fortnightly nutation in declination amounts theoretically to about ·1″, and it is so small as to have hitherto escaped observation. It probably would have been so large as to have been observed were the interior rigidity of the earth anything less than $\frac{1}{200000}$ of that of glass, always provided that the upper crust is rigid enough to prevent any change of form sensibly influencing the nutational motive couple. To understand the degree of rigidity meant by "$\frac{1}{200000}$ of the rigidity of glass," imagine a sheet of some such substance as gutta-percha or vulcanised india-rubber of a square metre area and a centimetre thickness. Let one side of the sheet be cemented to a perfectly hard plane vertical wall, and let a slab of lead 8·8 centimetres thick (weighing therefore a metrical ton)* be cemented to the other side of it. If the rigidity of the substance be $\frac{1}{200000}$ of the rigidity of glass†, and the range of its elasticity sufficient, the side of the sheet to which the lead is attached will be dragged down relatively to the other through a space of $\frac{1}{12}$ of a centimetre.

In the argument from tidal deformations of the solid part of the earth's material, which I communicated to the Royal Society ten years ago, and mentioned incidentally at the recent meeting of the British Association, I showed that though precession and

---

* The metrical ton, or the mass of a cubic metre of water at temperature of maximum density, is ·9842 of the British ton. The thickness of a slab of lead of a square metre area, weighing a metrical ton, is, of course, equal to a metre divided by the specific gravity of lead.

† Everett's measurements give 244×10^6 centimetres weight per square centimetre for the rigidity of the glass on which he experimented. Instead of this I take 240×10^6, for simplicity.

nutation would be augmented by want of rigidity in the interior, they would be diminished by want of rigidity in the upper crust, and that on no probable hypothesis can we escape the conclusion that the earth as a whole is less yielding than a globe of glass of the same dimensions and exposed to the same forces. That argument, therefore, proves about 200,000 times greater rigidity for the earth as a whole than what I have now written to you proves for the interior of the earth on the supposition of a thin preternaturally rigid crust.

I must apologise to you for having troubled you with so long a letter. I did not intend to make it so long when I commenced, but I have been led on by considerations of details, inevitable when such a subject is once entered upon.

I remain,

Yours very truly,

WILLIAM THOMSON.

G. Poulett Scrope, Esq., F.R.S.

143. INFLUENCE OF GEOLOGICAL CHANGES ON THE EARTH'S
ROTATION [PRESIDENTIAL ADDRESS].

[From *Glasg. Geol. Soc. Proc.* Vol. IV. [Feb. 12, 1874], pp. 311—313;
Nature, Vol. IX. March 5, 1874, pp. 345—346.]

AT the annual meeting of the Geological Society of Glasgow,
on Feb. 12, the president, Sir William Thomson, F.R.S., gave an
address on the above subject, of which the following is an
abstract:

He first briefly considered the rotation of rigid bodies in
general, defining a principal axis of rotation as one for which
the centrifugal forces balance while the body rotates around it.
He then took the case of the earth; and, having pointed out the
position of its present axis, showed that if from any cause it were
made to revolve round any other, that would be an "instantaneous
axis," changing every instant, and travelling through the solid,
from west to east, in a period of 296 days round the principal
axis. It would shift continually in the figure, owing to the varying
centrifugal force of two opposite portions of the body. This would
produce, by centrifugal force, a tide of peculiar distribution over
the ocean, having 296 days for period. An inclination of the axis
of instantaneous rotation to principal axis of 1″, or 100 ft. at the
earth's surface, would produce rise and fall of water in 45° latitude,
where the effect is greatest, amounting to ·17 of a foot above and
below mean level*.

* [The actual period is lengthened to about 430 days by elastic yielding of the
Earth, and is partly obscured by meteorological accumulations of matter. The
accompanying tide has been sought for by Bakhuysen. For modern records on
changes of latitude and their interpretation, see e.g. E. H. Hills and J. Larmor,
Monthly Notices Roy. Astron. Soc. Nov. 1906. See also Presidential Addresses to
the Roy. Soc. for 1891 and 1892, in *Popular Lectures and Addresses*, pp. 502, 523.]

He noticed, in passing, the application of these dynamical principles to the attraction which the sun and moon exercise on the protuberant parts of the earth, tending to bring the plane of the earth's equator into coincidence with the ecliptic. This causes an incessant change, to a certain limited extent, in the position of the axis of rotation, thereby occasioning what is known as the "precession of the equinoxes." Having illustrated these remarks by some interesting experiments, Sir William Thomson proceeded to consider more particularly the circumstances according to which the axis of the earth might become changed through geological influences, and the consequences of any such change. The possibility of such a change had been adduced to account for the great differences in climate which can be shown to have obtained at different periods in the same portion of the earth's surface. In the British Isles, for example, and in many other countries, there is clear evidence that at a comparatively recent period a very cold climate—much colder than at present—prevailed; while in the same places the remains of plants and animals belonging to several preceding eras indicate a high temperature and a comparatively tropical climate. The question arose, can changes in the earth s axis account for these changes of climate? In the present condition of the earth, any change in the axis of rotation could not be permanent, because the instantaneous axis would travel round the principal axis of the solid in a period of 296 days, as already stated. Maxwell had pointed out that this shifting of the instantaneous axis in the solid would constitute in its period a periodic variation everywhere of "latitude," ranging above and below the mean value, to an extent equal to the angular deviation of the instantaneous axis of rotation from the principal axis; and, by comparing observations of the altitude of the Pole-star during three years at Greenwich, had concluded that there may possibly be as much as $\frac{1}{2}''$ of such deviation, but not more.

In very early geologic ages, if we suppose the earth to have been plastic, the yielding of the surface might have made the new axis a principal axis. But certain it is that the earth at present is so rigid that no such change is possible. The precession of the equinoxes shows that the earth at present moves as a rigid body; and during the whole period of geologic history, or while it has been inhabited by plants and animals, it has been practically rigid. Changes of climate, then, have not been produced by changes of the axis of the earth. The learned professor then inquired what influences great subsidences or great elevations in different parts of the earth might have on the axis of rotation. No doubt the removal of a large quantity of solid matter from one part of the globe to another would sensibly alter the principal axis, as well as the axis of rotation, which so nearly coincides with it; but it could be shown that it would produce in the latter only about $\frac{1}{300}$th part of the change produced in the former. We know too little of the changes in the interior of the earth accompanying such changes on its surface to be able to state results with certainty. But he estimated that an elevation, for example, of 600 feet on a tract of the earth's surface 1000 miles square and 10 miles in thickness, would only alter the position of the principal axis by *one-third of a second*, or 34 feet. He called attention to the effect of tidal friction and subterranean viscosity in reducing any such deviation, and pointed out that it must be exceedingly slow; using for evidence the observationally proved slowness of the diminution of the earth's rotational velocity, and of the inclination of its equator to the ecliptic. It therefore seemed probable that geological changes had not produced any perceptible change in the principal axis or in the axis of rotation within the period of geological history.

144. Review of Evidence Regarding the Physical Condition of the Earth; its Internal Temperature; the Fluidity or Solidity of its Interior Substance; the Rigidity, Elasticity, Plasticity of its External Figure; and the Permanence or Variability of its Period and Axis of Rotation.

[Presidential Address to the Mathematical and Physical Science Section of the British Association at Glasgow. From *British Association Report*, 1876, pt. ii. pp. 1—12; *Nature*, Vol. xiv. Sept. 14, 1876, pp. 426—431; *Archives Sci. Phys. Nat.* Vol. lvii. 1876, pp. 138—161; *Amer. Journ. Sci.* Vol. xii. 1876, pp. 336—354. Reprinted in *Math. and Phys. Papers*, Vol. iii. art. xcv. pp. 320—335; *Popular Lectures and Addresses*, Vol. ii. pp. 238—272.]

145. Geological Climate.

[From *Glasg. Geol. Soc. Trans.* Vol. v. [Feb. 22, 1877], pp. 238—250. Reprinted in *Popular Lectures and Addresses*, Vol ii. pp. 273—298.]

146. The Internal Condition of the Earth, as to Temperature, Fluidity, and Rigidity.

[From *Glasg. Geol. Soc. Trans.* Vol. vi. 1882 [read Feb. 14, 1878], pp. 38—49. Reprinted in *Popular Lectures and Addresses*, Vol. ii. pp. 299—318.]

147. PROBLEMS RELATING TO UNDERGROUND TEMPERATURE :
A FRAGMENT *.

[From *Phil. Mag.* Vol. v. May, 1878, pp. 370—374; *Journ. de Phys.* Vol. VII. 1878, pp. 397—402; *Amer. Journ. of Science*, Vol. XVI. 1878, pp. 132—135.

Problem I. A fire is lighted on a small portion of an un-interrupted plane boundary of a mass of rock of the precise quality of that of Calton Hill, and after burning for a certain time is removed, the whole plane area of rock being then freely exposed to the atmosphere. It is required to determine the consequent conduction of heat through the interior.

Problem II. It is required to trace the effect of an un-usually hot day on the internal temperature of such a mass of rock.

Problem III. It is required to trace the secular effect con-sequent on a sudden alteration of mean temperature.

Problem IV. It is required to determine the change of temperature within a ball of the rock consequent upon suddenly removing it from a fluid of one constant temperature and plunging it into a fluid maintained at another constant temperature.

Problems I., II., and III. In solving each of these problems,

---

* An old MS., written eighteen years ago and found to-day. It was kept back until the time should be found to write out the solutions of Problems II., III., and IV. The time was never found; but as mere synthesis from the solution of Problem I. suffices for II. and III. (surface integration of the solution for I. over the medial plane solves II., and the time-integral from $t = -\infty$ to $t=0$, of the solution of II., gives that of III.), and as IV. is merely an example of Fourier's now well-known solution for the globe (see Professors Ayrton and Perry's paper, " On the Heat-conductivity of Stone," *Philosophical Magazine* for April, 1878), with numerical results calculated for trap-rock according to its thermal conductivity, as determined by the Edinburgh observations referred to in the fragment now pub-lished, the non-completion of the original proposal need not be much regretted.— W. T., March 25, 1878.

we shall suppose the air in contact with the rock to be not sensibly influenced in its temperature by the conduction of heat inwards or outwards through the solid substance. In reality, the stratum of air in immediate contact with the rock must always have precisely the same temperature as the rock itself at its bounding surface; and the continual mixing up of the different strata, whether by wind or by local convective currents due to differences of temperature, tends to bring the whole superincumbent mass of air to one temperature. Our supposition therefore amounts to assuming that the rate of variation of temperature from point to point in the rock near its surface, owing to the special cause under consideration, is much less than the ordinary changing variations from day to night. Hence, in Problems I., II., and III., the solutions will not be applicable until so much time has been allowed to elapse as will leave only a residual variation, small in comparison with the ordinary diurnal maximum rates of increase and diminution of temperature from point to point inwards in the immediate neighbourhood of the surface.

In the case of Problem II. these conditions will be practically fulfilled, and continue to be fulfilled, very soon after the day of extraordinary temperature of which the effect is to be considered, and we shall have a perfectly practical solution illustrative of the consequences experienced several days or weeks later at the 3-foot and 6-foot deep thermometers of the observing-station. The solution of Problem I., which we now proceed to work out, will show clearly what dimensions as to space, time, and temperature may be chosen for a really practical illustration of its conclusions.

Problem I., subject to the limitations we have just stated, is equivalent to the following:—*An infinitely small area of an infinite plane terminating on one side a mass of uniform trap-rock which extends up indefinitely in all directions on the other side, is infinitely heated for an infinitely short time, and the whole surface is instantly and for ever after maintained at a constant temperature. It is required to determine the constant internal variations of temperature.*

Let the solid be doubled so as to extend to an infinite distance on both sides of the plane mentioned in the enunciation. This

plane, when no longer a boundary, we shall call *the medial plane.*
Let P, P' be two points equidistant from the medial plane in a
line perpendicular to it, on each side of the portion heated
according to enunciation. Let a certain quantity of heat Q be
suddenly created in an infinitely small portion of the solid round
P, and at the same instant let an equal quantity be abstracted
from an infinitely small portion of the solid round P'. The
consequent variations of temperature on the two sides of the
medial plane of reference will be equal and opposite, being a
heating effect which spreads from the medial plane in one
direction, and a symmetrical cooling effect spreading from the
same plane through the matter which we have imagined placed
on its other side. The heating effect on the first side will, as is
easily seen, be precisely the same as that proposed for investigation
in Problem I.; and the thermal action of the mass we have
supposed added on the other side will merely have the effect of
maintaining the temperature of the bounding plane unvaried.
Now if a quantity Q of heat be placed at one point (α, β, γ) of an
infinite homogeneous solid, the effect at any subsequent time t at
any point x, y, z of the solid will be expressed by the formula

$$\frac{Q}{8\sqrt{k^3\pi^3}} . t^{-\frac{3}{2}} \epsilon^{-\frac{(x-\alpha)^2+(y-\beta)^2+(z-\gamma)^2}{4kt}},$$

discovered by Fourier; and the effect of simultaneously placing
other quantities of heat, positive or negative, at other points will,
as he has shown, be determined by finding the effect of each
source separately by proper application of the same formula, and
adding the results in accordance with the principle of the superposition of thermal conductions stated above. Hence the effect
of simultaneously placing equal positive and negative quantities,
$+Q$ and $-Q$, at two points, (α, β, γ), $(\alpha', \beta', \gamma')$, will for any
subsequent time t be expressed by the formula

$$\frac{Q}{8\sqrt{k^3\pi^3}} t^{-\frac{3}{2}} \left\{ \epsilon^{-\frac{(x-\alpha)^2+(y-\beta)^2+(z-\gamma)^2}{4kt}} - \epsilon^{-\frac{(x-\alpha')^2+(y-\beta')^2+(z-\gamma')^2}{4kt}} \right\}.$$

If in this expression we take $\alpha = \frac{1}{2}a$, $\alpha' = -\frac{1}{2}a$, $\beta = 0$, $\beta' = 0$, $\gamma = 0$,
$\gamma' = 0$, and suppose a to be infinitely small, we find what it
becomes by differentiating the first term with reference to α,
writing a instead of $d\alpha$, and taking $\alpha = 0$, $\beta = 0$, $\gamma = 0$. The result
constitutes the solution of the proposed problem; and thus, if v

denote the required temperature at time t and point (x, y, z) of the solid, we find

$$v = \frac{Qa}{16\pi^{\frac{3}{2}} k^{\frac{5}{2}}} \cdot xt^{-\frac{5}{2}} \epsilon^{-\frac{x^2+y^2+z^2}{4kt}}.$$

A more convenient formula* to express the solution will be obtained by putting $x^2 + y^2 + z^2 = r^2$ and $x = r\cos\theta$. We thus have

$$v = \frac{Qa}{16\pi^{\frac{3}{2}}} (kt)^{-\frac{5}{2}} \cos\theta \cdot r\epsilon^{-\frac{r^2}{4kt}},$$

which expresses the temperature assumed at a time t after the application of the fire, by a point of the solid at a distance r from the point of the surface where the fire was applied, and situated in a direction inclined at an angle θ to the vertical through this point. From this expression we conclude :—

(1) The simultaneous temperatures at different points equidistant from the position of the fire are simply proportional to the distances of these points from the plane surface.

(2) The law of variation of temperature with distance in any one line from the place where the fire was applied is the same at all times.

---

* In this formula k denotes what I have called the thermal diffusivity of the substance—that is to say, its thermal conductivity divided by the thermal capacity of unit bulk of the substance. Diffusivity is essentially reckoned in units of area per unit of time; or, as Maxwell puts it, its dimensions are $[L^2 T^{-1}]$. Its value (267 square British feet per annum for the trap-rock of Calton Hill, used further on in the text) was taken from my paper on the "Reduction of Observations of Underground Temperature," published in the *Transactions of the Royal Society of Edinburgh* for April 1860, where it was found by the application of Fourier's original formula to a harmonic reduction of Forbes's observations of underground temperature. Reducing this number to square centimetres per second, and expressing similarly the results of my own reduction of Forbes's observations for two other localities in the neighbourhood of Edinburgh, and of Professor Everett's reductions of the Greenwich Underground Observations, we have the following Table of diffusivities:—

Diffusivities.

Trap-rock of Calton Hill	·00786	of a square centim. per second.		
Sand of experimental garden	·00872	,,	,,	,,
Sandstone of Craigleith Quarry	·02311	,,	,,	,,
Gravel of Greenwich Observatory Hill	·01249	,,	,,	,,

These numbers were first published by Everett, in his *Illustrations of the Centimetre-Gramme-Second (C. G. S.) System of Units*, published by the Physical Society of London (1875), a most opportune and useful publication.

(3) The law of variation of temperature with time is the same at all points of the solid.

(4) Corresponding distances in the law of variation with distance increase in proportion to the square root of the time from the application and removal of the fire; and therefore, of course, corresponding times in the law of variation with time are proportional to the squares of the distances.

(5) The maximum value of the temperature, in the law of variation with distance, diminishes inversely as the square of the increasing time.

(6) The maximum value of the temperature in the law of variation with time, at any one point of the rock, is inversely as the fourth power of the distance from the place where the fire was applied.

(7) At any one time subsequent to the application of the fire, the temperature increases in any direction from the place where the fire was applied to a maximum at a distance equal to $\sqrt{2kt}$, and beyond that falls to zero at an infinite distance in every direction. The value of k for the trap-rock of Calton Hill being 267 *, when a year is taken as the unit of time and a British foot the unit of space, the radius of the hemispherical surface of maximum temperature is therefore $23\cdot1 \times \sqrt{t}$ feet. Thus at the end of one year it is $23\cdot1$ feet, at the end of 10,000 years it is 2310 feet, from the origin. The curve of fig. 1 shows graphically the law of variation of temperature with distance. The ordinates of the curve are proportional to the temperatures, and the corresponding abscissas to the distances from the origin or place of application of the fire.

(8) At any one point at a finite distance within the solid (which, by hypothesis, is at temperature zero at the instant when the fire is applied and removed) the temperature increases to a maximum at a certain time, and then diminishes to zero again after an infinite time; the ultimate law of diminution being inversely as the square root of the fifth power of the time. The time when the maximum temperature is acquired at a distance r from the place where the fire was applied, is $r^2/10k$, or, according

---

* [This number has been changed to agree with *Math. and Phys. Papers*, Vol. III. p. 289, and the following ones altered in consequence.]

to the value we have found for the trap-rock, $r^2/2351^*$ of a year. Thus it appears that at one French foot from the place of the fire, the maximum temperature is acquired about four hours (more exactly ·155 day) after the application and removal of the

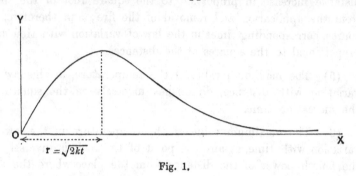

$$r = \sqrt{2kt}$$

Fig. 1.

fire. At 48·5 French feet from the fire the maximum temperature is reached just a year from the beginning; and at 4850 feet the maximum is reached in 10,000 years. The law of variation of temperature with time is shown by the curve of fig. 2, the ordinates of which represent temperatures, and the abscissas times.

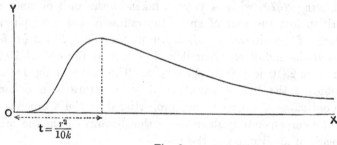

$$t = \frac{r^2}{10k}$$

Fig. 2.

From these results we can readily see how the circumstances of the proposed problem may be actually realized, if not rigorously, yet to any desired degree of approximation, in the manner supposed—namely, by keeping a fire burning for a certain time over a small area of rock, and then removing it and cooling the surface.

* [This number has been changed to suit the French foot and year as units, involving change in subsequent numbers.]

148. On a New Form of Portable Spring Balance for the Measurement of Terrestrial Gravity.

[From *Edinb. Roy. Soc. Proc.* Vol. XIII. [April 19, 1886], pp. 683—686; *British Association Report*, 1886, pp. 534, 535.]

THE design and construction of the instrument now to be described was undertaken on the suggestion of General Walker, of the East Indian Trigonometrical Survey. At the Aberdeen Meeting of the British Association in 1885, General Walker obtained the appointment of a committee to examine into the whole question of the present methods and instruments for the measurement of gravitational force, and to promote investigation, having for its object the production of gravitation measuring instruments of a more reliable and accurate character than those now in use.

The secretary of this committee, Professor Poynting, has already issued a circular note to the members of the committee (of whom the author is one), stating the conditions which must be fulfilled by any gravimeter laid before the committee for examination and report.

An instrument, constructed according to the following description, promises to fulfil all the conditions mentioned in Professor Poynting's circular. Its sensibility is amply up to the specified degree. It is, of necessity, largely influenced by temperature, and it is not certain that the allowance for temperature, or the means which may be worked out for bringing the instrument always to one temperature, may prove satisfactory. It is almost certain, although not quite certain, that the constancy of the latent zero of the spring will be sufficient, after the instrument has been kept for several weeks or months under the approximately constant stress under which it is to act in regular use.

The instrument consists of a thin flat plate of springy German silver of the kind known as "doctor," used for scraping the colour off the copper rollers in calico printing The piece used was 75 centimetres long, and was cut to a breadth of about 2 centimetres. A brass weight of about 200 grammes was securely soldered to one end of it, and the spring was bent like the spring of a hanging bell, to such a shape that when held firmly by one end the spring stood out approximately in a straight line, having the weight at the other end. If the spring had no weight the curvature, when free from stress, must be in simple proportion to the distance along the curve from the end at which the weight is attached, in order that when held by one end it may be straightened by the weight fixed at the other end.

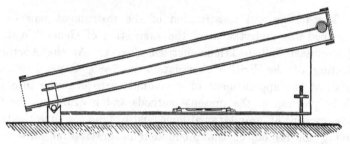

Front elevation, with one-half of Tube removed.

The weight is about 2 per cent. heavier than that which would keep the spring straight when horizontal; and the fixed end of it is so held that the spring stands, not horizontal, but inclined at a slope of about 1 in 5, with the weighted end above the level of the fixed end. In this position the equilibrium is very nearly unstable. A definite sighted position has been chosen for the weight, relatively to a mark rigidly connected to the fixed end of the spring, fulfilling the condition that in this position the equilibrium is stable at all the temperatures for which it has hitherto been tested; while the position of unstable equilibrium is only a few millimetres above it for the highest temperature for which the instrument has been tested, which is about 16° C.

The fixed end is rigidly attached to one end of a brass tube, about 8 centimetres diameter, surrounding the spring and weight, and closed by a glass plate at the upper end of the incline, through which the weight is viewed. The tube is fixed to the hypotenuse of a right-angled triangle of sheet brass, of which one leg, inclined to it at an angle of about $\frac{1}{5}$ radian, is approximately horizontal, and is supported by a transverse trunnion resting on fixed V-s under the lower end of the tube, and a micrometer screw under the short, approximately vertical, leg of the triangle.

The observation consists in finding the number of turns and parts of a turn of the micrometer screw, required to bring the instrument from the position at which the bubble of the spirit-level is between its proper marks, to the position which equilibrates the spring-borne weight, with a mark upon it exactly in line with a chosen divisional line on a little scale of 20 half-millimetres, fixed in this tube in the vertical plane perpendicular to its length.

The instrument is, as is to be expected, exceedingly sensitive to changes of temperature. An elevation of temperature of 1° C. diminishes the Young's modulus of the German silver so much, that about a turn and a half of the micrometer screw (lowering the upper end of the tube at the rate of 2/3 millimetre per turn) produced the requisite change of adjustment for the balanced position of the movable weight. About $1\frac{1}{8}$ turn of the screw corresponds to a difference of 1/5000 in the force of gravity, and the sensibility of the instrument is amply valid for 1/40 of this amount; that is to say, for 1/200,000 difference in the force of gravity. Hence it is not want of sensibility in the instrument that can prevent its measuring differences of gravity to the 1/100,000; but to attain this degree of minuteness it will be necessary to know the temperature of the spring to within 1/20° C. I do not see that there can be any great difficulty in achieving the thermal adjustment by the aid of a water jacket and a delicate thermometer. To facilitate the requisite thermal adjust-ment, I propose, in a new instrument of which I shall immediately commence the construction, to substitute for the brass tube a long double girder of copper (because of the high thermal con-ductivity of copper), by which sufficient uniformity of temperature along the spring throughout the mainly effective portion of its length and up to near the sighted end, shall be secured. The water jacket will secure a slight enough variation of temperature to allow the absolute temperature to be indicated by the thermo-meter with, I believe, the required accuracy.

149. The Sun's Heat.

[From *Roy. Instit. Proc.* Vol. XII. 1889 [Jan. 21, 1887], pp. 1—21; *Nature*, Vol. XXXV. Jan 27, 1887, pp. 297—300; *Ciel et Terre*, Vol. III. 1887—1888, pp. 79—80, 281—288; *Good Words*, March and April, 1887. Reprinted in *Popular Lectures and Addresses*, Vol. I. pp. 369—422.]

150. On the Equilibrium of a Gas under its own Gravitation only.

From *Edinb. Roy. Soc. Proc.* Vol. xiv. [read Feb. 21, 1887], pp. 111—118, pt. iii. p. 118 [title only]; *Phil. Mag.* Vol. xxii. March 1887, pp. 287—292.]

This problem, for the case of uniform temperature, was first, I believe, proposed by Tait in the following highly interesting question, set in the Ferguson Scholarship Examination (Glasgow, October 2, 1885):—"Assuming Boyle's Law for all pressures, form the equation for the equilibrium-density at any distance from the centre of a spherical attracting mass, placed in an infinite space filled originally with air. Find the special integral which depends on a power of the distance from the centre of the sphere alone."

The answer (in examination style!) is:—Choose units properly; we have

$$\frac{d\rho}{dr} = -r^{-2}\rho \int_0^r \rho r^2 dr \quad \dots\dots\dots\dots\dots(1),$$

where ρ is the density at distance r from the centre. Assume

$$\rho = Ar^\kappa \quad \dots\dots\dots\dots\dots\dots(2).$$

We find $A = 2$, $\kappa = -2$; and therefore

$$\rho = 2r^{-2} \dots\dots\dots\dots\dots\dots\dots(3)$$

satisfies the equation in the required form.

---

* *Note of February* 22, 1887.—Having yesterday sent a finally revised proof of this paper for press, I have to-day received a letter from Prof. Newcomb, calling my attention to a most important paper by Mr J. Homer Lane, "On the Theoretical Temperature of the Sun," published in the *American Journal of Science* for July 1870, p. 57, in which precisely the same problem as that of my article is very powerfully dealt with, mathematically and practically. It is impossible now, before going to press, for me to do more than to refer to Mr Lane's paper; but I hope to profit by it very much in the continuation of my present work which I intended, and still intend, to make.—W. T.

Tait informs me that this question occurred to him while writing for *Nature* a review of Stokes's Lecture* "On Inferences from the Spectrum Analysis of the Lights of Sun, Stars, Nebulæ, and Comets"; and in the *Proceedings of the Mathematical Society* he has given some transformations of the Equation of Equilibrium. The same statical problem has recently been forced on myself by considerations which I could not avoid in connection with a lecture which I recently gave in the Royal Institution of London, on "The Probable Origin, the Total Amount, and the Possible Duration of the Sun's Heat."

Helmholtz's explanation, attributing the Sun's heat to condensation under mutual gravitation of all parts of the Sun's mass, becomes not a hypothesis but a statement of fact, when it is admitted that no considerable part of the heat emitted from the Sun is produced by present in-fall of meteoric matter from without. The present communication is an instalment towards the gaseous dynamics of the Sun, Stars, and Nebulæ.

To facilitate calculation of practical results, let a kilometre be the unit of length; and the terrestrial-surface heaviness of a cubic kilometre of water at unit density, taken as the maximum density under ordinary pressure, be the unit of force (or, approximately, a thousand million tons heaviness at the earth's surface). If p be the pressure, ρ the density, and t the temperature from absolute zero, we have, by Boyle and Charles's laws,

$$p = H\rho t \quad \dots\dots\dots\dots\dots\dots(4);$$

where t denotes absolute (thermodynamic†) temperature, with $0°$ Cent. taken as unit; and H denotes what is commonly, in technical language, called "the height of the homogeneous atmosphere" at $0°$ C. For dry common air, according to Regnault's determination of density,

$$H = 7\cdot985 \text{ kilometres} \quad \dots\dots\dots\dots(4').$$

---

* Lecture III. of Second Course of *Burnet Lectures*, Aberdeen, Dec. 1884; published, London 1885 (Macmillan).

† The notation of the text is related to temperature Centigrade on the thermodynamic principle (which is approximately temperature Centigrade by the air-thermometer), as follows :—$t = \frac{1}{273}$ (temperature Centigrade + 273); see my *Collected Mathematical and Physical Papers*, Vol. I. Arts. xxxix. and xlviii. Part VI. §§ 99, 100 ; and article "Heat," §§ 35—38 and 47—67, *Encyc. Brit.*, and Vol. III. (soon to be published) of *Collected Papers*.

Let β be the gravitational coefficient proper to the units chosen; so that $\beta mm'/D^2$ is the force between m, m' at distance D. The earth's mean density being 5·6, and radius 6370 kilometres, we have

$$\tfrac{4}{3}\pi \,.\,6370\,.\,5\cdot6\beta = 1\,; \text{ and therefore } 4\pi\beta = 1/11890 \ldots(5).$$

Let now the p, ρ, t of (4) be the pressure, density, and temperature at distance r from the centre of a spherical shell containing gas in gross-dynamic* equilibrium. We have, by elementary hydrostatics,

$$\frac{dp}{dr} = -\rho\left(M + \int_a^r dr\,4\pi r^2 \rho\right)\beta/r^2 \ldots\ldots\ldots\ldots(6),$$

whence $$\frac{d}{dr}\left(\frac{r^2}{\rho}\frac{dp}{dr}\right) = -4\pi\beta r^2 \rho \ldots\ldots\ldots\ldots\ldots(7),$$

where M denotes the whole quantity of matter within radius a from the centre; which may be a nucleus and gas, or may be all gas.

If the gas is enclosed in a rigid spherical shell, impermeable to heat, and left to itself for a sufficiently long time, it settles into the condition of gross-thermal equilibrium, by "conduction of heat," till the temperature becomes uniform throughout. But if it were stirred artificially all through its volume, currents not considerably disturbing the static distribution of pressure and density will bring it approximately to what I have called convective equilibrium† of temperature—that is to say, the condition in which the temperature in any part P is the same as that which any other part of the gas would acquire if enclosed in an impermeable cylinder with piston, and dilated or expanded to the same density as P. The *natural stirring* produced in a great free fluid mass like the Sun's, by the cooling at the surface, must, I believe, maintain a somewhat close approximation to convective equilibrium throughout the whole mass. The known relations between temperature, pressure, and density for the ideal "perfect gas," when condensed or allowed to expand in a cylinder and piston of material impermeable to heat, are‡

$$p = HT\rho^k, \quad t = T\rho^{k-1} \ldots\ldots\ldots\ldots(8),(9);$$

---

* Not in molecular equilibrium of course; and not in gross-thermal equilibrium, except in the case of t uniform throughout the gas.

† See "On the Convective Equilibrium of Temperature in the Atmosphere," *Manchester Phil. Soc.* Vol. II., 3rd series, 1862; and Vol. III. of *Collected Papers*.

‡ See my *Collected Mathematical and Physical Papers*, Vol. I. Art. xlviii. note 3.

where k denotes the ratio of the thermal capacity of the gas, pressure constant, to its thermal capacity, volume constant, which is approximately equal to 1·41 or 1·40 (we shall take it 1·4) for all gases, and all temperatures, densities, and pressures; and T denotes the temperature corresponding to unit density in the particular gaseous mass under consideration.

Using (8) to eliminate p from (7) we find

$$\frac{d}{dr}\left[r^2\frac{d(\rho^{k-1})}{dr}\right] = -\frac{4\pi\beta(k-1)}{HTk}r^2\rho \quad \ldots\ldots\ldots(10);$$

which, if we put

$$\rho^{k-1}=u,\quad 1/(k-1)=\kappa\ldots\ldots\ldots\ldots(11),\ (12),$$

and

$$r^{-1}\sqrt{\frac{HTk}{4\pi\beta(k-1)}}=x \quad \ldots\ldots\ldots\ldots\ldots(13)$$

takes the remarkably simple form

$$\frac{d^2u}{dx^2}=-\frac{u^\kappa}{x^4}\ldots\ldots\ldots\ldots\ldots\ldots\ldots\ldots(14).$$

Let $f(x)$ be a particular solution of this equation; so that

and therefore
$$\left.\begin{array}{l} f''(x)=-[f(x)]^\kappa x^{-4} \\[2mm] f''(mx)=-[f(mx)]^\kappa m^{-4}x^{-4} \end{array}\right\} \quad \ldots\ldots\ldots(15).$$

From this we derive a general solution with one disposable constant, by assuming

$$u=Cf(mx) \quad \ldots\ldots\ldots\ldots\ldots(16);$$

which, substituted in (14), yields, in virtue of (15),

$$m^2=C^{-\kappa+1} \quad \ldots\ldots\ldots\ldots\ldots(17);$$

so that we have, as a general solution,

$$u=Cf\left[xC^{-\frac12(\kappa-1)}\right] \quad \ldots\ldots\ldots\ldots\ldots(18).$$

Now the class of solutions of (14) which will interest us most is that for which the density and temperature are finite and continuous from the centre outwards, to a certain distance, finite as we shall see presently, at which both vanish. In this class of cases u increases from 0 to some finite value, as x increases from some finite value to ∞. Hence if $u=f(x)$ belongs to this class, $u=Cf(mx)$ also belongs to it; and (18) is the general solution for the class. We have therefore, immediately, the following conclusions:—

(1)   The diameters of different globular* gaseous stars of the same kind of gas are inversely as the $\frac{1}{2}(\kappa-1)$th powers (or $\frac{3}{4}$ powers) of their central temperatures, at the times when, in the process of gradual cooling, their temperatures at places of the same densities are equal (or "T" the same for the different masses). Thus, for example, one sixteenth central temperature corresponds to eight-fold diameter: one eighty-first central temperature corresponds to twenty-seven fold diameter.

(2)   Under the same conditions as (1) (that is, H and T the same for the different masses), the central densities are as the κth powers (or $\frac{5}{2}$ powers) of the central temperatures; and therefore, inversely as the $2\kappa/(\kappa-1)$, or $2/(2-k)$, or $10/3$, powers of the diameters.

(3)   Under still the same conditions as (1) and (2), the quantities of matter in the two masses are inversely as the $[2/(2-\kappa)-3]$th powers (inversely as the cube roots), of their diameters.

(4)   The diameters of different globular gaseous stars, of the same kind of gas, and of the same central densities, are as the square roots of their central temperatures.

(5)   The diameters of different globular gaseous stars of different kinds of gas, but of the same central densities and temperatures, are inversely as the square roots of the specific densities of the gases.

(6)   A single curve [$y=f(r^{-1})$] with scale of ordinate (r) and scale of abscissa (y) properly assigned according to (18), (17), and (11) shows for a globe of any kind of gas in molecular equilibrium, of given mass and given diameter, the absolute temperature at any distance from the centre. Another curve, $\{y=[f(r^{-1})]^\kappa\}$, with scales correspondingly assigned, shows the distribution of density from surface to centre.

---

* This adjective excludes stars or nebulæ *rotating steadily* with so great angular velocities as to be much flattened, or to be annular; also nebulæ revolving circularly with different angular velocities at different distances from the centre, as may be approximately the case with spiral nebulæ. It would approximately enough include the sun, with his small angular velocity of once round in 25 days, were the fluid not too dense through a large part of the interior to approximately obey gaseous law. It no doubt applies very accurately to earlier times of the sun's history, when he was much less dense than he is now.

It is easy to find, with any desired degree of accuracy, the particular solution of (14), for which

$$u = A, \text{ and } \frac{du}{dx} = A', \text{ where } x = a \quad \dots\dots(19),$$

a denoting any chosen value of x, and A and A' any two arbitrary numerics, by successive applications of the formula

$$u_{i+1} = A - \int_a^x dx \left(A' - \int_a^x dx \frac{u^{\kappa_i}}{x^4} \right) \quad \dots\dots\dots(20);$$

the quadratures being performed with labour moderately proportional to the accuracy required, by tracing curves on "section"-paper (paper ruled with small squares) and counting the squares and parts of squares in their areas. To begin, u_0 may be taken arbitrarily; but it may conveniently be taken from a hasty graphic construction by drawing, step by step, successive arcs [*] of a curve with radii of curvature calculated from (14) with the value of du/dx found from the step-by-step process. If this preliminary construction is done with care, by aid of good drawing-instruments, u_1 calculated from u_0 by quadratures will be found to agree so closely with u_0, that u_0 itself will be seen to be a good solution. If any difference is found between the two, u_1 is the better: u_2 is a closer approximation than u_1; and so on, with no limit to the accuracy attainable.

Mr Magnus Maclean, my official assistant in the University of Glasgow, has made a successful beginning of the working-out this process for the case $u = 16$ where $x = \infty$; and has already obtained a somewhat approximate solution, of which the produce useful for our problem is expressed in the following table.

---

[*] This method of graphically integrating a differential equation of the second order, which first occurred to me many years ago as suitable for finding the shapes of particular cases of the capillary surface of revolution, was successfully carried out for me by Prof. John Perry, when a student in my laboratory in 1874, in a series of skilfully executed drawings representing a large variety of cases of the capillary surface of revolution, which have been regularly shown in my Lectures to the Natural Philosophy Class of the University of Glasgow. These curves were recently published in the *Proc. Roy. Instit.* (Lecture of Jan. 29, 1886), and *Nature*, July 22 and 29, and Aug. 19, 1886; also to appear in a volume of Lectures now in the press, to be published in the *Nature* series.

Numerical Solution of $\dfrac{d^2u}{dx^2} + x^{-4}u^{2\cdot5} = 0$.

Distance from centre $= r = 1/x$	Reciprocal of distance from centre $= x = 1/r$	Temperature $= u$	Density $= u^{2\cdot5}$	Mass within distance r from the centre $= du/dx$ $= \displaystyle\int_x^\infty dx\, u^{2\cdot5}x^{-4}$
0	∞	16·00	1024	·00
·100	10	14·46	795·2	·28
·111	9	14·14	751·6	·38
·125	8	13·71	695·8	·52
·143	7	13·10	621·2	·731
·167	6	12·20	520·0	1·056
·200	5	10·92	394·1	1·566
·250	4	9·00	243·0	2·336
·333	3	6·15	93·81	3·436
·500	2	2·25	7·595	4·366
·667	1·5	0	0	4·49

The deduction from these numbers, of results expressing in terms of convenient units the temperature and density at any point of a given mass of a known kind of gas, occupying a sphere of given radius, must be reserved for a subsequent communication*.

One interesting result which I can give at present, derived from the first and last numbers of the several columns of the preceding table, is, that the central density of a globular gaseous star is 22½ times its average density.

* [See no. 160 *infra*.]

151. POLAR ICE-CAPS AND THEIR INFLUENCE IN CHANGING SEA-LEVELS.

[From *Glasg. Geol. Soc.* Vol. VIII. [Feb. 16, 1888], pp. 322—340. Reprinted in *Popular Lectures and Addresses*, Vol. II. pp. 319—359.]

152. Comment le Soleil a Commencé à Brûler.

[From *L'Astronomie*, Vol. XI. Oct. 1892, pp. 361—367; abstracted in *Nature*, Vol. XLVI. Oct. 20, 1892, p. 597. Incorporated in *Popular Lectures and Addresses*, Vol. I. pp. 369—422.]

La question de savoir si le Soleil va actuellement en se refroidissant ou en s'échauffant est excessivement compliquée et, en fait, la poser ou y répondre est un paradoxe à moins que nous ne définissions exactement où la température doit être comptée. Si nous demandons comment la température de régions d'égale densité du globe solaire varie d'âge en âge, la réponse est certainement que la matière du Soleil dont la densité a une valeur determinée, soit par exemple la densité ordinaire de notre atmosphère, va constamment en se refroidissant, quelle que soit sa place dans le fluide, et quelle que soit la loi de compression de ce dernier. Mais la distance à l'intérieur du globe à laquelle on trouve une densité constante diminue avec la contraction du globe, et il peut se faire qu'à des profondeurs égales au-dessous de la surface la chaleur devienne de plus en plus élevée. Tel serait certainement le cas, si la loi de la condensation gazeuse se continuait partout; mais alors, la température rayonnante effective en vertu de laquelle le Soleil répand sa chaleur au dehors peut s'abaisser parce que les températures de portions de même densité vont en diminuant dans toutes les circonstances.

Laissons cette question difficile et compliquée aux investigateurs spéciaux de la physique solaire, et posons simplement le problème sous ce titre : Quelle est la température du centre du Soleil ? Augmente-t-elle ou diminue-t-elle ?

Si nous remontons dans le passé à quelques millions d'années en arrière, époque à laquelle le Soleil était entièrement gazeux jusqu'à son centre, alors, certainement la température centrale allait en augmentant. D'autre part, comme il est possible quoique non probable pour le temps présent (mais ce qui arrivera sans doute dans l'avenir), s'il y a là un noyau solide, dans ce cas la température centrale va aussi en augmentant, parce que la conductibilité de la chaleur au dehors de ce noyau solide est trop lente pour compenser l'augmentation de pression due à l'accroissement de gravité dans le fluide se condensant autour du centre solide. Mais à une certaine époque dans l'histoire d'un globe entièrement fluide et primitivement gazeux, se contractant sous l'influence de sa propre gravitation, et du rayonnement de sa chaleur, dans l'espace extérieur glacé, lorsque les régions centrales sont devenues assez denses pour ne plus subir de condensation plus grande que celle qui s'effectue suivant la loi gazeuse des simples proportions, il me semble certain que l'accroissement de chaleur doit cesser, et que la température centrale doit commencer à diminuer en raison du refroidissement par le rayonnement de la surface et du mélange de ce fluide refroidi avec celui des régions intérieures.

Nous arrivons maintenant à l'aspect le plus intéressant de notre sujet : l'histoire ancienne du Soleil.

Il y a cinq à dix millions d'années, le globe solaire a dû avoir un diamètre double de celui qu'il possède actuellement et une densité égale au huitième de sa densité actuelle, c'est-à-dire à 0,174 de celle de l'eau ; mais nous ne pouvons pas, avec quelques probabilités d'argument ou de spéculation, remonter beaucoup plus haut dans la physique solaire.

Une question s'impose pourtant : quel était l'état de la matière constitutive du Soleil avant qu'elle fût réunie en une seule masse et devînt chaude ? Nous pouvons imaginer dans l'espace deux masses solides froides s'attirant mutuellement, et tombant l'une sur l'autre en vertu de cette attraction. Ou bien, mais ceci est beaucoup moins probable, il peut avoir existé deux masses se heurtant avec des vitesses considérablement plus grandes que celles qui auraient été dues à leur attraction mutuelle.

Cette dernière hypothèse implique que, si nous appelons les deux corps A et B, le mouvement du centre d'inertie de B relativement à A doit, lorsque la distance entre eux était grande, avoir été dirigé justement vers le centre d'inertie de A. Cette hypothèse est fort improbable. D'un autre côté, il est certain que les deux corps A et B en repos dans l'espace, abandonnés à eux-mêmes sans perturbations dues à des corps extérieurs et uniquement influencés par leur gravitation mutuelle, se heurteraient directement sans mouvement de leur centre d'inertie et sans rotation consécutive après la collision.

La probabilité d'une rencontre entre deux astres voisins, appartenant à un grand nombre de corps s'attirant mutuellement et disséminés dans l'espace, est beaucoup plus grande si ceux-ci sont en repos que s'ils se meuvent dans quelque direction que ce soit, et avec des vitesses beaucoup plus grandes que celles qu'ils acquerraient en tombant du repos les uns sur les autres.

L'Astronomie stellaire, si splendidement aidée par le spectroscope, nous montre, du reste, que les mouvements propres des étoiles et de notre Soleil sont généralement très petits comparativement à la vitesse (612 kilomètres par seconde) qu'une masse acquerrait en tombant sur le Soleil, et sont comparables au mouvement très modéré de la Terre sur son orbite (29 kilomètres et demi par seconde).

Pour fixer les idées, imaginons deux globes solides et froids, chacun d'un diamètre égal à la moitié de celui du Soleil et d'une densité moyenne égale à celle de la Terre, ces globes étant en repos ou à peu près et éloignés l'un de l'autre de deux fois la distance de la Terre au Soleil. Ils tomberont l'un sur l'autre et se heurteront juste après une demi-année de chute. La collision durera une demi-heure, dans le cours de laquelle ces deux corps seront transformés en une masse fluide, incandescente, violemment agitée, flottant en dehors de la ligne de mouvement avant le choc, et agrandie à des dimensions plusieurs fois supérieures à la somme des volumes primitifs des deux globes.

De combien cette masse fluide s'écartera-t-elle autour de la surface de collision ? Il est impossible de le dire. Le mouvement est trop compliqué pour être analysé par aucune méthode mathématique connue. Mais un calculateur armé d'une patience suffisante pourrait le déterminer avec une certaine approximation. La distance atteinte par la frange circulaire externe de la masse fluide serait probablement beaucoup moindre que la distance parcourue par chaque globe avant le choc, parce que la vitesse de translation des molécules, constituant la chaleur en laquelle l'énergie totale de la chute originaire des globes aurait été transformée, prendrait sans doute les trois cinquièmes de la quantité totale de cette énergie. La durée de la dispersion serait moindre qu'une demi-année, et alors le fluide commencerait à retomber de nouveau vers la surface. Un an environ après le premier choc, la masse fluide serait de nouveau dans un état de pression maximum autour du centre, et cette fois peut-être plus violemment agitée encore que la première, et de nouveau elle se disperserait encore, mais cette fois dans le sens de l'axe, et vers les positions d'où les deux globes seraient tombés. Une nouvelle chute vers le centre succéderait encore aux précédentes, et après une série d'oscillations de plus en plus rapides, cette masse finirait probablement, après deux ou trois ans, par former un soleil sphérique ayant environ la même masse, la même chaleur et le même éclat que notre Soleil actuel, mais sans mouvement de rotation.

Nous avons supposé les deux globes en repos au moment où ils tombaient l'un sur l'autre, en vertu de leur attraction mutuelle, d'une distance égale au diamètre de l'orbite terrestre. Admettons maintenant qu'au lieu d'avoir été en repos, ils aient été mus transversalement en direction contraire, avec une vitesse relative de 2 mètres par seconde, ou plus exactement, de $1^m,89$. Le moment de rotation de ces mouvements autour d'un axe passant par le centre de gravité des deux globes perpendiculaires à leur ligne de mouvement est juste égal au moment de la rotation du Soleil autour de son axe. La vitesse transversale dont nous parlons est si petite qu'aucun des résultats établis précédemment sur la grandeur, la chaleur, et l'éclat du Soleil ainsi créé ne serait sensiblement altéré; seulement, au lieu d'être sans rotation, il tournerait maintenant sur lui-même en vingt-cinq jours, et serait par conséquent à tous les points de vue identique à notre Soleil actuel.

Si, au lieu d'être primitivement en repos ou de se mouvoir avec les faibles vitesses transversales dont nous avons parlé, chaque globe avait une vitesse transversale de 710 mètres par seconde, les deux globes éviteraient le choc et tourneraient en ellipses autour de leur centre commun de gravité en une période d'un an, s'effleurant juste à la surface l'un de l'autre chaque fois qu'ils passeraient au point le plus proche de leurs orbites.

Si la vitesse initiale transversale de chaque globe était légèrement inférieure à 710 mètres par seconde, il y aurait une violente collision à l'effleurement et deux soleils brillants, globes solides baignés dans une atmosphère de feu, prendraient naissance en quelques heures, et commenceraient à tourner autour de leur centre commun de gravité, suivant de longues orbites elliptiques en une période d'un peu moins d'un an; l'action réciproque de leurs marées diminuerait les excentricités de leurs orbites et finirait par faire tourner les deux corps circulairement, à une distance de 6,44 diamètres de chacun d'eux, d'une surface à l'autre.

Imaginons maintenant, en choisissant encore un cas particulier pour fixer nos idées, que 29 millions de globes solides et froids, chacun ayant environ la même masse que la Lune, et dont la masse totale équivaudrait à celle du Soleil, soient disséminés aussi uniformément que possible sur une surface sphérique d'un rayon égal à cent fois le rayon de l'orbite terrestre, et que tous ces globes soient absolument en repos. Ils se mettront tous à tomber vers le centre de la sphère, qu'ils atteindront au bout de 250 ans. Chacun de ces 29 millions de globes sera alors, dans l'espace d'une demi-heure, fondu et élevé à la température de plusieurs centaines de mille ou de millions de degrés centigrades. La masse fluide ainsi engendrée par cette prodigieuse chaleur s'épandra extérieurement tout autour du centre à l'état de gaz ou de vapeur, mais elle n'atteindra pas les dimensions de la sphère primitive que nous avons imaginée.

Après une série d'expansions et de condensations consécutives, le globe incandescent finira, au bout de trois ou quatre cents ans, par former une nébuleuse gazeuse mesurant quarante fois le rayon de l'orbite terrestre. La densité moyenne de cette nébuleuse serait $1/(215 \times 40)^3$, ou le six cent trente-six mille millionième, de la densité moyenne du Soleil, ou le quatre cent cinquante-quatre mille millionième de la densité de l'eau, ou le cinq cent soixante-dix millionième de la densité de l'air, à la température de 10° et au niveau de la mer.

Dans les régions centrales de cette immense nébuleuse la densité, sensiblement uniforme à travers plusieurs millions de kilomètres, serait un vingt mille millionième de celle de l'eau, ou un vingt-cinq millionième de celle de l'air. Si c'était de l'oxygène ou de l'azote, ou quelque gaz simple ou composé, de densité spécifique égale à celle de notre atmosphère, la température centrale serait de 51200 degrés centigrades, et la vitesse moyenne des molécules serait de 6700 mètres par seconde.

La nébuleuse gazeuse ainsi constituée serait, dans le cours de quelques millions d'années, par le rayonnement extérieur de sa chaleur, condensée au volume actuel du Soleil, avec la même chaleur et la même lumière, mais sans mouvement de rotation.

Le moment de rotation du système solaire tout entier est environ dix-huit fois celui de la rotation du Soleil, et, dans cette quantité, Jupiter représente les $\frac{17}{18}$ et le Soleil $\frac{1}{18}$; les autres corps du système solaire peuvent être négligés.

Maintenant, au lieu d'être absolument en repos à l'origine, admettons que nos 29 millions de lunes soient animées chacune d'un léger mouvement représentant en tout une quantité de moment de rotation autour d'un certain axe, égal au moment de rotation du système solaire, ou bien beaucoup plus considérable, pour tenir compte du milieu résistant, toutes ces lunes tomberont aussi vers le centre, en 250 ans, et quoiqu'elles ne se rencontrent pas juste en ce point comme dans l'hypothèse précédente, elles se rencontreront néanmoins et subiront les unes par les autres des myriades de chocs, de sorte que chacun de ces 29 millions de globes sera fondu et réduit en vapeur par la chaleur résultant de ces chocs. Il s'en suivra un mouvement de rotation; la nébuleuse sera aplatie et son rayon équatorial s'étendra fort au delà de l'orbite de Neptune, avec un moment de rotation égal ou supérieur à celui du système solaire.

Telle est précisément l'origine demandée par Laplace pour sa théorie nébulaire de l'évolution du système solaire, théorie qui, fondée sur l'histoire de l'Univers sidéral tel qu'Herschel l'avait observé, et complétée dans ses détails par le profond jugement dynamique et le génie imaginatif de Laplace, paraît aujourd'hui une vérité démontrée par la Thermodynamique.

Ainsi, il semble ne plus rester en réalité de grand mystère ni de grande difficulté dans l'évolution automatique du système solaire, venu de matériaux froids répandus à travers l'espace, jusqu'à son ordre et sa beauté actuels, illuminé et échauffé par un brillant soleil central, pas plus qu'il n'y en a dans la marche d'une horloge jusqu'à ce qu'elle s'arrête. Il serait superflu d'ajouter que l'origine et l'entretien de la vie sur la Terre est absolument et infiniment au delà des limites des spéculations sûres de la science dynamique. La seule contribution de la Dynamique à la biologie théorique est la négation absolue d'un commencement automatique ou d'un entretien automatique de la vie.

Je ne tirerai de ce qui précède qu'une conclusion. Supposons que la masse du Soleil soit composée de matériaux qui étaient primitivement très éloignés les uns des autres et froids, l'antécédent immédiat de l'incandescence solaire doit être cherché, soit dans deux corps différents seulement dans les détails des cas que nous avons considérés comme exemples, ou dans un plus grand nombre qui, malgré sa grandeur, peut cependant être défini, attendu que le nombre des atomes qui constituent la masse actuelle du Soleil n'est pas infini, et doit être compris entre 4×10^{57} et 140×10^{57}. L'antécédent immédiat à l'incandescence peut avoir été une subdivision extrême, telle que des atomes séparés, ou bien quelque chose de plus grand, comme des groupes d'atomes ou de petits cristaux, ce qu'on pourrait appeler des flocons de matière; ou bien on peut imaginer des morceaux plus gros, tels que des pierres ordinaires que l'on peut lancer à la main, ou bien encore des uranolithes plus considérables, tels que ceux que nous possédons dans nos collections et qui sont véritablement tombés du ciel sur la Terre.

Pour la théorie du Soleil, il est indifférent d'adopter l'un ou l'autre de ces états de la matière, depuis l'atome ou la poussière jusqu'aux plus grandes masses. Mais ici se pose une question plus primordiale encore. Pouvons-nous imaginer que ces uranolithes aient été tels dès l'origine des choses? Nous pouvons penser que le Soleil est le résultat de l'agglomération de pierres météoriques; mais nous ne pouvons guère remonter plus haut.

Sûrement ces pierres tombées du ciel ont une histoire pleine d'événements, mais il est difficile de la conjecturer; ce sont les fragments brisés de masses plus considérables, mais quelles étaient ces masses, nous ne le savons pas.

153. SOLAR RELATIONS OF MAGNETIC STORMS: CHANGES OF LATITUDE.

[*Presidential Address*, from *Roy. Soc. Proc.* Vol. LII. [Nov. 30, 1892], 1893, pp. 300—310; *Nature*, Vol. XLVII. Dec. 1, 1892, pp. 107—111. Reprinted in *Popular Lectures and Addresses*, Vol. II. pp. 508—529.]

154. On the Temperature Variation of the Thermal Conductivity of Rocks. By Lord Kelvin and J. R. Erskine Murray, B.Sc.

[From *Glas. Phil. Soc. Proc.* Vol. xxvi. [March 27, 1895], pp. 227—232; *Roy. Soc. Proc.* Vol. lviii. [May 30, 1895], pp. 162—167; *Nature*, Vol. lii. May 16, p. 70, June 20, 1895, pp. 182—184; *Amer. Journ. Sci.* Vol. l. (3rd Ser.), 1895, pp. 419—423.]

1. The experiments described in this communication were undertaken for the purpose of finding temperature variation of thermal conductivity of some of the more important rocks of the earth's crust.

2. The method which we adopted was to measure, by aid of thermoelectric junctions, the temperatures at different points of a flux line in a solid, kept unequally heated by sources (positive and negative) applied to its surface, and maintained uniform for a sufficiently long time to cause the temperature to be as nearly constant at every point as we could arrange for. The shape of the solid and the thermal sources were arranged to cause the flux lines to be, as nearly as possible, parallel straight lines; so that, according to Fourier's elementary theory and definition of thermal conductivity, we should have

$$\frac{k(M, B)}{k(T, M)} = \frac{[v(M) - v(T)] \div MT}{[v(B) - v(M)] \div BM},$$

where T, M, B denote three points in a stream line (respectively next to the top, at the middle, and next to the bottom in the slabs and columns which we used); $v(T)$, $v(M)$, $v(B)$ denote the steady temperatures at these points; and $k(T, M)$, $k(M, B)$, the mean conductivities between T and M, and between M and B respectively.

3. The rock experimented on in each case consisted of two equal and similar rectangular pieces, pressed with similar faces together. In one of these faces three straight parallel grooves are

cut, just deep enough to allow the thermoelectric wires and junctions to be embedded in them, and no wider than to admit the wires and junctions (see diagram, § 8 below). Thus, when the two pieces of rock are pressed together, and when heat is so applied that the flux lines are parallel to the faces of the two parts, we had the same result, so far as thermal conduction is concerned, as if we had taken a single slab of the same size as the two together, with long fine perforations to receive the electric junctions. The compound slab was placed with the perforations horizontal, and their plane vertical. Its lower side, when thus placed, was immersed under a bath of tin, kept melted by a lamp below it. Its upper side was flooded over with mercury in our later experiments (§§ 6, 7, 8), as in Hopkins' experiments on the thermal conductivity of rock. Heat was carried off from the mercury by a measured quantity of cold water poured upon it once a minute, allowed to remain till the end of a minute, and then drawn off and immediately replaced by another equal quantity of cold water. The chief difficulty in respect to steadiness of temperature was the keeping of the gas lamp below the bath of melted tin uniform. If more experiments are to be made on the same plan, whether for rocks or metals, or other solids, it will, no doubt, be advisable to use an automatically regulated gas flame, keeping the temperature of the hot bath in which the lower face of the slab or column is immersed at as nearly constant a temperature as possible, and to arrange for a perfectly steady flow of cold water to carry away heat from the upper surface of the mercury resting on the upper side of the slab or column. It will also be advisable to avoid the complication of having the slab or column in two parts, when the material and the dimensions of the solid allow fine perforations to be bored through it, instead of the grooves which we found more readily made with the appliances available to us.

4. Our first experiments were made on a slate slab, 25 cm. square and 5 cm. thick, in two halves, pressed together, each 25 cm. by 12·5, and 5 cm. thick. One of these parts cracked with a loud noise in an early experiment, with the lower face of the composite square resting on an iron plate heated by a powerful gas burner, and the upper face kept cool by ice in a metal vessel resting upon it. The experiment indicated, very decidedly, less conductivity in the hotter part below the middle than in the cooler part above the

middle of the composite square slab. We supposed this might possibly be due to the crack, which we found to be horizontal and below the middle, and to be complete across the whole area of 12½ cm. by 5, across which the heat was conducted in that part of the composite slab; and to give rise to palpably imperfect fitting together of the solid above and below it. We therefore repeated the experiment with the composite slab turned upside down, so as to bring the crack in one half of it now to be above the middle, instead of below the middle, as at first. We still found for the composite slab less conductivity in the hot part below the middle than in the cool part above the middle. We inferred that, in respect to thermal conduction through slate across the natural cleavage planes, the thermal conductivity diminishes with increase of temperature.

5. We next tried a composite square slab of sandstone of the same dimensions as the slate, and we found for it also decisive proof of diminution of thermal conductivity with increase of temperature. We were not troubled by any cracking of the sandstone, with its upper side kept cool by an ice-cold metal plate resting on it, and its lower side heated to probably as much as 300° or 400° C.

6. After that we made a composite piece, of two small slate columns, each 3·5 cm. square and 6·2 cm. high, with natural cleavage planes vertical, pressed together with thermoelectric junctions as before; but with appliances (§ 10 below) for preventing loss or gain of heat across the vertical sides, which the smaller horizontal dimensions (7 cm., 3·5 cm.) might require, but which were manifestly unnecessary with the larger horizontal dimensions (25 cm., 25 cm.) of the slabs of slate and sandstone used in our former experiments. The thermal flux lines in the former experiments on slate were perpendicular to the natural cleavage planes, but now, with the thermal flux lines parallel to the cleavage planes, we still find the same result, smaller thermal conductivity at the higher temperatures. Numerical results will be stated in § 12 below.

7. Our last experiments were made on a composite piece of Aberdeen granite, made up of two columns, each 6 cm. high and 7·6 cm. square, pressed together, with appliances similar to those described in § 6; and, as in all our previous experiments on slate

and sandstone, we found less thermal conductivity at higher temperatures. The numerical results will be given in § 12 below.

8. The accompanying diagram represents the thermal appliances and thermoelectric arrangement of §§ 6, 7. The columns of slate or granite were placed on supports in a bath of melted tin with about 0·2 cm. of their lower ends immersed. The top of each column was kept cool by mercury, and water changed once a minute, as described in § 3 above, contained in a tank having the top of the stone column for its bottom and completed by four vertical metal walls fitted into grooves in the stone and made tight against wet mercury by marine glue.

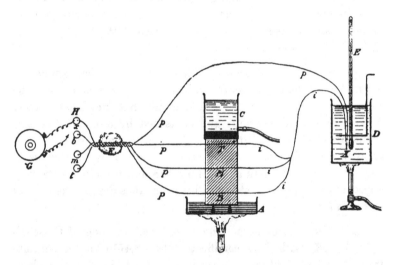

Iron wires are marked *i*.   Platinoid wires are marked *p*.

B, T, M. Thermoelectric junctions in slab.

X.   ,,   ,,   oil bath.

A. Bath of molten tin.   *C*. Tank of cold water.

D. Oil bath.   *E*. Thermometer.

F. Junctions of platinoid and copper wires. The wires are insulated from one another, and wrapt all together in cotton wool at this part, to secure equality of temperature between these four junctions, in order that the current through the galvanometer shall depend solely on differences of temperature between whatever two of the four junctions, X, T, M, B, is put in circuit with the galvanometer.

G. Galvanometer.

H. Four mercury cups, for convenience in connecting the galvonometer to any pair of thermoelectric junctions.

x, *b*, *m*, *t* are connected, through copper and platinoid, with X, B, M, T, respectively.

9. The temperatures, $v(B)$, $v(M)$, $v(T)$, of B, M, T, the hot, intermediate, and cool points in the stone, were determined by equalising to them successively the temperature of the mercury thermometer placed in the oil-tank, by aid of thermoelectric circuits and a galvanometer used to test equality of temperature by nullity of current through its coil when placed in the proper circuit, all as shown in the diagram. The steadiness of temperature in the stone was tested by keeping the temperature of the thermometer constant, and observing the galvanometer reading for current when the junction in the oil-tank and one or other of the three junctions in the stone were placed in circuit. We also helped ourselves to attaining constancy of temperature in the stone by observing the current through the galvanometer, due to differences of temperature between any two of the three junctions B, M, T placed in circuit with it.

10. We made many experiments to test what appliances might be necessary to secure against gain or loss of heat by the stone across its vertical faces, and found that *kieselguhr*, loosely packed round the columns and contained by a metal case surrounding them at a distance of 2 cm. or 3 cm., prevented any appreciable disturbance due to this cause. This allowed us to feel sure that the thermal flux lines through the stone were very approximately parallel straight lines on all sides of the central line BMT.

11. The thermometer which we used was one of Cassella's (No. 64,168) with Kew certificate (No. 48,471) for temperature from 0° to 100°, and for equality in volume of the divisions above 100°. We standardised it by comparison with the constant volume

Reading		Correction to be subtracted from reading of mercury thermometer
Air thermometer	Mercury thermometer	
0°	1·9°	1·9
120·2	122·2	2·0
166·8	168·6	1·8
211·1	212·7	1·6
265·7	267·5	1·8

air thermometer* of Dr Bottomley with the above result. This is satisfactory as showing that when the zero error is corrected the greatest error of the mercury thermometer, which is at 211° C., is only 0·3°.

12. Each experiment on the slate and granite columns lasted about two hours from the first application of heat and cold; and we generally found that after the first hour we could keep the temperatures of the three junctions very nearly constant. Choosing a time of best constancy in our experiments on each of the two substances, slate and granite, we found the following results:—

Slate: flux lines parallel to cleavage

$$v(T) = 50°·2 \text{ C.} \quad v(M) = 123°·3 \text{ C.} \quad v(B) = 202°·3 \text{ C.}$$

The distances between the junctions were $BM = 2·57$ cm. and $MT = 2·6$ cm. Hence by the formula of § 2,

$$\frac{k(M, B)}{k(T, M)} = \frac{73·1 \div 2·6}{79·0 \div 2·57} = \frac{28·1}{30·7} = 0·91.$$

Aberdeen granite:

$$v(T) = 81°·1 \text{ C.} \quad v(M) = 145°·6 \text{ C.} \quad v(B) = 214°·6 \text{ C.}$$

The distances between the junctions were $BM = 1·9$ cm. and $MT = 2·0$ cm.

$$\frac{k(MB)}{k(TM)} = \frac{64·5 \div 2·0}{69·0 \div 1·9} = \frac{32·2}{36·3} = 0·88.$$

13. Thus we see, that for slate, with lines of flux parallel to cleavage planes, the mean conductivity in the range from 123° C. to 202° C. is 91 per cent. of the mean conductivity in the range from 50° C. to 123° C., and for granite, the mean conductivity in the range from 145° C. to 214° C. is 88 per cent. of the mean conductivity in the range from 81° C. to 145° C. The general plan of apparatus, described above, which we have used only for comparing the conductivities at different temperatures, will, we believe, be found readily applicable to the determination of conductivities in absolute measure.

---

* *Phil. Mag.*, August, 1888, and *Edinb. Roy. Soc. Proc.*, January 6, 1888.

155. On the Fuel-Supply and the Air-Supply of the Earth.

[From *British Association Report*, 1897, pp. 553, 554.]

ALL known fuel on the earth is probably residue of ancient vegetation. One ton average fuel takes three tons oxygen to burn it, and therefore its vegetable origin, decomposing carbonic acid and water by power of sunlight, gave three tons oxygen to our atmosphere. Every square metre of earth's surface bears ten tons of air, of which two tons is oxygen. The whole surface is 126 thousand millions of acres, or 510 million millions of square metres. Hence there is not more than 340 million million tons of fuel on the earth, and this is probably the exact amount, because probably all the oxygen in our atmosphere came from primeval vegetation.

The surely available coal supply of England and Scotland was estimated by the Coal Supply Commission of 1871, which included Sir Roderick Murchison and Sir Andrew Ramsay among its members, as being 146 thousand million tons. This is approximately six-tenths of a ton per square metre of area of Great Britain. To burn it all would take one and eight-tenths of a ton of oxygen, or within two-tenths of a ton of the whole oxygen of the atmosphere resting on Great Britain. The Commission estimated fifty-six thousand million tons more of coal as probably existing at present in lower and less easily accessible strata. It may therefore be considered as almost quite certain that Great Britain could not burn all its own coal with its own air, and therefore that the coal of Britain is considerably in excess of fuel supply of rest of world reckoned per equal areas, whether of land or sea.

156. The Age of the Earth as an Abode fitted for Life*.

[From *Journ. Victoria Institute, London,* Vol. XXXI. 1899, pp. 11—35, *Presidential Address* (1897) ; *Ann. Rep. Smithsn. Inst.* 1897, 1898, pp. 337—357 ; *Phil. Mag.* Vol. XLVII. Jan. 1899, pp. 66—90 [with additions †].]

1.   The age of the earth as an abode fitted for life is certainly a subject which largely interests mankind in general.   For geology it is of vital and fundamental importance—as important as the date of the battle of Hastings is for English history—yet it was very little thought of by geologists of thirty or forty years ago ; how little is illustrated by a statement‡, which I will now read, given originally from the presidential chair of the Geological Society by Professor Huxley in 1869, when for a second time, after a seven years' interval, he was president of the Society.

"I do not suppose that at the present day any geologist would be found …to deny that the rapidity of the rotation of the earth *may* be diminishing, that the sun *may* be waxing dim, or that the earth itself *may* be cooling. Most of us, I suspect, are Gallios, 'who care for none of these things,' being of opinion that, true or fictitious, they have made no practical difference to the earth, during the period of which a record is preserved in stratified deposits."

2.   I believe the explanation of how it was possible for Professor Huxley to say that he and other geologists did not care for things on which the age of life on the earth essentially depends, is because he did not know that there was valid foundation for any estimates worth considering as to absolute magnitudes. If science did not allow us to give any estimate whatever as to whether 10,000,000 or 10,000,000,000 years is the age of this earth

---

* [Lord Kelvin inserted in his press-copy of this paper the account of " The Lava Lake of Kilauea," signed S. E. Bishop, which appeared in *Nature*, September 4, 1902.]

† Communicated by the Author, being the 1897 Annual Address of the Victoria Institute with additions written at different times from June 1897 to May 1898.

‡ In the printed quotations the italics are mine in every case, not so the capitals in the quotation from Page's *Text-Book*.

as an abode fitted for life, then I think Professor Huxley would
have been perfectly right in saying that geologists should not
trouble themselves about it, and biologists should go on in their
own way, not enquiring into things utterly beyond the power of
human understanding and scientific investigation. This would
have left geology much in the same position as that in which
English history would be if it were impossible to ascertain whether
the battle of Hastings took place 800 years ago, or 800 thousand
years ago, or 800 million years ago. If it were absolutely impossible to find out which of these periods is more probable than
the other, then I agree we might be Gallios as to the date of the
Norman Conquest. But a change took place just about the time
to which I refer, and from then till now geologists have not considered the question of absolute dates in their science as outside
the scope of their investigations.

3. I may be allowed to read a few extracts to indicate how
geological thought was expressed in respect to this subject, in
various largely used popular text-books, and in scientific writings
which were new in 1868, or not so old as to be forgotten. I have
several short extracts to read and I hope you will not find them
tedious.

The first is three lines from Darwin's *Origin of Species*, 1859
edition, p. 287 :—

"In all probability a far longer period than 300,000,000 years has elapsed
since the latter part of the secondary period."

Here is another still more important sentence, which I read to
you from the same book :—

"He who can read Sir Charles Lyell's grand work on the Principles of
Geology, which the future historian will recognize as having produced a
revolution in natural science, yet does not admit how *incomprehensibly vast*
have been the past periods of time, *may at once close this volume.*"

I shall next read a short statement from Page's *Advanced
Students' Text-Book of Geology*, published in 1859 :—

"Again where the FORCE seems unequal to the result, the student
should never lose sight of the element TIME : *an element to which we can
set no bounds in the past,* any more than we know of its limit in the future."

"It will be seen from this hasty indication that there are two great
schools of geological causation—the one ascribing every result to the ordinary
operations of Nature, combined with the element of *unlimited time,* the other

appealing to agents that operated during the earlier epochs of the world with greater intensity, and also for the most part over wider areas. *The former belief is certainly more in accordance with the spirit of right philosophy*, though it must be confessed that many problems in geology seem to find their solution only through the admission of the latter hypothesis."

4. I have several other statements which I think you may hear with some interest. Dr Samuel Haughton, of Trinity College, Dublin, in his *Manual of Geology*, published·in 1865, p. 82, says :—

" The infinite time of the geologists is in the past; and *most of their speculations regarding this subject seem to imply the absolute infinity of time*, as if the human imagination was unable to grasp the period of time requisite for the formation of a few inches of sand or feet of mud, and its subsequent consolidation into rock." (This delicate satire is certainly not overstrained.)

" Professor Thomson has made an attempt to calculate the length of time during which the sun can have gone on burning at the present rate, and has come to the following conclusion :—'It seems, on the whole, most probable that the sun has not illuminated the earth for 100,000,000 years, and almost certain that he has not done so for 500,000,000 years. As for the future, we may say with equal certainty, that the inhabitants of the earth cannot continue to enjoy the light and heat essential to their life for many million years longer, unless new sources, now unknown to us, are prepared in the great storehouse of creation.' "

I said that in the sixties and I repeat it now; but with charming logic it is held to be inconsistent with a later statement that the sun has not been shining 60,000,000 years; and that both that and this are stultified by a still closer estimate which says that probably the sun has not been shining for 30,000,000 years! And so my efforts to find some limit or estimate for Geological Time have been referred to and put before the public, even in London daily and weekly papers, to show how exceedingly wild are the wanderings of physicists, and how mutually contradictory are their conclusions, as to the length of time which has actually passed since the early geological epochs to the present date.

Dr Haughton further goes on :—

" *This result* (100 to 500 million years) of Professor Thomson's, *although very liberal in the allowance of time, has offended geologists, because, having been accustomed to deal with time as an infinite quantity at their disposal, they feel naturally embarrassment and alarm at any attempt of the science of Physics to place a limit upon their speculations.* It is quite possible that even a hundred million of years may be greatly in excess of the actual time during which the sun's heat has remained constant."

5.   Dr Haughton admitted so much with a candid open mind; but he went on to express his own belief (in 1865) thus:—

"Although I have spoken somewhat disrespectfully of the geological calculus in my lecture, yet I believe that the time during which organic life has existed on the earth is practically infinite, because it can be shown to be so great as to be inconceivable by beings of our limited intelligence."

Where is inconceivableness in 10,000,000,000?   There is nothing inconceivable in the number of persons in this room, or in London.   We get up to millions quickly.   Is there anything inconceivable in 30,000,000 as the population of England, or in 38,000,000 as the population of Great Britain and Ireland, or in 352,704,863 as the population of the British Empire?   Not at all. It is just as conceivable as half a million years or 500 millions.

6.   The following statement is from Professor Jukes's *Students' Manual of Geology*:—

"The time required for such a slow process to effect such enormous results must of course be taken to be inconceivably great.   The word 'inconceivably' is not here used in a vague but in a literal sense, to indicate that the lapse of time required for the denudation that has produced the present surfaces of some of the older rocks, is vast beyond any idea of time which the human mind is capable of conceiving."

"Mr Darwin, in his admirably reasoned book on the origin of species, so full of information and suggestion on all geological subjects, estimates the time required for the denudation of the rocks of the Weald of Kent, or the erosion of space between the ranges of chalk hills, known as the North and South Downs, at *three hundred millions of years*.   The grounds for forming this estimate are of course of the vaguest description.   It may be possible, perhaps, that the estimate is a hundred times too great, and that the real time elapsed did not exceed three million years, but, on the other hand, it is just as likely that the time which actually elapsed since the first commencement of the erosion till it was nearly as complete as it now is, was really a hundred times greater than his estimate, or thirty thousand millions of years."

7.   Thus Jukes allowed estimates of anything from 3 millions to 30,000 millions as the time which actually passed during the denudation of the Weald.   On the other hand Professor Phillips in his Rede lecture to the University of Cambridge (1860), decidedly prefers one inch per annum to Darwin's one inch per century as the rate of erosion; and says that most observers would consider even the one inch per annum too small for all but the most invincible coasts !   He thus, on purely geological grounds, reduces Darwin's estimate of the time to less than one-hundredth.

And, reckoning the actual thicknesses of all the known geological strata of the earth, he finds 96 million years as a possible estimate for the antiquity of the base of the stratified rocks; but he gives reasons for supposing that this may be an over-estimate, and he finds that from stratigraphical evidence alone, we may regard the antiquity of life on the earth as possibly between 38 millions and 96 millions of years. Quite lately a very careful estimate of the antiquity of strata containing remains of life on the earth has been given by Professor Sollas, of Oxford, calculated according to stratigraphical principles which had been pointed out by Mr Alfred Wallace. Here it is*:—"So far as I can at present see, the lapse of time since the beginning of the Cambrian system is probably less than 17,000,000 years, even when computed on an assumption of uniformity, which to me seems contradicted by the most salient facts of geology. Whatever additional time the calculations made on physical data can afford us, may go to the account of pre-Cambrian deposits, of which at present we know too little to serve for an independent estimate."

8. In one of the evening Conversaziones of the British Association during its meeting at Dundee in 1867 I had a conversation on geological time with the late Sir Andrew Ramsay, almost every word of which remains stamped on my mind to this day. We had been hearing a brilliant and suggestive lecture by Professor (now Sir Archibald) Geikie on the geological history of the actions by which the existing scenery of Scotland was pro-duced. I asked Ramsay how long a time he allowed for that history. He answered that he could suggest no limit to it. I said, "You don't suppose things have been going on always as they are now? You don't suppose geological history has run through 1,000,000,000 years?" "Certainly I do." "10,000,000,000 years?" "Yes." "The sun is a finite body. You can tell how many tons it is. Do you think it has been shining on for a million million years?" "I am as incapable of estimating and understanding the reasons which you physicists have for limiting geological time as you are incapable of understanding the geological reasons for our unlimited estimates." I answered, "You can understand physicists' reasoning perfectly if you give your mind to it." I ventured also to say that physicists were not wholly incapable

* "The Age of the Earth," *Nature*, April 4th, 1895.

of appreciating geological difficulties; and so the matter ended, and we had a friendly agreement to temporarily differ.

9. In fact, from about the beginning of the century till that time (1867), geologists had been nurtured in a philosophy originating with the Huttonian system: much of it substantially very good philosophy, but some of it essentially unsound and misleading: witness this, from Playfair, the eloquent and able expounder of Hutton:—

"How often these vicissitudes of decay and renovation have been repeated is not for us to determine; they constitute a series of which as the author of this theory has remarked, we neither see the beginning nor the end; a circumstance that accords well with what is known concerning other parts of the economy of the world. In the continuation of the different species of animals and vegetables that inhabit the earth, we discern neither a beginning nor an end; in the planetary motions where geometry has carried the eye so far both into the future and the past we discover no mark either of the commencement or the termination of the present order."

10. Led by Hutton and Playfair, Lyell taught the doctrine of eternity and uniformity in geology; and to explain plutonic action and underground heat, invented a thermo-electric "perpetual" motion on which, in the year 1862, in my paper on the "Secular Cooling of the Earth*," published in the *Transactions of the Royal Society of Edinburgh*, I commented as follows:—

"To suppose, as Lyell, adopting the chemical hypothesis, has done†, that the substances, combining together, may be again separated electrolytically by thermo-electric currents, due to the heat generated by their combination, and thus the chemical action and its heat continued in an endless cycle, violates the principles of natural philosophy in exactly the same manner, and to the same degree, as to believe that a clock constructed with a self-winding movement may fulfil the expectations of its ingenious inventor by going for ever."

It was only by sheer force of reason that geologists have been compelled to think otherwise, and to see that there was a definite beginning, and to look forward to a definite end, of this world as an abode fitted for life.

11. It is curious that English philosophers and writers should not have noticed how Newton treated the astronomical problem. Playfair, in what I have read to you, speaks of the planetary

* Reprinted in Thomson and Tait, *Treatise on Natural Philosophy*, 1st and 2nd Editions, Appendix D (g).

† *Principles of Geology*, chap. xxxi. ed. 1853.

system as being absolutely eternal, and unchangeable: having had no beginning and showing no signs of progress towards an end. He assumes also that the sun is to go on shining for ever, and that the earth is to go on revolving round it for ever. He quite overlooked Laplace's nebular theory; and he overlooked Newton's counterblast to the planetary "perpetual motion." Newton, commenting on his own *First Law of Motion*, says, in his terse Latin, which I will endeavour to translate, "But the greater bodies of planets and comets moving in spaces less resisting, keep their motions longer." That is a strong counterblast against any idea of eternity in the planetary system.

12. I shall now, without further preface, explain, and I hope briefly, so as not to wear out your patience, some of the arguments that I brought forward between 1862 and 1869, to show strict limitations to the possible age of the earth as an abode fitted for life.

Kant* pointed out in the middle of last century, what had not previously been discovered by mathematicians or physical astronomers, that the frictional resistance against tidal currents on the earth's surface must cause a diminution of the earth's rotational speed. This really great discovery in Natural Philosophy seems to have attracted very little attention,—indeed to have passed quite unnoticed,—among mathematicians, and astronomers, and naturalists, until about 1840, when the doctrine of energy began to be taken to heart. In 1866, Delaunay suggested that tidal retardation of the earth's rotation was probably the cause of an outstanding acceleration of the moon's mean motion reckoned according to the earth's rotation as a timekeeper found by Adams in 1853 by correcting a calculation of Laplace which had seemed to prove the earth's rotational speed to be uniform†. Adopting

---

* In an essay first published in the Kœnigsberg *Nachrichten*, 1754, Nos. 23, 24; having been written with reference to the offer of a prize by the Berlin Academy of Sciences in 1754. Here is the title-page, in full, as it appears in Vol. vi. of Kant's *Collected Works*, Leipzig, 1839:—Untersuchung der Frage: Ob die Erde in ihrer Umdrehung um die Achse, wodurch sie die Abwechselung des Tages und der Nacht hervorbringt, einige Veränderung seit den ersten Zeiten ihres Ursprunges erlitten habe, welches die Ursache davon sei, und woraus man sich ihrer versichern könne? welche von der Königlichen Akademie der Wissenschaften zu Berlin zum Preise aufgegeben worden, 1754.

† *Treatise on Natural Philosophy* (Thomson and Tait), § 830, ed. 1, 1867, and

Delaunay's suggestion as true, Adams, in conjunction with Professor Tait and myself, estimated the diminution of the earth's rotational speed to be such that the earth as a timekeeper, in the course of a century, would get 22 seconds behind a thoroughly perfect watch or clock rated to agree with it at the beginning of the century. According to this rate of retardation the earth, 7,200 million years ago, would have been rotating twice as fast as now: and the centrifugal force in the equatorial regions would have been four times as great as its present amount, which is $\frac{1}{289}$ of gravity. At present the radius of the equatorial sea-level exceeds the polar semi-diameter by $21\frac{1}{2}$ kilometres, which is, as nearly as the most careful calculations in the theory of the earth's figure can tell us, just what the excess of equatorial radius of the surface of the sea all round would be if the whole material of the earth were at present liquid and in equilibrium under the influence of gravity and centrifugal force with the present rotational speed, and $\frac{1}{4}$ of what it would be if the rotational speed were twice as great. Hence, if the rotational speed had been twice as great as its present amount when consolidation from approximately the figure of fluid equilibrium took place, and if the solid earth, remaining absolutely rigid, had been gradually slowed down in the course of millions of years to its present speed of rotation, the water would have settled into two circular oceans round the two poles: and the equator, dry all round, would be 64·5 kilometres above the level of the polar sea bottoms. This is on the supposition of absolute rigidity of the earth after primitive consolidation. There would, in reality, have been some degree of yielding to the gravitational tendency to level the great gentle slope up from each pole to equator. But if the earth, at the time of primitive consolidation, had been rotating twice as fast as at present, or even 20 per cent. faster than at present, traces of its present figure must have been left in a great preponderance of land, and probably no sea at all, in the equatorial regions. Taking into account all uncertainties, whether in respect to Adams' estimate of the rate of frictional retardation of the earth's rotatory speed, or to the conditions as to the rigidity of the earth once consolidated, we may safely conclude that the earth was certainly

later editions; also *Popular Lectures and Addresses*, Vol. II. (Kelvin), "Geological Time," being a reprint of an article communicated to the Glasgow Geological Society, February 27th, 1868.

not solid 5,000 million years ago, and was probably not solid 1,000 million years ago*.

13. A second argument for limitation of the earth's age, which was really my own first argument, is founded on the consideration of underground heat. To explain a first rough and ready estimate of it I shall read one short statement. It is from a very short paper that I communicated to the Royal Society of Edinburgh on the 18th December, 1865, entitled, "The Doctrine of Uniformity in Geology briefly refuted."

"The 'Doctrine of Uniformity' in Geology, as held by many of the most eminent of British Geologists, assumes that the earth's surface and upper crust have been nearly as they are at present in temperature, and other physical qualities, during millions of millions of years. But *the heat which we know, by observation, to be now conducted out of the earth yearly* is so great, that if this action had been going on with any approach to uniformity for 20,000 million years, the amount of heat lost out of the earth would have been about as much as would heat, by 100° C., a quantity of ordinary surface rock of 100 times the earth's bulk. This would be more than enough to melt a mass of surface rock equal in bulk to the whole earth. No hypothesis as to chemical action, internal fluidity, effects of pressure at great depth, or possible character of substances in the interior of the earth, possessing the smallest vestige of probability, can justify the supposition that the earth's upper crust has remained nearly as it is, while from the whole, or from any part, of the earth, so great a quantity of heat has been lost."

14. The sixteen words which I have emphasized in reading this statement to you (italics in the reprint) indicate the matter-of-fact foundation for the conclusion asserted. This conclusion suffices to sweep away the whole system of geological and biological speculation demanding an "inconceivably" great vista of past time, or even a few thousand million years, for the history of life on the earth, and approximate uniformity of plutonic action throughout that time; which, as we have seen, was very generally prevalent thirty years ago among British Geologists and Biologists; and which, I must say, some of our chiefs of the present day have not yet abandoned. Witness the Presidents of the Geological and Zoological Sections of the British Association at its meetings of 1893 (Nottingham), and of 1896 (Liverpool).

* " The fact that the continents are arranged along meridians, rather than in an equatorial belt, affords some degree of proof that the consolidation of the earth took place at a time when the diurnal rotation differed but little from its present value. It is probable that the date of consolidation is considerably more recent than a thousand million years ago."—Thomson and Tait, *Treatise on Natural Philosophy*, 2nd ed., 1883, § 830.

Mr Teall : Presidential Address to the Geological Section, 1893. "The good old British ship 'Uniformity,' built by Hutton and refitted by Lyell, has won so many glorious victories in the past, and appears still to be in such excellent fighting trim, that I see no reason why she should haul down her colours either to 'Catastrophe' or 'Evolution.' Instead, therefore, of acceding to the request to 'hurry up' we make a demand for more time."

Professor Poulton : Presidential Address to the Zoological Section, 1896. "Our argument does not deal with the time required for the origin of life, or for the development of the lowest beings with which we are acquainted from the first formed beings, of which we know nothing. Both these processes may have required an immensity of time; but as we know nothing whatever about them and have as yet no prospect of acquiring any information, we are compelled to confine ourselves to as much of the process of evolution as we can infer from the structure of living and fossil forms—that is, as regards animals, to the development of the simplest into the most complex Protozoa, the evolution of the Metazoa from the Protozoa, and the branching of the former into its numerous Phyla, with all their Classes, Orders, Families, Genera, and Species. But we shall find that this is quite enough to necessitate *a very large increase in the time estimated by the geologist.*"

15. In my own short paper from which I have read you a sentence, the rate at which heat is at the present time lost from the earth by conduction outwards through the upper crust, as proved by observations of underground temperature in different parts of the world, and by measurement of the thermal conductivity of surface rocks and strata, sufficed to utterly refute the Doctrine of Uniformity as taught by Hutton, Lyell, and their followers; which was the sole object of that paper.

16. In an earlier communication to the Royal Society of Edinburgh*, I had considered the cooling of the earth due to this loss of heat; and by tracing backwards the process of cooling had formed a definite estimate of the greatest and least number of million years which can possibly have passed since the surface of the earth was everywhere red hot. I expressed my conclusion in the following statement† :—

"We are very ignorant as to the effects of high temperatures in altering the conductivities and specific heats and melting temperatures of rocks, and as to their latent heat of fusion. We must, therefore, allow very wide limits in such an estimate as I have attempted to make; but I think we may with

---

* "On the Secular Cooling of the Earth," *Trans. Roy. Soc. Edinburgh*, Vol. XXIII. April 28th, 1862, reprinted in Thomson and Tait, Vol. III. pp. 468—485, and *Math. and Phys. Papers*, Art. xciv. pp. 295--311.

† "On the Secular Cooling of the Earth," *Math. and Phys. Papers*, Vol. III. § 11 of Art. xciv.

much probability say that the consolidation cannot have taken place less than 20 million years ago, or we should now have more underground heat than we actually have; nor more than 400 million years ago, or we should now have less underground heat than we actually have. That is to say, I conclude that Leibnitz's epoch of emergence of the *consistentior status* [the consolidation of the earth from red hot or white hot molten matter] was probably between those dates."

17. During the 35 years which have passed since I gave this wide-ranged estimate, experimental investigation has supplied much of the knowledge then wanting regarding the thermal properties of rocks to form a closer estimate of the time which has passed since the consolidation of the earth, and we have now good reason for judging that it was more than 20 and less than 40 million years ago ; and probably much nearer 20 than 40.

18. Twelve years ago, in a laboratory established by Mr Clarence King in connexion with the United States Geological Survey, a very important series of experimental researches on the physical properties of rocks at high temperatures was commenced by Dr Carl Barus, for the purpose of supplying trustworthy data for geological theory. Mr Clarence King, in an article published in the *American Journal of Science**, used data thus supplied, to estimate the age of the earth more definitely than was possible for me to do in 1862, with the very meagre information then available as to the specific heats, thermal conductivities, and temperatures of fusion, of rocks. I had taken 7000° F. (3871° C.) as a high estimate of the temperature of melting rock. Even then I might have taken something between 1000° C. and 2000° C. as more probable, but I was most anxious not to *under-estimate* the age of the earth, and so I founded my primary calculation on the 7000° F. for the temperature of melting rock. We know now from the experiments of Carl Barus† that diabase, a typical basalt of very primitive character, melts between 1100° C. and 1170° C., and is thoroughly liquid at 1200° C. The correction from 3871° C. to 1200° C. or 1/3·22 of that value, for the temperature of solidification, would, with no other change of assumptions, reduce my estimate of 100 million to $1/(3\cdot22)^2$ of its amount, or a little less than 10 million years; but the effect of pressure on the temperature of solidification must also be taken into account, and

---

* "On the Age of the Earth," Vol. XLV. January, 1893.

† *Phil. Mag.* 1893, first half-year, pp. 186, 187, 301—305.

Mr Clarence King, after a careful scrutiny of all the data given him for this purpose by Dr Barus, concludes that without further data "we have no warrant for extending the earth's age beyond 24 millions of years."

19.   By an elaborate piece of mathematical book-keeping I have worked out the problem of the conduction of heat outwards from the earth, with specific heat increasing up to the melting-point as found by Rücker and Roberts-Austen and by Barus, but with the conductivity assumed constant; and, by taking into account the augmentation of melting temperature with pressure in a somewhat more complete manner than that adopted by Mr Clarence King, I am not led to differ much from his estimate of 24 million years.   But, until we know something more than we know at present as to the probable diminution of thermal conductivity with increasing temperature, which would shorten the time since consolidation, it would be quite inadvisable to publish any closer estimate.

20.   All these reckonings of the history of underground heat, the details of which I am sure you do not wish me to put before you at present, are founded on the very sure assumption that the material of our present solid earth all round its surface was at one time a white hot liquid.   The earth is at present losing heat from its surface all round from year to year and century to century. We may dismiss as utterly untenable any supposition such as that a few thousand or a few million years of the present regime in this respect was preceded by a few thousand or a few million years of heating from without.   History, guided by science, is bound to find, if possible, an antecedent condition preceding every known state of affairs, whether of dead matter or of living creatures. Unless the earth was created solid and hot out of nothing, the regime of continued loss of heat must have been preceded by molten matter all round the surface.

21.   I have given strong reasons* for believing that *immediately* before solidification at the surface, the interior was solid close up to the surface: except comparatively small portions of lava or melted rock among the solid masses of denser solid rock which had sunk through the liquid, and possibly a somewhat large space

---

* "On the Secular Cooling of the Earth," Vol. III. *Math. and Phys. Papers*, §§ 19—33.

around the centre occupied by platinum, gold, silver, lead, copper, iron, and other dense metals, still remaining liquid under very high pressure.

22.   I wish now to speak to you of depths below the great surface of liquid lava bounding the earth before consolidation; and of mountain heights and ocean depths formed probably a few years after a first emergence of solid rock from the liquid surface (see § 24, below), which must have been quickly followed by complete consolidation all round the globe.   But I must first ask you to excuse my giving you all my depths, heights, and distances, in terms of the kilometre, being about six-tenths of that very inconvenient measure the English statute mile, which, with all the other monstrosities of our British metrical system, will, let us hope, not long survive the legislation of our present Parliamentary session destined to honour the sixty years' Jubilee of Queen Victoria's reign by legalising the French metrical system for the United Kingdom.

23.   To prepare for considering consolidation at the surface let us go back to a time (probably not more than twenty years earlier as we shall presently see—§ 24) when the solid nucleus was covered with liquid lava to a depth of several kilometres; to fix our ideas let us say 40 kilometres (or 4 million centimetres). At this depth in lava, if of specific gravity 2·5, the hydrostatic pressure is 10 tons weight (10 million grammes) per square centimetre, or ten thousand atmospheres approximately.   According to the laboratory experiments of Clarence King and Carl Barus* on Diabase, and the thermodynamic theory† of my brother, the late Professor James Thomson, the melting temperature of diabase is 1170° C. at ordinary atmospheric pressure, and would be 1420° C. under the pressure of ten thousand atmospheres, if the rise of temperature with pressure followed the law of simple proportion up to so high a pressure.

24.   The temperature of our 40 kilometres deep lava ocean of melted diabase may therefore be taken as but little less than 1420° C. from surface to bottom.   Its surface would radiate heat out

---

* *Phil. Mag.* 1893, first half-year, p. 306.

† *Trans. Roy. Soc. Edinburgh*, Jan. 2, 1849; *Cambridge and Dublin Mathematical Journal*, Nov. 1850.   Reprinted in *Math. and Phys. Papers* (Kelvin), Vol. I. p. 156.

into space at some such rate as two (gramme-water) thermal units Centigrade per square centimetre per second*. Thus, in a year (31½ million seconds) 63 million thermal units would be lost per square centimetre from the surface. This is, according to Carl Barus, very nearly equal to the latent heat of fusion abandoned by a million cubic centimetres of melted diabase in solidifying into the glassy condition (pitch-stone) which is assumed when the freezing takes place in the course of a few minutes. But, as found by Sir James Hall in his Edinburgh experiments† of 100 years ago, when more than a few minutes is taken for the freezing, the solid formed is not a glass but a heterogeneous crystalline solid of rough fracture; and if a few hours or days, or any longer time, is taken, the solid formed has the well known rough crystalline structure of basaltic rocks found in all parts of the world. Now Carl Barus finds that basaltic diabase is 14 per cent. denser than melted diabase, and 10 per cent. denser than the glass produced by quick freezing of the liquid. He gives no data, nor do Rücker and Roberts-Austen, who have also experimented on the thermo-dynamic properties of melted basalt, give any data, as to the latent heat evolved in the consolidation of liquid lava into rock of basaltic quality. Guessing it as three times the latent heat of fusion of the diabase pitch-stone, I estimate a million cubic centimetres of liquid frozen per square centimetre per three years. This would diminish the depth of the liquid at the rate of a million centimetres per three years, or 40 kilometres in twelve years.

25.   Let us now consider in what manner this diminution of depth of the lava ocean must have proceeded, by the freezing of portions of it; all having been at temperatures very little below the assumed 1420° C. melting temperature of the bottom, when the depth was 40 kilometres. The loss of heat from the white-hot surface (temperatures from 1420° C. to perhaps 1380° C. in different parts) at our assumed rate of two (gramme-water Centigrade) thermal units per sq. cm. per sec. produces very rapid cooling of the liquid within a few centimetres of the surface (thermal capacity

* This is a very rough estimate which I have formed from consideration of J. T. Bottomley's accurate determinations in absolute measure of thermal radiation at temperatures up to 920° C. from platinum wire and from polished and blackened surfaces of various kinds in receivers of air-pumps exhausted down to one ten-millionth of the atmospheric pressure. *Phil. Trans. Roy. Soc.*, 1887 and 1893.

† *Trans. Roy. Soc. Edinburgh.*

36 per gramme, according to Barus) and in consequence great downward rushes of this cooled liquid, and upwards of hot liquid, spreading out horizontally in all directions when it reaches the surface. When the sinking liquid gets within perhaps 20 or 10 or 5 kilometres of the bottom, its temperature* becomes the freezing-point as raised by the increased pressure; or, perhaps more correctly stated, a temperature at which some of its ingredients crystallize out of it. Hence, beginning a few kilometres above the bottom, we have a snow shower of solidified lava or of crystalline flakes, or prisms, or granules of felspar, mica, hornblende, quartz, and other ingredients: each little crystal gaining mass and falling somewhat faster than the descending liquid around it, till it reaches the bottom. This process goes on until, by the heaping of granules and crystals on the bottom, our lava ocean becomes silted up to the surface.

Probable Origin of Granite.

26. Upon the suppositions we have hitherto made we have, at the stage now reached, all round the earth at the same time a red hot or white hot surface of solid granules or crystals with interstices filled by the mother liquor still liquid, but ready to freeze with the slightest cooling. The thermal conductivity of this heterogeneous mass, even before the freezing of the liquid part, is probably nearly the same as that of ordinary solid granite or basalt at a red heat, which is almost certainly† somewhat less than the thermal conductivity of igneous rocks at ordinary temperatures. If you wish to see for yourselves how quickly it would cool when wholly solidified take a large macadamising stone, and heat it red hot in an ordinary coal fire. Take it out with a pair of tongs and leave it on the hearth, or on a stone slab at a distance from the fire, and you will see that in a minute or two, or perhaps in less than a minute, it cools to below red heat.

27. Half an hour‡ after solidification reached up to the

---

* The temperature of the sinking liquid rock rises in virtue of the increasing pressure : but much less than does the freezing point of the liquid or of some of its ingredients. (See Kelvin, *Math. and Phys. Papers*, Vol. III. pp. 69, 70.)

† *Proc. R. S.*, May 30, 1895.

‡ Witness the rapid cooling of lava running red hot or white hot from a volcano, and after a few days or weeks presenting a black hard crust strong enough and cool enough to be walked over with impunity.

surface in any part of the earth, the mother liquor among the granules must have frozen to a depth of several centimetres below the surface and must have cemented together the granules and crystals, and so formed a crust of primeval granite, comparatively cool at its upper surface, and red hot to white hot, but still all solid, a little distance down; becoming thicker and thicker very rapidly at first; and after a few weeks certainly cold enough at its outer surface to be touched by the hand.

Probable Origin of Basaltic Rock*.

28. We have hitherto left, without much consideration, the mother liquor among the crystalline granules at all depths below the bottom of our shoaling lava ocean. It was probably this interstitial mother liquor that was destined to form the basaltic rock of future geological time. Whatever be the shapes and sizes of the solid granules when first falling to the bottom, they must have lain in loose heaps with a somewhat large proportion of space occupied by liquid among them. But, at considerable distances down in the heap, the weight of the superincumbent granules must tend to crush corners and edges into fine powder. If the snow shower had taken place in air we may feel pretty sure (even with the slight knowledge which we have of the hardnesses of the crystals of felspar, mica and hornblende, and of the solid granules of quartz) that, at a depth of 10 kilometres, enough of matter from the corners and edges of the granules of different kinds, would have been crushed into powder of various degrees of fineness, to leave an exceedingly small proportionate volume of air in the interstices between the solid fragments. But in reality the effective weight of each solid particle, buoyed as it was by hydrostatic pressure of a liquid less dense than itself by not more than 20 or 15 or 10 per cent., cannot have been more than from about one-fifth to one-tenth of its weight in air, and therefore the same degree of crushing effect as would have been experienced at 10 kilometres with air in the interstices, must have been experienced only at depths of from 50 to 100 kilometres below the bottom of the lava ocean.

29 A result of this tremendous crushing together of the solid granules must have been to press out the liquid from among

* See Addendum at end of Lecture.

them, as water from a sponge, and cause it to pass upwards through the less and less closely packed heaps of solid particles, and out into the lava ocean above the heap. But, on account of the great resistance against the liquid permeating upwards 30 or 40 kilometres through interstices among the solid granules, this process must have gone on somewhat slowly; and, during all the time of the shoaling of the lava ocean, there may have been a considerable proportion of the whole volume occupied by the mother liquor among the solid granules, down to even as low as 50 or 100 kilometres below the top of the heap, or bottom of the ocean, at each instant. When consolidation reached the surface, the oozing upwards of the mother liquor must have been still going on to some degree. Thus, probably for a few years after the first consolidation at the surface, not probably for as long as one hundred years, the settlement of the solid structure by mere mechanical crushing of the corners and edges of solid granules, may have continued to cause the oozing upwards of mother liquor to the surface through cracks in the first formed granite crust and through fresh cracks in basaltic crust subsequently formed above it.

Leibnitz's Consistentior Status.

30. When this oozing everywhere through fine cracks in the surface ceases, we have reached Leibnitz's *consistentior status*; beginning with the surface cool and permanently solid and the temperature increasing to 1150° C. at 25 or 50 or 100 metres below the surface.

Probable Origin of Continents and Ocean Depths of the Earth.

31. If the shoaling of the lava ocean up to the surface had taken place everywhere at the same time, the whole surface of the consistent solid would be the dead level of the liquid lava all round, just before its depth became zero. On this supposition there seems no possibility that our present-day continents could have risen to their present heights, and that the surface of the solid in its other parts could have sunk down to their present ocean depths, during the twenty or twenty-five million years which may have passed since the *consistentior status* began or

during any time however long. Rejecting the extremely improbable hypothesis that the continents were built up of meteoric
matter tossed from without, upon the already solidified earth, we
have no other possible alternative than that they are due to
heterogeneousness in different parts of the liquid which constituted
the earth before its solidification. The hydrostatic equilibrium of
the rotating liquid involved only homogeneousness in respect to
density over every level surface (that is to say, surface perpendicular to the resultant of gravity and centrifugal force): it
required no homogeneousness in respect to chemical composition.
Considering the almost certain truth that the earth was built up
of meteorites falling together, we may follow in imagination the
whole process of shrinking from gaseous nebula to liquid lava and
metals, and solidification of liquid from central regions outwards,
without finding any thorough mixing up of different ingredients,
coming together from different directions of space—any mixing
up so thorough as to produce even approximately chemical homogeneousness throughout every layer of equal density. Thus we
have no difficulty in understanding how even the gaseous nebula,
which at one time constituted the matter of our present earth,
had in itself a heterogeneousness from which followed by dynamical
necessity Europe, Asia, Africa, America, Australia, Greenland,
and the Antarctic Continent, and the Pacific, Atlantic, Indian,
and Arctic Ocean depths, as we know them at present.

32. We may reasonably believe that a very slight degree of
chemical heterogeneousness could cause great differences in the
heaviness of the snow shower of granules and crystals on different
regions of the bottom of the lava ocean when still 50 or 100 kilometres deep. Thus we can quite see how it may have shoaled
much more rapidly in some places than in others. It is also
interesting to consider that the solid granules, falling on the
bottom, may have been largely disturbed, blown as it were into
ridges (like rippled sand in the bed of a flowing stream, or like
dry sand blown into sand-hills by wind) by the eastward horizontal
motion which liquid descending in equatorial regions must acquire,
relatively to the bottom, in virtue of the earth's rotation. It is
indeed not improbable that this influence may have been largely
effective in producing the general configuration of the great ridges
of the Andes and Rocky Mountains and of the West Coasts of
Europe and Africa. It seems, however, certain that the main

determining cause of the continents and ocean-depths was chemical differences, perhaps very slight differences, of the material in different parts of the great lava ocean before consolidation.

33. To fix our ideas let us now suppose that over some great areas such as those which have since become Asia, Europe, Africa, Australia, and America, the lava ocean had silted up to its surface, while in other parts there still were depths ranging down to 40 kilometres at the deepest. In a very short time, say about twelve years according to our former estimate (§ 24) the whole lava ocean becomes silted up to its surface.

34. We have not time enough at present to think out all the complicated actions, hydrostatic and thermodynamic, which must accompany, and follow after, the cooling of the lava ocean surrounding our ideal primitive continent. By a hurried view, however, of the affair we see that in virtue of, let us say, 15 per cent. shrinkage by freezing, the level of the liquid must, at its greatest supposed depth, sink six kilometres relatively to the continents: and thus the liquid must recede from them; and their bounding coast-lines must become enlarged. And just as water runs out of a sandbank, drying when the sea recedes from it on a falling tide, so rivulets of the mother liquor must run out from the edges of the continents into the receding lava ocean. But, unlike sandbanks of incoherent sand permeated by water remaining liquid, our uncovered banks of white-hot solid crystals, with interstices full of the mother liquor, will, within a few hours of being uncovered, become crusted into hard rock by cooling at the surface, and freezing of the liquor, at a temperature somewhat lower than the melting temperatures of any of the crystals previously formed. The thickness of the wholly solidified crust grows at first with extreme rapidity, so that in the course of three or four days it may come to be as much as a metre. At the end of a year it may be as much as 10 metres; with a surface, almost, or quite, cool enough for some kinds of vegetation. In the course of the first few weeks the regime of conduction of heat outwards becomes such that the thickness of the wholly solid crust, as long as it remains undisturbed, increases as the square root of the time; so that in 100 years it becomes 10 times, in 25 million years 5000 times, as thick as it was at the end of one year; thus, from one year to 25 million years after the time of surface freezing

the thickness of the wholly solid crust might grow from 10 metres to 50 kilometres. These definite numbers are given merely as an illustration; but it is probable they are not enormously far from the truth in respect to what has happened under some of the least disturbed parts of the earth's surface.

35. We have now reached the condition described above in § 30, with only this difference, that instead of the upper surface of the whole solidified crust being level we have in virtue of the assumptions of §§ 33, 34, inequalities of 6 kilometres from highest to lowest levels, or as much more than 6 kilometres as we please to assume it.

36. There must still be a small, but important, proportion of mother liquor in the interstices between the closely packed un-cooled crystals below the wholly solidified crust. This liquor, differing in chemical constitution from the crystals, has its freezing-point somewhat lower, perhaps very largely lower, than the lowest of their melting-points. But, when we consider the mode of formation (§ 25) of the crystals from the mother liquor, we must regard it as still always a solvent ready to dissolve, and to redeposit, portions of the crystalline matter, when slight variations of temperature or pressure tend to cause such actions. Now as the specific gravity of the liquor is less, by something like 15 per cent., than the specific gravity of the solid crystals, it must *tend* to find its way upwards, and will actually do so, however slowly, until stopped by the already solidified impermeable crust, or until itself becomes solid on account of loss of heat by conduction outwards. If the upper crust were everywhere continuous and perfectly rigid the mother liquor must, inevitably, if sufficient time be given, find its way to the highest places of the lower boundary of the crust, and there form gigantic pockets of liquid lava tending to break the crust above it and burst up through it.

37. But in reality the upper crust cannot have been infinitely strong; and, judging alone from what we know of properties of matter, we should expect gigantic cracks to occur from time to time in the upper crust tending to shrink as it cools and prevented from lateral shrinkage by the non-shrinking uncooled solid below it. When any such crack extends downwards as far as a pocket of mother liquor underlying the wholly solidified crust, we should have an outburst of trap rock or of volcanic lava just such as have

been discovered by geologists in great abundance in many parts of the world. We might even have comparatively small portions of high plateaus of the primitive solid earth raised still higher by outbursts of the mother liquor squeezed out from below them in virtue of the pressure of large surrounding portions of the superincumbent crust. In any such action, due to purely gravitational energy, the centre of gravity of all the material concerned must sink, although portions of the matter may be raised to greater heights; but we must leave these large questions of geological dynamics, having been only brought to think of them at all just now by our consideration of the earth, antecedent to life upon it.

38. The temperature to which the earth's surface cooled within a few years after the solidification reached it, must have been, as it is now, such that the temperature at which heat radiated into space during the night exceeds that received from the sun during the day, by the small difference due to heat conducted outwards from within*. One year after the freezing of the granitic interstitial mother liquor at the earth's surface in any locality, the average temperature at the surface might be warmer, by 60° or 80° Cent., than if the whole interior had the same average temperature as the surface. To fix our ideas, let us suppose, at the end of one year, the surface to be 80° warmer than it would be

---

* Suppose, for example, the cooling and thickening of the upper crust has proceeded so far, that at the surface and therefore approximately for a few decimetres below the surface, the rate of augmentation of temperature downwards is one degree per centimetre. Taking as a rough average ·005 c.g.s. as the thermal conductivity of the surface rock, we should have for the heat conducted outwards ·005 of a gramme water thermal unit Centigrade per sq. cm. per sec. (Kelvin, *Math. and Phys. Papers*, Vol. III. p. 226). Hence if (*ibid.* p. 223) we take $\frac{1}{8000}$ as the radiational emissivity of rock and atmosphere of gases and watery vapour above it radiating heat into the surrounding vacuous space (æther), we find $8000 \times ·005$, or 40 degrees Cent. as the excess of the mean surface temperature above what it would be if no heat were conducted from within outwards. The present augmentation of temperature downwards may be taken as 1 degree Cent. per 27 metres as a rough average derived from observations in all parts of the earth where underground temperature has been observed. (See *British Association Reports* from 1868 to 1895. The very valuable work of this Committee has been carried on for these twenty-seven years with great skill, perseverance, and success, by Professor Everett, and he promises a continuation of his reports from time to time.) This with the same data for conductivity and radiational emissivity as in the preceding calculation makes 40°/2700 or 0·0148° Cent. per centimetre as the amount by which the average temperature of the earth's surface is at present kept up by underground heat.

with no underground heat: then at the end of 100 years it would be 8° warmer, and at the end of 10,000 years it would be ·8 of a degree warmer, and at the end of 25 million years it would be ·016 of a degree warmer, than if there were no underground heat.

39.   When the surface of the earth was still white-hot liquid all round, at a temperature fallen to about 1200° Cent., there must have been hot gases and vapour of water above it in all parts, and possibly vapours of some of the more volatile of the present known terrestrial solids and liquids, such as zinc, mercury, sulphur, phosphorus.   The very rapid cooling which followed instantly on the solidification at the surface must have caused a rapid downpour of all the vapours other than water, if any there were; and a little later, rain of water out of the air, as the temperature of the surface cooled from red heat to such moderate temperatures as 40° and 20° and 10° Cent., above the average due to sun-heat and radiation into the ether around the earth.   What that primitive atmosphere was, and how much rain of water fell on the earth in the course of the first century after consolidation, we cannot tell for certain; but Natural History and Natural Philosophy give us some foundation for endeavours to discover much towards answering the great questions,—Whence came our present atmosphere of nitrogen, oxygen, and carbonic acid? Whence came our present oceans and lakes of salt and fresh water?   How near an approximation to present conditions was realized in the first hundred centuries after consolidation of the surface?

40.   We may consider it as quite certain that nitrogen gas, carbonic acid gas, and steam, escaped abundantly in bubbles from the mother liquor of granite, before the primitive consolidation of the surface, and from the mother liquor squeezed up from below in subsequent eruptions of basaltic rock; because all, or nearly all, specimens of granite and basaltic rock, which have been tested by chemists in respect to this question*, have been found to contain, condensed in minute cavities within them, large quantities of nitrogen, carbonic acid, and water.   It seems that in no specimen of granite or basalt tested has chemically free oxygen been discovered, while in many, chemically free hydrogen has been found;

---

* See, for example, Tilden, *Proc. R. S.*, February 4th, 1897.   "On the Gases enclosed in Crystalline Rocks and Minerals."

and either native iron or magnetic oxide of iron in those which do not contain hydrogen. From this it might seem probable that there was no free oxygen in the primitive atmosphere, and that if there was free hydrogen, it was due to the decomposition of steam by iron or magnetic oxide of iron. Going back to still earlier conditions we might judge that, probably, among the dissolved gases of the hot nebula which became the earth, the oxygen all fell into combination with hydrogen and other metallic vapours in the cooling of the nebula, and that although it is known to be the most abundant material of all the chemical elements constituting the earth, none of it was left out of combination with other elements to give free oxygen in our primitive atmosphere.

41. It is, however, possible, although it might seem not probable, that there was free oxygen in the primitive atmosphere. With or without free oxygen, however, *but with sunlight*, we may regard the earth as fitted for vegetable life as now known in some species, wherever water moistened the newly solidified rocky crust cooled down below the temperature of 80° or 70° of our present Centigrade thermometric scale, a year or two after solidification of the primitive lava had come up to the surface. The thick tough velvety coating of living vegetable matter, covering the rocky slopes under hot water flowing direct out of the earth at Banff (Canada)*, lives without help from any ingredients of the atmosphere above it, and takes from the water and from carbonic acid or carbonates, dissolved in it, the hydrogen and carbon needed for its own growth by the dynamical power of sunlight; thus leaving free oxygen in the water to pass ultimately into the air. Similar vegetation is found abundantly on the terraces of the Mammoth hot springs and on the beds of the hot water streams flowing from the Geysers in the Yellowstone National Park of the United States. This vegetation, consisting of confervæ, all grows under flowing water at various temperatures, some said to be as high as 74° Cent. We cannot doubt but that some such confervæ, if sown or planted in a rivulet or pool of warm water in the early years of the first century of the solid earth's history, and if favoured with sunlight, would have lived, and grown, and multiplied, and would have made a beginning of oxygen in the air, if there had been none of it before their contributions. Before the end of the century, if sun-

---

* Rocky Mountains Park of Canada, on the Canadian Pacific Railway.

heat, and sunlight, and rainfall, were suitable, the whole earth not under water must have been fitted for all kinds of land plants which do not require much or any oxygen in the air, and which can find, or make, place and soil for their roots on the rocks on which they grow; and the lakes or oceans formed by that time must have been quite fitted for the life of many or all of the species of water plants living on the earth at the present time. The moderate warming, both of land and water, by underground heat, towards the end of the century, would probably be favourable rather than adverse to vegetation, and there can be no doubt but that if abundance of seeds of all species of the present day had been scattered over the earth at that time, an important proportion of them would have lived and multiplied by natural selection of the places where they could best thrive.

42. But if there was no free oxygen in the primitive atmosphere or primitive water, several thousands, possibly hundreds of thousands, of years must pass before oxygen enough for supporting animal life, as we now know it, was produced. Even if the average activity of vegetable growth on land and in water over the whole earth was, in those early times, as great in respect to evolution of oxygen as that of a Hessian forest, as estimated by Liebig* 50 years ago, or of a cultivated English hayfield of the present day, a very improbable supposition, and if there were no decay (*eremacausis*, or gradual recombination with oxygen) of the plants or of portions such as leaves falling from plants, the rate of evolution of oxygen, reckoned as three times the weight of the wood or the dry hay produced, would be only about 6 tons per English acre per annum or 1½ tons per square metre per thousand years. At this rate it would take only 1533 years, and therefore in reality a much longer time would almost certainly be required, to produce the 2·3 tons of oxygen which we have at present resting on every square metre of the earth's surface, land and sea†. But probably quite a moderate number of hundred thousand years may have sufficed. It is interesting at all events to remark that, at any time, the total amount of combustible material on the earth, in the form of living

---

* Liebig, *Chemistry in its application to Agriculture and Physiology*, English, 2nd ed., edited by Playfair, 1842.

† In our present atmosphere, in average conditions of barometer and thermometer we have, resting on each square metre of the earth's surface, ten tons total weight, of which 7·7 is nitrogen and 2·3 is oxygen.

plants or their remains left dead, must have been just so much that to burn it all would take either the whole oxygen of the atmosphere, or the excess of oxygen in the atmosphere at the time, above that, if any, which there was in the beginning. This we can safely say, because we almost certainly neglect nothing considerable in comparison with what we assert when we say that the free oxygen of the earth's atmosphere is augmented only by vegetation liberating it from carbonic acid and water, in virtue of the power of sunlight, and is diminished only by virtual burning* of the vegetable matter thus produced. But it seems improbable that the average of the whole earth—dry land and sea-bottom—contains at present coal, or wood, or oil, or fuel of any kind originating in vegetation, to so great an amount as ·767 of a ton per square metre of surface; which is the amount at the rate of one ton of fuel to three tons of oxygen, that would be required to produce the 2·3 tons of oxygen per square metre of surface, which our present atmosphere contains. Hence it seems probable that the earth's primitive atmosphere must have contained free oxygen.

43. Whatever may have been the true history of our atmosphere it seems certain that if sunlight was ready, the earth was ready, both for vegetable and animal life, if not within a century, at all events within a few hundred centuries after the rocky consolidation of its surface. But was the sun ready? The well founded dynamical theory of the sun's heat carefully worked out and discussed by Helmholtz, Newcomb, and myself†, says NO if the consolidation of the earth took place as long ago as 50 million years; the solid earth must in that case have waited 20 or 50 million years for the sun to be anything nearly as warm as he is at present. If the consolidation of the earth was finished 20 or 25 million years ago, the sun was probably ready,—though probably not then quite so warm as at present, yet warm enough to support some kind of vegetable and animal life on the earth.

44. My task has been rigorously confined to what, humanly speaking, we may call the fortuitous concourse of atoms, in the

---

* This " virtual burning " includes eremacausis of decay of vegetable matter, if there is any eremacausis of decay without the intervention of microbes or other animals. It also includes the combination of a portion of the food with inhaled oxygen in the regular animal economy of provision for heat and power.

† See *Popular Lectures and Addresses*, Vol. I. pp. 376—429, particularly p. 397.

preparation of the earth as an abode fitted for life, except in so far
as I have referred to vegetation, as possibly having been concerned
in the preparation of an atmosphere suitable for animal life as we
now have it. Mathematics and dynamics fail us when we con-
template the earth, fitted for life but lifeless, and try to imagine
the commencement of life upon it. This certainly did not take
place by any action of chemistry, or electricity, or crystalline
grouping of molecules under the influence of force, or by any
possible kind of fortuitous concourse of atoms. We must pause,
face to face with the mystery and miracle of the creation of living
creatures.

ADDENDUM.—*May* 1898.

Since this lecture was delivered I have received from Professor Roberts-Austen the following results of experiments on
the melting-points of rocks which he has kindly made at my
request:

	Melting-point	Error
Felspar . . .	1520° C.	± 30°
Hornblende .	about 1400°	
Mica	1440°	± 30°
Quartz . . .	1775°	± 15°
Basalt . . .	about 880°	

These results are in conformity with what I have said in
§§ 26—28 on the probable origin of granite and basalt, as they
show that basalt melts at a much lower temperature than felspar,
hornblende, mica, or quartz, the crystalline ingredients of granite.
In the electrolytic process for producing aluminium, now practised
by the British Aluminium Company at their Foyers works, alumina,
of which the melting-point is certainly above 1700° C. or 1800° C.,
is dissolved in a bath of melted cryolite at a temperature of about
800° C. So we may imagine melted basalt to be a solvent for
felspar, hornblende, mica, and quartz at temperatures much below
their own separate melting-points; and we can understand how
the basaltic rocks of the earth may have resulted from the solidification of the mother liquor from which the crystalline ingredients
of granite have been deposited.

157. BLUE RAY* OF SUNRISE OVER MONT BLANC.

[From Letter to *Nature*, Vol. LX. Aug. 31, 1899 [letter dated
Aug. 27], p. 411.]

LOOKING out at 4 o'clock this morning from a balcony of this
hotel, 1545 metres above sea-level, and about 68 kilometres W.
18° S. from Mont Blanc, I had a magnificent view of Alpine
ranges of Switzerland, Savoy, and Dauphiné; perfectly clear and
sharp on the morning twilight sky. This promised me an opportunity for which I had been waiting five or six years; to see the
earliest instantaneous light of sunrise through very clear air, and
find whether it was perceptibly blue. I therefore resolved to
watch an hour till sunrise, and was amply rewarded by all the
splendours I saw. Having only vague knowledge of the orientation of the hotel, I could not at first judge whereabouts the sun
would rise; but in the course of half an hour rosy tints on each
side of the place of strongest twilight showed me that it would
be visible from the balcony; and I was helped to this conclusion
by Haidinger's brushes when the illumination of the air at greater
altitudes by a brilliant half-moon nearly overhead, was overpowered
by sunlight streaming upwards from beyond the mountains. A
little later, beams of sunlight and shadows of distant mountains
converged clearly to a point deep under the very summit of Mont
Blanc. In the course of five or ten minutes I was able to watch
the point of convergence travelling obliquely upwards till in an
instant I saw a blue light against the sky on the southern profile
of Mont Blanc; which, in less than one-twentieth of a second
became dazzlingly white, like a brilliant electric arc-light. I had
no dark glass at hand, so I could not any longer watch the rising
sun.

Hotel du Mont-Revard, above Aix-les-Bains.
August 27.

* The "Rayon Vert" of Jules Verne is the corresponding phenomenon at
sunset; which I first saw about six years ago.

158. ON THE CLUSTERING OF GRAVITATIONAL MATTER
IN ANY PART OF THE UNIVERSE.

[From *British Association Report*, 1901, pp. 563—578 ; *Nature*, Vol. LXIV.
Oct. 24, 1901, pp. 626—629; *Phil. Mag.* Vol. III. Jan. 1902, pp. 1—9 ;
Baltimore Lectures, Appendix D, 1904, pp. 532—540.]

159. On Homer Lane's Problem of a Spherical Gaseous Nebula.

[From *Edinb. Roy. Soc. Proc.* Vol. XXVII. [read Jan. 21, 1907, title only], p. 375; *Nature*, Vol. LXXV. Feb. 1907, pp. 368, 369.]

1. A HIGHLY interesting problem of pure mathematics was brought before the world in the *American Journal of Science*, July, 1870, by the late Mr Homer Lane, who, as we are told by Mr T. J. J. See*, was for many years connected with the U.S. Coast and Geodetic Survey at Washington. Lane's problem is the convective equilibrium, of density, of pressure, and of temperature, in a rotationless spherical mass of gaseous fluid†, hot in its central parts, and left to itself in waveless quiescent ether.

2. For the full discussion of this problem we must, according to the evolutionary philosophy of the physics of dead matter, try to solve it for all past and future time. But we may first, after the manner of Fourier, consider the gaseous globe as being at any time given with any arbitrarily assumed distribution of temperature, subject only to the condition that it is uniform throughout every spherical surface concentric with the boundary. And our subject might be the absolutely determinate problem of finding the density and pressure at every point necessary for

---

* "Researches on the Physical Constitution of the Heavenly Bodies," *Astr. Nachr.*, November, 1895.

† By a gaseous fluid I here mean what is commonly called a "perfect gas," that is, a gas which fulfils two laws:—(1) Boyle's law. At constant temperature it exerts pressure exactly in proportion to its density, or in inverse proportion to the volume of a given homogeneous mass of it. (2) A given mass of it, kept at constant pressure, has its volume exactly proportional to its temperature, according to the absolute thermodynamic definition of temperature (Preston's *Theory of Heat*, Article 290). According to the "Kinetic Theory of Gases," every gas or vapour approximates more and more closely to the fulfilment of these two laws, the smaller is the proportion of the sum of times in collision to the sum of times of moving approximately in straight lines between collisions.

dynamical equilibrium. But for stability of this equilibrium Homer Lane assumed, rightly as I believe is now generally admitted, that it must be of the kind which eight years earlier* I called convective equilibrium.

3.   If the fluid globe were given with any arbitrary distribution of temperature, for example uniform temperature throughout, the cooling, and consequent augmentation of density of the fluid at its boundary, by radiation into space, would immediately give rise to an instability according to which some parts of the outermost portions of the globe would sink, and upward currents would consequently be developed in other portions. In any real fluid, whether gaseous or liquid, or liquid with an atmosphere of vapour around it, this kind of automatic stirring would tend to go on until a condition of approximate equilibrium is reached, in which any portion of the fluid descending or ascending would, by the thermodynamic action involved in change of pressure, always take the temperature corresponding to its level, that is to say, its distance from the centre of the globe.

4.   The condition thus reached, when heat is continually being radiated away from the spherical boundary, is not perfect equilibrium. It is only an approximation to equilibrium, in which the temperature and density are each approximately uniform at any one distance from the centre, and vary slowly with time, the variable irregular convective currents being insufficient to cause any considerable deviation of the surfaces of equal density and temperature from sphericity.

5.   A very interesting and important theorem was given by Prof. Perry, on p. 252 of *Nature* for July, 1899, according to which, for cosmical purposes, it is convenient to divide gases into two species—species P, gases for which the ratio (k) of thermal capacity, pressure constant, to thermal capacity, volume constant, is greater than $1\frac{1}{3}$; species Q, gases for which k is less than $1\frac{1}{3}$. On looking at the page of *Nature* referred to, it will be seen that Perry questioned or even denied the possibility of a gas of species Q. His theorem is:—*A finite spherical globe of gas, given in equilibrium with any arbitrary distribution of temperature having*

---

* "On the Convective Equilibrium of Temperature in the Atmosphere," *Literary and Philosophical Society of Manchester*, January 21, 1862 ; re-published as Appendix E, *Math. and Phys. Papers*, Vol. III. pp. 255—260.

isothermal surfaces spherical, has less heat if the gas is of species P, and more heat if of species Q, than the thermal equivalent of the work which would be done by the mutual gravitational attraction between all its parts, in ideal shrinkage from an infinitely rare distribution of the whole mass to the given condition of density.

6.   From this we see that if a globe of gas Q is given in a state of convective equilibrium, with the requisite heat given to it, no matter how, and left to itself in waveless quiescent ether, it would, through gradual loss of heat, immediately cease to be in equilibrium, and would begin to fall inwards towards its centre, until in the central regions it becomes so dense that it ceases to obey Boyle's law; that is to say, ceases to be a gas.   Then, notwithstanding Perry's theorem, it can come to approximate convective equilibrium as a cooling liquid globe surrounded by an atmosphere of its own vapour.

7.   But if, after being given as in § 6, heat be properly and sufficiently supplied to the globe of Q-gas at its boundary, and the interior be kept stirred by artificial stirrers, the whole gaseous mass can be brought into the condition of convective equilibrium.

8.   In the course of the communication to the Royal Society of Edinburgh, curves were shown representing the distributions of density and temperature in convective equilibrium for four different gases, corresponding to the four values of k:

Gas (1) $k = 1\frac{2}{3}$ (approximately the value of k for the monatomic gases, mercury vapour according to Kundt and Warburg, argon, helium, neon, krypton, and xenon).

Gas (2) $k = 1\frac{2}{5}$ (approximately the value of k for seven known diatomic gases, hydrogen, nitrogen, oxygen, carbon monoxide, nitric oxide, hydrochloric acid, hydrogen bromide).

Gas (3) $k = 1\frac{1}{3}$ (approximately the value of k for water vapour, chlorine, marsh gas, bromine iodide, chlorine iodide).

Gas (4) $k = 1\frac{1}{4}$ (approximately the value of k for sulphur dioxide).

Four of these curves agree practically with curves given by Homer Lane for $k = 1\frac{2}{3}$ and $k = 1\frac{2}{5}$, in his original paper to the *American Journal of Science*, July, 1870.

9. In a communication to the Edinburgh Royal Society of February, 1887, "On the Equilibrium of a Gas under its own Gravitation only," I indicated a graphical treatment of Lane's problem by successive quadratures, which facilitated the accurate calculation of numerical results, and was worked out fully for the case $k = 1\frac{2}{3}$ by Mr Magnus Maclean, with results shown in a table on p. 117 of the *Proceedings of the Royal Society of Edinburgh*, Vol. XIV., and on p. 292 of the *Phil. Mag.*, March, 1887. The numbers in that table expressing temperature and density are represented by two of the curves now laid before the society. The other curves represent numerical results calculated by Mr George Green, according to a greatly improved process which he has found, giving the result by step by step calculation without the aid of graphical constructions.

The mathematical interpretation of the solution for Perry's critical case of $k = 1\frac{1}{3}$, and for gases of the Q-species, is exceedingly interesting.

The communication included also fully worked out examples of the general solution of Lane's problem for gases of class P of different total quantities and of different specific densities.

10. In my communication to the Royal Society of Edinburgh, of February, 1887 I pointed out that Homer Lane's problem gives no approximation to the present condition of the sun, because of his great average density (1·4). This was emphasised by Prof. Perry in the seventh paragraph, headed "Gaseous Stars," of his letter to Sir Norman Lockyer on "The Life of a Star" (*Nature*, July 13, 1899), which contains the following sentence:

"It seems to me that speculation on this basis of perfectly gaseous stuff ought to cease when the density of the gas at the centre of the star approaches 0·1 or one-tenth of the density of ordinary water in the laboratory."

160.　On the Formation of Concrete Matter from Atomic Origins.

[From *Phil. Mag.* Vol. xv. April, 1908, pp. 397—413.]

THE following paper was written by Lord Kelvin some months before his death, and the subject with which it is concerned was occupying his attention down to the last few days of his life, in fact, till the commencement of his last illness. Very shortly before his death he wrote out in detail a paper on "Homer Lane's Problem," which he had communicated to the Royal Society of Edinburgh in January 1907, under the title "The Problem of a Spherical Gaseous Nebula"; and it is proposed to reprint this paper, along with an appendix, written at Lord Kelvin's desire by his Private Secretary, Mr George Green, in the next number of the *Philosophical Magazine*.

The present paper has been carefully corrected by Mr Green and myself, but it was left in a perfectly finished state, and little or no editing was necessary.　J. T. BOTTOMLEY.

1.　COALITION, due to gravitational attraction between materials given originally in small parts widely distributed through space, is probably the most ancient history of all the bodies in the universe.　What the primitive forms or magnitudes of those pieces of matter may have been can never be made known to us by historical evidence.　If they had been all globes, or irregular broken solids, of diameters a few kilometres, or a few thousandths of a millimetre, and if among them all there existed all the atoms which at present exist, the present condition of the universe might be very much the same as it is.　Towards a speculative answer we might be guided, and perhaps wrongly guided, by what we see of meteorites, stony or iron.　We can hardly regard it as probable that those broken looking lumps of solid matter, with their corners and edges rounded off by melting in their final rush through our atmosphere before arriving at the Earth's surface, were primitive forms in which matter either was created, or existed through all infinity of past time.　On the contrary, it may seem to us quite probable that the primitive condition of matter was atomic; perhaps every primitive particle was a separate indivisible atom; or perhaps some of the primitive particles were atoms, and some of them doublets such as the O_2, N_2, H_2, which we know as the molecular constituents of gaseous oxygen, gaseous nitrogen, gaseous hydrogen, according to modern chemical doctrine. Or perhaps some of the primitive particles may have been given in groups consisting of a moderate number of atoms ready for building into crystals; or they may have been given as very small complete crystals each consisting of a very large number of atoms.

2. To illustrate the dynamics of the real conglomeration, which we believe to be an event of ancient history, consider an ideal case of 1083 million million million cubic metres of solid matter; the sum of their volumes being equal to the Earth's volume. Let the density of the material of each cube be equal to the Earth's mean density, 5·67. The sum of their masses will be 6·14 thousand million million million metric tons, being equal to the Earth's mass. Place them at rest in cubic order, equally distributed through a vast spherical space, of radius one thousand times the Earth's radius, and therefore of volume equal to a thousand million times the Earth's volume. Let every one of the cubes be oriented with its faces and edges parallel to the planes and lines of the cubic order. In this order, the lines of shortest distance between the centres of constituents are perpendicular to the three pairs of parallel faces of the cubes. The distance from centre to centre would be one metre, if the cubes were given in contact, occupying a sphere equal in bulk to the Earth. The distance between the centres of nearest neighbours is therefore a kilometre, when they are given in their wide-spread initial arrangement.

3. Leave now the cubes all free to fall inwards in virtue of mutual gravitation. Each one of those on the bounding surface of the whole group will commence falling towards the centre of the sphere, with acceleration one millionth of the acceleration of a body falling freely near the Earth's surface: that is to say 9·8 millionths of a metre *per second per second*. The centreward acceleration of all the others will be in simple proportion to their distances from the centre of the assemblage. This is easily seen

by remarking that, according to Newtonian principles, the resultant force on each cube is the same as if all the cubes at less distances than its own from the centre were condensed in a point there, and all the cubes at greater distances than its own were annulled. Hence as long as the cubes all fall freely inwards their distribution remains uniform, and their boundary spherical.

4.   Hence the cubes, if all similarly oriented as in § 2, will, at one instant of time, all come into contact, each with all its six nearest neighbours; and for an instant they will be fitted together making a great globe of the same size as the Earth, but with angular projections at the spherical boundary according to their cubic arrangement.   All the parts of this composite globe are, at the instant of first contact, moving inwards at speeds simply proportional to their distances from the centre.   If the cubes were perfectly elastic, they would rebound outwards with velocities equal to those they had at the instant of first contact; and a periodic falling inwards and rebounding outwards would follow; which would continue for ever, if there were no ether to resist this gigantic oscillatory movement.   But what would be done by cubes of real matter, with its true molecular constitution and imperfect elasticity, in and after such a prodigious, purely pressural, collision, is not a subject for profitable conjecture.

5.   Let now the cubes be placed initially at rest, with their centres arranged as in § 2, but with orientations given at random, and let them fall freely.   Contacts between some neighbours will begin to occur when the shortest distances between centres are equal to $3^{\frac{1}{3}}$, or 1·732, metres, being the length of the body diagonal of each cube.   At this instant the whole assemblage occupies $3^{\frac{3}{2}}$, or 5·196, times the Earth's bulk: and the average density is somewhat greater than the density of water.

6.   The velocity of the outermost cubes at the instant of impact, in the case of the similar orientations of § 2, is 11,175 metres per second.   If the cubes were reduced to infinitely small material points at their centres, and so could continue to fall freely to the centre of the sphere, they would all reach that point in 897 of a year from the time of our ideally assumed initial state of rest.   As the material points fall towards the centre of the sphere, the mean density of the assemblage continuously increases.   The following table shows the number of minutes

during which the system must continue to fall to reach the centre, after it has attained the densities indicated:

Time before reaching the centre *	Mean Density
47·7 minutes	0·1
21·3 „	0·5
15·1 „	1·0
6·3 „	5·67

If the beginning is an assemblage of cubes oriented at random, as in § 5, and with their centres in cubic order, as in § 2, contacts would commence at a time about 8 minutes earlier than the time of coming to fit exactly on the supposition of uniform orientation.

7. In the case of cubes initially oriented at random, collisions will not begin simultaneously as in § 4; but will begin with a crushing to powder at colliding edges or corners. The stupendous system of collisions which follows the commencement would, if the material of the cubes is of any known substance, metallic or rocky, cause, in the course of a few minutes, melting of the whole mass: unless the prodigious pressure in the central parts should have the effect of preventing fluidity in those parts, which does not seem probable.

8. The same general description is applicable to the ideal case of a vast number of large and small fragments of any shapes, instead of our equal cubic metres of homogeneous matter; provided only that the initial distribution through the great spherical space is of uniform average density all through.

9. Let us now, instead of masses large or small of concrete matter, begin with a vast number of atoms; or of atoms, and doublets such as O_2, N_2, H_2, given at rest distributed uniformly in respect to average density through a sphere of a thousand million times the Earth's bulk: and having the sum of their masses equal to the Earth's mass. Every particle (atom or doublet) will have the same centreward velocity at the same time, as that found for the ideal cubes in § 6; until some of the atoms or doublets get into touch with neighbours, that is, come so near one another that mutual molecular forces become effective. This

---

* Calculated by means of a formula given on page 538 of Lord Kelvin's *Baltimore Lectures*, Appendix D.

must be the case when the mean density is considerably less than one-tenth of the density of water, as we see by considering the known properties of gases and vapours, and of liquids of small density. Hence the time during which the atoms will continue to fall freely without jostling one another must be a few minutes less than the time of getting into touch at mean density somewhat greater than that of water, on the supposition of cubes randomly oriented at rest, as defined in § 5.

10. When the density becomes so great that the atoms begin to seriously jostle one another, we have the first step to the formation of concrete matter from atomic origins. Our present knowledge of the properties of matter does not suffice to allow us to follow, with definite and complete understanding, the progress after this step. We can see a prodigious cloud of atoms crowding turbulently around the centre. Atoms coming from all directions meet and collide. The energy of the relative motions of the atoms is still, let us suppose, sufficient to prevent them from ever remaining in contact except during very short times when any two of them are in collision. And let us suppose the central mean density to be still considerably less than 1 of the density of water. The whole assemblage now constitutes the gaseous fluid mass which forms the subject of Homer Lane's celebrated problem*. As long as the whole mass remains thus in gaseous condition, loss of heat to the surrounding ether allows mutual gravitation to condense the whole assemblage, doing its work by increasing the kinetic energy of the relative motions. A result, as found by the mathematical solution of Homer Lane's problem, is that all of the assemblage outside a certain distance from the centre sinks in temperature, while all within that distance rises in temperature. I may here explain that the temperature of a "perfect gas" means the kinetic energy per unit mass of all the relative translatory motions of its molecules in free paths from collision to collision.

11.   During the whole time of the gaseous stage which we

* J. Homer Lane, *American Journal of Science*, July, 1870; A. Ritter, *Wiedemann's Annalen*, 1878—1882; Prof. A. Schuster, *Brit. Assoc. Report*, 1883; Sir William Thomson, *Phil. Mag.* Vol. xxiii. p. 287, 1887; Prof. J. Perry, *Nature*, Vol. lx. 1899; T. J. J. See, *Astr. Nachr.* No. 4053, Bd. 169, 1905; T. J. J. See, *Astr. Nachr.* No. 4104, Bd. 171, 1906; Lord Kelvin, *Royal Society of Edinburgh*, Jan. 21, 1907; Lord Kelvin, *Nature*, Feb. 14, 1907.

have been considering, the crowding becomes denser and denser
in the central regions, until *there* every atom comes to be always
in collision with all its nearest neighbours. Throughout all the
space around the centre in which this condition has been reached
the crowd constitutes a fluid of the species called liquid. In the
outlying parts the crowd is much less dense; each atom, instead
of being always in collision, is in collision only during comparatively
short intervals of time; and, for the rest of its time, is moving in
approximately straight lines, not perceptibly disturbed by other
atoms far or near. The less dense crowd of atoms in those parts
constitutes a fluid of the species called gas or vapour. Every
collision between two atoms in those outlying parts, and in the
central region every change of speed and of direction of atomic
motions, due to mutual forces between neighbouring atoms, sets
ether locally in motion. The motion thus produced in ether gives
rise to etherial waves which travel outwards through the outlying
parts of the assemblage; and continue their outward course into
void ether all around the assemblage. These etherial waves carry
away gradually into infinite space the kinetic energy of the atoms,
originally given to them by gravitation between all parts of the
contracting assemblage. A first effect of this loss of energy would
be to continue the raising of temperature in the central region,
which in § 10 was said to take place during the whole of the
gaseous stage of the evolution. As time advances, the dense gas
or liquid in the central parts comes to a maximum of temperature.
After this there is a general diminution of temperature, by the
conduction and radiation of heat outwards: the whole mass goes
on cooling, and is automatically kept largely stirred by irregular
convection-currents, of cooled liquid flowing downwards from the
surface, and of hotter and less dense liquid rising from below.

12. If, as would be the case were the liquid melted iron, or
water, the solidified material is less dense than the liquid at the
same temperature and pressure, a continuous crust would form
all over the surface of the globe; which would grow thicker and
thicker inwards, by freezing of the interior liquid in contact with
it, in virtue of conduction of heat outwards through the crust.
This solid crust, completely enclosing a liquid, would be burst by
the liquid expanding as it freezes (just as a closed water-pipe is
burst by water freezing inside it). If the spherical shell bursts
into many fragments these would all float. The freezing of the

liquid exposed in the openings thus produced in the crust would quickly fill up the gaps; and thus the process of freezing all around would spread inwards to the centre. Such may possibly be the history of the earliest solidification of part of the Earth's mass, forming a metallic central nucleus by coalition of primeval atoms of iron, nickel, gold, platinum, or other dense metals.

13. But the Earth, while not improbably metallic in its central parts, is probably in the main of "earthy" or rocky materials. It seems highly probable that, unlike the materials mentioned in § 12 which expand in freezing, all the "earthy" materials of the Earth contract in freezing. Bischoff*, in experiments made about eighty years ago, found that melted granite, slate, and trachyte, all contract by more than ten per cent. in freezing; and sixteen years ago, Carl Barus† found that diabase (a partially crystalline basaltic rock) is fourteen per cent. denser than melted diabase. He found the melting temperature of diabase to be about 1170° C.; and the late Professor Roberts-Austen, by experiments which he kindly made at my request, found the melting temperatures of several different rocks under ordinary atmospheric pressure‡ to be as follows:

	Melting-point	Error
Felspar	1520° C.	± 30°
Hornblende ....about	1400° „	
Mica	1440° „	± 30°
Quartz	1775° „	± 15°
Basalt ....about	880° „	

14. Go back now to the end of § 11, and consider the Earth after the gaseous stage of its evolution has ended in liquefaction at the centre. The temperature will sink throughout, in virtue of the automatic stirring, and of outward thermal conduction; and the amount of material in the thoroughly liquid condition will increase until the whole globe becomes liquid, except a vaporous or gaseous atmosphere of comparatively small total mass, resting on the liquid all around its spherical surface. The density of this liquid increases from surface to centre, in its earlier stages

---

* *Bulletin de la Soc. Géol.*, 2nd series, Vol. IV. p. 1312.

† *Phil. Mag.* 1893, first half-year, pp. 173—175.

‡ See Addendum to "The Age of the Earth as an Abode fitted for Life," *Phil. Mag.* 1899, first half-year, p. 89.

probably only because of the greater pressure; but ultimately also in consequence of subsidence of the denser chemical ingredients, after the convective currents have become too slack for thorough mixing up of all the materials.

15. Crystalline freezing may begin at the surface, because of the rapid loss of heat by radiation outwards, but each solid crystal sinks because its density is greater than that of the fluid in contact with it. In sinking, it is melted and redissolved by the hotter fluid below. This process goes on until the temperature at every part of the liquid is reduced to that at which some of the ingredients such as quartz, felspar, hornblende, mica, crystallize out of the liquid, under the hydrostatic pressure at the depth of the portion considered. This formation of crystals leaves a mother liquor consisting of ingredients which freeze at a lower temperature. At this stage the portions frozen at the surface do not melt in sinking through the liquid, and they fall down to the centre, or to the central nucleus if there is one. Thus the main rigidification commences in the central region, by conglomeration into a granite of crystals descending through a vast surrounding lava ocean. The solid granite thus formed extends outwards till it comes to the surface.

16. The views regarding the solidification of the Earth, described in § 15, were first published in the *Proceedings* of the Victoria Institute, for 1897, in an article on "The Age of the Earth as an Abode fitted for Life," republished in the *Phil. Mag.*, Jan., 1899. From this article the following passage is quoted:— "If the shoaling of the lava ocean up to the surface had taken place everywhere at the same time, the whole surface of the consistent solid would be the dead level of the liquid lava all round, just before its depth became zero. On this supposition there seems no possibility that our present-day continents could have risen to their present heights, and that the surface of the solid in its other parts could have sunk down to their present ocean depths."

17. Our question is:—How can we explain why the Earth is not at present a mass of solid granite of approximately spherical surface, deviating from sphericity just so much that, if it were covered with water, the water would be at the same depth in every part, when in equilibrium under the combined influence of

gravitation and centrifugal force due to its diurnal rotation ?   A possible, but it seems to me an almost infinitely improbable, explanation is that ocean depths are scars due to collisions with outside bodies, and mountain heights are due to matter left on the Earth by such collisions.   When we look at the scarred surface of the Moon, we cannot but feel that it would be pushing possibilities beyond the verge of absurdity to attribute the geographical features of the Earth, and the corresponding features of the Moon, all to blows received by them, or matter shot down on them, from without.

18.   After solidification, as described in § 15, contraction by loss of heat would almost certainly produce abundance of vertical cracks, proceeding inwards from all parts of the spherical surface (on the same dynamical principle as that which explains the well-known "crackling" seen on the glaze of many kinds of pottery). But there seems no possibility that the wide-spread hollows of the Antarctic, Pacific, Atlantic, and Indian Oceans, and the great areas of elevation in the continents, Europe and Asia, Africa, America, and Australia, and the seven to ten kilometre heights in the Andes and Himalayas, can have followed, by any natural causes, merely from the condition described in § 15.   I have come to this conclusion after careful consideration of the dynamics of cooling and shrinkage, and possible cavitations, through two hundred kilometres below the surface all round the globe.   It seems indeed quite certain that when the Earth came to be almost wholly solid, it must have had in itself some great heterogeneousness of constitution or figure, from which its present geographical condition had its origin.   This heterogeneousness must have had *its* origin in some heterogeneousness of the primordial distributions of atoms: and we must abandon the uniform distribution which we chose in § 9 merely as an illustration.   But we have much more than geography to account for.   We have to account for:—(1) the diurnal rotation of the Earth; (2) the Earth's motion through space, at about thirty kilometres per second, relatively to the Sun; (3) the Sun's motion through space towards a point in the constellation Hercules, first indicated by Sir William Herschel, and more recently estimated at about nineteen kilometres per second, relatively to the average of sufficiently well observed stars.   These three deviations from the spherical and irrotational conditions of §§ 2—16 are, it seems to

me, essentially connected with the explanation of merely geographical heterogeneousness demanded in §§ 16, 17.

19. Any system of bodies large or small, or of atoms, given at rest, and left subject only to mutual gravitational and collisional forces, fulfils throughout all time two laws :—

Law (1). The centre of inertia of the whole system remains at rest.

Law (2). The sum of moments of momentum* of the motions of all the parts, relatively to any axis through the centre of inertia, is zero.

The corresponding laws for a system set in motion in any manner, and left to move under the action of mutual forces only, are as follows :—

Law (1′). The centre of inertia of the whole system moves uniformly in a straight line.

Law (2′). The sum of moments of momentum of the motions of all the parts, relatively to any axis through the centre of inertia, parallel to any fixed line in space, is constant.

Law (3). There is a certain definite line, fixed in direction, through the centre of inertia of the system, such that the sum of moments of momentum round it is greater than that round any other axis through the centre of inertia; and the moment of momentum round every axis perpendicular to it, through the centre of inertia, is zero.

That maximum moment of momentum is called the resultant moment of momentum of the system. Its axis may be called the rotational axis of the system.

---

* (1) The momentum (a name first given in the seventeenth century when mathematicians wrote in Latin, and retained in the nineteenth and twentieth centuries) of a moving particle is the product of its mass into its velocity.

(2) The moment (a nineteenth century name) of momentum of a particle round any axis is the product of its momentum into the shortest distance of its line of motion from that axis, into the sine of the inclination of its line of motion to that axis.

(3) The moment of momentum round any axis of any number of moving particles is the name given to the sum of their moments of momentum round that axis.

It makes no difference to this definition if any set or sets of the particles are rigidly connected to make a rigid body or rigid bodies.

20. Consider a vast assemblage of atoms, or of small bodies, given at rest at any time, distributed in any manner, uniformly or non-uniformly, through any finite volume of space. According to § 19, Law (1), the centre of inertia of the whole assemblage would remain at rest, and the total moment of momentum round every axis through it would remain zero, whatever motions the atoms receive in virtue of mutual gravitational attractions, and mutual repulsions in collision.

21. Consider now separately some part of the whole assemblage, which to avoid circumlocution I shall call part S, including all the primitive atoms or particles which at present form the Solar System, but not including any other great quantity of matter. Part S has, at each instant, a definite resultant moment of momentum round a definite axis through its centre of inertia; and its centre of inertia is, at each instant, moving with a definite velocity in a definite direction. In a vast assemblage such as we were considering in § 20, which may be the whole matter of the universe (finite* in quantity as it may with all probability be supposed to be), let there be denser parts and less dense parts. In the denser parts, there will be gravitational coalition; in the less dense parts, there will consequently be rarefaction. The present existence of the Sun is undoubtedly due to gravitational coalition in some of the denser parts. The velocity of the centre of inertia of the Solar System is due to the gravitational attractions of matter outside S, so also is the moment of momentum of the Solar System round any axis through its centre of inertia. The rarefaction of the distribution of particles, large or small, around S, leaves the matter belonging to S more and more nearly free from force acting on it from without; and it becomes more and more nearly subject to Laws (1′) and (2′) of § 19.

22. The approximately constant momentum of the Solar System in its motion through space is chiefly the momentum of the Sun's motion, because his mass is much greater than the sum of the masses of Jupiter and all the other planets and satellites. On the other hand, the resultant moment of momentum of the motions of all parts of the Solar System, relatively to the fixed line of greatest moment of momentum through the inertial centre of all, is chiefly due to the orbital motion of Jupiter and Saturn, and but in small part (about $\frac{1}{60}$ of the whole) due to the

* See Lord Kelvin's *Baltimore Lectures*, Lect. XVI. § 15.

Sun's axial rotation; as we see by the following table of moments
of momentum, given by Mr See* in his paper of 1905, "Researches
on the physical constitution of the heavenly bodies."

	M. of m. of orbital motion, that of Sun's axial rotation being unity.
Sun	1·0000000
Mercury	0·00069654
Venus	0·035444
Earth	0·0517385
Mars	0·00676526
Jupiter	36·98288
Saturn	14·98374
Uranus	3·26959
Neptune	4·83260

We have not now the simple and direct gravitational coalition,
by motions towards the centre of a spherical assemblage, which
we had in §§ 1—16, and which gave us Homer Lane's beautiful
problem of a spherical gaseous nebula. Our vast assemblage S
has moment of momentum; and its main condensation has led to
the formation of our rotating Sun. Local condensations of smaller
portions of S, each having some share of the moment of momentum
of the whole, have produced the planets, all revolving round the
Sun, and rotating round their axes, in the same general direction:
anti-clockwise, when viewed from the northern side of the general
plane of their orbits.

23. In Kant's and Laplace's Nebular Theory the local condensations, from which have been evolved the planets moving in
their orbits round the Sun, and the satellites moving in *their*
orbits round the planets, were, according to the suggestion
presented to us by Saturn's rings, supposed to begin as rings of
detached particles, which later became gravitationally drawn
together into spheroidal groups, and formed ultimately liquid or
solid approximately spherical bodies. This may probably be the
true history of many of the planets and satellites; but Sir George
Darwin† has given the very important suggestion that the

---

* *Astr. Nachr.* Bd. 169, Nov. 1905.

† *Phil. Trans.* 1879, p. 447, "On the Precession of a Viscous Liquid and on
the remote History of the Earth;" *Phil. Trans.* 1880, p. 713, "On the Secular
Changes in the Elements of the Orbit of a Satellite revolving about a tidally
distorted Planet;" *British Association Report*, 1905, p. 3, Presidential Address.

separation from a planet of material to become a satellite may in some cases have been a single portion of the mass, breaking away from what was earlier a rotating mass of liquid, in the shape of a figure of revolution, contracting by loss of heat, and therefore rotating with greater and greater angular velocity as it became denser.

24. Suppose for example the mass which is now Earth and Moon to have been at one time a single oblate spheroid of revolution. Its figure would then have been exactly elliptic, if its rotational angular velocity, and its density, were each equal from surface to centre. If it was denser in the central regions, its figure would have been an oblate figure of revolution, but not in general exactly elliptic in its meridional sections. While the spheroid shrinks in cooling it becomes more and more oblate, till, all round the equator, gravity is exactly balanced by centrifugal force; or till the spheroid becomes lopsided, as suggested in § 25 below Farther continued shrinkage cannot give a stable oblate figure of revolution. It might cause an equatorial belt to be detached from the main body; or the result might be as suggested in §§ 25, 26, below.

25. Poincaré's "pear-shaped" figure of equilibrium of a rotating liquid suggests the idea that the first instability produced by cooling and shrinkage, with constant moment of momentum, may possibly give rise to a stable figure with a protrusion on one side of the centre of what was the equatorial circle, and a flattening of the surface on the other side of the centre of inertia. This idea is to some degree supported by the elaborate and powerful mathematical investigations of Poincaré* and Darwin† on "pear-shaped" figures of liquids rotating in stable equilibrium, under the influence of gravity and centrifugal force.

26. Continued cooling and shrinkage would produce more and more protrusion on one side of the centre of inertia, until the protrusion becomes unstable, and a comparatively small portion

---

* H. Poincaré, "Sur la Stabilité de l'Equilibre des Figures Pyriformes," *Phil. Trans.* 1902.

† Sir George Darwin, "On the Pear-shaped form of Equilibrium of a rotating Mass of Liquid," *Phil. Trans.* 1902 ; "On the Stability of the Pear-shaped figure of Equilibrium of a rotating Mass of Liquid," *Phil. Trans.* 1903 ; "On the Figure and Stability of a Liquid Satellite," *Phil. Trans.* 1906.

of the whole liquid breaks away from the main mass, at the thin end of the "pear."

That separation must have been a sudden and violent catastrophe, however gradual may have been the changes of figure and distribution of matter which led to it. If at the time when it took place, the whole material was perfectly liquid, the act of separation would leave no permanent marks on either of the two bodies. After some moderate time of subsidence from the violent oscillations suddenly produced by the catastrophe, the Earth and Moon would have subsided into the comparatively tranquil conditions of rotating liquid spheroids, revolving round the centre of inertia of the two; disturbed from hydrostatic equilibrium only by the interior convective currents due to cooling at the surfaces; and with no prospect of ever freezing into the largely unsymmetrical shapes which we now see on the visible half of the Moon's surface, and over the whole surface of the Earth.

27. To account for the evolution of present configurations and conditions, it seems to me that we must suppose the material of Moon and Earth, at the time of the separational catastrophe, to have reached some such condition as that described in § 15 :— a conglomeration of crystals with still liquid lava filling all the interstices between them. Such a conglomeration would have plasticity enough to pass through the changes of figure which, according to Darwin's theory, the material of Moon and Earth must have experienced before the separational catastrophe: and yet may have possessed sufficient subpermanent or permanent resistance against change of shape to allow them to keep permanent traces of the wounds left on the two bodies by the convulsive separation.

28. The scar, and subsequent surgings, left on the semi-plastic Earth by the tearing away of the Moon from it might account for persisting deviation from rotational symmetry, and from equilibrium of gravity and centrifugal force, as great as that which is presented by the elevations of Africa, Asia, Europe, America, and the depths of the Atlantic, Pacific, and Indian Oceans. If, at the time when the Moon left the Earth, the material was all in the semi-solid semi-plastic condition of granite conglomerate, with a mother liquor of melted basalt in the interstices among the crystals, this *quasi* unset Portland cement,

constituting the two bodies, might well in the course of fifty or one hundred million years become as nearly solid as we know both the Earth and Moon to be at present. It must however be quite understood that the present features of the Earth, with mountains and ravines, and ocean depths, have been produced by long continued geological actions of upheavals and erosion.

29. Immediately after the separation, the Moon, about $\frac{1}{81}$ of the Earth's mass, would begin moving from the perigee of a somewhat approximately elliptic orbit round the centre of inertia of Earth and Moon, much disturbed on account of the great and violently changing deviations from sphericity of the two masses. The period of this orbital motion of the two bodies round their centre of inertia would be longer than the rotational period of the Earth, which would be but little changed by the catastrophe. In becoming rounded into a spheroidal form, the Moon would come to rotate round its own axis in a somewhat shorter period than that of the whole mass before the separation. Thus, in the beginning of the new regime, we have three different periods; the shortest being the rotational period of the Moon round an axis through her centre of inertia; somewhat longer than this the rotational period of the Earth; and considerably longer than it, the orbital period of the two bodies round their centre of inertia.

30. The changes of shape of the two semi-plastic spheroids in their subsequent motions under the influence of mutual gravitation between all their parts, would give rise to a loss of energy: while the total moment of momentum would remain unchanged. The main action would be loss of kinetic energy of the Earth and Moon, by transformation into heat of quasi-tidal work within the two bodies. Essentially concomitant features would be augmentation of the distance between them, involving work done against mutual gravity, and gradual transformation of the moment of momentum of their rotations into augmentation of the moment of momentum of their orbital motions round the centre of inertia of the two. The energy of the initial rotation of the Moon would be small compared with that of the Earth. The whole kinetic energy of the rotations, and the motions of centres of inertia, of the two bodies, at the present time exceeds by a relatively small quantity the present kinetic energy of the Earth's rotation. The

work done in separating the Moon to its present distance from the Earth, and in giving it the kinetic energy of its orbital motion, has been wholly drawn from the Earth's rotational kinetic energy at the time of the disruption, with the exception of a small contribution derived from the Moon's initial kinetic energy of rotation. A comparatively early result of the motions of the two bodies must have been to bring the Moon to keep always the same face to the Earth as she does at the present time.

31. Sir George Darwin, with comprehensively penetrating dynamical insight, has traced the course of events following the stage reached in § 29. He has given a reasonable account of the evolution of the present eccentricity of the Moon's orbit; and he has made the remarkable and important discovery that the axis of the Earth's rotation could not remain as it is at the stage of § 29, perpendicular to the plane of the orbital motion of Earth and Moon. We might readily enough work out the general character of the motions and transformations that would follow the stage of § 29, if the Earth's and Moon's rotational axes did in reality remain perpendicular to the plane of their orbital motions. But Darwin finds that this possible and easily understood association of motions would be unstable; and that the slightest deviation from exact perpendicularity of the Earth's axis to the orbital plane would become, not diminished but, augmented by the Earth's viscous resistance against change of shape. With this hint it is almost as easy for us to see, by dynamical reasoning, that the Earth's axis must, through millions of years, have become more and more oblique to the orbital plane, as it is, for us, with Archibald Smith's hint*, to see that a "teetotum" or a boy's spinning-top, having a well rounded bearing point, and set to spin at a sufficiently high speed round an axis oblique to the vertical, and dropped on a hard horizontal plane, will in a short time, perhaps less than a minute, be found spinning round a fixed exactly vertical axis ("sleeping"), and will go on so for ever, if the materials of the top and plane are perfectly hard, and if there is no resistance of the air.

32. Darwin's theory of the birth of the Moon might seem improbable; might seem even an extravagant attempt in evolu-

---

* "Note on the Theory of the Spinning Top," *Camb. Math. Journal*, 1839, Vol. I. p. 42.

tionary philosophy, insufficiently founded on knowledge. In reality it is rendered highly probable, it is indeed forced upon us, by tracing backwards to earlier and earlier times the dynamical antecedents of the present conditions of Earth, Moon, and Sun.

33. A hundred years before the doctrine of energy thoroughly entered the minds of mathematicians and naturalists, Kant made known the truth that the Earth's rotational velocity is diminished by tidal friction. When we consider the dynamics on which this statement is founded, we see that it implies reactive forces gravitationally exerted on the Sun and Moon by terrestrial waters. Ignoring the Sun, as less influential than the Moon in respect to the tides, we "see that the mutual action between the Moon and the Earth must tend, in virtue of the tides, to diminish the rapidity of the Earth's rotation, and increase the moment of the Moon's motion round the Earth*."..."The tidal deformation of the water exercises the same influence on the Moon as if she were attracted not precisely in the line towards the Earth's centre, but in a line slanting very slightly, relatively to her motion, in the direction forwards. The Moon, then, continually experiences a force forward in her orbit by reaction from the waters of the sea. Now, it might be supposed for a moment that a force acting forwards would quicken the Moon's motion; but, on the contrary, the action of that force is to retard her motion. It is a curious fact easily explained, that a force continually acting forward with the Moon's motion will tend, in the long run, to make the Moon's motion slower, and increase her distance from the Earth*."

34. Thus we see that in the present regime the Moon is getting farther and farther from the Earth, and the Earth's rotational velocity is becoming less and less; the sum of moments of momentum being thus kept constant. Hence in more and more ancient times, the Moon must have been nearer and nearer to the Earth, and the Earth's rotational velocity must have been greater and greater. Trace then the course of motions backwards for a sufficient number of millions of years, and we find the Earth much hotter than at present, and in the semi-solid semi-plastic condition described in § 28 above. The distance of the Moon

---

* Quoted from Sections 7 and 14 respectively of an address by Sir William Thomson, to the Geological Society of Glasgow, " On Geological Time," Feb. 27, 1868; republished in Lord Kelvin's *Popular Lectures and Addresses*, Vol. II; see pp. 21 and 33.

from the Earth must then as now have been increasing, and the
Earth's rotational velocity then as now diminishing.   But these
two changes must have been much more rapid then than now,
because of the viscosity of the semi-solid material of the Earth,
and because of the Moon's shorter distance from the Earth and
therefore greater gravitational influence on the Earth, then
than now.   Sir George Darwin had perfect right to trace the
regime backwards, until there was contact and continuity between
the Earth and Moon.   The continuity of the whole mass must
have come to an end with the sudden and violent catastrophe
described in § 28 above, to which Sir George Darwin boldly went
back with sober truthfulness.   It is conceivable that meteorites
large or small may have at various times produced disturbances;
but I cannot see any probability for any other history of Earth
and Moon, differing materially from that which Darwin has
given us.

35.   Returning now to § 1, an unanswerable question occurs:—
Were the primordial atoms relatively at rest in the most ancient
time, or were they moving with velocities, relative to fixed axes
through the centre of inertia of the whole, sufficiently great to
give any considerable contribution to the present kinetic energy
of the universe?   It is conceivable that all the atoms were
relatively at rest in the most ancient time, and that "the potential
energy of gravitation may be in reality the ultimate created
antecedent of all the motion, heat, and light at present in the
universe *."

---

* Quoted from "On Mechanical Antecedents of Motion, Heat, and Light,"
Brit. Assoc. Report, Part II. 1854; *Edin. New Phil. Journal*, Vol. I. 1855 ; *Comptes
Rendus*, Vol. XL. 1855 ; republished in Sir William Thomson's *Math. and Phys.
Papers*, Vol. II. p. 34.

161. THE PROBLEM OF A SPHERICAL GASEOUS NEBULA*.

[From *Edinb. Roy. Soc. Proc.* Vol. XXVIII. [MS. recd. March 9, 1908], Issued
separately May 6, 1908, pp. 259—302; *Phil. Mag.* Vol. XV. June 1908,
pp. 687—711; Vol. XVI. July 1908, pp. 1—23.]

THIS paper was begun about the close of 1906, in order to fulfil a promise
given at the end of the paper "On the Convective Equilibrium of a Gas under
its own Gravitation only," published in the *Philosophical Magazine*, 1887;
and part of it was communicated by Lord Kelvin to the Royal Society of
Edinburgh at its meeting on the 21st January, 1907. Since then, however,
important additions have been made to it, and the subject has been dealt
with more fully than was originally intended. Unfortunately the manuscript
was left incomplete at Lord Kelvin's death. It ended with § 35.

However, from information which I received from Lord Kelvin while
carrying out the earlier work connected with the paper, I have been able to
write the sections from § 36 to the end. These complete all that Lord Kelvin
desired to include in this communication; and they express, I believe, the
views he held while writing the earlier sections.

The statement of mathematical solutions and numerical results separately,
as an Appendix to the paper, under my own name, is in accordance with Lord
Kelvin's wishes.

GEORGE GREEN,
Secretary.

1.† IF a fluid globe were given with any arbitrary distribution
of temperature, subject only to the condition that it is uniform
throughout every spherical surface concentric with the boundary,
the cooling, by radiation into space, and consequent augmentation
of density of the fluid at its boundary, would immediately give
rise to an instability according to which some parts of the outermost portions of the globe would sink, and upward currents would

* Communicated by Dr J. T. Bottomley, F.R.S.
† § 1 is extracted from "On Homer Lane's Problem of a Spherical Gaseous
Nebula," *Nature*, Feb. 14, 1907.

consequently be developed in other portions. In any real fluid,
whether gaseous or liquid, this kind of automatic stirring would
tend to go on until a condition of approximate equilibrium is
reached, in which any portion of the fluid descending or ascending
would, by the thermodynamic action involved in change of
pressure, always take the temperature corresponding to its level,
that is to say, its distance from the centre of the globe. The
condition thus reached, when heat is continually being radiated
away from the spherical boundary, is not perfect equilibrium. It
is only an approximation to equilibrium, in which the temperature
and density are each approximately uniform at any one distance
from the centre, and vary slowly with time, the variable irregular
convective currents being insufficient to cause any considerable
deviation of the surfaces of equal density and temperature from
sphericity.

2. The problem of the convective equilibrium of temperature,
pressure, and density, in a wholly gaseous, spherical fluid mass,
kept together by mutual gravitation of its parts, was first dealt
with by the late Mr Homer Lane, who, as we are told by Mr
T. J. J. See, was for many years connected with the U.S. Coast
and Geodetic Survey at Washington. His work was published in
the *American Journal of Science*, July 1870, under the title " On
the Theoretical Temperature of the Sun*."

In a letter to Joule, which was read before the Literary and

* The real subject of this paper is that stated in the text above. The application of the theory of gaseous convective equilibrium to sun heat and light is very
largely vitiated by the greatness of the sun's mean density (1·4 times the standard
density of water). Common air, oxygen, and carbonic acid gas show resistance to
compression considerably in excess of the amount calculated according to Boyle's
Law, when compressed to densities exceeding four, or five, or six, tenths of the
standard density of water. There seems strong reason to believe that every fluid
whose density exceeds a quarter of the standard density of water resists compression much more than according to Boyle's Law, whatever be the temperature
of the fluid, however high, or however low. We may consider it indeed as quite
certain that a large proportion of the sun's interior, if not indeed the whole of the
sun's mass within the visible boundary, resists compression much more than
according to Boyle's Law. It seems indeed most probable that the boundary,
which we see when looking at the sun through an ordinary telescope, is in reality
a surface of separation between a liquid and its vapour ; and that all the fluid
within this boundary resists compression so much more than according to Boyle's
Law that it does not even approximately satisfy the conditions of Homer Lane's
problem ; and that in reality its density increases inwards to the centre vastly less
than according to Homer Lane's solution (see § 56 below).

Philosophical Society of Manchester, January 21, 1862, and published in the Memoirs of the Society under the title, "On the Convective Equilibrium of Temperature in the Atmosphere*," it was shown that natural up and down stirring of the earth's atmosphere, due to upward currents of somewhat warmer air, and return downward flow of somewhat cooler air, in different localities, causes the average temperature of the air to diminish from the earth's surface upwards to a definite limiting height, beyond which there is no air. It was also shown that, were it not for radiation of heat across the air, outwards from the earth's surface, and inwards from the sun, the temperature of the highly rarefied air close to the bounding surface would be just over absolute zero; that is to say, temperature and density would come to zero at the same height as we ideally rise through the air to the boundary of the atmosphere. Homer Lane's problem gives us a corresponding law of zero density and zero temperature, at an absolutely defined spherical bounding surface (see § 27 below). In fact it is clear that if in Lane's problem we first deal only with a region adjoining the spherical boundary, and having all its dimensions very small in comparison with the radius, we have the same problem of convective equilibrium as that which was dealt with in my letter to Joule.

3. According to the definition of "convective equilibrium" given in that letter, any fluid under the influence of gravity is said to be in convective equilibrium if density and temperature are so distributed throughout the whole fluid mass that the surfaces of equal temperature, and of equal pressure, remain unchanged when currents are produced in it by any disturbing influence so gentle that changes of pressure due to inertia of the motions are negligible. The essence of convective equilibrium is that if a small spherical or cubic portion of the fluid in any position P is ideally enclosed in a sheath impermeable to heat, and expanded or contracted to the density of the fluid at any other place P', its temperature will be altered, by the expansion or contraction, from the temperature which it had at P, to the actual temperature of the fluid at P'. The formulas to express this condition were first given by Poisson. They are now generally known as the equations of adiabatic expansion or contraction, so

* Republished in Sir William Thomson's *Math. and Phys. Papers*, Vol. III. p. 255.

named by Rankine. They may be written as follows, for the ideal case of a perfect gas:

$$\frac{p}{p'} = \left(\frac{\rho}{\rho'}\right)^k, \quad \frac{\rho}{\rho'} = \left(\frac{t}{t'}\right)^{\frac{1}{k-1}}, \quad \frac{p}{p'} = \left(\frac{t}{t'}\right)^{\frac{k}{k-1}} \quad \ldots\ldots(1), (2), (3);$$

where (t, ρ, p), (t', ρ', p') denote the temperatures, densities, and pressures, at any two places in the fluid (temperatures being reckoned from absolute zero); and k denotes the ratio of the thermal capacity of the gas when kept at constant pressure to its thermal capacity at constant volume, which, according to a common usage, is for brevity called "the ratio of specific heats." For dry air, at any temperature, and at any density within the range of its approximate fulfilment of the gaseous laws, we have

$$k = 1\text{·}41, \quad \frac{k-1}{k} = \text{·}291, \quad \frac{k}{k-1} = 3\text{·}44\ldots\ldots\ldots\ldots(4).$$

For monatomic gases we have

$$k = \frac{5}{3}, \quad \frac{k-1}{k} = \frac{2}{5}, \quad \frac{k}{k-1} = \frac{5}{2} \quad \ldots\ldots\ldots\ldots\ldots(5).$$

For real gases, we learn from the Kinetic Theory of Gases, and by observation, that k may have any value between 1 and $1\frac{2}{3}$, but that it cannot have any value greater than $1\frac{2}{3}$, or less than 1.

4. To specify fully the quality of any gas, so far as concerns our present purpose, we need, besides k, the ratio of its specific heats, just one other numerical datum, the volume of a unit mass of it at unit temperature and unit pressure. This, which we shall denote by S, is commonly called the specific volume; and its reciprocal, $1/S$, we shall call the specific density (D) of the gas. In terms of this notation, the Boyle and Charles gaseous laws are expressed by either of the equations

$$pv = St, \quad \text{or} \quad p = \rho St\ldots\ldots\ldots\ldots(6), (6');$$

where p, v, ρ, denote respectively the pressure, the volume of unit mass, and the density of the gas at temperature t, reckoned from absolute zero. Our unit of temperature throughout the present paper will be 273° C. Thus the Centigrade temperature corresponding to t in our notation is $273(t-1)$.

5. In virtue of § 4, what is expressed by (1), (2), (3), equivalent as they are to two equations, may now, for working

258   COSMICAL AND GEOLOGICAL PHYSICS   [161

purposes, be expressed much more conveniently by the single
formula (6), together with the following equation—

$$p = A\rho^k \quad \dots\dots\dots\dots\dots\dots(7);$$

where A denotes what we may call the Adiabatic Constant, which
is what the pressure would be, in adiabatic convective equilibrium,
at unit density, if the fluid could be gaseous at so great a density
as that.

6.   Looking to (6), remark that p being pressure per unit of
area, the dimensions of pv are $L^{-2} \times L^3$ or L, if we express force in
terms of an arbitrary unit, as in § 10 below; therefore S, though
we call it specific volume, is a length. It is in fact, as we see by
(9) below, equal to the height of the homogeneous atmosphere
at unit temperature, in a place for which the heaviness of a unit
mass is the force which we call unity in the reckoning of p.

7.   In the definition of what is commonly called the "height
of the homogeneous atmosphere," and denoted by H, an idea very
convenient for our present purpose is introduced. Let p be the
pressure and ρ the density, at any point P within a fluid, liquid
or gaseous, homogeneous or heterogeneous, in equilibrium under
the influence of mutual gravitation between its parts; and let g
be the gravitational attraction on a unit of mass at the position
P. Let

$$g\rho H = p \quad \dots\dots\dots\dots\dots\dots(8).$$

This means that H is the height to which homogeneous liquid,
of uniform density ρ, ideally under the influence of uniform
gravity equal to g, must stand in a vertical tube to give pressure
at its foot equal to p.

8.   The idea expressed by (8) is useful in connection with
questions connected with internal pressure throughout a spherical
liquid mass, such as the sun. It is also useful when we are
considering pressure and temperature in gaseous fluids, such as
our terrestrial atmosphere, or the outermost parts of the sun;
which may be regarded as practically gaseous where the density
is anything less than ·1.

9.   For a perfect gas, (8) divided by ρ, becomes

$$gH = St \quad \dots\dots\dots\dots\dots\dots(9).$$

By this we see, what is interesting to remark, that for the same
temperature and same gaseous material, the "height of the

homogeneous atmosphere" is the same for the air at the earth's surface and for the air at any height above the surface; and is the same for different barometric pressures. For different temperatures, it varies as the absolute temperature. For different gases at the same temperature, it is proportional to their specific volumes. For different forces of gravity, it is inversely proportional to them.

10. Even for cosmical reckonings in respect to our present subject, and in many and varied terrestrial reckonings, it is convenient to take as unit of force the heaviness in mid-latitudes of the unit of mass. The unit of mass, for all nations and peoples of the earth, must for general convenience be founded on the existing French Metrical System. The unit may, according to the particular magnitude or character of substance of which the mass or quantity is to be specified, be conveniently taken as a milligram, or a gram, or a kilogram, or a metric ton (one thousand kilograms), or 10^9 tons.

11. The choice of unit force as mean terrestrial heaviness of unit mass is very convenient for ordinary earthly purposes, but language in which it is adopted is, unless properly guarded and tacitly understood, always liable to ambiguity as to whether force or quantity of matter is meant. Thus if (using for a moment the moribund British Engineering reckonings in pounds, inches, etc.) we speak of 73 pounds of lead, there is no doubt that we mean quantity of a particular kind of matter; but if we speak of 73 pounds per square inch (which might be 73 pounds of lead, or of iron, or of stone) we mean a force. If we call the pressure on the boiler of a ship 73 pounds per square inch, we mean a somewhat greater pressure when the ship is in middle or northern latitudes than when she is on the equator; though the difference is, for pressures on safety-valves, practically negligible, being for example three-tenths per cent. between the equator and the latitude of Glasgow or Edinburgh.

12. In the present paper we shall take as unit of mass the mass of a cubic kilometre of water at standard density (which is 10^9 metric tons); and we shall take its heaviness in mid-latitudes as unit of force. This means taking for g in (8) and (9), and in all future formulas, the ratio of gravity at the place under consideration to terrestrial gravity in mid-latitudes. Hence (remembering that in § 4 we have chosen for our unit temperature

reckoned from absolute zero the temperature of melting ice, being equal to 273° Centigrade above absolute zero) we see by (8) that S is simply the height in kilometres of the Homogeneous Atmosphere in mid-latitudes, at the freezing temperature. Thus, from known measurements of densities, we have the following table* of values of S for several different gases:—

Gas				S
Air	...	...	...	7·988 kilometres
Ammonia	...	...	...	13·414 ,,
Argon	...	...	...	5·767 ,,
Carbon dioxide		...	...	5·232 ,,
Carbon monoxide		...	...	8·370 ,,
Chlorine	...	...	...	3·297 ,,
Helium	...	...	...	58·354 ,,
Hydrogen	...	...	...	114·76 ,,
Nitrogen	...	...	...	8·256 ,,
Oxygen	...	...	...	7·233 ,,
Sulphur dioxide		...	...	3·709 ,,

13. Consider now convective equilibrium in any part of a wholly gaseous globe, or in any part of a fluid globe so near the boundary as to have density small enough to let it fulfil the gaseous laws. Let z be depth measured inwards from any convenient point of reference. The differential equation of fluid equilibrium is

$$dp = g\rho\, dz \quad\ldots\ldots\ldots\ldots\ldots\ldots(10).$$

Now, if the equilibrium is convective, we have by (3)

$$dp = \frac{k}{k-1}\frac{p'}{t'}\left(\frac{t}{t'}\right)^{\frac{1}{k-1}} dt \quad\ldots\ldots\ldots\ldots\ldots(11).$$

Using this, and (2), in (10), and dividing both members by $(t/t')^{\frac{1}{k-1}}$, we find

$$\frac{dt}{dz} = \frac{k-1}{k}\frac{g\rho' t'}{p'} \quad\ldots\ldots\ldots\ldots\ldots(12).$$

Whence, by (6), we find $\quad \dfrac{dt}{dz} = \dfrac{k-1}{k}\dfrac{g}{S} \quad\ldots\ldots\ldots\ldots\ldots(13);$

and, ((2) repeated) $\quad \dfrac{\rho}{\rho'} = \left(\dfrac{t}{t'}\right)^{\frac{1}{k-1}} \quad\ldots\ldots\ldots\ldots\ldots(14).$

---

* If instead of taking 10^9 tons as our unit of mass we take a gram, the numbers in this table must each be multiplied by 10^6, and they will then be the values of S in centimetres instead of in kilometres.

14. These are exceedingly important and interesting results. By (13) we see that in any part of a wholly gaseous spherical nebula, or in a gaseous atmosphere around a solid or liquid nucleus, in convective equilibrium, sufficiently stirred to have the same chemical constitution throughout, the temperature-gradient of increase inwards is in simple proportion to the force of gravity at different distances from the centre. We also see that in gaseous spherical nebulas of different chemical constitutions, or in gaseous atmospheres of different chemical constitutions, around solid or liquid nucleuses, the temperature-gradients at places of the same gravity are simply proportional to the values of $(k-1)/(kS)$ for the different gases or gaseous mixtures.

15. For the terrestrial atmosphere we have by (4) $k/(k-1) = 3.44$, and by the table in § 12, $S = 7.988$ kilometres. The temperature-gradient according to (13) is therefore, at the rate of our unit of temperature, or 273 degrees Centigrade, per 27·5 kilometres; or 1° C. in 100·6 metres. This is much greater than the temperature-gradient found by Welsh, in balloon ascents of about fifty years ago, which was only 1° C. in 161 metres*. Joule, with whom I had been in discussion on the subject in 1862, suggested to me that the discrepancy might be accounted for by the condensation of vapour in upward currents of air. In endeavouring to test this suggestion, I made some calculations of which results are shown in the following table, extracted from a table given in my paper of 1862, referred to in § 2 above.

Temperature centigrade or $t - 273.7$	Elevation from Earth's surface required to cool moist air by 1° C.
	$\dfrac{dx}{-dt}$
0	Metres 152
5	168
10	186
15	207
20	229
25	252
30	274
35	284

* Mr Shaw informs me that much investigation in later times gives a general average mean gradient of 1° C. per 164 metres. This is very nearly the same as it would be with no disturbance from radiation in air saturated with moisture, at 4° C.

16.   From this we see that an ascending current of moist air at 3° C. would sink in temperature at about the rate of 1° C. in 161 metres of ascent.   This is exactly Welsh's gradient; "and we may conclude that at the times and places of his observations the lowering of temperature upwards was nearly the same as that which air saturated with moisture [at 3° C.] would experience in ascending*."   But it is not to be supposed, indeed it cannot have been the case, that his observations were made in a single ascent through cloud.   "It is to be remarked that except when the air is saturated, and when, therefore, an ascending current will always keep forming cloud, the effect of vapour of water, however near saturation, will be scarcely sensible on the cooling effect of expansion †."

17.   But, considering our terrestrial atmosphere as a whole, and the complicated circumstances of winds, and rain, and snow, and its heatings by radiations from the sun, and its coolings by radiation into starlit space, and its heatings and coolings by radiations to land and sea in different latitudes, we may feel sure that Joule's suggestion shows a cause contributing importantly to the general average temperature-gradient being less than it would be in dry air in convective equilibrium.

18.   For the solar atmosphere, we have approximately, $g = 28$ (28 times middle latitude gravity at the earth's surface).   By way of example, we may take S and k the same as for the terrestrial atmosphere, as we have not sufficient knowledge from spectrum analysis to allow us to guess other probable values of S and k for the mixture of gases constituting the upper parts of the sun's atmosphere, than those we know for the mixture of Oxygen, Nitrogen, Argon, and Carbonic Acid, which in the main constitutes our terrestrial atmosphere.   Thus in the upper atmosphere of the sun, if in purely convective equilibrium, and undisturbed by radiations and other complications, the temperature would increase at the rate of 280 degrees Centigrade per kilometre downwards, and, looking forward to § 27 below, we see that the increase of temperature would start from absolute zero at the boundary, where density, pressure, and temperature, are all zero.   It would

---

* Quoted from the Manchester paper above referred to, *Math. and Phys. Papers*, Vol. III. p. 260.

 † *Ibid.*

require very robust faith in the suggestion of convective equilibrium for the gaseous atmosphere of the sun to believe in $+7°$ C.
being the actual temperature of the sun's atmosphere at one
kilometre below the boundary. I am afraid I cannot quite profess
that faith. It seems to me that the enormous radiation from
below would, if the upward and downward currents were moderately
tranquil, overheat the air in the uppermost kilometre of the sun's
atmosphere to far above the temperatures ranging from $-273°$
Centigrade to $+7°$ Centigrade, calculated as above from the adiabatic convective theory.

19. Keeping, however, for the present by way of example,
to the calculated results of this theory, with the data for S and k
chosen in § 15, we find at ten and at fifty kilometres below the
boundary, the temperatures, reckoned in Centigrade degrees above
absolute zero, would be respectively 2800 and 14000. Calling
these temperatures t' and t, and the densities at the same places
ρ' and ρ, we find by (14)

$$\frac{\rho}{\rho'} = \left(\frac{14000}{2800}\right)^{\frac{5}{2}} = 55\text{·}9 \quad \dots\dots\dots\dots\dots(15).$$

Suppose for example ρ' to be ·001 (1/1000 of the density of water),
we shall have $\rho = ·056$. This last is nearly but not quite too
great a density for approximate fulfilment of the gaseous laws for
the same gaseous mixture as our air. Thus, if not too much
disturbed by radiation of heat from below, the uppermost fifty
kilometres of the sun's atmosphere might be quite approximately
in gaseous convective equilibrium; with density and temperature
augmenting from zero at the boundary, to density ·056, and
temperature 14000 Centigrade degrees above absolute zero, at
the fifty kilometres depth. But, going down fifty kilometres
deeper, we find that the temperature at one hundred kilometres
depth would be 28000°, and the density would be 316. This
density is much too great to allow even an approximate fulfilment
of the gaseous laws, by any substance known to us, even if its
temperature were as high as 28000°. This single example is
almost enough to demonstrate that the approximately gaseous
outer shell of the sun cannot be as much as 100 kilometres thick
—a conclusion which may possibly be tested, demonstrated, or
contradicted, by sufficiently searching spectroscopic analysis. The
character of the test would be to find the thickness of the

COSMICAL AND GEOLOGICAL PHYSICS

outermost layer from which the bright spectrum lines proceed. If it were ·1″ as seen from the earth, it would be 73 kilometres thick.

20. Considering the great force of gravity at the sun's surface (about 28 times terrestrial gravity), it is scarcely possible to conceive that any fluid, composed of the chemical elements known to us, could be gaseous in the sun's atmosphere at depths exceeding one hundred kilometres. I am forced to conclude that the uppermost luminous bright-line-emitting layer of our own sun's atmosphere, and of the atmosphere of any other sun of equal mass, and of not greater radius, cannot probably be as much as one hundred kilometres thick.

21. There must have been a time, now very old, in the history of the sun when the gravity at his boundary was much less than 28, and the thickness of his bright-line-emitting outermost layer very much greater than one hundred kilometres. Going far enough back through a sufficient number of million years, in all probability we find a time when the sun was wholly a gaseous spherical nebula from boundary to centre, and a splendid realization of Homer Lane's problem.   The mathematical solution of Homer Lane's problem will, for a spherical gaseous nebula of given mass, tell exactly what, under the condition of convective equilibrium, the density and temperature were at any point within the whole gaseous mass, when the central density was of any stated amount less than ·1 ; on the assumption that we know the specific volume (S) and the ratio of specific heats (k) for the actual mixture of gases constituting the nebula. It will also allow us to find, at the particular time when any stated quantity of heat has been radiated from the gaseous nebula into space, exactly what its radius was, what its central temperature and density were, and what were the temperature and density at any distance from the centre.   Thus, on the assumption of S and k known, we have a complete history of the sun (or any other spherical star) for all the time before the central density had come to be as large as ·1.

22. To pass from the case of convective equilibrium in a gaseous atmosphere so thin that the force of gravity is practically constant throughout its thickness, to the problem of convective equilibrium through any depth, considerable in comparison with

the radius, or through the whole depth down to the centre, provided the fluid is gaseous so far, we have only to use (13) and (14), with the proper value of g, varying according to distance from the centre. Remembering that we are taking g in terms of terrestrial gravity, and that the mean density of the earth is 5·6, in terms of the standard density of water, which we are taking as our unit density, we have the following expression for g, in any spherical mass, m, having throughout equal densities, ρ, at equal distances, r, from the centre:—

$$g = \frac{m/r^2}{E/e^2} = \frac{3}{5\cdot6 \cdot e} r^{-2} \int_0^r dr\, r^2 \rho \quad \dots\dots\dots\dots(16),$$

where E denotes the earth's mass, and e the earth's radius. This expression we find by taking g as the force of gravity due to matter within the sphere of radius r, according to Newton's gravitational theorem, which tells us that a spherical shell of matter having equal density throughout each concentric spherical surface exerts no attraction on a point within it. Using this in (13) of § 13, with $dz = -dr$; multiplying both members by r^2, and introducing m to denote the mass of matter within the spherical surface of radius r, we find

$$-r^2 \frac{dt}{dr} = \frac{3}{5\cdot6 \cdot e} \frac{k-1}{kS} \int_0^r dr\, r^2 \rho = \frac{3}{5\cdot6 \cdot e} \frac{k-1}{kS} \frac{m}{4\pi} \quad \dots(17).$$

Differentiating (17) with reference to r, we find

$$-\frac{d}{dr}\left[r^2 \frac{dt}{dr}\right] = \frac{3}{5\cdot6 \cdot e} \frac{k-1}{kS} r^2 \rho \quad \dots\dots\dots\dots(18).$$

23. By (6), and (7), of §§ 4, 5, we find

$$\rho = \left(\frac{St}{A}\right)^{\kappa}, \text{ where } \kappa = \frac{1}{k-1} \quad \dots\dots\dots(19),(20).$$

Eliminating ρ from (18) by (19), we find

$$-\frac{d}{dr}\left[r^2 \frac{dt}{dr}\right] = \frac{r^2 t^{\kappa}}{\sigma^2}, \text{ where } \sigma^2 = \frac{5\cdot6 \cdot e\,(\kappa+1)\,A^{\kappa}}{3S^{\kappa-1}} \quad \dots(21),(22).$$

24. By putting

$$r = \sigma/x \quad \dots\dots\dots\dots\dots\dots\dots(23),$$

we reduce (21) to the very simple form,

$$\frac{d^2 t}{dx^2} = -\frac{t^{\kappa}}{x^4} \quad \dots\dots\dots\dots\dots\dots(24);$$

the equation of the first and third members of (17), modified by (20) and (23), gives

$$\frac{m}{E} = \frac{(\kappa + 1) S\sigma}{e^2} \frac{dt}{dx} \quad \dots\dots\dots\dots\dots(25).$$

25.   Let $t = \mathfrak{F}(x)$ be any particular solution of this equation; we find as a general solution, with one disposable constant C,

$$t = C\mathfrak{F}\left[xC^{-\frac{1}{2}(\kappa-1)}\right] \quad \dots\dots\dots\dots\dots(26),$$

which we may immediately verify by substitution in (24).   Here $\mathfrak{F}(x)$ may denote a solution for a gaseous atmosphere around a solid or liquid nucleus, or it may be the solution for a wholly gaseous globe, in which case $\mathfrak{F}(x)$ will be finite, and $\mathfrak{F}'(x)$ will be zero, when $x = \infty$.   Each solution $\mathfrak{F}(x)$ must belong to one or other of two classes:—

Class A: that in which the density increases continuously from the spherical boundary to a finite maximum at the centre. In this class we have $d\rho/dr = 0$ $(dt/dr = 0)$, when $r = 0$; or, which amounts to the same, $d\rho/dx = 0$ $(dt/dx = 0)$, when $x = \infty$.

Class B: that in which, in progress from the boundary inwards, we come to a place at which the density begins to diminish, or is infinite; or that in which the density increases continuously to an infinite value at the centre.

With units chosen to make $\mathfrak{F}(\infty) = 1$, we shall denote the function $\mathfrak{F}$ of class A by Θ_κ, and call it Homer Lane's Function; because he first used it, and expressed in terms of it all the features of a wholly gaseous spherical nebula in convective equilibrium, and calculated it for the cases, $\kappa = 1{\cdot}5$ and $\kappa = 2{\cdot}5$ $(k = 1\frac{2}{3}$ and $k = 1{\cdot}4)$.   He did not give tables of numbers, but he represented his solutions by curves.*   He did give some of his numbers for three points of each curve, and Mr Green, by very different methods of calculation, has found numbers for the case $\kappa = 2{\cdot}5$, agreeing with them to within $\frac{1}{10}$th per cent.

26.   By improvements which Mr Green has made on previous methods of calculation of Homer Lane's Function, and which he describes in an Appendix to the present paper, he has calculated values of the function $\Theta_\kappa(x)$, and of its differential coefficient $\Theta'_\kappa(x)$, which are shown in five tables corresponding to the following

* *American Journal of Science*, July 1870, p. 69.

five values of κ, 1·5, 2·5, 3, 4, ∞. For the four finite values of κ the practical range of each table is from $x = q$ to $x = \infty$, q denoting the value of x which makes $t = 0$.

27. There is such a value of x which is real in every case in which κ is positive and less than 5. This we see exemplified in the four diminishing values of q found by Mr Green (·2737, ·1867, 1450, ·0667)* for the four finite values of κ, 1·5, 2·5, 3, 4, and in the zero value of q for $\kappa = 5$, the case described in § 29 below. In this case equation (24) has a solution in finite terms, which gives $t = \sqrt{3} \cdot x$ for infinitely small values of x, and therefore makes $q = 0$ for $x = 0$.

28. Two interesting cases, $\kappa = 1$ and $\kappa = 5$, for each of which the differential equation (24) is soluble in finite terms, have been noticed, the former by Ritter†, the latter by Schuster‡. Ritter's case yields in reality Laplace's celebrated law§ of density for the earth's interior ($\sin nr/r$), which Laplace suggested as a consequence of supposing the earth to be a liquid globe, having pressure increasing from the surface inwards in proportion to the augmentation of the square of the density. With Ritter, however, the value of n is taken equal to π/R, so as to make the density zero at the bounding surface ($r = R$). With Laplace, n is taken equal to $\frac{5}{6}\pi/R$ to fit terrestrial conditions, including a ratio of surface density to mean density which is approximately 1/2·5. The ratio of surface density to mean density given by Laplace's law, with $n = \frac{5}{6}\pi/R$, is in fact 1/2·4225, which is as near to 1/2·5 as our imperfect knowledge of the surface density of the earth requires.

29. For the case $\kappa = 5$, Schuster found a solution in finite terms, which with our present notation may be written as follows:—

$$\frac{A}{S}\rho^{\frac{1}{5}} = t = \Theta_5(x) = \frac{x\sqrt{3}}{\sqrt{(3x^2 + 1)}} \quad \dots\dots\dots\dots(27)$$

This makes $t = 1$ at the centre ($\sigma/r = x = \infty$). At very great distances from the centre ($x = 0$) it makes

$$t = x\sqrt{3} = \frac{\sigma\sqrt{3}}{r}, \text{ and } \rho = \left(\frac{St}{A}\right)^5 = \left(\frac{S\sqrt{3}}{A}\right)^5 x^5 = \left(\frac{S\sqrt{3}}{A}\right)^5 \frac{\sigma^5}{r^5} \dots(28).$$

---

* See Appendix to the present paper, Tables I.—IV.
† Wiedemann's *Annalen*, Bd. xi. 1880, p. 338.
‡ *Brit. Assoc. Report*, 1883, p. 428.
§ *Mécanique Céleste*, Vol. v. livre xi. p. 49.

Using (27) in (25), we find

$$\frac{m}{E} = \frac{(\kappa+1)\,S\sigma}{e^2}\;\frac{\sqrt{3}}{(3x^2+1)^{3/2}} \quad \dots\dots\dots\dots(29);$$

and if in this we put $x = 0$, we find

$$\frac{M}{E} = \frac{(\kappa+1)\,S\sigma\,\sqrt{3}}{e^2} \quad\dots\dots\dots\dots\dots(30),$$

where M denotes the whole mass of the fluid. Thus we see that while the temperature and density both diminish to zero at infinite distance from the centre, the whole mass of the fluid is finite.

30. It is both mathematically and physically very interesting to pursue our solutions beyond $\kappa = 5$, to larger and larger values of κ up to $\kappa = \infty$: though we shall see in § 43 below, that, for all values of κ greater than 3 (or $k < 1\frac{1}{3}$), insufficiency of gravitational energy causes us to lose the practical possibility of a natural realisation of the convective equilibrium on which we have been founding. But notwithstanding this large failure of the convective approximate equilibrium, we have a dynamical problem of true fluid equilibrium, continuous through the whole range of κ from -1 to $-\infty$, and from $+\infty$ to 0; that is to say, for all values of k from 0 to ∞. In fact, looking back to the hydrostatic equation (10), and the physical equations (1), or (7), and (16), we have the whole foundations of equations (17) to (26), in which we may regard t merely as a convenient mathematical symbol defined by (6') in § 4. Any positive value of k is clearly admissible in (1), if we concern ourselves merely with a conceivable fluid having any law of relation between pressure and density which we please to give it, subject only to the condition that pressure is increased by increase of density. It is interesting to us now to remark, what is mathematically proved in § 44 below, that unless $k > 1\frac{1}{3}$, the repulsive quality in the fluid represented by k in equation (1) is not vigorous enough to give stable equilibrium to a very large globe of the fluid. in balancing the conglomerating effect of gravity.

31. As to the range of cases in which κ has finite values greater than 5, we leave it for the present and pass on to $\kappa = \infty$, or $k = 1$. In this case equation (1) becomes

$$\frac{\rho}{\rho'} = \frac{p}{p'} \quad\dots\dots\dots\dots\dots(31);$$

which is simply Boyle's law of the "Spring of air," as he called it.
It was on this law that Newton founded his calculation of the
velocity of sound, and got a result that surprised him by being
much too small. It was not till more than a hundred years later
that the now well-known cause of the discrepance was discovered
by Laplace, and a perfect agreement obtained between observation
and dynamical theory. But at present we are only concerned
with an ideal fluid which, irrespectively of temperature, exerts
pressure in simple proportion to its density. This ideal fluid we
shall call for brevity a Boylean gas.

32. For this extreme case of $\kappa = \infty$, our differential equation
(24) fails; but we deal with the failure by expressing t in terms
of ρ by (19), and then modifying the result by putting $\kappa = \infty$.
We thus find

$$\frac{d^2 \log \rho}{dx^2} = -\frac{\rho}{x^4}: \text{ where } x = \frac{\sigma}{r} \dots\dots\dots\dots(32);$$

σ denoting a linear constant given by (37) below. Equation (32)
is the equation of equilibrium of any quantity of Boylean gas,
when contained within a fixed spherical shell, under the influence
of its own gravity, but uninfluenced by the gravitational attraction
of any matter external to it. The value of σ might, but not
without considerable difficulty, be found from (22) by putting
$\kappa = \infty$. But it is easier and more clear to work out afresh, as in
§ 33 below, the equation of equilibrium of a Boylean gas, unen-
cumbered by the exuviæ of the adiabatic principle from which
our present problem emerges.

33. Let $\qquad\qquad p = B\rho \qquad\dots\dots\dots\dots\dots\dots(33),$

where B denotes what we may call the Boylean constant for the
particular gas considered; being its pressure at unit density.
According to our units, as explained in §§ 10, 11, 12, B is a linear
quantity. The analytical expression of the hydrostatic equilibrium is

$$dp = -g\rho \, dr \qquad\dots\dots\dots\dots\dots(34),$$

where [(16) repeated]

$$g = \frac{m/r^2}{E/e^2} = \frac{3}{5 \cdot 6 \cdot e} r^{-2} \int_0^r dr \, r^2 \rho \dots\dots\dots\dots(35).$$

Eliminating p from (34) by (33), and multiplying both members by r^2, we find

$$-r^2 \frac{d \log \rho}{dr} = \frac{3}{5 \cdot 6 \cdot e \cdot B} \int_0^r dr\, r^2 \rho = \frac{e^2 m}{BE} \quad \ldots\ldots\ldots(36).$$

Differentiating this with reference to r, and then transforming from r to x as in equations (21)...(24) above, we find (32), with the following expression for σ:—

$$\sigma^2 = \frac{5 \cdot 6}{3} eB \quad \ldots\ldots\ldots\ldots\ldots\ldots(37).$$

The equation of the first and third members of (36) gives

$$\frac{m}{E} = \frac{B\sigma}{e^2} \frac{d \log \rho}{dx} \quad \ldots\ldots\ldots\ldots\ldots\ldots(38).$$

34.   Let now $\rho = F(x)$ be any particular solution of (32); we find as a general solution with one disposable constant C,

$$\rho = CF\left(\frac{x}{\sqrt{C}}\right) \quad \ldots\ldots\ldots\ldots\ldots\ldots(39),$$

which we may immediately verify by substitution in (32) (compare § 25 above). The particular solution F must belong to one or other of the two classes, class A and class B, defined in § 25 above.

35.   We shall denote by $\Psi(x)$ what $F(x)$ of § 34 becomes, when the particular solution of (32), denoted by F, is of class A, with units so adjusted as to make $\Psi(\infty) = 1$; that is to say, central density unity. Mr Green in his Appendix to the present paper has calculated $\Psi(x)$ and $\Psi'(x)/\Psi(x)$, through the range from $x = \infty$ to $x = \cdot 1$. His results are shown in Table V. of the Appendix. Thus we may consider $\Psi(x)$ and its differential co-efficient $\Psi'(x)$ as known for all values of x through that range.

36.   Using this solution, $\Psi(x)$, instead of F in (39) above, we find that the solution of class A, which makes the central density C, is

$$\rho = C\Psi\left(\frac{x}{\sqrt{C}}\right) \quad \ldots\ldots\ldots\ldots\ldots(40);$$

and when we insert this expression for ρ in equation (38) we obtain

$$\frac{m}{E} = \frac{B\sigma}{e^2} \frac{1}{\sqrt{C}} \Psi'\left(\frac{x}{\sqrt{C}}\right) \Big/ \Psi\left(\frac{x}{\sqrt{C}}\right) \quad \ldots\ldots\ldots(41).$$

37. From equations (40) and (41), with values of $\Psi(x/\sqrt{C})$ and $\Psi'(x/\sqrt{C})/\Psi(x/\sqrt{C})$ obtained from the curves of $\Psi(x)$ and $\Psi'(x)/\Psi(x)$ in the range from $x = \infty$ to $x = \cdot 1$, and with the relation $r = \sigma/x$ where σ is given by (37) above, we can tell exactly the density at any point of a spherical mass of an ideal Boylean gas, and the mass of gas within each spherical surface of radius r, when the gas is in equilibrium under its own gravitation only, and has a density at its centre of any stated amount C. It is interesting to examine by means of these solutions the changes in ρ and m at any given distance from the centre when the central density C increases by any small amount dC; and to find also the changes in the radius of the spherical cell enclosing a given mass m, required to allow the mass to continue in equilibrium when the central density is increasing or diminishing continuously. The following table shows the values of ρ or $C\Psi(x/\sqrt{C})$, and $e^2 m/EB\sigma$ or $\Psi'(x/\sqrt{C})/\sqrt{C}\Psi(x/\sqrt{C})$, for several of the larger values of r, corresponding to the central densities 1 and 1·21 respectively.

$\dfrac{\sigma}{r}$	ρ	$\dfrac{e^2}{EB\sigma}m$	ρ	$\dfrac{e^2}{EB\sigma}m$
∞	1	0	1·21	0
·275	·2491	6·697	·2511	7·19
·250	·2076	7·905	·2069	8·39
·225	·1673	9·38	·1647	9·86
·200	·1295	11·20	·1260	11·64
·195	·1223	11·61	·1189	12·03
·190	·1153	12·04	·1118	12·46
·185	·1085	12·50	·1048	12·89
·180	·1017	12·97	·0982	13·35
·175	·0952	13·47	·0918	13·83
·170	·0889	13·99	·0855	14·34
·165	·0828	14·53	·0795	14·86
·160	·0769	15·10	·0738	15·40
·155	·0712	15·71	·0681	16·10
·150	·0657	16·34	·0528	16·59
·145	·0605	17·01	·0577	17·22
·140	·0554	17·71	·0529	18·04
·135	·0506	18·45	·0483	18·61
·130	·0461	19·23	·0439	19·35
·125	·0418	20·06	·0398	20·14
·120	·0377	20·95	·0359	20·98
·115	·0339	21·89	·0322	21·88
·110	·0303	22·88	·0288	22·82
·105	·0269	23·95	·0257	23·83
·100	·0238	25·10	·0227	24·94

38. From this table we see that it is possible to have the same mass of an ideal Boylean gas ($e^2m/EB\sigma \doteqdot 21\cdot9$) distributed in two different equilibrium conditions within a given sphere ($\sigma/r \doteqdot \cdot115$). We see also that in all smaller spheres the mass has increased, and in greater spheres it has decreased, through the alteration of density at the centre from 1 to $1\cdot21$. Indeed, when we trace the changes in the condition of any stated mass of a Boylean gas as its central density ideally increases from very small to very great values, we find that its radius diminishes till a certain central density has been reached, after which it increases till it becomes infinite.

39. By taking any two values of C in equation (26) above, and comparing the two solutions thus obtained as in § 37, it may be verified that results similar to those found in the case of a finite mass of an ideal Boylean gas, are found also in the case of a finite mass of any gas for which $\kappa > 3$, or $k < 1\frac{1}{3}$; while for any finite mass of a gas for which $\kappa < 3$, an increase in the density at the centre is always accompanied by a decrease in the radius of the shell enclosing the mass in equilibrium. These differences in the behaviour of the Boylean gas from that of gases for which $\kappa < 3$, and the resemblances of the Boylean gas and of gases for which $\kappa > 3$ (of which it may be regarded as the limiting case, $\kappa = \infty$), become of interest when we come to the question of the possibility of equilibrium of a mass of gas which is gradually losing energy by radiation into space. The result found above that there are two equilibrium conditions of a mass of any gas for which $\kappa > 3$, and one equilibrium condition of a mass of any gas for which $\kappa < 3$, within a given sphere, makes it desirable to investigate the nature of the equilibrium in each case, and leads us to the consideration of the energy required to maintain a mass of gas in equilibrium, within a sphere of radius R, in balancing the condensing influence of gravity.

40. Let K_v denote the thermal capacity at constant volume of the particular gas considered. The energy within unit volume of the gas at temperature t is $K_v\rho t$; and the total energy I, within a sphere of radius R, is given by

$$I = 4\pi K_v \int_0^R dr\, r^2 \rho t = K_v \int_0^R dm\, t \ \ldots\ldots\ldots\ldots(42).$$

By using equation (6), and then integrating by parts, we obtain

$$I = \frac{4\pi K_v}{S} \int_0^R dr\, r^2 p = \frac{4\pi K_v}{S} \cdot \left[\left(\frac{1}{3} r^3 p \right)_0^R - \frac{1}{3} \int_0^R dr\, r^3 \frac{dp}{dr} \right] \quad ...(43);$$

and since $p = 0$ at the outer boundary of the sphere and $r = 0$ at the centre, we have

$$I = - \frac{4\pi K_v}{3S} \int_0^R dr\, r^3 \frac{dp}{dr} \quad ....(44).$$

Substituting now the expression given for $-dp/dr$ in the equation of hydrostatic equilibrium (34), we obtain finally

$$I = \frac{4\pi K_v}{3S} \int_0^R dr\, r^3 g \rho ....(45).$$

41. The work which is done by the gravitational attraction of the matter within any layer of gas $4\pi r^2 \rho dr$ in bringing that layer from an infinite distance to its final position in the sphere is given by

$$dw = 4\pi r^2 \rho\, dr \cdot gr ....(46);$$

and the work done by gravity in collecting the whole sphere of radius R is therefore

$$W = 4\pi \int_0^R dr\, r^3 g \rho = \frac{e^2}{E} \int_0^R dm\, \frac{m}{r} \quad ....(47).$$

42. From equations (45) and (47) we obtain, as the ratio of the intrinsic energy within the sphere of gas to the work done by gravity in collecting the whole mass from an infinite distance,

$$\frac{I}{W} = \frac{K_v}{3S} \quad ....(48).$$

If K_p be the specific heat of the gas at constant pressure, we have $S = K_p - K_v$, and equation (48) may now be written in the form

$$\frac{I}{W} = \frac{K_v}{3(K_p - K_v)} = \frac{1}{3(k-1)} = \frac{\kappa}{3} \quad ....(49).$$

43. According to this theorem, it is convenient to divide gases into two species: species P, gases for which the ratio (k) of thermal capacity pressure constant to thermal capacity volume constant is greater than $1\frac{1}{3}$; species Q, gases for which k is less than $1\frac{1}{3}$. And the theorem expressed mathematically in equations (48) and (49) may be stated thus:—" A spherical globe of gas, given in equilibrium, with any arbitrary distribution of temperature having isothermal surfaces spherical, has less heat if the gas is of species P, and more heat if of species Q, than the thermal equivalent of the work which would be done by the mutual gravitational attraction between all its parts, in ideal shrinkage from an infinitely rare distribution of the whole mass to the given condition of density*."

44. It is easy to show from the theorem of §§ 42, 43 that the equilibrium of a globe of Q gas is essentially unstable. Let us first suppose for a moment that by a slight disturbance of the equilibrium condition the ratio I/W for the globe of Q gas becomes greater than that required for equilibrium by equation (49). Unless the excess of internal energy were quickly radiated away, the repulsive force which the globe of gas possesses by virtue of its internal energy would more than balance the condensing influence of gravity, and the globe would tend to expand. Since the internal energy lost in expansion is exactly equivalent to the work done against gravity, we see that the ratio I/W would continue to increase and the globe would become farther from an equilibrium condition than before. The expansion of the globe would therefore go on at an ever increasing speed till the density of the gas becomes infinitely small throughout.

If, on the other hand, through a slight disturbance of the equilibrium condition, the ratio I/W becomes less than that required for equilibrium, the globe of gas would in this case tend to contract. The increase in the internal energy due to any slight condensation would be exactly equal to the thermal equivalent of the work done by gravitation; and the ratio I/W would therefore go on diminishing instead of increasing, as it would require to do if the gas is to be restored to a condition of equilibrium.

45. "From this we see that if a globe of gas Q is given in a state of equilibrium, with the requisite heat given to it no matter how, and left to itself in waveless quiescent ether, it would, through gradual loss of heat, immediately cease to be in equilibrium, and would begin to fall inwards towards its centre, until in the central regions it becomes so dense that it ceases to obey Boyle's Law; that is to say, ceases to be a gas. Then, notwithstanding the

* Quoted from "On Homer Lane's Problem of a Spherical Gaseous Nebula," *Nature*, Feb. 14, 1907.

above theorem, it can come to approximate convective equilibrium as a cooling liquid globe surrounded by an atmosphere of its own vapour*."

46.   But if, after being given in convective equilibrium as in § 45, heat be properly and sufficiently supplied to the globe of Q gas at its centre, the whole gaseous mass can be kept in the condition of convective equilibrium.

47.   The theorem of §§ 42, 43 is given by Professor Perry on page 252 of *Nature* for July 13, 1899; and in the short article "On Homer Lane's Problem of a Spherical Gaseous Nebula," published in *Nature*, February 14, 1907, I have referred to it as Perry's theorem. Since this was written, however, I have found the same theorem given by A. Ritter on pp. 160—162 of *Wiedemann's Annalen*, Bd. 8, 1879, with the same conclusion from it as that stated in § 44 above, namely, that when $k < 1\frac{1}{3}$ the equilibrium of the spherical gaseous mass is unstable.

48.   In the theorem of Ritter and of Perry, given in § 42, *convective* equilibrium is not assumed. For the purposes of our problem, indicated in § 21, it is desirable to obtain expressions for the energy and the gravitational work of a mass M in equilibrium with a stated density at its centre, in terms of the notation of §§ 23—25 above. Thus, taking as our solution with central temperature C (equation 26),

$$t = C\Theta(z) \qquad \qquad \dots\dots\dots\dots\dots\dots(50),$$

where
$$z = xC^{-\frac{1}{2}(\kappa-1)}; \quad r = \sigma C^{-\frac{1}{2}(\kappa-1)}/z;$$

and where σ is given in terms of the Adiabatic Constant, A, by (22); we have from equations (25) and (50)

$$\frac{m}{E} = \frac{(\kappa+1)\,S\sigma C^{-\frac{1}{2}(\kappa-3)}}{e^2}\,\Theta'(z) \qquad \dots\dots\dots\dots(51),$$

and by differentiating this we obtain

$$\frac{dm}{E} = \frac{(\kappa+1)\,S\sigma C^{-\frac{1}{2}(\kappa-3)}}{e^2}\,\Theta''(z)\,dz \qquad \dots\dots\dots\dots(52).$$

---

* Quoted from " On Homer Lane's Problem of a Spherical Gaseous Nebula," *Nature*, Feb. 14, 1907.

49. With these values of t and dm substituted in the third member of equation (42), the expression for the internal energy, i, of the gas within a sphere of radius r becomes

$$i = K_v \int_0^r dm\, t = -\frac{K_v E (\kappa + 1) S\sigma C^{-\frac{1}{2}(\kappa - 5)}}{e^2} \int_z^\infty dz\, \Theta''(z)\, \Theta(z) \dots (53).$$

By putting $\Theta''(z) = -[\Theta(z)]^\kappa / z^4$ in this, and then integrating by parts as in § 40, equation (43), we may write i in the form

$$i = \frac{K_v E (\kappa + 1) S\sigma C^{-\frac{1}{2}(\kappa - 5)}}{e^2} \left[\frac{[\Theta(z)]^{\kappa+1}}{3z^3} + \frac{\kappa+1}{3} \int_z^\infty dz\, \frac{[\Theta(z)]^\kappa}{z^3}\, \Theta'(z) \right]$$
$$\dots\dots\dots (54).$$

Similarly, from the third member of (47), with the values of m and dm given in (51) and (52) above, we obtain the following expression for the gravitational work, w, done in collecting the gas within a sphere of radius r from infinite space

$$w = \frac{E(\kappa + 1)^2 S^2 \sigma C^{-\frac{1}{2}(\kappa - 5)}}{e^2} \int_z^\infty dz\, \frac{[\Theta(z)]^\kappa}{z^3}\, \Theta'(z) \dots (55).$$

It is easy to verify from these equations for i and w that with $S = K_p - K_v$, as in § 42,

$$i = \frac{K_v E (\kappa + 1) S\sigma C^{-\frac{1}{2}(\kappa - 5)}}{e^2} \frac{[\Theta(z)]^{\kappa+1}}{3z^3} + \frac{\kappa}{3} w \dots (56).$$

50. For the complete mass of gas, M, which can be in convective equilibrium under the influence of its own gravitation only, with central temperature C, we have the following results:

$$\frac{M}{E} = \frac{(\kappa + 1) S\sigma C^{-\frac{1}{2}(\kappa - 3)}}{e^2}\, \Theta'_\kappa(q) \dots\dots\dots\dots (57);$$

$$R = \frac{\sigma C^{-\frac{1}{2}(\kappa - 1)}}{q} \dots\dots\dots\dots\dots (58);$$

$$I = \frac{K_v E (\kappa + 1)^2 S\sigma C^{-\frac{1}{2}(\kappa - 5)}}{3e^2} \int_q^\infty dz\, \frac{[\Theta(z)]^\kappa}{z^3}\, \Theta'(z) \dots (59);$$

$$W = \frac{E(\kappa + 1)^2 S^2 \sigma C^{-\frac{1}{2}(\kappa - 5)}}{e^2} \int_q^\infty dz\, \frac{[\Theta(z)]^\kappa}{z^3}\, \Theta'(z) \dots (60);$$

with $\qquad \sigma^2 = \frac{5 \cdot 6 \cdot e (\kappa + 1) A^\kappa}{3 \cdot S^{\kappa - 1}} \dots\dots\dots\dots [(22) \text{ repeated}].$

The two equations (59) and (60) give as before

$$\frac{I}{W} = \frac{\kappa}{3} \quad \dots\dots\dots\dots\dots\dots\dots(61).$$

51. The equations of §§ 48—50, with equation (19), give the solution of Homer Lane's problem for all values of κ for which the function $\Theta_\kappa(z)$ and its derivative $\Theta'_\kappa(z)$ have been completely determined, namely for $\kappa = 1$ and $\kappa = 5$, referred to in §§ 28, 29 above, and for the values 1·5, 2·5, 3, 4, for which the Homer Lane functions and their derivatives are given in the Appendix to the present paper (Tables I.—IV.). It is important to remark that these equations indicate clearly the critical case $\kappa = 3$, and that they also reveal some interesting peculiarities of the case $\kappa = 5$; which we have found to be the smallest value of κ for which a finite mass of gas is unable to arrange itself in equilibrium within a finite boundary (see §§ 27, 29).

Equation (57) shows that in spherical nebulas for whose gaseous stuff $\kappa = 3$ the total mass of any gas which can exist in the equilibrium condition corresponding to a definite central temperature, when so distributed throughout its whole volume that the temperature and density at every point are related to each other in accordance with a chosen value of the adiabatic constant A, can also be brought into the equilibrium condition corresponding to any smaller central temperature, through gradual loss of energy, without disturbing the relation of temperature and density at any point of the mass.

Equations (59) and (60) show that in spherical gaseous nebulas for whose gaseous stuff $\kappa = 5$ the total internal energy, and the gravitational work, corresponding to each equilibrium distribution of gas, has the same value, whatever be the central temperature or total mass, provided temperature and density at each point within the mass are related to each other in accordance with the same value of the adiabatic constant in each case.

52. We may now apply the above equations to obtain the complete solution of our problem of § 21:—to determine for any spherical gaseous nebula of given mass, initially in convective equilibrium, exactly what its radius was, what its central temperature and density were, and what were the temperature and density at any distance from the centre, at the time when a stated quantity of heat has been radiated into space. Looking

to equation (57), we see that throughout all approximate equilibrium conditions of a constant total mass the relation

$$\sigma C^{-\frac{1}{2}(\kappa-3)} = \mathscr{M} \text{ (a constant)} \quad \dots\dots\dots\dots(62)$$

holds: and, with this condition, equation (51) shows that, during the gradual loss of heat from the nebula, the value of z for each stated mass m, concentric with the boundary, is constant. We have accordingly for the mass m

$$r = \frac{\sigma C^{-\frac{1}{2}(\kappa-1)}}{z} = \frac{\mathscr{M}}{z}\cdot\frac{1}{C} \quad \dots\dots\dots\dots(63),$$

where C varies slowly as time goes on. If we suppose C_1 to be the initial central temperature of the nebula, and C_2 its central temperature after a quantity of heat H has been lost by radiation, by applying (62) in the equations given above, we easily find (with suffixes 1 and 2 referring to the initial and final conditions respectively) the following results:

$$\begin{matrix} t_2 = \frac{C_2}{C_1}t_1; & \rho_2 = \left(\frac{C_2}{C_1}\right)^3\rho_1 \\ r_2 = \frac{C_1}{C_2}r_1; & R_2 = \frac{C_1}{C_2}R_1 \\ I_2 = \frac{C_2}{C_1}I_1; & W_2 = \frac{C_2}{C_1}W_1 \end{matrix} \right\} \dots\dots\dots(64);$$

in which t_2, t_1, ρ_2, ρ_1, r_2, r_1, all refer to points on the spherical surface enclosing a stated mass m. The total heat lost by radiation may now be written

$$H = (W_2 - W_1) - (I_2 - I_1) = \frac{C_2 - C_1}{C_1}(W_1 - I_1) \dots(65);$$

and for an infinitesimal change in the condition of the whole mass at any time this becomes

$$\delta H = \frac{\delta C}{C}(W - I) \quad \dots\dots\dots\dots\dots(66).$$

53. These are interesting results. Remembering that $I_1 = \kappa/3 . W_1$, we see by (65) and (66) that the central temperature of a globe of gas P in equilibrium increases, through gradual loss of heat by radiation into space. We then see also by (64) that the internal energy of a globe of gas P, continuing in a condition of approximate equilibrium while heat is being radiated away across its boundary, would go on increasing, and the work done

by the mutual gravitation of its parts would go on increasing, till the gas in the central regions became too dense to obey Boyle's Law. At the same time the radius of the globe would diminish. In other words, the repulsive power which the globe of gas P possesses by virtue of its internal energy, while in approximate equilibrium, is, owing to gradual loss of energy by radiation, at each instant just insufficient to exactly balance the attractive force due to the mutual gravitation of its parts. The globe is therefore compelled to contract: and, as the heat due to the contraction is not radiated away so quickly as it is produced, the shrinkage of the globe is accompanied by augmentation of its internal energy.

In figures 1 and 2 curves are shown illustrating five successive stages, numbered 1, 2, 3, 4, 5 respectively, in the history of a constant mass of any monatomic gas ($\kappa = 1\cdot5$; $k = 1\frac{2}{3}$) in approxi-

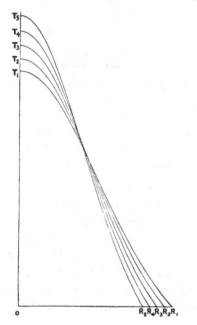 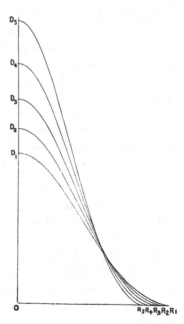

Fig. 1. $T_1 R_1, T_2 R_2, T_3 R_3, T_4 R_4, T_5 R_5$, are temperature curves for a constant mass of monatomic gas in equilibrium, at five stages of its history, numbered 1, 2, 3, 4, 5, in order of time, while it is losing heat by radiation into space.

Fig. 2. $D_1 R_1, D_2 R_2, D_3 R_3, D_4 R_4, D_5 R_5$ are density curves, corresponding respectively to the temperature curves $T_1 R_1, T_2 R_2, T_3 R_3, T_4 R_4, T_5 R_5$, of figure 1.

mate convective equilibrium while heat is being radiated from it into space. The abscissas represent distance from the centre. The ordinates in figure 1 represent temperature reckoned from absolute zero; $OT_1...OT_5$ being proportional to 1, 1·052, 1·108, 1·169, 1·235: figure 2 gives the corresponding density curves.

54. The remarkable result we have arrived at for P gases (for which alone, as we have seen, convective equilibrium can be realised), that the internal energy of a given mass in approximate convective equilibrium increases through gradual loss of heat by radiation into space, was first suggested as a possibility by Homer Lane; the suggestion being given in his paper referred to in § 2. To understand it more fully, go back to equation (62), and observe that in the case of P gases σ is continually diminishing, while the globe is shrinking through loss of heat. The adiabatic constant A, which determines the relation between temperature and density throughout the fluid at any instant, must therefore also continually diminish as time goes on [see (22) above]. Thus, we find from equation (19) that, although the density and temperature of the gas near the centre of the sphere are increasing, as we see from figures 1 and 2, and the total energy is increasing, in reality the temperatures at places of the same density are continually diminishing. And this diminution of temperature at places of the same density causes a diminution of the elastic resistance of the gas to compression which allows the gravitational forces to effect a contraction of the gaseous mass.

55. It seems certain that, as the condensation illustrated in figures 1 and 2 continues with increasing total energy, a time must come when the resistance to compression of the matter in the central regions must become much more than in accordance with the laws of perfect gases; and after that occurs, the cooling at the surface, with continual mixing of cooled fluid throughout the interior mass, must ultimately check the process of becoming hotter in the central regions, and bring about a gradual cooling of the whole mass.

56. The application of the above theory of approximate convective equilibrium to the sun, regarded as a mass of matter in the monatomic state, requires that the law of increase of density from the surface inwards should be such that the density at the centre is about six times the mean density (see Appendix,

§ 16). The mean density of the sun is about 1·4, the density of water being taken as unity. From this fact itself it seems certain that the sun is not gaseous as a whole. Disregarding, therefore, the high velocities which we know to exist in portions of the sun's atmosphere, and which are, according to the definition given in § 3, inconsistent with a condition of convective equilibrium, we are still forced to conclude that Homer Lane's exquisite mathematical theory gives no approximation to the present condition of the sun, because of his great average density. "This was emphasised by Professor Perry in the seventh paragraph, headed 'Gaseous Stars,' of his letter to Sir Norman Lockyer on 'The Life of a Star' (*Nature*, July 13, 1899), which contains the following sentence:

'It seems to me that speculation on this basis of perfectly gaseous stuff ought to cease when the density of the gas at the centre of the star approaches 0·1, or one-tenth of the density of ordinary water in the laboratory*.'"

57. According to a promise in the 1887 paper to the *Philosophical Magazine* "On the equilibrium of a gas under its own gravitation only," I now give examples of the application of this theory of convective equilibrium to spherical masses of argon and of nitrogen; choosing, for illustration, amounts of matter equal to masses of the sun, earth, and moon, with density at the centre 0·1 in each case.

Assuming $$t = C\Theta \left[xC^{-\frac{1}{2}(\kappa-1)} \right] \quad \ldots\ldots\ldots\ldots\ldots(67)$$
as the solution of (24), which gives central density 0·1, we find from equation (19)
$$0\cdot1 = \left(\frac{SC}{A} \right)^{\kappa} \quad \ldots\ldots\ldots\ldots\ldots(68);$$

and, as in this case we suppose the total mass M of the nebula to be known, we can determine C by applying equation (25) above. Thus
$$\frac{M}{E} = \frac{(\kappa+1) S\sigma C^{-\frac{1}{2}(\kappa-3)}}{e^2} \Theta'_{\kappa}(q) \ldots\ldots\ldots\ldots(69),$$

where q denotes the value of x for which $\Theta_{\kappa}(x) = 0$. Eliminating A and σ by means of equations (22) and (68), we obtain
$$C = \left(\frac{e}{S} \right) \cdot 3770 \cdot \left[\Theta'_{\kappa}(q) \right]^{-\frac{2}{3}} \frac{1}{\kappa+1} \left(\frac{M}{E} \right)^{\frac{2}{3}} \ldots\ldots\ldots(70).$$

* Quoted from "On Homer Lane's Problem of a Spherical Gaseous Nebula," *Nature*, Feb. 14, 1907.

From equations (68) and (22) we can determine the following expressions for A and σ:

$$\frac{A}{e} = \cdot 8122 \left[\Theta'_\kappa (q) \right]^{-\frac{2}{3}} \cdot 1^{\frac{\kappa-3}{3\kappa}} \frac{1}{\kappa+1} \left(\frac{M}{E} \right)^{\frac{2}{3}} \quad \dots\dots\dots(71),$$

$$\frac{\sigma}{\sqrt{(eS)}} = \left(\frac{e}{S} \right)^{\frac{\kappa}{2}} 1 \cdot 366 \left[1 \cdot 366 \cdot \Theta'_\kappa (q) \right]^{-\frac{\kappa}{3}} \frac{1^{\frac{\kappa-3}{6}}}{(\kappa+1)^{\frac{1}{2}(\kappa-1)}} \left(\frac{M}{E} \right)^{\frac{\kappa}{3}} \quad \dots(72).$$

The radius of the outer boundary of the nebula is given by

$$R = \sigma/x, \quad \text{where} \quad xC^{-\frac{1}{2}(\kappa-1)} = q \quad \dots\dots(73), (74).$$

We have therefore $R = \sigma q^{-1} C^{-\frac{1}{2}(\kappa-1)}$, which, by means of equations (70) and (72), may be written in the form

$$\frac{R}{e} = 2 \cdot 6527 \left[\Theta'_\kappa (q) \right]^{-\frac{1}{3}} q^{-1} \left(\frac{M}{E} \right)^{\frac{1}{3}} \quad \dots\dots\dots\dots(75).$$

For argon we have $k = 1\frac{2}{3}$, or $\kappa = 1 \cdot 5$; and $S = 5 \cdot 767$ kilometres; and for nitrogen we have $k = 1 \cdot 4$, or $\kappa = 2 \cdot 5$; and $S = 8 \cdot 256$ kilometres. With these values of S and κ, inserted in the above formulas, we obtain the results shown in the following table:

Matter in nebula	Total mass, that of	Central density	Central temperature in Centigrade degrees above absolute zero	Central pressure in metric tons per sq. kilometre	Radius of boundary in kilometres	Adiabatic Constant in kilometres
Argon	Sun	·1	$1 \cdot 105 \times 10^8$	$2 \cdot 33 \times 10^{14}$	$3 \cdot 04 \times 10^6$	$1 \cdot 08 \times 10^7$
,,	Earth	·1	$2 \cdot 342 \times 10^4$	$4 \cdot 95 \times 10^{10}$	$4 \cdot 42 \times 10^4$	$2 \cdot 30 \times 10^3$
,,	Moon	·1	$1 \cdot 243 \times 10^3$	$2 \cdot 63 \times 10^9$	$1 \cdot 02 \times 10^4$	$1 \cdot 22 \times 10^2$
Nitrogen	Sun	·1	$6 \cdot 383 \times 10^7$	$1 \cdot 92 \times 10^{14}$	$4 \cdot 79 \times 10^6$	$4 \cdot 82 \times 10^6$
,,	Earth	·1	$1 \cdot 353 \times 10^4$	$4 \cdot 07 \times 10^{10}$	$6 \cdot 97 \times 10^4$	$1 \cdot 02 \times 10^3$
,,	Moon	·1	$7 \cdot 185 \times 10^2$	$2 \cdot 16 \times 10^9$	$1 \cdot 61 \times 10^4$	$54 \cdot 3$

58. The curves of figures 3 and 4 represent temperature and density at different distances from the centres of nebulas for which κ has the values $1 \cdot 5, 2 \cdot 5, 3$, and 4. The temperature curves are drawn from the numbers given in the third columns of Tables I.—IV. of the Appendix: the density curves, from the numbers given in the fourth columns. With properly chosen scales of ordinates and abscissas, the curves shown may represent the condition of any gaseous mass, corresponding to any of the

solutions (26) above. Thus, with scales so chosen that $OR_\kappa = R = \sigma q^{-1} C^{-\frac{1}{2}(\kappa-1)}$, and $OT = C$, each curve, TR_κ, represents the temperature reckoned from absolute zero; and with $OD_\kappa = (SC/A)^\kappa$, each curve $D_\kappa R_\kappa$, represents the density, in a nebula composed of gas for which κ has one of the values given above, when the central temperature is C.

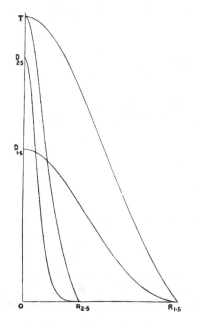

Fig. 3. $T\,R_{1\cdot5}$ is the curve of tempera-ture, and $D_{1\cdot5}\,R_{1\cdot5}$ is the curve of density, for a monatomic gas $(k=1\tfrac{2}{3})$. $T\,R_{2\cdot5}$, and $D_{2\cdot5}\,R_{2\cdot5}$ are corresponding curves for a diatomic gas $(k=1\cdot4)$.

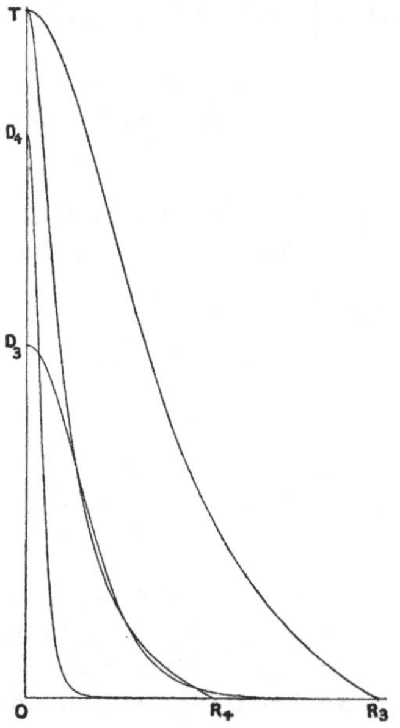

Fig. 4. $T\,R_3$ is the curve of tempera-ture, and $D_3\,R_3$ is the curve of density, for a gas for which $k=1\tfrac{1}{3}$. $T\,R_4$ and $D_4\,R_4$ are corresponding curves for a gas for which $k=1\tfrac{1}{4}$.

Each curve shown meets the axis of R at a finite angle; this angle being so small for the density curves that they appear to meet OR tangentially.

APPENDIX.

By George Green.

[From *Edinb. Roy. Soc. Proc.* Vol. xxviii. 1908, pp. 289—302; *Phil. Mag.* Vol. xvi. July 1908, pp. 10—23.]

MOLECULAR AND CRYSTALLINE THEORY*.

162. NOTE ON GRAVITY AND COHESION.

[From *Edinb. Roy. Soc. Proc.* Vol. IV. [read April 21, 1862], pp. 604—606; *Edinb. New Phil. Journ.* Vol. XVI. 1862, pp. 146—148; *Popular Lectures*, Vol. I. pp. 59—63.]

163. VOLTAIC POTENTIAL DIFFERENCES AND ATOMIC SIZES.

[From *Manchester Phil. Soc. Proc.* Vol. IX. [March 22, 1870], pp. 136—141; *Les Mondes*, Vol. XXII. 1870, pp. 701—708, Vol. XXVII. 1872, pp. 616—623; *Annal. Chem. Pharm.* Vol. CLVII. 1871, pp. 54—66; *Amer. Journ. Sci.* Vol. L. (2nd ser.) 1870, pp. 258—261; *Nature*, Vol. II. May 19, 1870, pp. 56, 57.]

THE following extract of a letter, dated March 21, 1870, from Sir Wm. Thomson, D.C.L., F.R.S., Hon. Member of the Society, to the President, [Dr Joule,] was read :—

I have now at last got into good working order measurements of electrostatic capacity (which, perhaps, you may remember I was working on the first time you ever came to see me, and more or less almost ever since). I have two students of last year, junior assistants in my laboratory, measuring electrostatic capacities of condensers, and variations of specific inductive capacities of dielectric, with sensibility of $\frac{1}{10}$ per cent., and with constancy in spite of accidental variations, generally within $\frac{1}{2}$ or $\frac{1}{3}$ per cent.

---

* See also *Baltimore Lectures*, 1904, *passim*.

My occupation on the Kinetic Theory of gases has led me at last to come to definite terms as to the size of molecules. Ever since about the first year of my professorship I have taught my students that Cauchy's theory of Dispersion proves heterogeneousness, or molecular structure, to become sensible in contiguous portions of glass or water, of dimensions moderately small in comparison with the wave lengths of ordinary light. I have spoken to you also, I think, of the argument deducible from the contact electricity of metals. This, I now find, proves a limit to the dimensions of the molecules in metals quite corresponding to that established for transparent solids and liquids by the dynamics of dispersion. In experiments made about ten years ago, of which a slight sketch is published in the *Proceedings* of the Literary and Philosophical Society of Manchester, I found that a plate of zinc and a plate of copper kept in metallic connection with one another (by a fine wire or otherwise) act electrically upon electrified bodies in their neighbourhood, and upon one another, as they would if they were of the same metal and kept at a difference of potentials equal to about three-quarters of that produced by a single cell of Daniell's. Hence, and from my measurement of the electrostatic effects of a Daniell's battery, published in the *Proceedings* of the Royal Society, for February and April, 1860, I find that plates of zinc and copper held parallel to one another at any distance, D, apart which is a small fraction of the linear dimensions of their opposed surfaces, and kept in metallic communication with one another, exercise a mutual attraction equal to

$$2 \times 10^{-10} \times A/D^2 \text{ grammes weight.}$$

Hence if they were allowed to approach from any greater distance, D', to the distance D, the work done by their mutual attraction is

$$2 \times 10^{-10} \times \frac{A\,(D'-D)}{D'D} \text{ centimetre grammes;}$$

which, if D is very small in comparison with D', is very approximately equal to

$$2 \times 10^{-10} \times A/D.$$

Now suppose a pile to be made of a great number $(N+1)$ of very thin plates alternately of zinc and copper, kept in metallic connection while they are brought towards one another. Let their positions in the pile be parallel, with narrow spaces intervening. For simplicity let the thickness of each metal plate and intervening space be D. The whole work done will be

$$2 \times 10^{-10} \times NA/D.$$

The whole mass of the pile (if we neglect that of one of the end plates) is

$$NAD\rho,$$

where ρ denotes the mean of the densities of zinc and copper. Hence, if h be the height to which the whole mass must be raised against a constant force equal to its weight at the earth's surface, to do the same amount of work, we have

$$NAD\rho h = 2 \times 10^{-10} \times NA/D,$$

which gives $\qquad h = 2 \times 10^{-10}/\rho D^2,$

or, as $\rho = 8$, nearly enough for the present rough estimate,

$$h = 1/(200000D)^2.$$

Hence if

$$D = \tfrac{1}{200000} \text{ centimetre,} \qquad h = 1 \text{ centimetre.}$$

The amount of energy thus calculated is not so great as to afford any argument against the conclusion which general knowledge of divisibility, electric conductivity, and other properties of matter indicates as probable: that, down to thicknesses of $\tfrac{1}{200000}$ of a centimetre for the metal plates and intervening spaces, the contact electrification, and the attraction due to it, follow with but little if any sensible deviation the laws proved by experiment for plates of measurable thickness with measurable intervals between them. But let D be a two-hundred-millionth of a centimetre. If the preceding formulæ were applicable to plates and spaces of this degree of thinness we should have

$$h = 1{,}000{,}000 \text{ centimetres or 10 kilometres.}$$

The thermal equivalent of the work thus represented is about 248 times the quantity of heat required to warm the whole mass (composed of equal masses of zinc and copper) by 1° Cent. This is probably much more than the whole heat of combination of equal masses of zinc and copper melted together. For it is not probable that the compound metal when dissolved in an acid would show anything approaching to so great a deficiency in the heat evolved below that evolved when the metallic constituents are separately dissolved*, and their solutions mixed; but the experiment should be made. Without any such experiment however we may safely say that the fourfold amount of energy indicated by the preceding formula, for a value of D yet twice as small, is very much greater than any estimate which our present knowledge allows us to accept for the heat of combination of zinc and copper. For something much less than the thermal equivalent of that amount of energy would melt the zinc and copper; and therefore if in combining they generated by their mutual attraction any such amount of energy, a mixture of zinc and copper filings would rush into combination (as the ingredients of gunpowder do) on being heated enough in any small part of the whole mass to melt together there. Hence we may infer that the electric attraction between metallically-connected plates of zinc and copper of only $\frac{1}{400000000}$ of a centimetre thickness, at a distance of only $\frac{1}{400000000}$ of a centimetre asunder must be greatly less than that calculated from the magnitude of the force and the law of its variation observed for plates of measurable thickness, at measurable distances asunder. In other words, plates of zinc and copper so thin as a four-hundred-millionth of a centimetre, and placed at as short a distance as a four-hundred-millionth of a centimetre from one another, form a mixture closely approaching to a molecular combination, if indeed plates so thin could be made without splitting atoms.

Wishing to avoid complication, I have avoided hitherto noticing one important question as to the energy concerned in the electric attraction of metallically connected plates of zinc and copper. Is there not a change of temperature in molecularly thin strata of the two metals adjoining to the opposed surfaces, when they are allowed to approach one another, analogous to the heat produced by the condensation of a gas, the changes of temperature produced by the application of stresses to elastic solids which you have investigated experimentally, and the cooling effect I have proved to be produced by drawing out a liquid film which I shall have to notice particularly below? Easy enough experiments on the contact electricity of metals will answer this question. If the contact-difference diminishes as the temperature is raised, it will follow from the Second Law of Thermodynamics, by reasoning

---

* Will you try this experiment? You would easily make a good thing of it.

precisely corresponding with that which I applied to the liquid film in my letters to you of February 2 and February 3, 1858*, that plates of the two metals kept in metallic communication and allowed to approach one another will experience an elevation of temperature. But if the contact difference increases with temperature, the effect of mutual approach will be a lowering of temperature. On the former supposition, the diminution of intrinsic energy in quantities of zinc and copper, consequent on mutual approach with temperature kept constant, will be greater, and on the latter supposition less, than I have estimated above. Till the requisite experiments are made, farther speculation on this subject is profitless: but whatever be the result, it cannot invalidate the conclusion that a stratum of $\frac{1}{200000000}$ of a centimetre thick cannot contain in its thickness many, if so much as one, molecular constituent of the mass.

Besides the two reasons for limiting the smallness of atoms or molecules which I have now stated, two others are afforded by the theory of capillary attraction, and Clausius's and Maxwell's magnificent working out of the Kinetic theory of gases.

In my letters to you already referred to, I showed that the dynamic value of the heat required to prevent a bubble from cooling when stretched is rather more than half the work spent in stretching it. Hence if we calculate the work required to stretch it to any stated extent, and multiply the result by $\frac{3}{2}$, we have an estimate, near enough for my present purpose, of the augmentation of energy experienced by a liquid film when stretched and kept at a constant temperature. Taking 08 of a gramme weight per centimetre of breadth as the capillary tension of a surface of water, and therefore ·16 as that of a water bubble, I calculate (as you may verify easily) that a quantity of water extended to a thinness of $\frac{1}{200000000}$ of a centimetre would, if its tension remained constant, have more energy than the same mass of water in ordinary condition by about 1,100 times as much as suffices to warm it by 1° Cent. This is more than enough (as Maxwell suggested to me) to drive the liquid into vapour. Hence if a film of $\frac{1}{200000000}$ of a centimetre thick can exist as liquid at all, it is *perfectly certain* that there cannot be many molecules in its thickness.

* *Proceedings of the Royal Society* for April, 1858.

The argument from the Kinetic theory of gases leads me to quite a similar conclusion.

I need not trouble you with it at present, as I am writing a short sketch of those of the results of Maxwell and Clausius which I use in it, to form part of an article on the Size of Atoms for *Nature*.

ON THE SIZE OF ATOMS.

[From *Nature*, Vol. I. March 31, 1870, pp. 551—553; *Amer. Journ. Sci.* Vol. L. 1870, pp. 38—44. Reprinted in Thomson and Tait's *Natural Philosophy*, Part II. pp. 495—502.]

THE idea of an atom has been so constantly associated with incredible assumptions of infinite strength, absolute rigidity, mystical actions at a distance, and indivisibility, that chemists and many other reasonable naturalists of modern times, losing all patience with it, have dismissed it to the realms of metaphysics, and made it smaller than "anything we can conceive." But if atoms are inconceivably small, why are not all chemical actions infinitely swift? Chemistry is powerless to deal with this question, and many others of paramount importance, if barred by the hardness of its fundamental assumptions, from contemplating the atom as a real portion of matter occupying a finite space, and forming a not immeasurably small constituent of any palpable body.

More than thirty years ago naturalists were scared by a wild proposition of Cauchy's, that the familiar prismatic colours proved the "sphere of sensible molecular action" in transparent liquids and solids to be comparable with the wave-length of light. The thirty years which have intervened have only confirmed that proposition. They have produced a large number of capable judges; and it is only incapacity to judge in dynamical questions that can admit a doubt of the substantial correctness of Cauchy's conclusion. But the "sphere of molecular action" conveys no very clear idea to the non-mathematical mind. The idea which it conveys to the mathematical mind is, in my opinion, irredeemably false. For I have no faith whatever in attractions and repulsions acting at a distance between centres of force according to various laws. What Cauchy's mathematics really proves is this: that in palpably homogeneous bodies such as glass or water, contiguous portions are not similar when their dimensions are moderately small fractions of the wave-length. Thus in water contiguous cubes, each of one one-thousandth of a centimetre breadth are sensibly similar. But contiguous cubes of one ten-millionth of a centimetre must be very sensibly different. So in a solid mass of brickwork, two adjacent lengths of 20,000 centimetres each, may contain, one of them nine hundred and ninety-nine bricks and two half bricks, and the other one thousand bricks: thus two contiguous cubes of 20,000 centimetres breadth may be considered as sensibly similar. But two adjacent lengths of forty centimetres each might contain one of them, one brick, and two half bricks, and the other two whole bricks; and contiguous cubes of forty centimetres would be very sensibly dissimilar. In short, optical dynamics leaves no alternative but to admit that the diameter of a molecule, or the distance from the centre of a molecule to the centre of a contiguous molecule in glass, water, or any other of our transparent liquids and solids, exceeds a ten-thousandth of the wave-length, or a two-hundred-millionth of a centimetre.

By experiments on the contact electricity of metals made eight or ten years ago, and described in a letter to Dr Joule, which was published in the *Proceedings* of the Literary and Philosophical Society of Manchester, I found that plates of zinc and copper connected with one another by a fine wire attract one another, as would similar pieces of one metal connected with the two plates of a galvanic element, having about three-quarters of the electro-motive force of a Daniell's element.

Measurements published in the *Proceedings* of the Royal Society for 1860 showed that the attraction between parallel plates of one metal held at a distance apart small in comparison with their diameters, and kept connected with such a galvanic element, would experience an attraction amounting to two ten-thousand-millionths of a gramme weight per area of the opposed surfaces equal to the square of the distance between them. Let a plate of zinc and a plate of copper, each a centimetre square and a hundred-thousandth of a centimetre thick, be placed with a corner of each touching a metal globe of a hundred-thousandth of a centimetre diameter. Let the plates, kept thus in metallic communication with one another be at first wide apart, except at the corners touching the little globe, and let them then be gradually turned round till they are parallel and at a distance of a hundred-thousandth of a centimetre asunder. In this position they will attract one another with a force equal in all to two grammes weight. By abstract dynamics and the theory of energy, it is readily proved that the work done by the changing force of attraction during the motion by which we have supposed this position to be reached, is equal to that of a constant force of two grammes weight acting through a space of a hundred-thousandth of a centimetre; that is to say, to two hundred-thousandths of a centimetre gramme. Now let a second plate of zinc be brought by a similar process to the other side of the plate of copper; a second plate of copper to the remote side of this second plate of zinc, and so on till a pile is formed consisting of 50,001 plates of zinc and 50,000 plates of copper, separated by 100,000 spaces, each plate and each space one hundred-thousandth of a centimetre thick. The whole work done by electric attraction in the formation of this pile is two centimetre grammes.

The whole mass of metal is eight grammes. Hence the amount of work is a quarter of a centimetre-gramme per gramme of metal. Now 4,030 centimetre-grammes of work, according to Joule's dynamical equivalent of heat is the amount required to warm a gramme of zinc or copper by one degree Centigrade. Hence the work done by the electric attraction could warm the substance by only $\frac{1}{16120}$ of a degree. But now let the thickness of each piece of metal and of each intervening space be a hundred-millionth of a centimetre instead of a hundred-thousandth. The work would be increased a millionfold unless a hundred-millionth of a centimetre approaches the smallness of a molecule. The heat equivalent would therefore be enough to raise the temperature of the material by 62° C. This is barely, if at all, admissible, according to our present knowledge, or, rather, want of knowledge, regarding the heat of combination of zinc and copper. But suppose the metal plates and intervening spaces to be made yet four times thinner, that is to say, the thickness of each to be a four hundred millionth of a centimetre. The work and its heat equivalent will be increased sixteen-fold. It would therefore be 990 times as much as that required to warm the mass by 1° Cent., which is very much more than can possibly be produced by zinc and copper in entering into molecular combination. Were there in reality anything like so much heat of combination as this, a mixture of zinc and copper powders would, if melted in any one spot, run together, generating more than heat enough to melt each throughout; just as a large quantity of gunpowder if ignited in any one spot burns throughout without fresh application of heat. Hence plates of zinc and copper of a three hundred-millionth of a centimetre thick, placed close together alternately, form a near approximation to a chemical combination, if indeed such thin plates could be made without splitting atoms.

The theory of capillary attraction shows that when a bubble—a soap-bubble for instance—is blown larger and larger, work is done by the stretching of a film which resists extension as if it were an elastic membrane with a constant contractile force. This contractile force is to be reckoned as a certain number of units of force per unit of breadth. Observation of the ascent of water in capillary tubes shows that the contractile force of a thin film of water is about sixteen milligrammes weight per millimetre of breadth. Hence the work done in stretching a water film to any degree of thinness, reckoned in millimetre-milligrammes, is equal to sixteen times the number of square millimetres by which the area is augmented, provided the film is not made so thin that there is any sensible diminution of its contractile force. In an article "On the Thermal effect of drawing out a Film of Liquid," published in the *Proceedings* of the Royal Society for April 1858, I have proved from the second law of thermodynamics that about half as much more energy, in the shape of heat, must be given to the film to prevent it from sinking in temperature while it is being drawn out. Hence the intrinsic energy of a mass of water in the shape of a film kept at constant temperature increases by twenty-four millimetre-milligrammes for every square millimetre added to its area.

Suppose then a film to be given with a thickness of a millimetre, and suppose its area to be augmented ten thousand and one fold: the work done per square millimetre of the original film, that is to say per milligramme of the mass, would be 240,000 millimetre-milligrammes. The heat equivalent of this is more than half a degree Centigrade of elevation of temperature of the substance. The thickness to which the film is reduced on this supposition is very approximately a ten-thousandth of a millimetre. The commonest observation on the soap-bubble (which in contractile force differs no doubt very little from pure water) shows that there is no sensible diminution of contractile force by reduction of the thickness to the ten-thousandth of a millimetre; inasmuch as the thickness which gives the first maximum brightness round the black spot, seen where the bubble is thinnest, is only about an eight-thousandth of a millimetre.

The very moderate amount of work shown in the preceding estimates is quite consistent with this deduction. But suppose now the film to be farther stretched until its thickness is reduced to a twenty-millionth of a millimetre. The work spent in doing this is two thousand times more than that which we have just calculated. The heat equivalent is 1,130 times the quantity required to raise the temperature of the liquid by one degree Centigrade. This is far more than we can admit as a possible amount of work done in the extension of a liquid film. A smaller amount of work spent on the liquid would convert it into vapour at ordinary atmospheric pressure. The conclusion is unavoidable, that a water-film falls off greatly in its contractile force before it is reduced to a thickness of a twenty-millionth of a millimetre. It is scarcely possible, upon any conceivable molecular theory, that there can be any considerable falling off in the contractile force as long as there are several molecules in the thickness. It is therefore probable that there are not several molecules in a thickness of a twenty-millionth of a millimetre of water.

The kinetic theory of gases suggested a hundred years ago by Daniel Bernouilli has, during the last quarter of a century, been worked out by Herapath, Joule, Clausius, and Maxwell, to so great perfection that we now find in it satisfactory explanations of all non-chemical properties of gases. However difficult it may be to even imagine what kind of thing the molecule is, we may regard it as an established truth of science that a gas consists of moving molecules disturbed from rectilineal paths and constant velocities by collisions or mutual influences, so rare that the mean length of nearly rectilineal portions of the path of each molecule is many times greater than the average distance from the centre of each molecule to the centre of the molecule nearest it at any time. If, for a moment, we suppose the molecules to be hard elastic globes all of one size, influencing one another only through actual contact, we have for each molecule simply a zigzag path composed of rectilineal portions, with abrupt changes of direction. On this supposition Clausius proves, by a simple application of the calculus of probabilities, that the average length of the free path of a particle from collision to collision bears to the diameter of each globe, the ratio of the whole space in which the globes move, to eight times the sum of the volumes of the globes. It follows that the number of the globes in unit volume is equal to the square of this ratio divided by the volume of a sphere whose radius is equal to that average length of free path. But we cannot believe that the individual molecules of gases in general, or even of any one gas, are hard elastic globes. Any two of the moving particles or molecules must act upon one another somehow, so that when they pass very near one another they shall produce considerable deflexion of the path and change in the velocity of each. This mutual action (called force) is different at different distances, and must vary, according to variations of the distance so as to fulfil some definite law. If the particles were hard elastic globes acting upon one another only by contact, the law of force would be—zero force when the distance from centre to centre exceeds the sum of the radii, and infinite repulsion for any distance less than the sum of the radii. This hypothesis, with its "hard and fast" demarcation between no force and infinite force, seems to require mitigation. Without entering on the theory of vortex atoms at present, I may at least say that soft elastic solids, not necessarily globular, are more promising than infinitely hard elastic globes. And, happily, we are not left merely to our fancy as to what we are to accept for probable in respect to the law of force. If the particles were hard elastic globes the average time from collision to collision would be inversely as the average velocity of the particles. But Maxwell's experiments on the variation of the viscosities of gases with change of temperature prove that the mean time from collision to collision is independent of the velocity if we give the name collision to those mutual actions only which produce something more than a certain specified degree of deflection of the line of motion. This law could be fulfilled by soft elastic particles (globular or not globular); but, as we have seen, not by hard elastic globes. Such details, however, are beyond the scope of our present argument.

What we want now is rough approximations to absolute values, whether of time or space or mass—not delicate differential results. By Joule, Maxwell, and Clausius we know that the average velocity of the molecules of oxygen or nitrogen or common air, at ordinary atmospheric temperature and pressure, is about 50,000 centimetres per second, and the average time from collision to collision a five-thousand-millionth of a second. Hence the average length of path of each molecule between collisions is about $\frac{1}{100000}$ of a centimetre. Now, having left the idea of hard globes, according to which the dimensions of a molecule and the distinction between collision and no collision are perfectly sharp, something of apparent circumlocution must take the place of these simple terms.

First, it is to be remarked that two molecules in collision will exercise a mutual repulsion in virtue of which the distance between their centres, after being diminished to a minimum, will begin to increase as the molecules leave one another. This minimum distance would be equal to the sum of the radii, if the molecules were infinitely hard elastic spheres; but in reality we must suppose it to be very different in different collisions. Considering only the case of equal molecules, we might, then, define the radius of a molecule as half the average shortest distance reached in a vast number of collisions. The definition I adopt for the present is not precisely this, but is chosen so as to make as simple as possible the statement I have to make of a combination of the results of Clausius and Maxwell. Having defined the radius of a gaseous molecule, I call the double of the radius the diameter; and the volume of a globe of the same radius or diameter I call the volume of the molecule.

The experiments of Cagniard de la Tour, Faraday, Regnault, and Andrews, on the condensation of gases do not allow us to believe that any of the ordinary gases could be made forty thousand times denser than at ordinary atmospheric pressure and tempera-ture, without reducing the whole volume to something less than the sum of the volumes of the gaseous molecules, as now defined. Hence, according to the grand theorem of Clausius quoted above, the average length of path from collision to collision cannot be more than five thousand times the diameter of the gaseous molecule; and the number of molecules in unit of volume cannot exceed 25,000,000 divided by the volume of a globe whose radius is that average length of path. Taking now the preceding estimate, $\frac{1}{100000}$ of a centimetre, for the average length of path from collision to collision we conclude that the diameter of the gaseous molecule cannot be less than $\frac{1}{500000000}$ of a centimetre; nor the number of the molecules in a cubic centimetre of the gas (at ordinary density) greater than 6×10^{21} (or six thousand million million million).

The densities of known liquids and solids are from five hundred to sixteen thousand times that of atmospheric air at ordinary pressure and temperature; and, therefore, the number of molecules in a cubic centimetre may be from 3×10^{24} to 10^{26} (that is, from three million million million million to a hundred million million million million). From this (if we assume for a moment a cubic arrangement of molecules), the distance from centre to nearest centre in solids and liquids may be estimated at from $\frac{1}{140000000}$ to $\frac{1}{460000000}$ of a centimetre.

The four lines of argument which I have now indicated lead all to substantially the same estimate of the dimensions of molecular structure. Jointly they establish with what we cannot but regard as a very high degree of probability the conclusion that, in any ordinary liquid, transparent solid, or seemingly opaque solid, the mean distance between the centres of contiguous molecules is less than the hundred-millionth, and greater than the two thousand-millionth of a centimetre.

To form some conception of the degree of coarse-grainedness indicated by this conclusion, imagine a rain drop, or a globe of glass as large as a pea, to be magnified up to the size of the earth, each constituent molecule being magnified in the same proportion. The magnified structure would be coarser grained than a heap of small shot, but probably less coarse grained than a heap of cricket-balls.

164. ON THE SIZE OF ATOMS.

[From *Roy. Institut. Proc.* Vol. x. 1884 [Feb. 2, 1883], pp. 185—213; *Nature*, Vol. xxviii. June 28, 1883, pp. 203—205, July 12, 1886, pp. 250—254, July 19, 1883, pp. 274—278. Reprinted in *Popular Lectures*, Vol. i. pp. 147—217.]

165. On the Division of Space with Minimum Partitional Area.

[From *Phil. Mag.* Vol. XXIV. Dec. 1887, pp. 503—514; *Acta Math.*
Vol. XI. 1887–1888, pp. 121—134.]

1. THIS problem is solved in foam, and the solution is interestingly seen in the multitude of film-enclosed cells obtained by blowing air through a tube into the middle of a soap-solution in a large open vessel. I have been led to it by endeavours to understand, and to illustrate, Green's theory of "extraneous pressure" which gives, for light traversing a crystal, Fresnel's wave-surface, with Fresnel's supposition (strongly supported as it is by Stokes and Rayleigh) of velocity of propagation dependent, not on the distortion normal, but solely on the line of vibration. It has been admirably illustrated, and some elements towards its solution beautifully realised in a manner convenient for study and instruction, by Plateau, in the first volume of his *Statique des Liquides soumis aux seules Forces Moléculaires.*

2. The general mathematical solution, as is well known, is that every interface between cells must have constant curvature* throughout, and that where three or more interfaces meet in a curve or straight line their tangent-planes through any point of the line of meeting intersect at angles such that equal forces in these planes, perpendicular to their line of intersection, balance. The *minimax* problem would allow any number of interfaces to meet in a line; but for a pure minimum it is obvious that not

---

* By "curvature" of a surface I mean sum of curvatures in mutually perpendicular normal sections at any point; not Gauss's "curvatura integra," which is the product of the curvature in the two "principal normal sections," or sections of greatest and least curvature. (See Thomson and Tait's *Natural Philosophy*, Part I. §§ 130, 136.)

more than three can meet in a line, and that therefore, in the realisation by the soap-film, the equilibrium is necessarily unstable if four or more surfaces meet in a line. This theoretical conclusion is amply confirmed by observation, as we see at every intersection of films, whether interfacial in the interior of groups of soap-bubbles, large or small, or at the outer bounding-surface of a group, never more than three films, but, wherever there is intersection, always *just three films*, meeting in a line. The theoretical conclusion as to the angles for stable equilibrium (or pure minimum solution of the mathematical problem) therefore becomes, simply, that every angle of meeting of film-surfaces is exactly 120°.

3.   The rhombic dodecahedron is a polyhedron of plane sides between which every angle of meeting is 120°; and space can be filled with (or divided into) equal and similar rhombic dodeca-hedrons. Hence it might seem that the rhombic dodecahedron is the solution of our problem for the case of all the cells equal in volume, and every part of the boundary of the group either infinitely distant from the place considered, or so adjusted as not to interfere with the homogeneousness of the interior distribution of cells. Certainly the rhombic dodecahedron *is a solution of the minimax, or equilibrium-problem*; and certain it is that no other plane-sided polyhedron can be a solution.

4.   But it has seemed to me, on purely theoretical consideration, that the tetrahedral angles of the rhombic dodecahedron*, giving, when space is divided into such figures, twelve plane films meeting in a point (as twelve planes from the twelve edges of a cube meeting in the centre of the cube) are essentially unstable. That it is so is proved experimentally by Plateau (Vol. I. § 182, fig. 71) in his well-known beautiful experiment with his cubic skeleton frame dipped in soap-solution and taken out. His fig. 71 is re-produced here in fig. 1. Instead of twelve *plane* films stretched

---

* The rhombic dodecahedron has six tetrahedral angles and eight trihedral angles. At each tetrahedral angle the plane faces cut one another successively at 120°, while each is perpendicular to the one remote from it; and the angle between successive edges is $\cos^{-1}\frac{1}{3}$, or 70° 32'. The obtuse angles (109° 28') of the rhombs meet in the trihedral angles of the solid figure. The whole figure may be regarded as composed of six square pyramids, each with its alternate slant faces perpendicular to one another, placed on six squares forming the sides of a cube. The long diagonal of each rhombic face thus made up of two sides of pyramids conterminous in the short diagonal, is $\sqrt{2}$ times the short diagonal.

inwards from the twelve edges and meeting in the centre of the cube, it shows twelve films, of which eight are slightly curved and four are plane*, stretched from the twelve edges to a small central plane quadrilateral film with equal curved edges and four angles each of 109° 28′. Each of the plane films is an isosceles triangle with two equal curved sides meeting at a corner of the central curvilinear square in a plane perpendicular to its plane. It is in the plane through an edge and the centre of the cube. The angles of this plane curvilinear triangle are respectively 109° 28′, at the point of meeting of the two curvilinear sides, and each of the two others half of this, or 54° 44′.

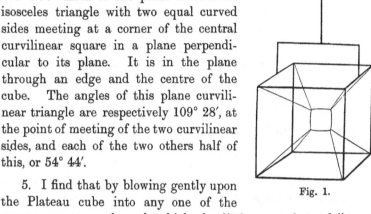

Fig. 1.

5. I find that by blowing gently upon the Plateau cube into any one of the square apertures through which the little central quadrilateral film is seen as a line, this film is caused to contract. If I stop blowing before it contracts to a point, it springs back to its primitive size and shape. If I blow still very gently but for a little more time, the quadrilateral contracts to a point, and the twelve films meeting in it immediately draw out a fresh little quadrilateral film similar to the former, but in a plane perpendicular to its plane and to the direction of the blast. Thus, again and again, may the films be transformed so as to render the little central curvilinear square parallel to one or other of the three pairs of square apertures of the cubic frame. Thus we see that the twelve plane films meeting in the centre of the cube is a configuration of unstable equilibrium which may be fallen from in three different ways.

6. Suppose now space to be filled with equal and similar ideal rhombic dodecahedrons. Draw the short diagonal of every rhombic face, and fix a real wire (infinitely thin and perfectly stiff) along each. This fills space with Plateau cubic frames.

---

* I see it inadvertently stated by Plateau that all the twelve films are "légèrement courbées."

Fix now, ideally, a very small rigid globe at each of the points
of space occupied by tetrahedral angles of the dodecahedrons, and
let the faces of the dodecahedrons be realised by soap-films. They
will be in *stable* equilibrium, because of the little fixed globes;
and the equilibrium would be stable without the rigid diagonals
which we require only to help the imagination in what follows.
Let an exceedingly small force, like gravity*, act on all the films
everywhere perpendicularly to one set of parallel faces of the
cubes. If this force is small enough it will not tear away the
films from the globes; it will only produce a very slight bending
from the plane rhombic shape of each film. Now annul the little
globes. The films will instantly jump (each set of twelve which
meet in a point) into the Plateau configuration (fig. 1), with the
little curve-edged square in the plane perpendicular to the determining force, which may now be annulled, as we no longer require
it. The rigid edges of the cubes may also be now annulled, as we
have done with them also; because each is (as we see by symmetry)
pulled with equal forces in opposite directions, and therefore is
not required for the equilibrium, and it is clear that the equilibrium
is stable without meeting them †.

* To do for every point of meeting of twelve films what is done by blowing in
the experiment of § 5.

† The corresponding two-dimensional problem is much more easily imagined,
and may indeed be realised by aid of moderately simple appliances.
Between a level surface of soap-solution and a horizontal plate of glass fixed at

Fig. 2.

7.   We have now space divided into equal and similar tetra-kaidecahedral cells* by the soap-film; each bounded by

(1)   Two small plane quadrilaterals parallel to one another;

(2)   Four large plane quadrilaterals in planes perpendicular to the diagonals of the small ones;

(3)   Eight non-plane hexagons, each with two edges common with the small quadrilaterals, and four edges common with the large quadrilaterals.

The films seen in the Plateau cube show one complete small quadrilateral, four halves of four of the large quadrilaterals, and eight halves of eight of the hexagons, belonging to six contiguous cells; all mathematically correct in every part (supposing the film and the cube-frame to be infinitely thin). Thus we see all the elements required for an exact construction of the complete tetra-kaidecahedron. By making a clay model of what we actually see, we have only to complete a symmetrical figure by symmetrically completing each half-quadrilateral and each half-hexagon, and putting the twelve properly together, with the complete small quadrilateral, and another like it as the far side of the 14-faced figure. We thus have a correct solid model.

8.   Consider now a cubic portion of space containing a large number of such cells, and of course a large but a comparatively small number of partial cells next the boundary. Wherever the boundary is cut by film, fix stiff wire; and remove all the film from outside, leaving the cubic space divided stably into cells by

a centimetre or two above it, imagine vertical film-partitions to be placed along the sides of the squares indicated in the drawing (fig. 2); these will rest in stable equilibrium if thick enough wires are fixed vertically through the corners of the squares. Now draw away these wires downwards into the liquid; the equilibrium in the square formation becomes unstable, and the films instantly run into the hexagonal formation shown in the diagram; provided the square of glass is provided with vertical walls (for which slips of wood are convenient), as shown in plan by the black border of the diagram. These walls are necessary to maintain the inequality of pull in different directions which the inequality of the sides of the hexagons implies. By inspection of the diagram we see that the pull is T/a per unit area on either of the pair of vertical walls which are perpendicular to the short sides of the hexagons; and on either of the other pair of walls $2 \cos 30° \times T/a$; where T denotes the pull of the film per unit breadth, and a the side of a square in the original formation. Hence the ratio of the pulls per unit of area in the two principal directions is as 1 to 1·732.

* The centres of these cells are in cubic order.

films held out against their tension by the wire network thus fixed in the faces of the cube. If the cube is chosen with its six faces parallel to the three pairs of quadrilateral films, it is clear that the resultant of the whole pull of film on each face will be perpendicular to the face, and that the resultant pulls on the two pairs of faces parallel to the greater quadrilaterals are equal to one another and less than the resultant pull on the pair of faces parallel to the smaller quadrilaterals. Let now the last-mentioned pair of faces of the cube be allowed to yield to the pull inwards, while the other two pairs are dragged outwards against the pulls on them, so as to keep the enclosed volume unchanged; and let the wirework fixture on the faces be properly altered, shrunk on two pairs of faces, and extended on the other pair of faces, of the cube which now becomes a square cage with distance between floor and ceiling less than the side of the square. Let the exact configuration of the wire everywhere be always so adjusted that the cells throughout the interior remain, in their altered configuration, equal and similar to one another. We may thus diminish, and if we please annul, the difference of pull per unit area, on the three pairs of sides of the cage. The respective shrinkage-ratio and extension-ratios, to exactly equalise the pulls per unit area on the three principal planes (and therefore on all planes), are $2^{-\frac{1}{3}}$, $2^{\frac{1}{6}}$, $2^{\frac{1}{6}}$, as is easily seen from what follows.

9. While the equalisation of pulls in the three principal directions is thus produced, work is done by the film, on the moving wire-work of the cage, and the total area of film is diminished by an amount equal to W/T, if W denote the whole work done, and T the pull of the film per unit breadth. The change of shape of the cage being supposed to be performed infinitely slowly, so that the film is always in equilibrium throughout, the total area is at each instant a minimum, subject to the conditions

(1) That the volume of each cell is the given amount;

(2) That every part of the wire has area edged by it; and

(3) That no portion of area has any free edge.

10. Consider now the figure of the cell (still of course a tetrakaidecahedron) when the pulls in the three principal directions are equalised, as described in § 8. It must be perfectly

isotropic in respect to these three directions.   Hence the pair of small quadrilaterals must have become enlarged to equality with the two pairs of large ones which must have become reduced, in the deformational process described in § 8.   Of each hexagon three edges coincide with edges of quadrilateral faces of one cell; and each of the three others coincides with edges of three of the quadrilaterals of one of the contiguous cells.   Hence the 36 edges of the isotropic tetrakaidecahedron are equal and similar plane arcs; each of course symmetrical about its middle point.   Every angle of meeting of edges is essentially 109° 28′ (to make trihedral angles between tangent planes of the films meeting at 120°). Symmetry shows that the quadrilaterals are still plane figures; and therefore, as each angle of each of them is 109° 28′, the change of direction from end to end of each arc-edge is 19° 28′. Hence each would be simply a circular arc of 19° 28′, if its curvature were equal throughout; and it seems from the complete mathematical investigation of §§ 16, 17, 18 below, that it is nearly so, but not exactly so even to a first approximation.

Of the three films which meet in each edge, in three adjacent cells, one is quadrilateral and two are hexagonal.

11.   By symmetry we see that there are three straight lines in each (non-plane) hexagonal film, being its three long diagonals; and that these three lines, and therefore the six angular points of the hexagon, are all in one plane.   The arcs composing its edges are not in this plane, but in planes making, as we shall see (§ 12), angles of 54° 44′ with it.   For three edges of each hexagon, the planes of the arcs bisect the angle of 109° 28′ between the planes of the six corners of contiguous hexagons; and for the other three edges are inclined on the outside of its plane of corners, at angles equal to the supplements of the angles of 125° 16′ between its plane of corners and the planes of contiguous quadrilaterals.

12.   The planes of corners of the eight hexagons constitute the faces of an octohedron which we see, by symmetry, must be a regular octohedron (eight equilateral triangles in planes inclined 109° 28 at every common edge).   Hence these planes, and the planes of the six quadrilaterals, constitute a plane-faced tetra-kaidecahedron obtained by truncating the six corners* of a regular

---

* This figure (but with probably indefinite extents of the truncation?) is given in books on mineralogy as representing a natural crystal of an oxide of copper.

octohedron each to such a depth as to reduce its eight original (equilateral triangular) faces to equilateral equiangular hexagons. An orthogonal projection of this figure is shown in fig. 3. It is to be remarked that space can be filled with such figures. For brevity we shall call it a plane-faced isotropic tetrakaidecahedron.

13.   Given a model of the plane-faced isotropic tetrakaidecahedron, it is easy to construct approximately a model of the *minimal tetrakaidecahedron*, thus:—Place on each of the six square faces a thin plane disk having the proper curved arcs of 19° 28′ for its edges. Draw the three long diagonals of each hexagonal face. Fill up by little pieces of wood, properly cut, the three sectors of 60° from the centre to the overhanging edges of the adjacent quadrilaterals. Hollow out symmetrically the other three sectors, and the thing is done. The result is shown in orthogonal projection, so far as the edges are concerned, in fig. 4; and as the orthogonal projections are equal and similar

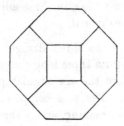

Fig. 3.

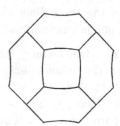

Fig. 4.

on three planes at right angles to one another, this diagram suffices to allow a perspective drawing from any point of view to be made by "descriptive geometry."

14.   No shading could show satisfactorily the delicate curvature of the hexagonal faces, though it may be fairly well seen on the solid model made as described in § 12. But it is shown beautifully, and illustrated in great perfection, by making a skeleton model of 36 wire arcs for the 36 edges of the complete figure, and dipping it in soap-solution to fill the faces with film, which is easily done for all the faces but one. The curvature of the hexagonal film on the two sides of the plane of its six long diagonals is beautifully shown by reflected light. I have made

these 36 arcs by cutting two circles, 6 inches diameter, of stiff wire, each into 18 parts of 20° (near enough to 19° 28'). It is easy to put them together in proper positions and solder the corners, by aid of simple devices for holding the ends of the three arcs together in proper positions during the soldering. The circular curvature of the arcs is not mathematically correct, but the error due to it is, no doubt, hardly perceptible to the eye.

15. But the true form of the curved edges of the quadri- lateral plane films, and of the non-plane surfaces of the hexagonal films, may be shown with mathematical exactness by taking, instead of Plateau's skeleton cube, a skeleton square cage with four parallel edges each 4 centimetres long: and the other eight, constituting the edges of two squares each $\sqrt{2}$ times as long, or

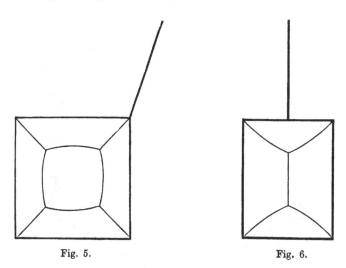

Fig. 5.   Fig. 6.

5·66 centim. Dipped in soap-solution and taken out it always unambiguously gives the central quadrilateral in the plane per- pendicular to the four short sides. It shows with mathematical accuracy (if we suppose the wire edges infinitely thin) a complete quadrilateral, four half-quadrilaterals, and eight half-hexagons of the minimal tetrakaidecahedron. The two principal views are represented in figs. 5 and 6.

16. The mathematical problem of calculating the forms of the plane arc-edges, and of the curved surface of the hexagonal faces, is easily carried out to any degree of approximation that

may be desired; though it would be very laborious, and not worth the trouble, to do so further than a first approximation, as given in § 17 below. But first let us state the rigorous mathematical problem; which by symmetry becomes narrowed to the consideration of a 60° sector BCB of our non-plane hexagon, bounded by straight lines CB, CB', and a slightly curved edge BEB', in a plane, Q, through BB', inclined to the plane BCB' at an angle of $\tan^{-1}\sqrt{2}$, or 54° 44'. The plane of the curved edge I call Q, because it is the plane of the contiguous quadrilateral. The mathematical problem to be solved is *to find the surface of zero curvature edged by BCB' and cutting at 120° the plane Q all along the intersectional curve* (fig. 7). It is obvious that this problem

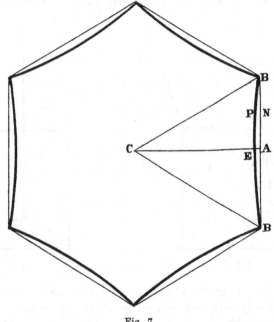

Fig. 7.

is determinate and has only one solution. Taking CA for axis of x, and z perpendicular to the plane BCB', and regarding z as a function of x, y, to be determined for finding the form of the surface, we have, as the analytical expression of the conditions

$$\frac{d^2z}{dx^2}\left(1+\frac{dz^2}{dy^2}\right) - 2\frac{dz}{dx}\frac{dz}{dy}\frac{d^2z}{dxdy} + \frac{d^2z}{dy^2}\left(1+\frac{dz^2}{dx^2}\right) = 0 \quad \dots(1);$$

and
$$\left(1 + \frac{dz^2}{dx^2} + \frac{dz^2}{dy^2}\right)^{-\frac{1}{2}} \left(\sqrt{\tfrac{1}{3}} - \frac{dz}{dx}\sqrt{\tfrac{2}{3}}\right) = \tfrac{1}{2} \Bigg\}$$
$$\text{where } z = (a - x)\sqrt{2} \qquad \text{........ (2).}$$

17. The required surface deviates so little from the plane BCB' that we get a good approximation to its shape by neglecting dz^2/dx^2, $dz/dx \cdot dz/dy$, and dz^2/dy^2, in (1) and (2), which thus become

$$\nabla^2 z = 0 \dots\dots\dots\dots\dots(3),$$

and
$$\frac{dz}{dx} = \frac{\sqrt{2}}{2} - \sqrt{\tfrac{3}{8}} = \cdot094735, \text{ where } x = a - z/\sqrt{2} \dots(4),$$

∇^2 denoting $(d/dx)^2 + (d/dy)^2$. The general solution of (3), in polar coordinates (r, ϕ) for the plane (x, y), is

$$\Sigma (A \cos m\phi + B \sin m\phi) r^m \dots\dots\dots\dots(5),$$

where A, B, and m are arbitrary constants. The symmetry of our problem requires $B = 0$, and $m = 3 \cdot (2i + 1)$, where i is any integer. We shall not take more than two terms. It seems not probable that advantage could be gained by taking more than two, unless we also fall back on the rigorous equations (1) and (2), keeping dz^2/dx^2 &c. in the account, which would require each coefficient A to be not rigorously constant but a function of r. At all events we satisfy ourselves with the approximation yielded by two terms, and assume

$$z = Ar^3 \cos 3\phi + A'r^9 \cos 9\phi \dots\dots\dots\dots(6);$$

with two coefficients A, A' to be determined so as to satisfy (4) for two points of the curved edge, which, for simplicity, we shall take as its middle, $E (\phi = 0)$; and end, $B (\phi = 30°)$. Now remark that, as z is small, even at E, where it is greatest, we have, in (4), $x = a$ or $r = a \sec \phi$. Thus, and substituting for dz/dx its expression in polar (r, ϕ) coordinates, which is

$$\frac{dz}{dx} = \frac{dz}{dr} \cos \phi - \frac{dz}{rd\phi} \sin \phi \dots\dots\dots\dots(7),$$

we find, from (4) with (6),

$$\text{(by case } \phi = 0) \qquad A + 3a^6 A' = \cdot031578a^{-2} \quad \dots(8),$$

$$\text{(and by case } \phi = 30°) \qquad A - \tfrac{64}{9}a^6 A' = \cdot031578 \cdot \tfrac{3}{2} \cdot a^{-2} \dots(9);$$

whence

$$A' = -\tfrac{1}{2} \times \tfrac{9}{91} \times \cdot 031578 . a^{-8} = -9 \times \cdot 0001735 . a^{-8}$$
$$= - \cdot 001561 . a^{-8},$$
$$A = \tfrac{1}{2}(3 - \tfrac{64}{91}) \times \cdot 031578 . a^{-2} = 209 \times \cdot 0001735 . a^{-2}$$
$$= \cdot 036261 . a^{-2};$$

and for required equation of the surface we have (taking $a = 1$ for brevity)

$$\left. \begin{aligned} z &= \cdot 03626 . r^3 \cos 3\phi - \cdot 001561 r^9 \cos 9\phi \\ &= \cdot 03626 . r^3 (\cos 3\phi - \cdot 043 . r^6 \cos 9\phi) \end{aligned} \right\} \quad \ldots\ldots(10).$$

18. To find the equation of the curved edge BEB', take, as in (4),

$$x = 1 - z/\sqrt{2} = 1 - \xi, \text{ where } \xi \text{ denotes } z/\sqrt{2} \ \ldots\ldots(11).$$

Substituting in this, for z, its value by (10), with for r its approximate value $\sec\phi$, we find

$$\xi = \frac{1}{\sqrt{2}} (\ 03626 \sec^3\phi \cos 3\phi - \cdot 001561 \sec^9\phi \cos 9\phi)\ldots(12)$$

as the equation of the orthogonal projection of the edge, on the plane BCB', with

$$AN = y = \tan\phi; \text{ and } NP = \xi \ \ldots\ldots\ldots\ldots(13).$$

The diagram was drawn to represent this projection roughly, as a circular arc, the projection on BCB' of the circular arc of $20°$ in the plane Q, which, before making the mathematical investigation, I had taken as the form of the arc-edges of the plane quadrilaterals. This would give, of CA, for the sagitta, AE; which we now see is too great. The equation (12), with $y = 0$, gives for the sagitta

$$AE = \cdot 024 \times CA \ \ldots\ldots\ldots\ldots\ldots\ldots(14),$$

or, say, $\tfrac{1}{40}$ of CA. The curvature of the projection at any point is to be found by expressing $\sec^3\phi \cos 3\phi$ and $\sec^9\phi \cos 9\phi$ in terms of $y = \tan\phi$ and taking d^2/dy^2 of the result.

By taking $\sqrt{(3/2)}$ instead of $\sqrt{(1/2)}$ in (12), we have the equation of the arc itself in the plane Q.

19. To judge of the accuracy of our approximation, let us find the greatest inclination of the surface to the plane BCB'. For the tangent of the inclination at (r, ϕ) we have

$$\left(\frac{dz^2}{dr^2} + \frac{dz^2}{r^2 d\phi^2} \right)^{\frac{1}{2}} = \cdot 1088 . r^2 (1 - 2 \times \cdot 129 . r^6 \cos 6\phi + \cdot 129^2 r^{12})^{\frac{1}{2}} \ldots(15).$$

The greatest values of this will be found at the curved bounding edge, for which $r \doteqdot \sec \phi$. Thus we find

$$
\left(\frac{dz^2}{dr^2} + \frac{1}{r^2}\frac{dz^2}{d\phi^2} \right)^{\frac{1}{2}}
$$
$$
\left\{
\begin{aligned}
&= \cdot 0948, \text{ and therefore inclination} = 5° \; 25' \text{ at } E \\
&= \cdot 1894, \text{,,} \text{,,} = 10° \; 44' \text{ at } B
\end{aligned}
\right\} \quad ...(16).
$$

Hence we see that the inaccuracy due to neglecting the square of the tangent of the inclination in the mathematical work cannot be large. The exact value of the inclination at E is $\tan^{-1}(-\sqrt{2}) - 120°$, or $5° \; 16'$, which is less by $9'$ than its value by (16).

166. MOLECULAR CONSTITUTION OF MATTER.

[From *Edinb. Roy. Soc. Proc.* Vol. XVI. [read July 1 and 15, 1889], pp. 693—
724. Reprinted in *Math. and Phys. Papers*, Vol. III. art. xcvii.
pp. 395—427.]

167. ON THE MODULUSES OF ELASTICITY IN AN ELASTIC SOLID ACCORDING TO BOSCOVICH'S THEORY.

[From *Edinb. Math. Soc.*, Feb. 1890, *not printed*, but substance of the paper
is in *Edinb. Roy. Soc. Proc.* Vol. XVI. pp. 693—724. Reprinted in *Math.
and Phys. Papers*, Vol. III. art. xcvii. pp. 423—427. Continuation in
No. **176** *infra*.]

168. ON THE MOLECULAR TACTICS OF A CRYSTAL.

[*The Robert Boyle Lecture, delivered before the Oxford University
Junior Scientific Club, May 16, 1893.*]

[Reprinted in *Baltimore Lectures*, Appendix H, pp. 602—642, 1904.]

169. ON THE ELASTICITY OF A CRYSTAL ACCORDING TO BOSCOVICH.

[From *Roy. Soc. Proc.* Vol. LIV. 1894 [June 15, 1893], pp. 59—75; *Phil. Mag.*
Vol. XXXVI. Nov. 1893, pp. 414—430. Reprinted in *Baltimore Lectures*,
Appendix I, 1904, pp. 643—661.]

170. ON THE PIEZO-ELECTRIC PROPERTY OF QUARTZ.

[From *Brit. Assoc. Report*, 1893, p. 691 [title only]; *Phil. Mag.* Vol. XXXVI. pp. 331—342, Oct. 1893, and p. 384, Nov. 1893; *Lum. Elec.* Vol. L. Oct. 7, 1893, pp. 37—41, and Nov. 4, 1893, pp. 236, 237.]

1.   IN the present communication we are not concerned with the six-sided pyramid, or planes parallel to the sides of a six-sided pyramid, seen at the ends of a quartz crystal farthest from the matrix. Nor are we concerned with the transverse striæ generally seen on the sides of the prism, which are undoubtedly steps (probably having faces parallel to the faces of the terminal pyramid) by which the prism becomes less in transverse section from the matrix outwards. We shall consider only a perfect, and therefore an unstriated, hexagonal prism. The sides of the prism may be, and generally are, unequal in nature but the angles are all exactly 120°. For simplicity of reference to the natural crystalline form, I shall suppose the prism to be equilateral, which it may be in nature ; as well as equiangular, which it must be. Thus we have three planes of symmetry, which for brevity I shall call the diagonal planes, being planes through the opposite edges of the prism. We have also three other planes of symmetry, which for brevity I shall call the normal planes, being planes perpendicular to the pairs of parallel faces.

2.   In the brothers J. and P. Curie's beautiful instrument for showing their discovery of the piezo-electric property of quartz, a thin plate of the crystal about half a millimetre thick, I believe, is taken from a position with its sides parallel to any of the three normal planes of symmetry; its length perpendicular to the faces of the prisms, and its breadth parallel to the edges. The sides of this plate are, through nearly all their length, silvered by the

chemical process to render them conductive*, and are metallically connected with two pairs of quadrants of my quadrant electrometer. I find that the effect is also well shown by my portable electrometer; the two sides of the quartz plate being connected respectively to the outer case, and the insulated electrode, of the electrometer. In an instrument which has been made for me under Mr Curie's direction, the silvered part of the plate is 7 centimetres long and 1·8 broad. A weight of 1 kilogramme hung upon the plate placed with its length vertical causes one side to become positively electrified and the other negatively.

3. A plate parallel to any one of the three normal planes of symmetry will give the same result, of transverse electropolarization; but a plate cut parallel to any one of the three diagonal planes of symmetry will give no result in the mode of experimenting described in § 2. But with its sides unsilvered it would, if properly tested, show positive electrification at one end and negative at the other when stretched longitudinally, as we see by the hypothesis, and theoretical considerations which I now proceed to explain; and by § 11 below without any hypothesis.

4. Electric eolotropy of the molecule, and nothing but electric eolotropy of the molecule, can produce the observed phenomena. The simplest kind of electric eolotropy which I can imagine is as follows:—For brevity I shall explain it in relation to the chemical constitution which, according to present doctrine, is one atom of silicon to two atoms of oxygen. The chemical molecule may be merely SiO_2 for silica in solution or it may consist of several compound molecules of this type, grouped together: but it seems certain that, in crystallized silica (in order that the crystal may have the hexagonally eolotropic piezo-electric property which we know it has) the crystalline molecule must consist of three SiO_2 molecules clustered together; or must be *some* configuration of three atoms of silicon and three double atoms of oxygen combined. As a ready and simple way of attaining the desired result, take a cluster of three atoms of silicon and three double atoms of oxygen placed at equal distances of 60° in alternate order, silicon and oxygen, on the circumference of a circle.

The diagram, fig. 1, shows a crystalline molecule of this kind

* For a description and drawing of this part of their instrument, given by the brothers Curie, see Appendix to the present paper.

surrounded by six nearest neighbours in a plane perpendicular to
the axis of a quartz crystal. Each silicon atom is represented by
+ (plus) and each oxygen double atom by − (minus). The constituents of each cluster must be supposed to be held together
in stable equilibrium in virtue of their chemical affinities. The
different clusters, or crystalline molecules, must be supposed to be
relatively mobile before taking positions in the formation of a

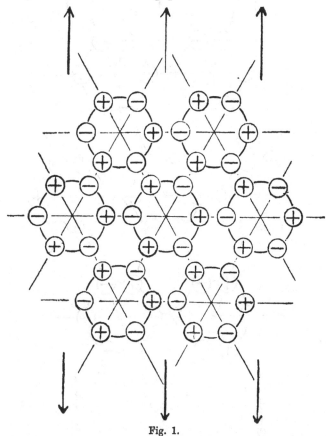

Fig. 1.

crystal. But we must suppose, or we may suppose, the mutual
forces of attraction (or chemical affinity), between the silicon of
one crystalline molecule and the oxygen of a neighbouring crystalline molecule, to be influential in determining the orientation of
each crystalline molecule, and in causing disturbance in the relative
positions of the atoms of each molecule, when the crystal is strained
by force applied from without.

5. Imagine now each double atom of oxygen to be a small negatively electrified particle, and each atom of silicon to be a particle electrified with an equal quantity of positive electricity.

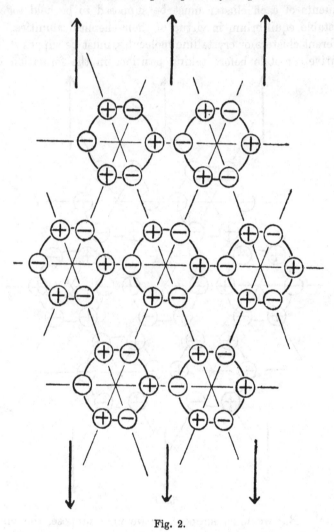

Fig. 2.

Suppose now such pressures, positive and negative, to be applied to the surface of a portion of crystal as shall produce a simple elongation in the direction perpendicular to one of the three sets of rows. This strain is indicated by the arrow-heads in fig. 1, and is realized to an exaggerated extent in fig. 2.

This second diagram shows all the atoms and the centres of all the crystalline molecules in the positions to which they are brought by the strain. Both diagrams are drawn on the supposition that the stiffness of the relative configuration of atoms of each molecule is slight enough to allow the mutual attractions between the positive atoms and the negative atoms of neighbouring molecules to keep them in lines through the centres of the molecules, as fig. 1 shows for the undisturbed condition of the system, and fig. 2 for the system subjected to the supposed elongation. Hence two of the three diameters through atoms of each crystalline molecule are altered in direction, by the elongation, while the diameter through the third pair of atoms remains unchanged, as is clearly shown in fig. 2 compared with fig. 1.

6. Remark, first, that the rows of atoms, in lines through the centres of the crystalline molecules, perpendicular to the direction of the strain, are shifted to parallel positions with distances between the atoms in them unchanged. Hence the atoms in these rows contribute nothing to the electrical effect. But, in parallels to these rows, on each side of the centre of each molecule, we find two pairs of atoms whose distances are diminished.

7. This produces an electric effect which, for great distances from the molecule, is calculated by the same formula as the magnetic effect of an infinitesimal bar-magnet whose magnetic moment is numerically equal to the product of the quantity of electricity of a single atom into the sum of the diminutions of the two distances between the atoms of the two pairs under consideration. Hence, denoting by N the number of crystalline molecules per unit bulk of the crystal; by b the radius of the circle of each crystalline molecule; by q the quantity of electricity on each of the six atoms or double-atoms, whether positive or negative; by ϑ the change of direction of each of the two diameters through atoms which experience change of direction; and by μ the electric moment* developed per unit volume of the

---

* I do not know if this designation has hitherto been used. I introduce it with precisely the same significance relatively to electricity, as the well-known "magnetic moment" in reference to magnetism.

crystal, by the strain which we have been considering and which is shown in fig. 2; we have*

$$\mu = Nq \cdot 4b\vartheta \cos 30° = N \cdot 2bq\vartheta \sqrt{3} \quad \ldots\ldots\ldots\ldots(1).$$

It is of course understood that ϑ is a small fraction of a radian.

8. To test the sufficiency of our theory, let us first consider quantities of electricity which probably, we may almost say certainly, are present in the atoms in nature.

Instead of the silicon atoms marked + in the diagram let us substitute globes of polished zinc; and instead of the double-oxygen atoms marked − let us substitute little globes of copper well oxidized (polished copper, heated in air till it becomes of a dark slate-colour). Let us suppose all the six atoms of each compound molecule to be metallically connected, and all the molecules insulated from one another. We are not concerned with conceivable permeation of electricity by conductance through the crystal; and therefore we must suppose the total quantity of electricity on each crystalline molecule to be zero. Let the circle of each compound molecule in the diagram be a real exceedingly thin stiff ring of metal, no matter what kind of metal, and let each of the six atoms be a bead (a perforated spherule), whether of zinc or of copper, moving frictionlessly on it. Thus we have, in idea, a working model of an electrically eolotropic crystalline molecule.

9. I have found by experiment† that the difference of potentials in air, beside a polished surface of zinc and an oxidized surface of copper, is about ·004 of a c.g.s. electrostatic unit, provided the zinc and copper are metallically connected. Hence, if a be the radius of each spherule, we have approximately $q = ·002 \times a$; because we shall suppose for simplicity that, except the infinitely thin ring on which it is movable, no spherule has any metal within a distance from it of less than two or three times its diameter. Let now $N = 10^{21}$ per cubic centimetre; and let b be a quarter of $N^{-\frac{1}{3}}$, that is to say, $b = \frac{1}{4} \times 10^{-7}$ of a centimetre. Lastly, to give definiteness to our example, let $a = ·2 \times b$. Equation (1) becomes

$$\mu = 866\vartheta. \quad \ldots\ldots\ldots\ldots\ldots\ldots\ldots\ldots(2).$$

* It should be remarked that δq is of the second order [ϑ^2].

† *Electrostatics and Magnetism*, § 400, and experiments not hitherto published, by another method.

10. From the admirable statement of Messrs Curie of the result of their measurements quoted in the Appendix of the present paper, I find that a stretching force of 1 kilogramme per square centimetre, in their experiment described in § 2 above, produces an electric moment of ·063 c.g.s. electrostatic reckoning per cubic centimetre of the crystal. Thus about ⅓ of a c.g.s. unit of electric moment per cubic centimetre is produced by 5 kilogrammes per square centimetre of stretching force; and this, according to equation (2), requires ϑ to be 1/2598, which is an amount of change of direction among atoms quite such as might be expected in pieces of crystal stretched by forces well within the limits of their strength. A rough mechanical illustration of the theory of electric atoms to account for the piezo-electric properties of crystals, is presented in an electrically working model of a piezo-electric pile, submitted to Section A in a separate communication at the present meeting of the British Association.

11. I shall now prove, without any hypothetical assumption, the statement at the end of § 3 above. Consider first a simple elongation perpendicular to one of the three pairs of parallel sides of the hexagon in fig. 3, as indicated by the arrow-heads, *A A A A A A*, in fig. 3. Superimpose now two equal negative elongations, one of them in the direction of the original elongation, and the other in a direction perpendicular to it. These negative elongations, indicated by the twelve arrow-heads marked *C*, constitute a condensation equal in all directions; which produces no change on the electrical effect of the first simple elongation. But it leaves us with a simple negative elongation in the direction perpendicular to that of the original positive elongation; which therefore alone produces the same effect as that which was produced by the original one alone. Thus we see that if elongations perpendicular to the three pairs of parallel sides produce electro-polarizations with electric axis in each case perpendicular to the line of elongation, equal elongations in the directions of the diagonals produce electric polarizations equal to those, but having their axes along the lines of elongation instead of perpendicular to them.

12. Consider now two simple elongations in the directions shown by the arrow-heads *A A, B B*, in fig. 4. These two elongations

produce electro-polarizations with axes and signs indicated by *aa* and *bb* respectively.   The resultant of our present two simple elongations is clearly a dilatation equal in all directions in the plane of the diagram, compounded with a single simple elongation in the line *K'OK*, bisecting the angle between them, and of magnitude equal to $\sqrt{3}$ times the magnitude of each of them, as is easily proved by the elementary geometry of strain.   Hence an

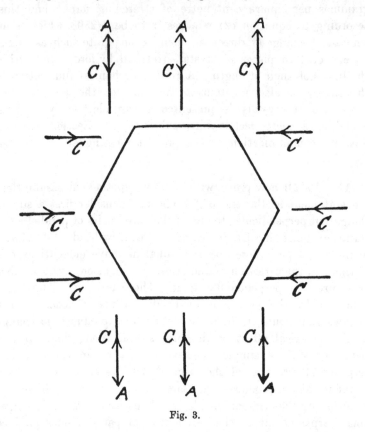

Fig. 3.

elongation in the direction *K'OK* does not produce zero of electric effect.   In fact, there are no "Axen fehlender Piëzo-electricität." The three axes so-called by Röntgen* are the lines of elongation in Curie's experiment.   Without sufficient consideration it might be imagined that the six lines corresponding to *K'OK* in fig. 4 are "Axen fehlender Piëzo-electricität."   On the contrary, elonga-

* Wiedemann's *Annalen*, 1883, Vol. xviii. p. 215.

tion in the line $K'OK$ produces an electro-polarization which is
the resultant of the equal polarizations indicated by aa and bb,
and which, as it bisects the angle bOa, is in a line inclined at 45°
to OK the line of the elongation. In fact simple elongation in
any direction perpendicular to the principal axis of a quartz

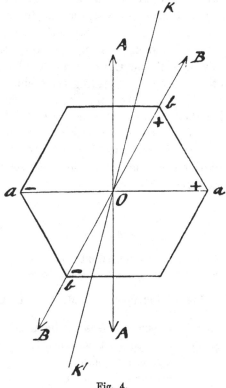

Fig. 4.

crystal produces electro-polarization; and it is only when the lines
of two simple elongations are coincident with one another, or are
perpendicular to one another, that the resultant of their electro-
polarizations can be zero.

13. A most important contribution to our knowledge of the
electric properties of crystals has been made by Röntgen*, and
by Friedel and J. Curie†, in independent investigations proving

* *Ber. der Oberrh. Ges. f. Natur- und Heilkunde*, Vol. xxii. [of date between
December 1882 and April 30, 1883].

† *Bulletin de la Société Minéralogique de France*, t. v. p. 282, Decembre 1882;
and *Comptes Rendus* of French Academy of Sciences, April 30, and May 14, 1883.

that the irregular electrifications of the corners of quartz crystals, which had been observed by many observers as consequences of heatings and of returns to lower temperatures, are wholly due to mechanical stresses developed by inequalities of temperature in different parts of the crystal. Those phenomena are therefore truly *piezo-electric*, and are not at all "pyro-electric" like the electric property of tourmaline which is due to change from one temperature to another, each the same throughout the crystal. The very important and interesting discovery thus made by Röntgen and by Friedel and Curie, is, as they have pointed out, available also to explain the perplexing and seemingly paradoxical statements regarding positive and negative electrifications of corners and hemihedral facets at opposite ends of the four long diagonals of crystals of the cubic class, and of cubes of boracite, which had been given by previous observers and writers, and which have not yet disappeared from elementary treatises on Mineralogy, Electricity, and General Physics.

APPENDIX.

[Extract from a pamphlet published by the "*Société Centrale de Produits Chimiques*," 42, 43 Rue des Écoles, Paris.]

Quartz Piézo-Électrique de MM. J. *et* P. CURIE.

Cet instrument se compose essentiellement d'une lame de quartz, sur laquelle on exerce des tractions à l'aide de poids placés dans un plateau. Cette action mécanique provoque un dégagement d'électricité sur les faces de la lame.

La lame de quartz *abc* (fig. 5) est montée solidement à ses extrémités dans deux garnitures métalliques *H* et *B*. Elle est suspendue, en *H*, à la partie supérieure; elle soutient, à son tour, en *B*, à la partie inférieure, le plateau et les poids, par l'intermédiaire d'une tige munie de crochets.

L'axe optique du quartz est dirigé horizontalement, suivant la largeur *ab* de la lame. Les faces sont normales à un des axes binaires (ou axes électriques) du cristal. On exerce les tractions dans le sens vertical, c'est-à-dire dans une direction à la fois normale à l'axe optique et à l'axe électrique.

Les deux faces de la lame sont argentées.

On a tracé dans l'argenture de chaque face deux traits fins, mn, $m'n'$, qui isolent des montures la plus grande partie de la surface. On recueille l'électricité, sur ces portions isolées, à l'aide de deux lames de cuivre faisant ressort (rr, rr), qui viennent s'appuyer sur les deux faces et communiquent avec les bornes de l'appareil.

Lorsque l'on place des poids dans le plateau, on provoque le dégagement de quantités d'électricité égales et de signes contraires sur les deux faces de la lame.

Lorsque l'on retire les poids, le dégagement se fait encore, mais avec inversion des signes de électricité dégagée sur chaque face.

La quantité d'électricité dégagée sur une face est rigoureusement proportionnelle à la variation de traction F.—On a

$$q = 0{\cdot}063 \frac{L}{e} F.$$

L est la longueur mm' de la partie argentée utilisée. e est l'épaisseur de la lame.

Fig. 5.

F est exprimé en kilogrammes et q est donné en unités C.G.S électrostatiques.

On a donc avantage, lorsque l'on veut avoir des effets très sensibles, à prendre une lame longue dans le sens de la traction et peu épaisse dans le sens de l'axe électrique. La dimension parallèle de l'axe optique n'a pas d'influence sur la quantité d'électricité dégagée*.

La lame de quartz est placée dans une enceinte métallique desséchée. Cette cage métallique, toutes les pièces métalliques de l'instrument et les montures de la lame de quartz sont mises en communication permanente avec la terre.

Le modèle no. 2 comporte encore un commutateur et un levier qui sert à soulever les plateaux et les poids. Nous reviendrons plus loin sur le rôle de ces organes.

---

* Double breadth, with doubled stretching force, would give double quantity.

Since my communication of a short article "On a Piezo-electric Pile" to the *Philosophical Magazine* [*infra*, p. 323], I have found a very important article* by Messrs J. and P. Curie, in which precisely the same combination is described, and the application of the principle illustrated by it to explain all the piezo-electric and pyro-electric properties of crystals is pointed out.

The discovery of the piezo-electric property in crystalline matter has been made known to the world by the experimental researches of the brothers Curie ; and it now interests me exceedingly to find that they have also given what seems to me undoubtedly the true electro-molecular theory of the constitution of crystals, explaining not only piezo-electricity, but the old known pyro-electricity ; and bringing these properties into relation with the electro-chemical constitution of the crystalline molecule.

* *Comptes Rendus* of the French Academy of Sciences for February 14, 1881.

171. On a Piezo-electric Pile.

[From *Brit. Assoc. Report*, 1893, pp. 691, 692; *Phil. Mag.* Vol. xxxvi. Oct. 1893, pp. 342, 343; *Electrician*, Vol. xxxi. Oct. 20, 1893, p. 664; *Lum. Élec.* Vol. l. Oct. 7, 1893, pp. 41, 42.]

THE application of pressure to a voltaic pile, dry or wet, has been suggested as an illustration of the piezo-electric properties of crystals, but no very satisfactory results have hitherto been obtained, whether by experiment or by theoretical considerations, so far as I know. Whatever effects of pressure have been observed have depended upon complex actions on the moist, or semi-moist, substances between the metals, and electrolytic or semi-electrolytic and semi-metallic conductances of these substances. Clearing away everything but air from between the opposed metallic surfaces of different quality, I have made the piezo-electric pile which accompanies this communication. It consists of twenty-four double plates, each 8 centimetres square, of zinc and copper soldered together, zinc on one side and copper on the other. Half a square centimetre is cut from each corner of each zinc plate, so that the copper square is left uncovered by the zinc at each of its four corners. Thus each plate presents on one side an uninterrupted copper surface, and on the other side a zinc surface, except the four uncovered half square centimetres of copper. A pile of these plates is made, resting one over the other on four small pieces of india-rubber at the four copper corners. The air-space between the opposed zinc and copper surfaces may be of any thickness from half a millimetre to 3 or 4 millimetres. Care must be taken that there are no minute shreds of fibre or dust bridging the air-space. In this respect so small an air-space as half a millimetre gives trouble, but with 3 or 4 millimetres no trouble is found.

The lowest and uppermost plates are connected by fine wires to the two pairs of quadrants of my quadrant electrometer, and it is generally convenient to allow the lowest to lie uninsulated on an ordinary table and to connect it metallically with the outer case of the electrometer.

To make an experiment, (1) connect the two fine wires metallically, and let the electrometer-needle settle to its metallic zero.

(2) Break the connection between the two fine wires, and let a weight of a few hektogrammes or kilogrammes fall from a height of a few millimetres above the upper plate and rest on this plate. A startlingly great deflexion of the electrometer-needle is produced. The insulation of the india-rubber supports and of the quadrants in the electrometer ought to be so good as to allow the needle to come to rest, and the steady deflexion to be observed, before there is any considerable loss.

If, for example, the plates are placed with their zinc faces up, the application of the weight causes positive electricity to come from the lower face of the uppermost plate and deposit itself over the upper surface of plate and weight, and on the electrode and pair of quadrants of the electrometer connected with it.

172. ON THE THEORY OF PYRO-ELECTRICITY AND
PIEZO-ELECTRICITY OF CRYSTALS.

[From *Comp. Rend.* Vol. CXVII. [Oct. 9, 1893], pp. 463—472 ; *Lum. Élec.*
Vol. L. Nov. 4, 1893, pp. 238—242 ; *Phil. Mag.* Vol. XXXVI. Nov. 1893,
pp. 453—459.]

1. THE doctrine of bodily electro-polarization masked by an
induced superficial electrification, which I gave thirty years ago
in Nichol's *Cyclopædia**, wanted a physical explanation of the
assumed molecular polarization to render it a satisfying physical
theory of pyro-electricity; and it was essentially defective, as has
been remarked by Röntgen† and by Voigt‡, in that it contained
no suggestion towards explaining the multiple electric polarities
irregularly produced by irregular changes of temperature in boracite,
in quartz, and in tourmaline itself; which had perplexed many
naturalists and experimenters. A short but very important paper
by MM. Jacques and Pierre Curie in the *Comptes Rendus* for
Feb. 14, 1881, supplies that want in a manner which suggests
what seems to me the true matter-of-fact electro-chemical theory
of a crystalline molecule, and at the same time makes easy the
extension of my slight primitive doctrine, to remedy its defect
in respect to multipolarity, and to render it available for explaining
not only the old-known pyro-electricity of crystals, but also the
piezo-electricity discovered by the brothers Curie§ themselves.
The element of zinc and copper soldered together and surrounded
only by air, which they suggest, represents perfectly a true electro-
chemical compound molecule such as H_2O or SiO_2 in a realizable

---

* Reprinted in *Collected Mathematical and Physical Papers* (Sir W. Thomson),
Vol. I. p. 315.
† Wiedemann's *Annalen*, 1883, Vol. XVIII. p. 213.
‡ " Allgemeine Theorie der Piëzo- und Pyro-electrischen Erscheinungen an
Krystallen," p. 6; separate publication from Vol. XXXVI. of *Abhand. König. Ges.
Wiss. Göttingen*, 1890.
§ *Comptes Rendus*, Aug. 2 and Aug. 16, 1880.

model, which indeed I actually made three weeks ago and described in a communication to the *Philosophical Magazine**** without knowing that I had been anticipated.

2. To represent pyro-electric and piezo-electric qualities in a crystal, take as crystalline molecule a rigid body of any shape, bounded by a surface made up of pieces of different metals, soldered together so as to constitute one metallic conductor. Arrange a large number of such molecules in order, as a Bravais homogeneous assemblage, not touching one another. Connect every molecule with neighbours by springs of non-conducting material (india-rubber may be taken if we wish to make a practically working model). We may, for example, suppose each molecule to be connected with only twelve neighbours; its two nearest, its two next-nearests, its two next-next-nearests in the plane of those four, and the three pairs of nearests, next-nearests, and next-next-nearests on the two sides of that plane. Thus we have a perfect mechanical model for the elasticity and the piezo-electricity of a crystal; and for pyro-electricity also, if we suppose change of temperature to produce either change of the contact-electricities of the metals, or change of configuration of the assemblage, whether by changing the shape of each molecule or by changing the forces of the springs.

3. The mathematical problem which this combination presents is as follows :—

Given a homogeneous assemblage of a large number of equal and similar closed surfaces, S, each composed of two or more different kinds of metal soldered together, all insulated in a large closed chamber, of which the bounding surface C is everywhere at a practically infinite distance from the assemblage, and is of the same metal as one of the metals of S, copper we shall suppose, to fix the ideas.

It is required to find :—

(1) The potential in the copper of every molecule when the total quantity of electricity on each is zero.

(2) The quantity of electricity on each molecule when all are metallically connected by infinitely fine wire.

---

* For October 1893, "On a Piezo-electric Pile."

4.   The mathematical expression of the conditions and requirements of the problem is as follows:—

Let $f(P)$ denote a given function of the position of a point P on the surface S of any one of the molecules; expressing the difference of the potential in the air infinitely near to P, from the potential in the air infinitely near to the copper parts of the surface S. This function is the same for corresponding points of all the molecules.

Let V_n be the potential at the copper of the molecule numbered n.

Let $D(P_i, P_n)$ be the distance between a point P_i on the molecule numbered i, and a point P_n on the molecule numbered n.

Let $\iint ds_i$ denote integration over the surface of molecule i, and ρ_i a function of the position of P_i on the surface of this molecule (the electric density at P_i).

Let Σ_i denote summation for all the molecules, including the case $i = n$.

Let q_n be the total quantity of electricity on molecule n.

The equation of electric equilibrium is

$$\Sigma_i \iint \frac{\rho_i ds_i}{D(P_i, P_n)} = f(P_n) + V_n \ \ldots\ldots\ldots\ldots\ldots(a),$$

and we have

$$\iint \rho_n ds_n = q_n \ \ldots\ldots\ldots\ldots\ldots\ldots\ldots(b).$$

It is required to find

(1) V_n, when $q_n = 0$, for every value of n;

and   (2) q_n, when $V_n = 0$,   „   „   .

5.   The problem thus proposed is of a highly transcendental character, unless the surface S is spherical. In this case it can be solved for any finite number of molecules by mere expenditure of labour; perhaps the work of the natural working-life of a competent mathematician, if the assemblage is a Bravais parallelepiped of 125 globes in 5 "réseaux" of 25 globes each, would suffice to give the solution for each item within one per cent. of accuracy; and not much more labour would be needed to solve the problem to the same degree of accuracy for each of 125×10^{21} spherical

molecules in a similar Bravais parallelepiped of 5×10^7 "réseaux," if the distance (h we shall call it) between the planes of corresponding points of two consecutive "réseaux" of nearest and next-nearest molecules is not greater than about twice the diameter of each molecule.

6. When this last condition is fulfilled, we can see, from general knowledge of the doctrine of electric screening, without solving the problem as proposed for every individual molecule, that the solution of the second part (2) of the requirements is $q_n \doteqdot 0$, very approximately for every molecule at any distance exceeding two or three times h from every part of the boundary of the assemblage; and this whether the molecules are spherical, or of any other shape not too wildly different. We see also that for all molecules not nearer than $3h$, or perhaps $4h$, or $5h$, from any part of the boundary of the assemblage, the distribution of electricity is similar. That is to say, the whole assemblage within a thin surface-layer (of some such thickness as $3h$ or $4h$) is homogeneous, not only geometrically and mechanically, but also electrically. The problem of finding, with moderate accuracy, the distribution of electricity on each molecule of the homogeneous assemblage thus constituted, is comparatively un-laborious if the shape of each molecule is spherical.

7. We also see, by the known elements of electrostatics, without solving the problem of finding the quantity of electricity on each molecule of the surface-layer in the circumstances described in § 6, that the sum of the quantities on all the molecules of this layer, per unit of the surface, is equal to the component, in the direction normal to the surface, of the electric moment per unit-volume of the homogeneous assemblage within the surface-layer.

8. The condition of the whole assemblage, surface-layer and homogeneous assemblage within it, at which we have arrived in §§ 6 and 7, may be regarded as representing the natural undisturbed condition of a crystal. Let now any homogeneous change of configuration of our assemblage be produced either by proper application of force to the molecules of the surface-layer, or by uniform change of temperature throughout the interior, or by both these causes acting simultaneously. We need not exclude the case of no change of shape or bulk of the boundary; that is

to say, the case of no change of the relative positions of corresponding points of the molecules; and our "change of configuration" only an infinitesimal rotation of each molecule. The inclusion of this case is important to guard against a tendency which I find in the writings both of MM. Curie and of Voigt:—a tendency to a hypothetical assumption unduly limiting the pyro-electric property to identity with the piezo-electric effect produced by force causing the same change of shape or bulk as that which is produced by the change of temperature. In nature, we may expect as a general possibility, and as a probable result in some cases, a bodily electro-polarization produced by change of temperature, even though change of bulk and shape are prevented by force applied to the surface. And, in our model, changes of forces of the springs would certainly cause rotation of the molecules, and so produce electro-polarization, even when the molecules of the boundary are held fixed, unless the springs are specially designed and constructed to annul this effect.

9. Solve now the electrical problem of finding the change of electric moment of each molecule of the homogeneous assemblage, produced by the change of configuration described in § 8, when the potential is zero throughout the surface-layer. To avoid tampering with the separate insulation of all or any of the molecules, and to conform our ideas to the realities of experiments on the electric properties of crystals, I suppose this equality of potential to be produced not by temporary metallic connexion between the molecules as in § 3 (2), but by a metal coat enclosing our model, and having its inner surface everywhere very near to the boundary of the assemblage; for example, everywhere within a distance of less than $2h$ or $3h$ in our model, or of less than $10^6 \times h$ if we are dealing with a real crystal in a real experiment.

10. To find experimentally the solution of the mathematical problem of § 9, divide the metal coat into two parts; one of them (corresponding to Coulomb's "proof-plane") we shall call for brevity E. It may be either so small that it is sensibly plane, or it may be a portion of the coat covering a finite plane part of the boundary of the assemblage. Commence now with the crystal in its natural undisturbed state, and having the metal coat on it with E insulated from the rest of the coat. Produce a change of

configuration as in § 8; and then measure how much electricity would need to pass from E to the rest of the coat to equalize the potential between them. This is wholly and exactly what MM. Curie do in their admirably designed measurement with their "quartz piézo-électrique," avoiding all need for consideration of the essentially transcendent problem of the distribution of electric potential at the surface of an uncoated crystal when there is either pyro-electric or piezo-electric disturbance of its interior.

The quantity of electricity thus measured, divided by the area of E, is equal to the component perpendicular to E of the interior electro-polarization when E and the rest of the coat are metallically connected.

11. In conclusion, following Voigt in his *Allgemeine Theorie*, already referred to, we see that there are essentially 18 independent coefficients for the piezo-electricity of a crystal in general; in three formulas expressing the three components of the electric moment per unit of its volume each as a linear function of the six components of the geometrical strain of the substance. To each of these expressions I add a term for the component of the electric moment due to change of temperature when force acting on the surface prevents change of volume or shape. Thus we have in all 21 independent coefficients for piezo-electricity and pyro-electricity; to be determined for a real crystal by observation. It is interesting to see how our model can be constructed to realize the piezo-electric and pyro-electric phenomena in accordance with any given values of these 21 coefficients, by experimental solution of as much of the mathematical problem of § 4 as is necessary for the purpose.

12. Choose any convenient shape, spherical or not wildly different from spherical, for each molecule. Divide the whole surface into 22 parts (not wildly unequal nor extravagantly different from squares or equilateral equi-angular hexagons), and number them 0, 1, 2, ... 21. Construct a trial molecule with part 0 always of copper; and with, for first trial, part 1 of zinc, and parts 2, 3, ... 21 of copper. Take a large number of such molecules and make of them a Bravais homogeneous assemblage with any arbitrarily chosen values for the six edges of the fundamental acute-angled tetrahedrons. Connect the molecules homogeneously

by springs of non-conducting material in the manner described in § 2 above. To provide fully for pyro-electricity, without hypothesis, we must now take care that these springs are such, that when the temperature is changed, and the border molecules are held fixed, all the interior molecules shall be caused to rotate round parallel axes through equal angles proportional to the difference of temperature. For this purpose the springs must be of two or more different materials; and when set in their proper positions between the molecules they must be under stress, some of them pushing and some pulling, in the undisturbed condition of the assemblage.

13. Subject now the assemblage successively to six different geometrical strain-components, e, f, g, a, b, c; and to one change of temperature, t, with the boundary molecules held fixed. With each of these seven configurations, measure, by three separate measurements conducted according to the method described in § 10, the three components of the sum of the electric moments of the molecules in unit volume.

14. Repeat the same 21 measurements with part 2 of the surface of each molecule zinc, and all the rest copper: next with part 3 zinc, and all the rest copper: and so on. Thus we have 21^2 distinct measurements, each giving independently one of the 21^2 multipliers $[x, e, 1]$, $[x, e, 2]$, &c., which appear in the following 21 equations:—

$$[x, e, 1] v_1 + [x, e, 2] v_2 + \ldots [x, e, 21] v_{21} = (x, e),$$
$$[x, f, 1] v_1 + [x, f, 2] v_2 + \ldots [x, f, 21] v_{21} = (x, f),$$
$$[x, g, 1] v_1 + [x, g, 2] v_2 + \ldots [x, g, 21] v_{21} = (x, g),$$
$$[x, a, 1] v_1 + [x, a, 2] v_2 + \ldots [\dot{x}, a, 21] v_{21} = (x, a),$$
$$[x, b, 1] v_1 + [x, b, 2] v_2 + \ldots [x, b, 21] v_{21} = (x, b),$$
$$[x, c, 1] v_1 + [x, c, 2] v_2 + \ldots [x, c, 21] v_{21} = (x, c),$$
$$[x, t, 1] v_1 + [x, t, 2] v_2 + \ldots [x, t, 21] v_{21} = (x, t),$$

&c., &c., &c., with y and z instead of x.

In these equations $v_1, v_2, \ldots v_{21}$ denote the volta-electric differences from copper which must be given to part 1, part 2, ... part 21 of the surface of the molecule in order that the 21 piezo-electric and pyro-electric coefficients may have their given values

$(x, e), (x, f), \ldots (z, t)$; the meaning of these coefficients being explained by the following three equations:—

$$X = (x, e)\, e + (x, f)\, f + (x, g)\, g + (x, a)\, a + (x, b)\, b + (x, c)\, c + (x, t)\, t,$$
$$Y = (y, e)\, e + \ldots\ldots\ldots\ldots\ldots\ldots\ldots\ldots\ldots\ldots\ldots\ldots\ldots\ldots + (y, t)\, t,$$
$$Z = (z, e)\, e + \ldots\ldots\ldots\ldots\ldots\ldots\ldots\ldots\ldots\ldots\ldots\ldots\ldots\ldots + (z, t)\, t,$$

where X, Y, Z denote the components of the electric moment per unit of volume, produced in the crystal by geometrical change, and change of temperature (e, f, g, a, b, c, t). Thus, the volta-electric difference of zinc from copper being taken as unity, the 21 volta-differences from copper, of parts 1 to 21 of the surface of each molecule, are determined by 21 linear equations.

15.   Thus we have, in idea, constructed a model for the piezo-electric and pyro-electric quality of a crystal in which each one of the 21 piezo-electric and pyro-electric coefficients has an arbitrarily given value.

173. On Homogeneous Division of Space.

[From *Roy. Soc. Proc.* Vol. LV. 1894, pp. 1—16 [Jan. 18, 1894]; *Nature*, Vol. XLIX. March 8, 1894, pp. 445—448, March 15, 1894, pp. 469—471.]

1. The homogeneous division of any volume of space means the dividing of it into equal and similar parts, or cells, as I shall call them, all sameways oriented. If we take any point in the interior of one cell or on its boundary, and corresponding points of all the other cells, these points form a homogeneous assemblage of single points, according to Bravais' admirable and important definition *. The general problem of the homogeneous partition of space may be stated thus :—Given a homogeneous assemblage of single points, it is required to find every possible form of cell enclosing each of them subject to the condition that it is of the same shape and sameways oriented for all. An interesting application of this problem is to find for a crystal (that is to say, a homogeneous assemblage of groups of chemical atoms) a homogeneous arrangement of partitional interfaces such that each cell contains all the atoms of one molecule. Unless we knew the exact geometrical configuration of the constituent parts of the group of atoms in the crystal, or crystalline molecule as we shall call it, we could not describe the partitional interfaces between one molecule and its neighbour.

Knowing as we do know for many crystals the exact geometrical character of the Bravais assemblage of corresponding points of its molecules, we could not be sure that any solution of the partitional problem we might choose to take would give a cell containing only the constituent parts of one molecule. For instance, in the case of quartz, of which the crystalline molecule is probably $3(SiO_2)$, a form of cell chosen at random might be

* *Journal de l'École Polytechnique*, tome 19, cahier 33, pp. 1—128 (Paris, 1850), quoted and used in my *Mathematical and Physical Papers*, Vol. III. Art. 97, p. 400.

such that it would enclose the silicon of one molecule with only some part of the oxygen belonging to it, and some of the oxygen belonging to a neighbouring molecule, leaving out some of its own oxygen, which would be enclosed in the cell of either that neighbour or of another neighbour or other neighbours.

2.   This will be better understood if we consider another illustration—a homogeneous assemblage of equal and similar trees planted close together in any regular geometrical order on a plane field either inclined or horizontal, so close together that roots of different trees interpenetrate in the ground, and branches and leaves in the air.   To be perfectly homogeneous, every root, every twig, and every leaf of any one tree must have equal and similar counterparts in every other tree.   So far everything is natural, except, of course, the absolute homogeneousness that our problem assumes; but now, to make a homogeneous assemblage of molecules in space, we must suppose plane above plane each homogeneously planted with trees at equal successive intervals of height.   The interval between two planes may be so large as to allow a clear space above the highest plane of leaves of one plantation and below the lowest plane of the ends of roots in the plantation above.   We shall not, however, limit ourselves to this case, and we shall suppose generally that leaves of one plantation intermingle with roots of the plantation above, always, however, subject to the condition of perfect homogeneousness.   Here, then, we have a truly wonderful problem of geometry—to enclose ideally each tree within a closed surface containing every twig, leaf, and rootlet belonging to it, and nothing belonging to any other tree, and to shape this surface so that it will coincide all round with portions of similar surfaces around neighbouring trees. Wonderful as it is, this is a perfectly easy problem if the trees are given, and if they fulfil the condition of being perfectly homogeneous.

In fact we may begin with the actual bounding surface of leaves, bark, and roots of each tree.   Wherever there is a contact, whether with leaves, bark, or roots of neighbouring trees, the areas of contact form part of the required cell-surface.   To complete the cell-surface we have only to swell out* from the untouched portions of surface of each tree homogeneously until the swelling

---

* Compare *Mathematical and Physical Papers*, Vol. III. Art. 97, § 5.

portions of surface meet in the interstitial air spaces (for simplicity we are supposing the earth removed, and roots, as well as leaves and twigs, to be perfectly rigid). The wonderful cell-surface which we thus find is essentially a case of the tetrakaidecahedronal cell, which I shall now describe for any possible homogeneous assemblage of points or molecules.

3. We shall find that the form of cell essentially consists of fourteen walls, plane or not plane, generally not plane, of which eight are hexagonal and six quadrilateral; and with thirty-six edges, generally curves, of meeting between the walls; and twenty-four corners where three walls meet. A cell answering this description must of course be called a tetrakaidecahedron, unless we prefer to call it a fourteen-walled cell. Each wall is an interface between one cell and one of fourteen neighbours. Each of the thirty-six edges is a line common to three neighbours. Each of the twenty-four corners is a point common to four neighbours. The old-known parallelepipedal partitioning is merely a very special case in which there are four neighbours along every edge, and eight neighbours having a point in common at every corner. We shall see how to pass (§ 4) continuously from or to this singular case, to or from a tetrakaidecahedron differing infinitesimally from it; and, still continuously, to or from any or every possible tetrakaidecahedronal partitioning.

4. To change from a parallelepipedal to a tetrakaidecahedronal cell, for one and the same homogeneous distribution of points, proceed thus:—Choose any one of the four body-diagonals of a parallelepiped and divide the parallelepiped into six tetrahedrons by three planes each through this diagonal, and one of the three pairs of parallel edges which intersect it in its two ends. Give now any purely translational motion to each of these six tetrahedrons. We have now the 4×6 corners of these tetrahedrons at twenty-four distinct points. These are the corners of a tetrakaidecahedron, such as that described generally in § 3. The two sets of six corners, which before the movement coincided in the two ends of the chosen diagonal, are now the corners of one pair of the hexagonal faces of the tetrakaidecahedron. When we look at the other twelve corners we see them as corners of other six hexagons, and of six parallelograms, grouped together as described in § 15 below. The movements of the six tetrahedrons may be

such that the groups of six corners and of four corners are in fourteen planes as we shall see in § 14; but, if they are made at random, none of the groups will be in a single plane. The fourteen faces, plane or not plane, of the tetrakaidecahedron are obtained by drawing arbitrarily any set of surfaces to constitute four of the hexagons and three of the quadrilaterals, with arbitrary curves for the edges between hexagon and hexagon and between hexagons and quadrilaterals, and then by drawing parallel equal and similar counterparts to these surfaces in the remaining four hexagonal and three quadrilateral spaces in the manner more particularly explained in § 8 below. It is clear, or at all events I shall endeavour to make it clear by fuller explanations and illustrations below, that the figure thus constituted fulfils our definition (§ 1) of the most general form of cell fitted to the particular homogeneous assemblage of points corresponding to the parallelepiped with which we have commenced. This will be more easily understood in general, if we first consider the particular case of *parallelepipedal* partitioning, and of the deviations which, without altering its corners, we may arbitrarily make from a plane-faced parallelepiped, or which we may be compelled by the particular figure of the molecule to make.

5. Consider, for example, one of the trees of § 2, or if you please a solid of less complex shape, which for brevity we shall call S, being one of a homogeneous assemblage. Let P [fig. 7 of § 9] be a point in unoccupied space (air, we shall call it for brevity), which, for simplicity we may suppose to be somewhere in the immediate neighbourhood of S, although it might really be anywhere far off among distant solids of the assemblage. Let PA PB, PC be lines parallel to any three Bravais rows not in one plane, and let A, B, C be the nearest points corresponding to P in these lines. Complete a parallelepiped on the lines PA, PB, PC, and let QD, QE, QF be the edges parallel to them through the opposite corner Q. Because of the homogeneousness of the assemblage, and because A, B, C, D, E, F, Q are points corresponding to P, which is in air, each of those seven points is also in air. Draw any line through air from P to A and draw the lines of corresponding points from B to F, D to Q, and C to E. Do the same relatively to PB, AF, EQ, CD; and again the same relatively to PC, AE, FQ, BD. These twelve lines are all in air, and they are the edges of our curved-faced parallelepiped. To

describe its faces take points infinitely near to one another along the line PC (straight or curved as may be): and take the corresponding points in BD. Join these pairs of corresponding points by lines in air infinitely near to one another in succession. These lines give us the face $PBDC$. Corresponding points in AE, FQ, and corresponding lines between them give us the parallel face $AFQE$. Similarly we find the other two pairs of the parallel faces of the parallelepiped. If the solids touch one another anywhere, either at points or throughout finite areas, we are to reckon the interface between them as air in respect to our present rules.

6.　We have thus found the most general possible parallelepipedal partitioning for any given homogeneous assemblage of solids. Precisely similar rules give the corresponding result for *any possible partitioning* if we first choose the twenty-four corners of the tetrakaidecahedron by finding six tetrahedrons and giving them arbitrary translatory motions according to the rule of § 4. To make this clear it is only now necessary to remark that the four corners of each tetrahedron are essentially corresponding points, and that if one of them is in air all of them are in air, whatever translatory motion we give to the tetrahedron.

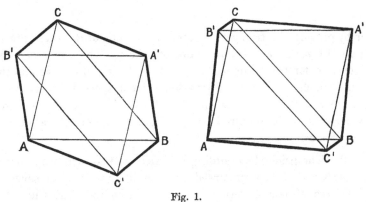

Fig. 1.

7.　The transition from the parallelepiped to the tetrakaidecahedron described in § 4 will be now readily understood, if we pause to consider the vastly simpler two-dimensional case of transition from a parallelogram to a hexagon. This is illustrated in figs. 1 and 2; with heavy lines in each case for the sides of the hexagon, and light lines for the six of its diagonals which are sides of constructional triangles. The four diagrams show different relative positions in one plane of two equal homochirally similar triangles ABC, $A'B'C'$; oppositely oriented (that is to say, with corresponding lines AB, $A'B'$ parallel but in inverted directions). The hexagon $AC'BA'CB'$, obtained by joining A with B' and C', B with C' and A', and C with A' and B', is clearly in each case a proper cell-figure for dividing plane space homogeneously according to the Bravais distribution of points defined by either triangle,

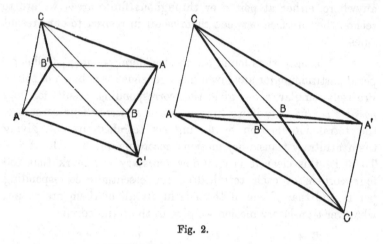

Fig. 2.

or by putting the triangles together in any one of the three proper ways to make a parallelogram of them. The corresponding operation for three-dimensional space is described in § 4: and the proof which is obvious in two-dimensional space is clearly valid for space of three dimensions, and therefore the many words which would be required to give it formal demonstration are superfluous.

8. The principle according to which we take arbitrary curved surfaces with arbitrary curved edges of intersection, for seven of the faces of our partitional tetrakaidecahedron, and the other seven correspondingly parallel to them, is illustrated in figs. 3, 4, 5, and 6, where the corresponding thing is done for a partitional hexagon suited to the homogeneous division of a plane. In these diagrams the hexagon is for simplicity taken equilateral and equi-angular. In drawing fig. 3, three pieces of paper were cut, to the shapes kl, mn, uv. The piece kl was first placed in the position shown relatively to AC', and a portion of the area of one cell to

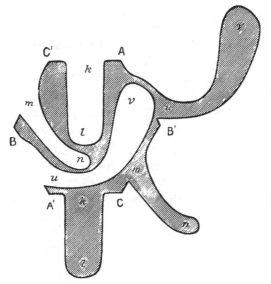

Fig. 3.

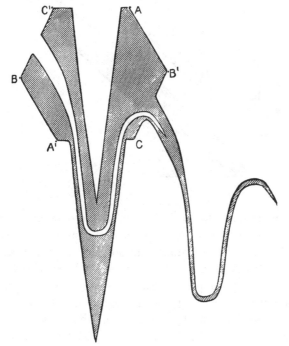

Fig. 4.

be given to a neighbour across the frontier $C'A$ on one side was marked off. It was then placed in the position shown relatively to $A'C$ and the equivalent portion to be taken from a neighbour on the other side was marked. Corresponding give-and-take delimitations were marked on the frontiers $C'B$ and $B'C$, according to the form mn; and on the frontiers BA', AB', according to the form uv. Fig. 4 was drawn on the same plan but with one pair of frontiers left as straight lines, and the two other pairs drawn by aid of two paper templets. It would be easy, but not worth the trouble, to cut out a large number of pieces of brass of the shapes shown in these diagrams and to show them fitted together like the pieces of a dissected map. Figs. 5 and 6 are drawn on the

Fig. 5.

same principle; fig. 6 showing, on a reduced scale, the result of putting pieces together precisely equal and similar to that shown in fig. 5. In these diagrams, unlike the cases represented in figs. 3 and 4, the primitive hexagon is, as shown clearly in fig. 5, divided into isolated parts. But if we are dealing with homogeneous division of solid space, the separating channels shown in fig. 5 might be sections, by the plane of the drawing, of perforations through the matter of one cell produced by the penetration of matter, rootlets for example, from neighbouring cells.

9.   Corresponding to the three ways by which two triangles can be put together to make a parallelogram, there are seven, and only seven, ways in which the six tetrahedrons of § 4 can be put together to make a parallelepiped, in positions parallel to those which they had in the original parallelepiped.   To see this, remark first that among the thirty-six edges of the six tetrahedrons seven

Fig. 6.

different lengths are found which are respectively equal to the three lengths of edges (three quartets of equal parallels); the three lengths of face-diagonals having ends in P or Q (three pairs of equal parallels); and the length of the chosen body-diagonal PQ.   (Any one of these seven is, of course, determinable from the other six if given.)

In the diagram, fig. 7, full lines show the edges of the primitive parallelepiped, and dotted lines show the body-diagonal PQ and two pairs of the face-diagonals, the other pair of face-diagonals (PF, QC), not being marked on the diagram to avoid confusion. Thus, the diagram shows, in the parallelograms $QDPA$ and $QEPB$, two of the three cutting planes by which it is divided into six tetrahedrons, and it so shows also two of the six tetrahedrons, $QPDB$ and $QPEA$. The lengths QP, QD, QE, QF are found in the edges of every one of the six tetrahedrons, the two other edges of each being of two of the three lengths QA, QB, QC. The six tetrahedrons may be taken in order of three pairs having edges of lengths respectively equal to QB and QC, QC and QA, QA and QB. It is the third of these pairs that is shown in fig. 7.

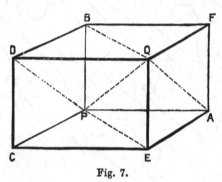

Fig. 7.

Remark now that the sum of the six angles of the six tetrahedrons at the edge equal to any one of the lengths QP, QD, QE, QF is four right angles. Remark also that the sum of the four angles at the edge of length QA in the two pairs of tetrahedrons in which the length QA is found is four right angles, and the same with reference to QB and QC. Remark lastly that the two tetrahedrons of each pair are equal and dichirally* similar, or enantiomorphs as such figures have been called by German writers.

10. Now, suppose any one pair of the tetrahedrons to be taken away from their positions in the primitive parallelepiped, and, by purely translational motion, to be brought into position with their edges of length QD coincident, and the same to be

* A pair of gloves are dichirally similar, or enantiomorphs. Equal and similar right-handed gloves are chirally similar.

done for each of the other two pairs. The sum of the six angles at the coincident edges being two right angles, the plane faces at the common edge will fit together, and the condition of parallelism in the motion of each pair fixes the order in which the three pairs come together in the new position, and shows us that in this position the three pairs form a parallelepiped essentially different from the primitive parallelepiped, provided that, for simplicity in our present considerations, we suppose each tetrahedron to be wholly scalene, that is to say, the seven lengths found amongst the edges to be all unequal. Next shift the tetrahedrons to bring the edges QE into coincidence, and next again to bring the edges QF into coincidence. Thus, including the primitive parallelepiped, we can make four different parallelepipeds in each of which six of the tetrahedrons have a common edge.

11. Now take the two pairs of tetrahedrons having edges of length equal to QA, and put them together with these edges coincident. Thus we have a scalene octahedron. The remaining pair of tetrahedrons placed on a pair of its parallel faces complete a parallelepiped. Similarly two other parallelepipeds may be made by putting together the pairs that have edges of lengths equal to QB and QC respectively with those edges coincident, and finishing in each case with the remaining pair of tetrahedrons. The three parallelepipeds thus found are essentially different from one another, and from the four of § 10; and thus we have the seven parallelepipeds fulfilling the statement of § 9. Each of the seven parallelepipeds corresponds to one and the same homogeneous distribution of points.

12. Going back to § 4, we see that, by the rule there given, we find four different ways of passing to the tetrakaidecahedron from any one chosen parallelepiped of a homogeneous assemblage. The four different cellular systems thus found involve four different sets of seven pairs of neighbours for each point. In each of these there are four pairs of neighbours in rows parallel to the three quartets of edges of the parallelepiped and to the chosen body-diagonal; and the other three pairs of neighbours are in three rows parallel to the face-diagonals which meet in the chosen body-diagonal. The second (§ 11) of the two modes of putting together tetrahedrons to form a parallelepiped which we have been considering suggests a second mode of dividing our primitive

parallelepiped, in which we should first truncate two opposite corners and then divide the octahedron which is left, by two planes through one or other of its three diagonals. The six tetra- hedrons obtained by any one of the twelve ways of effecting this second mode of division give, by their twenty-four corners, the twenty-four corners of a space-filling tetrakaidecahedronal cell, by which our fundamental problem is solved. But every solution thus obtainable is clearly obtainable by the simpler rule of § 4, commencing with some one of the infinite number of primitive parallelepipeds which we may take as representative of any homogeneous distribution of points.

13. The communication is illustrated by a model showing the six tetrahedrons derived by the rule § 4 from a symmetrical kind of primitive parallelepiped, being a rhombohedron of which the axial-diagonal is equal in length to each of the edges. The homogeneous distribution of points corresponding to this form of parallelepiped is the well-known one in which every point is surrounded by eight others at the corners of a cube of which it is the centre; or, if we like to look at it so, two simple cubical distributions of single points, each point of one distribution being at the centre of a cube of points of the other. [To understand the tactics of the single homogeneous assemblage constituted by these two cubic assemblages, let P be a point of one of the cubic assemblages, and Q any one of its eight nearest neighbours of the other assemblage. Q is at the centre of a cube of which P is at one corner. Let PD, PE, PF be three conterminous edges of this cube so that D, E, F are points of the first assemblage nearest to P. Again Q is a corner of a cube of which P is the centre; and if QA, QB, QC are three conterminous edges of this cube, A, B, C are points of the second assemblage nearest to Q. The rhombohedron of which PQ is body-diagonal and PA, PB, PC the edges conterminous in P, and QD, QE, QF the edges conter- minous in Q, is our present rhombohedron. The diagram of § 9 (fig. 7), imagined to be altered to proper proportions for the present case, may be looked to for illustration. Its three face- diagonals through P, being PD, PE, PF, are perpendicular to one another. So also are QA, QB, QC, its three face-diagonals through Q. The body-diagonal of the cube PQ, being half the body- diagonal of the cube whose edges are PD, PE, PF, is equal to $PD \times \frac{1}{2}\sqrt{3}$: and PA, PB, PC are also each of them equal to this,

because A, B, C are centres of other equal cubes, having P for a common corner.—January 30.]

14. The tetrahedrons used in the model are those into which the parallelepiped is cut by three planes through the axial diagonal, which in this case cut one another at angles of 60°. We wish to be able to shift the tetrahedrons into positions corresponding to those of the triangles in fig. 1, which we could not do if they were cut out of the solid. I, therefore, make a mere skeleton of each tetrahedron, consisting of a piece of wire bent at two points, one-third of its length from its ends, at angles of $70\frac{1}{2}°$, being $\sin^{-1}\frac{1}{3}\sqrt{8}$, in planes inclined at 60° to one another. The six skeletons thus made are equal and similar, three homochirals and the other three also homochirals, their enantiomorphs. In their places in the primitive parallelepiped they have their middle lines coincident in its axial diagonal PQ, and their other 6×2 arms coincident in three pairs in its six edges through P and Q. Looking at fig. 7 we see, for example, three of the edges CP, PQ, QE, of one of the tetrahedrons thus constituted; and DQ, QP, PB, three edges of its enantiomorph. In the model they are put together with their middle lines at equal distances around the axial diagonal and their arms symmetrically arranged round it. Wherever two lines cross they are tied, not very tightly, together by thin cord many times round, and thus we can slip them along so as to bring the six middle lines either very close together, nearly as they would be in the primitive parallelepiped, or farther and farther out from one another so as to give, by the four corners of the tetrahedrons, the twenty-four corners of all possible configurations of the plane-faced space-filling tetrakaidecahedron.

15. The six skeletons being symmetrically arranged around an axial line we see that each arm is cut by lines of other skeletons in three points. For an important configuration, let the skeletons be separated out from the axial line just so far that each arm is divided into four equal parts, by those three intersectional points. The tetrakaidecahedron of which the twenty-four corners are the corners of the tetrahedrons thus placed may conveniently be called the orthic tetrakaidecahedron. It has six equal square faces and eight equal equiangular and equilateral hexagonal faces. It was described in § 12 of my paper on "The Division of Space with

Minimum Partitional Area*," under the name of "plane-faced isotropic tetrakaidecahedron"; but I now prefer to call it orthic, because, for each of its seven pairs of parallel faces, lines joining corresponding points in the two faces are perpendicular to the faces, and the planes of its three pairs of square faces are perpendicular to one another. Fig. 8 represents an orthogonal projection

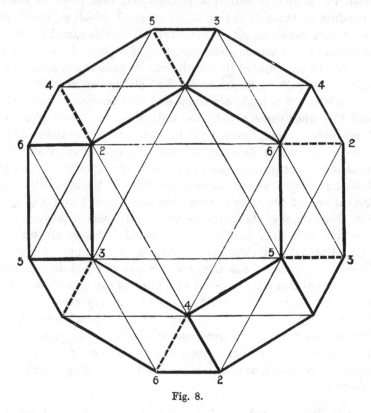

Fig. 8.

on a plane parallel to one of the four pairs of hexagonal faces. The heavy lines are edges of the tetrakaidecahedron. The light lines are edges of the tetrahedrons of § 13, or parts of those edges not coincident in projection with the edges of the tetrakaidecahedron. The figures 1, 1, 1; 2, 2, 2;...; 6, 6, 6 show corners belonging respectively to the six tetrahedrons, two of the four corners of each being projected on one point in the diagram.

---

* *Phil. Mag.* 1887, 2nd half-year, and *Acta Mathematica*, Vol. XI. pp. 121—134 [*supra*, p. 297].

Fig. 9 shows, on the same scale of magnitude with corresponding distinction between heavy and light lines, the orthogonal projection on a plane parallel to a pair of square faces.

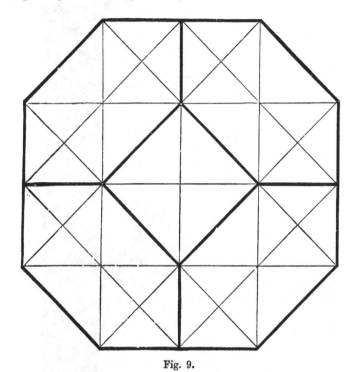

Fig. 9.

16. If the rule of § 15 with reference to the division of each arm of a skeleton tetrahedron into four equal parts by points in which it is cut by other lines of skeletons is fulfilled with all details of §§ 14 and 15 applied to any oblique parallelepiped, we find a tetrakaidecahedron which we may call orthoid, because it is an orthic tetrakaidecahedron, altered by homogeneous strain. Professor Crum Brown has kindly made for me the beautiful model of an orthoidal tetrakaidecahedron thus defined which is placed before the Royal Society as an illustration of the present communication.

Fig. 10 is a stereoscopic picture of an orthic tetrakaidecahedron, made by soldering together thirty-six pieces of wire, each 4 in. long, with three ends of wire at each of twenty-four corners.

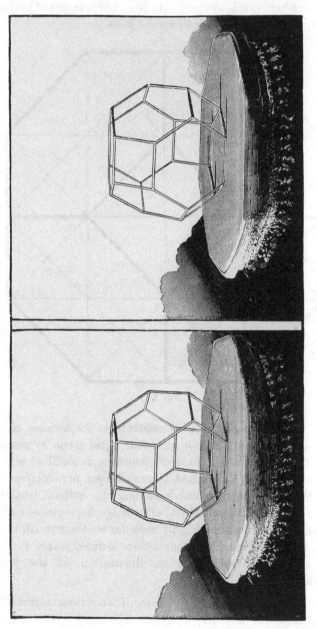

Fig. 10.

17. I cannot in the present communication enter upon the most general possible plane-faced partitional tetrakaidecahedron or show its relation to orthic and orthoidal tetrakaidecahedrons. I may merely say that the analogy in the homogeneous division of a plane is this:—an equilateral and equiangular hexagon (orthic); any other hexagon of three pairs of equal and parallel sides whose paracentric diagonals trisect one another (orthoidal). The angles of an orthoidal hexagon, other than equilateral, are not 120°. The angles of the left-hand hexagon fig. 1 (§ 7) are 120°, and its paracentric diagonals do not trisect one another, as the diagram clearly shows.

174. On the Molecular Dynamics of Hydrogen Gas, Oxygen Gas, Ozone, Peroxide of Hydrogen, Vapour of Water, Liquid Water, Ice, and Quartz Crystal.

[From *Brit. Assoc. Report*, 1896, pp. 721—724.]

In a communication " On the Different Crystalline Configurations possible with the same Law of Force according to Boscovich," to the last meeting (July 20) of the Royal Society of Edinburgh, a purely mathematical problem of fundamental importance for the physical theory of crystals—the equilibrium of any number of points acting on one another with forces in the lines joining them —was considered in the simplest case of Boscovichian statics: that in which the mutual force between every pair of atoms is the same for the same distance between any two atoms of the whole assemblage. The next simplest case is that in which there are two kinds of atom, h, o, with the distinction that the force between two h's and the force between two o's and the force between an h and an o are generally different at the same distance. The mutual force between two h's is, of course, always the same at the same distance. So also is the mutual force between two o's and between an h and an o.

The object of the present communication is to find how much of the known properties of the substances named in the title can be explained with no further assumption except the conferring of inertia upon a Boscovich atom.

The known chemical and physical properties to be provided for are :

1. That in each of the gases named, the molecule is divisible into two; which is the meaning of the symbols H_2, O_2, used to denote them in chemistry.

2. That Ozone (O_3) is a possible, though not a very stable, gaseous molecule, consisting of a group of Oxygen atoms of which

the constituents readily pass into the configuration (O_2) of Oxygen gas.

3. That Peroxide of Hydrogen (H_2O_2, or perhaps HO) is a possible, but not a very stable, combination, which, for all we know, may exist as a dry gas, but which is only generally known as a solution in water (of density 1·45 in the highest concentration hitherto reached), readily absorbing Hydrogen or parting with Oxygen so as to form H_2O.

4. That water (H_2O) is an exceedingly stable compound in the gaseous, in the liquid or in the crystalline form according to circumstances of temperature and pressure.

5. That dry mixtures of Hydrogen and Oxygen gases, and also mixtures of these gases with water in the same enclosure, have been kept by many experimenters for weeks or months, and perhaps for years, enclosed in glass vessels, without any combination of the two gases having been detected.

6. That Ice contracts by about 8 per cent. in melting, and that ice-cold water, when warmed, contracts till it reaches a maximum density at about 4° C., and expands on further elevation of temperature.

7. For Quartz crystal—

(a) The difference between neighbouring corners of the hexagonal prism;

(b) The similarity between each face and its neighbour on either side turned upside down (the axis of the prism supposed vertical).

(c) The right-handed and left-handed chiralities of different crystals in nature with, so far as known, an equal chance of one chirality or the other in any crystal that may be found.

In the present communication it is shown that all the properties stated in this schedule can be conceivably explained by making H consist of two Boscovich atoms (h, h), and O of two others (o, o). This essentially makes H_2 consist of four h's at the corners of an equilateral tetrahedron, and O_2 a similar configuration of four o's. It naturally shows Ozone as six o's at the corners of a regular octahedron: and peroxide of hydrogen as a tetrahedron of h's placed symmetrically within a tetrahedron of o's. It makes H_2O (the gaseous molecule of water) consist of two o's with two h's attached to one of them and two other h's attached to the other; the h's of each o getting as near to the other o as the mutual repulsion of the h's allows. This configuration and the modification it experiences in the formation of crystals of ice are illustrated by models which accompany the communication.

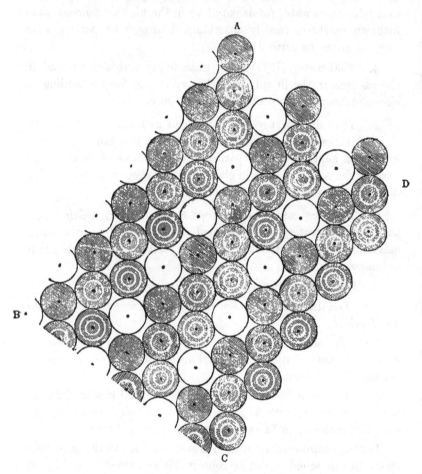

To understand what is probably the true configuration of ice-crystal, we are helped by first considering a double cubic assemblage of point-atoms, such that each point-atom is in the centre of a cube having eight point-atoms for its corners. This double cubic assemblage may be imagined as consisting of two simple cubic assemblages, so placed that one atom of each assemblage is in the centre of a cube of atoms of the other. The annexed diagram shows, in the centres of the circles which it contains, atoms of a double cubic assemblage, which lie in the plane of a pair of remote parallel edges, *AD*, *BC*, of one set of constituent cubes. It shows *all* the atoms in the lines of this plane which it contains except certain omissions in the lines *AD*, *DC*, made specially on account of the present application of the diagram. The circles of simple shading and of shading interrupted by two small concentric circles constitute one of the simple cubic assemblages; the unshaded and the circles with shading interrupted by one concentric circle constitute the other cubic assemblage. *AC*, *BD* are parallel to body diagonals, *AB*, *DC* are parallel to face diagonals, of the cubes. Annul now all the atoms at the centres of the blank circles*. Lastly, stretch the diagram perpendicularly to *AC* in some definite ratio of perhaps about 3 to 1. It then represents what we may believe to be probably the true molecular structure of ice-crystal: the circles with simple shading and with shading interrupted by two concentric circles denoting hydrogen atoms, and the circles with shading interrupted by single concentric circles the oxygen atoms.

The named properties of Quartz are explained by supposing the crystalline molecule to consist of three of the chemical molecules (OSiO) placed together in a manner readily imagined according to a suggestion which I communicated to the British Association at its Southport meeting in 1883. Models showing right-handed and left-handed specimens of these crystalline molecules and the configuration in which they must be placed to form a rock crystal ending in its well-known six-sided pyramid are shown to illustrate the present communication.

In a communication which I hope to make to the Royal Society of Edinburgh at an early meeting essential details of the configurations now suggested, and of the mutual forces between the atoms required by the conditions to be fulfilled, will be considered.

---

* The assemblage thus constituted is precisely that described in Section 24, and in footnote on Section 69 of "Molecular Constitution of Matter," *Proc. R. S. E.* July 1889, reprinted as Art. xcvii. of Vol. III. of *Mathematical and Physical Papers.* I was led to it in the course of my investigation of a Boscovichian elastic solid, having two independent moduluses of resistance to compression and of rigidity.— ("Elasticity of a Crystal according to Boscovich," *Proc. R. S.* June 1893 [*Baltimore Lectures*, Appendix I.].)

175. MAGNETISM AND MOLECULAR ROTATION *.

[From *Edinb. Roy. Soc. Proc.* Vol. XXII. [read July 17, 1899], pp. 631—635 ;
Phil. Mag. Vol. XLVIII. August, 1899, pp. 236—239.]

1.   CONSIDER the induction of an electric current in an endless wire when a magnetic field is generated around it.   For
simplicity, let the wire be circular and the diameter of its section
very small in comparison with that of the ring.   The time-
integral of the electromotive force in the circuit is $2AM$, if A
denote the area of the ring and M the component perpendicular
to its plane, of the magnetic force coming into existence.   This
is true whatever be the shape of the ring, provided it is all in
one plane.   Now, adopting the idea of two electricities, vitreous
and resinous, we must imagine an electric current of strength C
to consist of currents of vitreous and resinous electricities in
opposite directions, each of strength $\frac{1}{2}C$.   Hence the time-
integrals of the opposite electromotive forces on units of the
equal vitreous and resinous electricities are each equal to AM.

2.   Substitute now for our metal wire an endless tube of
non-conducting matter, vitreously electrified, and filled with an
incompressible non-conducting fluid, electrified with an equal
quantity, e, of resinous electricity.   The fluid and the containing
tube will experience equal and opposite tangential forces, of each
of which the time-integral of the line-integral round the whole
circumference is eAM, if the ring be a circle of radius r; and
the effect of the generation of the magnetic field will be to
cause the fluid and the ring to rotate in opposite directions with
moments of momentum each equal to $eAMr$, if neither fluid nor

* [Various corrections are needed in this tentative paper, marked by the author
as to be omitted from *Baltimore Lectures*.   The 2 should be omitted in line 5.   The
current in metals is known now to be carried mainly by negative electrions.   The
moment of momentum in § 2 and § 7 should be $eAM/2\pi$, and the following expressions changed accordingly.   The effects are of course excessively minute.]

ring is acted on by any other force than that of the electro-
magnetic induction. Their angular velocities are therefore
eAM/rw, eAM/rw', and their kinetic energies are $\frac{1}{2}e^2A^2M^2/w$,
$\frac{1}{2}e^2A^2M^2/w'$, where w, w' denote the masses of fluid and ring
respectively.

3. Suppose now for simplicity in the first place, the ring
to be embedded in ether, viewed as an incompressible solid, and
attached to the ether in contact with it firmly enough to prevent
slipping. The circuital impulse on the ring by the generation
of the magnetic field will give rise to a rapidly subsiding train
of waves of transverse vibration, of the kind which, in communications to Section A of the British Association* at its meeting
in Bristol last September, I described as a solitary wave of the
simplest possible kind in an elastic solid, and again, for periodic
motion, as a very simple and symmetrical case of a train of
periodic waves of transverse vibration. The work done by the
circuital force on the ring is spent on waves of this class travelling
outwards through ether, and in a very short time the ring comes
practically to rest. It does not come to perfect rest suddenly by
the departure from it of waves carrying away all its energy; it
subsides to absolute rest in an infinite time according to the law
$\epsilon^{-pt}\sin qt$. The resinously electrified fluid within the ring continues
revolving with unaltered energy as long as the force of the magnetic
field is maintained constant.

4. The simple molecular arrangement thus imagined supplies
the rotatory or revolutional motion, and the "moment of momentum," which, forty-three years ago, I pointed out† as wanted,
to explain "simply by inertia and pressure," the rotation of the
plane of polarization, then recently discovered by Faraday, for
light transmitted through heavy glass in a powerful magnetic
field along the lines of force. In my Baltimore Lectures I showed
that embedded gyrostats would in fact produce exactly the
rotation of the plane of polarization in a magnetic field discovered by Faraday. The idea which forms the subject of the
present communication shows how the fly-wheels of the gyro-

---

* "On the Simplest Possible ...;" "On Continuity in Undulatory Theory,...,"
B. A. Report, 1898.

† "Dynamical Illustrations of the Magnetic and the Helicoidal Rotatory Effects
of Transparent Bodies on Polarized Light," *Proc. R. S. L.* Vol. VIII. June 1856;
Phil. Mag. March 1857. [*Baltimore Lectures*, Appendix F.]

stats may be started into rotation in virtue of the generation of the magnetic field and stopped when the magnetic field is annulled.

5. The simply embedded gyrostat has not, however, the vibrational quality which is the essential of the Stokes-Maxwell*-Sellmeier vibratory molecule. For this a gyrostatic vibrator, capable of originating from a single blow on itself a subsidential train of at least 200,000 waves of light, must be connected with the surrounding ether by springs, having sufficient resilience to store up in themselves the total energy thus radiated out. Taking now as gyrostat our electric doublet of vitreously electrified rigid hollow ring filled with fluid resinously electrified, consider what must be the nature of the elastic communication between it and a rigid lining of a spherical hollow in ether around it, to fulfil some of the known conditions of radiant molecules.

6. (a) Let the spring connexion be equivalent to a simple force between I, the centre of inertia of ring and fluid, and O, the centre of the spherical sheath, varying directly as the distance between those points. The gyrostatic influence will be inoperative, and the result will be precisely the same as if we had a single Maxwell-Sellmeier material point at I, of mass equal to that of ring and fluid together.

(b) Let points on the ring be connected by springs with points on the sheath. Supposing now the sheath to be held fixed, the stiffnesses and the tensions of these springs may be adjusted to give 21 arbitrary values for the coefficients in the quadratic for the potential energy of any infinitesimal displacement, specified by three components of linear displacement of I, and three components of rotational displacement round axes through I. The well-known solution of the problem of infinitesimal vibrations about a position of equilibrium of a rigid body, modified in respect to moments of inertia to take into account the fluidity of the incompressible fluid in the ring, gives us immediately the periods and geometrical specifications of six fundamental modes of simple harmonic vibration. Hence our

---

* See Rayleigh, *Phil. Mag.* July 1899, quoting from *Camb. Univ. Calendar*, 1869.

combination, serving as a radiant molecule, without magnetic force, would give six bright lines (understood of course that each of the six periods is within the range of light-periods). Suppose now a vast number of such molecules, all equal and similar in every respect, but with different orientations, to be scattered through a flame. Each molecule, whatever its orientation, will give six lines of the same periods, though of different intensities when seen in any particular direction, according to the chances of orientation and of impulses. Hence each of the six bright lines will be perfectly sharp.

7. Now suppose a magnetic field to be suddenly instituted. The moment of momentum generated in any one of the molecules is $er AM \cos \theta$, where θ denotes the inclination of the axis of its ring to the lines of force. The gyrostatic influence will split each of our six fundamental modes of vibration into two, greater than it and less than it by equal very small differences. These differences will be different for different molecules, because of the different values of θ for their different orientations. Hence each bright line is not split into two sharp lines, but is broadened to an extreme breadth corresponding to the value $\theta = 0$. No simplifying suppositions as to the character of the molecule, such as symmetry of forces and moments of inertia round the axis of the ring, can possibly give Zeeman's normal results of the splitting of a bright line into two sharp lines circularly polarized in opposite directions, when the light is viewed from a direction parallel to the lines of magnetic force; and the dividing of each bright line into three, each plane-polarized, when the light is viewed from a direction perpendicular to the lines of force. Hence, although from 1856 till quite lately I felt satisfied in knowing that it sufficed to explain Faraday's magneto-optic discovery, I now, in the light of Zeeman's recent discovery, discard my old tempting gyrostatic hypothesis for an irrefragable reason, which is virtually the same as that stated by Larmor* in the following words:—"Hence a principal oscillator magnetically tripled must be capable of being excited with reference to any axis in the molecule; otherwise there would be merely hazy broadening or duplication instead of definite triplication."

* *Phil. Mag.* Vol. XLIV. 1897, "On the Theory of the Magnetic Influence on Spectra," p. 507.

8. It now seems to me that the theory of H. A. Lorentz (of Leyden), as expressed by equations (1) in Zeeman's first paper "On the Influence of Magnetism on the Nature of the Light emitted by a Substance*," is essentially true.

9. Though it cannot explain Zeeman's discovery, the molecular rotation caused by the institution of a magnetic field, which is the subject of the present communication, may, however, be considered as interesting not only because the idea of it seems to be new in electromagnetic theory, but also because it may conceivably constitute the explanation of Faraday's diamagnetism. Go back to §§ 2, 3 above, and remark that if a body containing a vast number of the molecules there described is situated between the poles of a steel magnet, the total energy will be greater than if there were nothing but ether between the poles, by a difference equal to the kinetic energy of the motion of the resinously electrified fluid. Hence if a body containing the supposed congregation of molecules is movable, it must be repelled from the place of strong magnetic force between the poles to places of weaker force further from them.

* *Phil. Mag.* Vol. XLIII. 1897, p. 226.

176. MOLECULAR DYNAMICS OF A CRYSTAL.

[From *Edinb. Roy. Soc. Proc.* Vol. XXIV. [read Jan. 20, 1902], pp. 205—224; *Phil. Mag.* Vol. IV. July, 1902, pp. 139—156; *Nature*, Vol. LXV. Feb. 27, 1902, p. 407. Reprinted in *Baltimore Lectures*, Appendix J, 1904, pp. 662—680.]

177. ÆPINUS ATOMIZED.

[*From the Jubilee Volume presented to Prof. Bosscha in Nov.* 1901.]

[*Phil. Mag.* Vol. III. March, 1902, pp. 257—283. Reprinted in *Baltimore Lectures*, Appendix E, 1904, pp. 541—568.]

ELECTRODYNAMICS.

178. ACCOUNT OF EXPERIMENTAL INVESTIGATIONS TO ANSWER QUESTIONS ORIGINATING IN THE MECHANICAL THEORY OF THERMO-ELECTRIC CURRENTS.

[From *Edinb. Roy. Soc. Proc.* Vol. III. [read May 1, 1854], p. 255. Abstract.]

IN this communication the mode of experimenting was described by which the experimental results quoted in the theoretical paper were obtained; and the principal parts of the special apparatus, which had been constructed and used in the investigation, were laid before the Royal Society.

179. ON THE HEAT PRODUCED BY AN ELECTRIC DISCHARGE.

[From *Phil. Mag.* Vol. VII. May 1854, pp. 347, 348.]

To the Editors of the Philosophical Magazine and Journal.

2 COLLEGE, GLASGOW,
April 19, 1854.

GENTLEMEN,

It has been pointed out by M. Clausius, in a letter addressed to you and published in the last Number of your Magazine, that the first discovery of the true relation between the generation of heat in the discharge of a Leyden phial and the quantity of the previous charge is not, as I had stated it to be, due to Joule, but that it had been given in a paper published about three years earlier by Riess. I may be allowed to explain, that, in making the statement in question, I considered the law of that relation as an evident corollary from the great principle, that the whole heat generated in any discharge of electricity is exactly the equivalent in thermal energy for the mechanical value of the electrical charge

which is lost*.   There is no doubt who is the discoverer of *this*,
and the originator, in your most valuable Magazine, of the theory
of mechanical equivalence among the electric, chemical, magnetic,
frictional, and pneumatic developments of energy, which has within
the last two or three years attracted so many investigators.

The mere law, that the heat generated by the discharge of a
Leyden phial or battery is proportional to the square of the
quantity of electricity in the previous charge, is not, as I inadvertently stated, due to Joule; neither is it, as M. Clausius seems
to suppose, due to Riess.   Becquerel, I find, in his *Traité de
l'Electricité* (Vol. III. p. 150, published in 1835, or two years earlier
than the paper referred to by M. Clausius), enunciates it quite
explicitly as having been established by "Cuthbertson and others,
who had used electrometers in measuring the calorific action of
the discharge of a battery."   Mr Joule, too, although in his first
publication he only referred to the researches of Snow Harris
which had recently appeared in the *Philosophical Transactions*,
remarks in a subsequent paper (On the Heat disengaged in
Chemical Combinations, *Phil. Mag.* June 1852), "that Brooke and
Cuthbertson found that the length of wire melted by an electrical
battery varied nearly with the square of its charge"; and at the
same time he refers to the researches of Riess on the calorific
effects of frictional electricity, acknowledging their priority to
his own researches on the heat generated by continuous electric
currents.

I remain, Gentlemen, Yours very faithfully,

WILLIAM THOMSON.

---

* The application of this principle to the discharge of a Leyden phial shows
that the whole heat generated must be equal to $\frac{1}{J} \cdot \frac{1}{2}Q^2 \frac{4\pi\tau}{IS}$, if J denote the
mechanical equivalent of the thermal unit, Q the amount of the charge, τ the
thickness of the glass, I its specific inductive capacity, and S the area of either
side of the coated surface; a conclusion which wants no other experimental verification than such as may be considered desirable for verifying that $I\frac{S}{4\pi\tau}$ is the
true expression for the capacity of a Leyden phial.

180. On Mr Whitehouse's Relay and Induction Coils
in Action on Short Circuit*.

[From *British Association Report*, 1857, pt. ii. p. 21. Abstract.]

THE peculiarities of Mr Whitehouse's induction coils, which
fit them remarkably for the purpose for which they are adapted,
as distinguished from the induction coils by which such brilliant
effects of high intensity are obtained, were described. The chief
part of the telegraphic receiving apparatus, the relay, was fully
described, and was shown in action, through thirty yards of the
Atlantic cable, after some remarks explaining the general nature
of a relay,—an electrical hair-trigger. The relation of Mr Whitehouse's relay to the Henley receiving instrument was pointed
out. The author expressed his conviction, that by using Mr
Whitehouse's system of taking advantage of each motion for a
single signal, instead of the to-and-fro motion, as in all systems
hitherto practised, the Henley single needle instrument might be
easily used, so as to give as great a speed on one line of wire
alone, as is at present attained by two with the double needle
instrument. The beautiful method of reading by bells would be
most ready and convenient for giving the indications to be interpreted as the messages, but the author believes that either by the
eye or ear, messages may be read off with the rapidity and ease
which will render the use of one telegraph wire in all respects as
satisfactory as that of two.

* [On the working of the early Atlantic Cable cf. *Math. and Phys. Papers*,
Vol. II., especially pp. 92—112.]

181. ON THE EFFECTS OF INDUCTION IN LONG SUBMARINE
LINES OF TELEGRAPH.

[From *British Association Report*, 1857, pt. ii. pp. 21, 22.   Abstract.]

A GENERAL explanation of the theory was given, and the "law
of squares" was proved to be rigorously true.   It was pointed out,
that when the resistances of the instruments employed to generate
and to receive the electric current are considerable in comparison
with the resistance of the line, the observed phenomena do not
fulfil the law of squares, because the conditions on which that law
is founded are deviated from.   The application of the theory to
the alternate "positive" and "negative" electrical actions used by
Mr Whitehouse for telegraphing was explained, and the circumstances which practically limit the speed of working were pointed
out.   Curves illustrating the enfeeblement of the current towards
the remote end of the telegraph line, and the consequent necessity
of the high pressure system introduced by Mr Whitehouse, were
shown.   The embarrassment occasioned by the great electrical
effect through the wire, which follows the commencement of a
series of uniform signals with a full strength of electrical force,
was illustrated in one diagram, which showed a succession of eight
impulses following one another at equal intervals of time, and
giving only one turn of the electrical tide at the remote end,
or two motions of the relay, including the initial effect.   The
remedy suggested by the author was illustrated by another diagram,
in which a succession of seven equal alternate applications
of positive and negative force, following a first impulse of half
strength, was shown to give seven turns of the tide at the remote
end, and therefore eight motions of the relay, following one another
at not very unequal intervals of time.

182. On [Fault] Phenomena of Submerged Atlantic Cable.

[From Letter to J. P. Joule, dated Sept. 25, 1858; *Manchester Phil. Soc. Proc.* Vol. I. [Oct. 5, 1858], pp. 60—62.]

INSTEAD of telegraphic work, which, when it has to be done through 2,400 miles of submarine wire, and when its effects are instantaneous interchange of ideas between the old and new worlds, possesses a combination of physical, and (in the original sense of the word) *metaphysical* interest, which I have never found in any other scientific pursuit—instead of this, to which I looked forward with so much pleasure, I have had, almost ever since I accepted a temporary charge of this Station, only the dull and heartless business of investigating the pathology of "faults" in submerged conductors. A good deal that I have learned in this time has, I believe, a close analogy with some curious phenomena you have described, and which you partially shewed me last winter, regarding intermittent effects of resistance to the passage of an electric current between two metal plates in a liquid. Thus I have been informed by Mr France, of the Submarine and Mediterranean Companies, who has had long experience in testing and working submarine cables, that he has frequently observed, when applying constant electromotive force to one end of a submerged cable in which there is a bad defect of insulation, that the indicating needle of his galvanometer has continued oscillating through nearly the whole range of its scale without any apparent cause. Phenomena of the same kind, to a greater or less degree are, I believe, familiar to all careful observers who have been engaged in submarine telegraphing. Another very remarkable feature of the insulation of gutta-percha-covered wire, is the difference in the effects of positive and negative electrifications.

It is well known that a fault of insulation in an actually submerged cable causes a much greater loss of current when the wire is negatively, than when it is positively electrified, and that if after the wire has been left to itself, or has been negatively electrified for some time, a positive electrification be applied and maintained, the insulating power (resistance to loss) gradually rises, and continues rising, minute after minute, sometimes even sensibly for hours; as is shown by the current from the battery at one end of the cable, gradually diminishing, while the current through the other end, if put to earth, gradually rises in strength. On the fourth day after the end of the cable was landed here, I found that a positive current entering from ten cells of a constant battery fell in the course of a few minutes to half strength. When the battery was next suddenly reversed, the negative current rose, and remained after that nearly constant, at about the same degree of strength as that at which the positive current had commenced. The same kind of action is, I have learned, certainly observed in cables actually submerged, and known to have faults in the gutta percha, by which the conductor becomes exposed to the water; and this has been attributed to electrolytic action upon the water giving rise to oxydation, or to the evolution of hydrogen at the surface of the copper, according as it is positively or negatively electrified, relatively to the earth at the spot.

I had observed the same difference as to insulating power for positive and negative charges, at Keyham, the cable being dry, and therefore think that the electrolytic explanation is either insufficient, or implies a very remarkable electrolytic action on gutta percha itself, or on pitch, or possibly moisture in crevasses.

In some experiments on artificial faults placed in basons of sea water, I have paid particular attention to the green and white incrustations, observed according as the current is from imperfectly protected wire to water or the reverse. The latter is very remarkable, and appears like an exudation on the bark of a tree, when the fault consists of a minute incision or aperture. In the last case there is always a fine passage or crater in the middle, by which bubbles of hydrogen escape.

183. ON PHOTOGRAPHED IMAGES OF ELECTRIC SPARKS.

[From *Glasg. Phil. Soc. Proc.* Vol. IV. [read Dec. 14, 1859],
pp. 266, 267. Abstract.]

PROFESSOR W. THOMSON exhibited photographed images of electric sparks reflected from a revolving mirror, which a few days since he had received from Mr Feddersen of Leipsic, and which afforded a remarkable illustration of the "oscillatory discharge" indicated by dynamical theory as occurring, when a Leyden phial of not too great electrostatic capacity is discharged, by a sufficiently easy conducting train, through a channel presenting sufficient "induction on itself" or "electro-dynamic capacity." The occurrence of an oscillatory discharge, under certain conditions, had been first anticipated by Helmholtz, in his *Erhaltung der Kraft* (Berlin, 1847). The law of discharge, when the discharging train possesses no sensible electrostatic capacity, had been fully investigated and the conditions under which it takes place with oscillations discriminated from those under which it takes place with continuous subsidence, had been determined, in a mathematical paper communicated to this Society about seven years ago "On Transient Electric Currents"—*Proceedings of Glasgow Philosophical Society*, January, 1853, and *Philosophical Magazine*, June of same year. [*Math. and Phys. Papers*, Vol. I. p. 534.]

At that time the numerical relation between electrostatic and electrodynamic units had not been determined, and therefore a certain coefficient in the mathematical formula was left for experimental investigation. The want has been since supplied by W. Weber, who has continued the system of absolute measurement inaugurated by himself and Gauss for terrestrial magnetism, and has extended it with the greatest advantage into every department of electric and magnetic science. The consequence is that the mathematical formula for an electric discharge can now be fully reduced to numbers, the criterion as to whether it is oscillatory or continuous applied, and, when it is oscillatory, the time of an oscillation determined for any stated dimensions and form of apparatus.   Professor Thomson further stated that he had calculated dimensions, &c., and found that arrangements could readily be made to give rise to oscillations in periods not less than $\frac{1}{10000}$ of a second, which could therefore be easily shown by Wheatstone's method of the revolving mirror, as he had anticipated might be possible when he first communicated his mathematical investigation to this Society.   The barred appearance of each of the photographic images now before the Society would, if the rate of rotation of the mirror, and the distance from it of the plate receiving the impression, were known, be enough to determine the period of the electric oscillation by which they had been produced.   He hoped soon to have particular information as to these and other details from Mr Feddersen, and so be able to make a thorough numerical verification of the theory, which he would not delay to lay before the Society.

184. REPORT OF COMMITTEE APPOINTED TO PREPARE A SELF-
RECORDING ATMOSPHERIC ELECTROMETER FOR KEW, AND
PORTABLE APPARATUS FOR OBSERVING ATMOSPHERIC ELEC-
TRICITY.

[From *British Association Report*, 1860, pp. 44, 45.]

YOUR Committee, acting according to your instructions, applied
to the Royal Society for £100 out of the Government grant for
scientific investigation, to be applied to the above-mentioned
objects. This application was acceded to, and the construction
of the apparatus was proceeded with. The progress was necessarily
slow, in consequence of the numerous experiments required to find
convenient plans for the different instruments and arrangements
to be made. An improved portable electrometer was first completed, and is now in a form which it is confidently hoped will be
found convenient for general use by travellers, and for electrical
observation from balloons. A house electrometer, on a similar
plan, but of greater sensibility and accuracy, was also constructed.
Three instruments of this kind have been made, one of which
(imperfect, but sufficiently convenient and exact for ordinary work)
is now in constant use for atmospheric observation in the laboratory
of the Natural Philosophy Class in the University of Glasgow.
The two others are considerably improved, and promise great ease,
accuracy, and sensibility for atmospheric observation, and for a
large variety of electrometric researches. Many trials of the
water-dropping collector, described at the last Meeting of the
Association, were also made, and convenient practical forms of the
different parts of the apparatus have been planned and executed.
A reflecting electrometer was last completed, in a working form,
and, along with a water-dropping collector and one of the improved common house electrometers, was deposited at Kew on
the 19th of May. A piece of clockwork, supplied by the Kew
Committee, completes the apparatus required for establishing the

self-recording system, with the exception of the merely photo-graphic part. It is hoped that this will be completed, under the direction of Mr Stewart, and the observations of atmospheric electricity commenced, in little more than a month from the present time. In the mean time preparations for observing the solar eclipse, and the construction of magnetic instruments for the Dutch Government, necessarily occupy the staff of the Observatory, to the exclusion of other undertakings. It is intended that the remaining one of the ordinary house electrometers, with a water-dropping collector, and the portable electrometer referred to above, will be used during the summer months for observation of atmospheric electricity in the Island of Arran. Your Com-mittee were desirous of supplying portable apparatus to Prof. Everett, of Windsor, Nova Scotia, and to Mr Sandiman, of the Colonial Observatory of Demerara, for the observation of atmo-spheric electricity in those localities; but it is not known whether the money which has been granted will suffice, after the expenses yet to be incurred in establishing the apparatus at Kew shall have been defrayed. In conclusion, it is recommended to you for your consideration by your Committee, whether you will not immediately take steps to secure careful and extensive obser-vations in this most important and hitherto imperfectly investigated branch of meteorological science. For this purpose it is suggested, —1, that, if possible, funds should be provided to supply competent observers in different parts of the world with the apparatus necessary for making precise and comparable observations in absolute measure; and 2, that before the conclusion of the present summer a commencement of the electrical observation from balloons should be made.

185. Velocity of Electricity.

[From Nichol's *Cyclopædia* [second edition], 1860. Reprinted in *Math. and Phys. Papers*, Vol. II. art. lxxxi, pp. 131—137; and in extended form in *Baltimore Lectures*, Appendix L, 1904, pp. 688—694.]

186. ON THE MEASUREMENT OF ELECTRIC RESISTANCE.

[From *Roy. Soc. Proc.* Vol. XI. 1860—1862 [read June 20, 1861], pp. 313—
328; *Annal. de Chimie*, Vol. LXVII. 1863, pp. 501—506; *Phil. Mag.*
Vol. XXIV. Aug. 1862, pp. 149—162.]

PART I. *New Electrodynamic Balance for resistances of short
bars or wires.*

IN measuring the resistances of short lengths of wire by
Wheatstone's Balance*, I have often experienced considerable
difficulty in consequence of the resistances presented by the
contacts between the ends of the several connected branches or
arcs. This difficulty may generally be overcome by soldering or
amalgamating the contacts, when allowable; but even with
soldered connexions there is some uncertainty relating to the
dimensions of the solder itself, when the wires tested are very
short. When soldering was not admissible, I have avoided being
led into error, by repeating the experiment several times with
slightly varied connexions; but I have in consequence sometimes
altogether failed to obtain results by either Wheatstone's or any
other method hitherto practised, as for instance in attempting to
measure the electric resistances of a number of metallic bars each
6 millimetres long and 1 millim. square section, which were
put into my hands by Mr Calvert of Manchester, being those of
which he and Mr Johnstone determined the relative thermal
conductivities in their investigation published in the *Transactions
of the Royal Society* for March 1858. I have thus been compelled
to plan a new method for measuring electric resistances in which
no sensible error can be produced by uncertainty of the connexions,
even though made with no extraordinary care.

---

* I have given this name to the beautiful arrangement first invented by Professor Wheatstone, and called by himself a "differential resistance measurer." It
is frequently called "Wheatstone's Bridge," especially by German writers. It is
sometimes also, but most falsely, called "Wheatstone's Parallelogram.'

Let AB and CD be the standard and the tested conductors respectively. Let the actual standard of resistance be the resistance of the portion of AB between marks* S, S' on it, and let it be required to find a portion TT' of CD which has a resistance either equal, or bearing a stated ratio, to that standard.

Join BC either by direct metallic contact between them, or by any ordinarily good metallic connexion with binding screws or otherwise; and join the two electrodes of a galvanic element to their other ends, A, D. Let GPH and KQL be two auxiliary conductors, which, to avoid circumlocutions, I shall call the primary

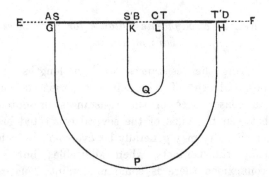

and the secondary testing-conductors respectively, with their ends applied to the marked points S, T', S', T. Let P and Q be points in these conductors to which the electrodes of the galvanometer are to be applied.

It is easily seen, and will be demonstrated below, that if the resistances of the testing-conductors be similarly divided in Q and P, and if their ends be in perfect conducting communication with the marked points of the main line to which they are applied, the condition that the galvanometer indication may be zero is that the ratio of the resistances of the standard and tested conductors must be the same as that in which the auxiliary conductors are each divided. Further, it is clear that by making the testing-conductors of incomparably greater resistances than any that can exist in the connexions at S, S', T, T', which can easily be done if these connexions are moderately good, the error arising from such imperfections as they must present may be made as small as is

---

* On the same principle as the "mètre à traits' instead of the "mètre à bouts" for a standard of length.

required*. To demonstrate the above and to form an accurate idea of the operation of this method, it is necessary to investigate the difference of potentials (electromotive force) produced between Q and P, when a stated difference of potentials, E, is maintained between S and T'.

Let SS', TT' denote the resistances between the marks, on the standard and tested conductors respectively. Let GPH, GP, PH, KQL, KQ, QL denote the resistances of the testing-conductors and their parts according to the diagram, implying that

$$GPH = GP + PH, \text{ and } KQL = KQ + QL.$$

Let SG, HT', $S'K$, LT be the resistances in the connexions at the marks; let

$$SG + GPH + HT'' \text{ be denoted by } SPT'',$$

---

* This method may be readily applied to Siemens's mercury standards (see *Phil. Mag.* Jan. 1861, or *Poggendorff's Annalen*, 1860, No. 5), by introducing platinum wires through holes in the glass tube near its ends, as electrodes for the testing-conductors, and wires or plates of platinum at the ends, as electrodes for one pole of the battery and for connexion with the conductor to be compared with it, respectively. It will then not be the whole line of mercury from end to end, but the portion of it between the two platinum wires first mentioned, that will be the actual standard. The objection against the use of mercury as a standard of resistance, urged by Matthiessen, that the amalgamated copper electrodes which Siemens found necessary to give very perfect *end* connexions must render the mercury impure and increase its resistance sensibly after a time, is thus completely removed. It must be shown, however, that different specimens of commercial mercury, dealt with in the manner prescribed by Siemens, to remove impurities, shall always be found to have equal specific resistances, before his proposal to produce independent standards by filling gauged tubes with mercury can be admitted as valid. But the *transportation and comparison of actual standards* between different experimenters in different places is, and probably must always be, the only way to obtain *the most accurate possible common system of measurement* : and when a proper mutual understanding between electricians and national scientific academies, in all parts of the world, has been arrived at, as it is to be hoped it may be soon, through the assistance of the British Association and Royal Society if necessary, the use of definite metallic standards, whether the liquid mercury as proposed by Siemens, on the one hand, or the solid wire, alloy of gold and silver, on the other hand, proposed by Matthiessen (*Phil. Mag.* Feb. 1861), would be essential only in the event of all existing standards being destroyed.

Weber's absolute system is often referred to as if its object were merely to fix standards of resistance, and the difficulty and expense of applying it independently have been objected to as fatal to its general adoption. In reality its great value consists in the dynamic conditions which it fulfils, with relation to electro-magnetic induction, and to the mechanical theories of heat and of electro-chemical action But it most probably will also be much more accurate than any definite metallic convention, for the re-establishment of a common metrical system, in case of the destruction of all existing standards.

and $\qquad S'K + KQL + LT$ be denoted by $S'QT$;

and let $S'BCT$ denote the resistance between S' and T composed of the resistance in the connexion and the resistances in the portions of the main conductors from the marks S' and T to their ends. Lastly, let R denote the resistance in the double channel $\begin{Bmatrix} S'BCT \\ S'QT \end{Bmatrix}$ between S' and T. By the well-known principles of electric conduction, we have

$$R = \left(\frac{1}{S'BCT} + \frac{1}{S'QT} \right)^{-1} \quad \dots\dots\dots\dots\dots(1)$$

for the resistance in the double arc between S' and T. Then, by addition, we have

$$SS' + R + TT'$$

for the resistance from S to T' by the channel $SS' \begin{Bmatrix} S'BCT \\ S'QT \end{Bmatrix} TT'$

This whole resistance is divided, by Q and its equipotential point in the direct channel $S'BCT$, into the parts

$$SS' + (S'Q/S'QT) . R \text{ and } (QT/S'QT) . R + TT'.$$

Hence if, for simplicity, we suppose the potential at S to be 0, and at T' to be E, and if we denote by q the potential at Q, we have

$$q = E \frac{SS' + (S'Q/S'QT) . R}{SS' + R + TT'} \quad \dots\dots\dots\dots(2).$$

Again, since P divides the resistance between S and T', along the channel SPT', into the parts SP and PT', we have

$$p = E (SP/SPT') \dots\dots\dots\dots\dots\dots(3),$$

if we denote by p the potential at P. Hence

$$q - p = E \frac{SS' + (S'Q/S'QT) . R - (SP/SPT')(SS' + R + TT')}{SS' + R + TT'};$$

or, since $1 - (SP/SPT') = (PT'/SPT')$,

$$q - p = E \frac{\dfrac{PT'}{SPT'} . SS' - \dfrac{SP}{SPT'} . TT' + R \left(\dfrac{S'Q}{S'QT} - \dfrac{SP}{SPT'} \right)}{SS' + R + TT'} \quad \dots(4).$$

Now let us suppose that, by varying one or more of the component arcs in the balance-circuit, we reduce the galvanometer indication to zero, that is to say, make $q - p = 0$. We shall have by equating the numerator of the preceding expression to zero, and resolving for TT',

$$TT' = (PT'/SP).SS' + R\{(SPT'/S'QT)(S'Q/SP) - 1\}...(5).$$

To interpret this expression, it may be remarked that if the second term vanishes, that is to say, if

$$R\{(SPT'/S'QT)(S'Q/SP) - 1\} = 0$$

we have $\qquad\qquad TT' = (PT'/SP).SS' \qquad\qquad$ ....(6);

and this is the condition aimed at in the arrangement. Now the connexions at S and T' must be made so good that the resistance SG in the first is inappreciable in comparison with GP, and the resistance HT', in the second, inappreciable in comparison with PH; so that we may have

$$PT'/SP \sqsupset\!\sqsubset PH/GP,$$

where $\sqsupset\!\sqsubset$ denotes an equality not perfect, but having no appreciable error; and hence

$$TT' \sqsupset\!\sqsubset (PH/GP).SS'.$$

The condition

$$R\{(SPT'/S'QT).(S'Q/SP) - 1\} \sqsupset\!\sqsubset 0$$

is to be secured by one or other of two ways or by both combined; that is, by making

$$R \sqsupset\!\sqsubset 0 \qquad\qquad...............................(a)$$

or $\qquad\qquad (SPT'/S'QT)(S'Q/SP) \sqsupset\!\sqsubset 1 \qquad...................(b),$

or each as nearly as possible. If the connexion BC were quite perfect and the marks S' and T were at the very ends of the conductors, the condition (a) would be fulfilled and there would be no necessity for the condition (b). We should then have a perfect Wheatstone balance,—the secondary testing-conductor $\left\{{K \atop L}\right\}Q$ becoming merely a part of the galvanometer electrode. Hence whenever the resistance $S'T$ can be made absolutely insensible, Wheatstone's balance leaves nothing to desire, provided the ends of the testing-conductor are applied to marked points on the standard and tested conductors, and the battery electrodes to their outer ends, or to points of them between their outer ends and those marked points. When, however, as very frequently is the case, $S'T$ may be made small but not absolutely insensible in comparison with the resistances of the standard and tested conductors, the addition of the "secondary testing-conductor" becomes valuable, even if it be only arranged to give a rough approximation to the condition $(SPT'/S'QT)\ (S'Q/SP) = 1$*, since it will reduce the error to the fraction $(SPT'/S'QT).(S'Q/SP) - 1$, of the small resistance R†. But further, when, as in experiments on short thick bars like those of Mr Calvert, $S'T$ cannot by any management be got to be small in comparison with TT', the use of the secondary testing-conductor becomes essential, and the most accurate possible fulfilment of the condition

$$(SPT'/S'QT).(S'Q/SP) \eqcirc 1$$

must be aimed at. This is to be done by dividing the secondary testing-conductor at Q, in very exactly the same ratio as the primary at P, and taking care that the resistances in the connexions $S'K$, LT are very small in comparison with KL and QL.

PART II. *Suggestions for carrying out these principles in practice.*

When high accuracy is not required, the two "testing-conductors" may be made of wires stretched straight in parallel lines, and the connexions for the galvanometer electrodes may be applied to them by means of a slide on a graduated scale—as in one of the common forms of Wheatstone's balance, with sliding contact on single testing-conductor. This form is very objectionable, however, whether for Wheatstone's balance or the method I now propose: (1) because it is impossible to secure that the different parts of each testing-conductor shall be accurately at the same temperature; (2) because the resistances at the ends of the fine stretched wire or wires are always sensible in comparison with the smallest measured differences produced by the slide; (3) because the stretched wire itself is never of absolutely equal gauge throughout, and, even if sensibly so when first put into the instrument, soon ceases to be so in consequence of the friction of

---

* This of course is equivalent to $SPT' : SP :: S'QT : S'Q$, and means that the secondary conductor is to be divided by one galvanometer electrode in the same proportion as the primary is divided by the other.

† In such cases R will, according to equation (1) above, be nearly equal to $S'BCT$, but somewhat less.

the sliding contact which it experiences in use*; (4) because, in even the hastiest experiments, provided a rationally planned galvanometer is used, a far higher proportional degree of accuracy is easily attained in measuring electrical resistances against a standard of resistance than can be at all attained, without very extraordinary precautions and the assistance of a microscope, in measuring lengths under a yard or two against a standard of length.

When the highest accuracy is required, I always use for primary testing-conductor the bisected conductor which I described to the British Association at its Glasgow meeting in 1855. This consists of a fine, very perfectly insulated wire, doubled on itself and wound on a bobbin, with very stout terminals soldered to its ends, and an electrode soldered to its middle, for joining to the galvanometer electrode. The two terminal and the middle electrodes thus attached to the testing-conductor, I have generally hitherto made flexible, either of thick wire, or strand of wires like the conductor of a submarine cable; but, for many applications, it is more convenient to make them solid metal blocks, with binding screws, insulated rigidly upon the bobbin which bears the conductor. The two halves into which the conductor is doubled must be very accurately equalized as to electric resistance when they are wound on the bobbin, and before the terminals are finally attached. This I find can be done with great accuracy; and when, after the terminals are soldered on, the electric bisection is once found perfect, it seems to remains so, without sensible change, for years. The close juxtaposition of the two branches of the testing-conductor on this plan ensures an almost absolute equality of temperature between them in all circumstances, and thus renders easy a degree of accuracy in the measurement of resistances quite unattainable with any other form of Wheatstone's balance. In the new method which I now propose for low resistances, I make the secondary conductor on exactly the same plan, and generally of about the same dimensions, as the primary. The bisected testing-conductors are only available when the resistances of the standard and of the tested conductor can be made equal; and with them the method which has been described above seems to be the most accurate possible for testing a perfect equality of resistance between two conductors.

---

* This defect I have remedied by frequently putting in a new wire for testing-conductor in working with a sliding-scale Wheatstone's balance.

The same plan of testing-conductors seems still the best, even when testing by equality cannot be practised,—with only this difference, that the two branches of each testing-conductor, instead of being made of equal resistance, must be adjusted to bear to one another very exactly the ratio which the tested resistance is to bear to the standard. By proper care, to prevent the bobbin of either testing-conductor from getting any non-uniform distribution of temperature, great accuracy may still be secured; but it is scarcely possible to maintain so very close an agreement of temperature, and therefore so constant a ratio of resistances, as when the two branches are equal lengths of one wire coiled side by side.

The use of this plan of conductors divided in a fixed ratio, whether for the single testing-conductor in Wheatstone's balance, or for the primary and secondary testing-conductors in the new method now proposed, requires that either the standard or the tested conductor can be varied so as to adjust the resistance of one to bear precisely that ratio to the resistance of the other. In certain cases this may be done advantageously by shifting one or other of the contacts S, S', T, T' along the standard or the tested conductor, as the case may be. If, for instance, T or T' can be shifted conveniently, the object of the measurement may be to find by trial on the tested conductor a portion TT' from mark to mark, of which the resistance bears a stated ratio to the fixed standard SS' from mark to mark. But by far the easiest working, and in most cases the most accurate also, is to be done by means of a well-arranged series of standards with terminals adapted for combining them in such a manner as to give to a minute degree of accuracy whatever resistance may be required. In a future communication on standards of electric resistance, I intend to describe plans for attaining this object through a wide range of magnitude (resistances from 10^5 to 10^{13} British absolute units of feet per second on Weber's invaluable system). In the mean time I shall merely say that I have formed a plan which I expect will prove very advantageous for low resistances, and which consists in combining the standards, whichever of them are required, in multiple arc (or "parallel" arcs, according to an expression sometimes used), so as to add their *conducting powers* *,

---

* The reciprocal of the resistance of a "conductor" or "arc" I call its *conducting power*. The conducting power of a bar or wire of any substance one foot long and weighing one grain, I call the *specific conductivity of its substance*.

—instead of in series, as in all arrangements of resistance coils hitherto used, by which the *resistances* of the component standards are added.

PART III. *General Remarks on Testing by Electro-dynamic Balance.*

I shall conclude by remarking that the sensibility of the method which has been explained, as well as of Wheatstone's balance, is limited solely by the heating effect of the current used for testing. To estimate the amount of this heating effect, let e and f be the parts of the whole electromotive force, E, which act in the standard SS', and tested conductor TT' respectively; so that, in accordance with the notation used above, we have

$$e = E\,\frac{SS'}{SS' + R + TT'}, \qquad f = E\,\frac{TT'}{SS' + R + TT'} \quad \text{...(7)}$$

of its substance. Following Weber, I define the resistance of a bar or wire one foot long, and weighing one grain, its *specific resistance*. It is much to be desired that the *weight-measure*, rather than the *diameter* or *the volume-measure*, should be generally adopted for accurately specifying the gauge of wires used as electric conductors.

With reference to either SS' or TT' (the first, for instance), let us use the following notation:—

l its length in feet;

w its mass per foot in grains;

s the specific heat of its substance;

σ the specific resistance of its substance.

Thus, since we have taken SS' to denote its actual resistance, we have

$$SS' = \sigma l/w.$$

Now, Weber's system of absolute measurement for electro-motive forces and for resistances being followed, I have shown* that *the mechanical value of the heat generated per unit of time in any fixed conductor of uniform metallic substance is equal to the*

---

* In a paper "On the Mechanical Theory of Electrolysis," *Philosophical Magazine*, Dec. 1851. [*Math. and Phys. Papers*, Vol. I. p. 472.]

square of the electromotive force between its extremities, divided by its resistance. This in the present case is equal to

$$e^2 w/(l\sigma);$$

and if J denote Joule's mechanical equivalent of the thermal unit, we therefore have

$$e^2 w/(Jl\sigma)$$

for the rate per second at which heat is generated in SS'. This will at first go entirely to raise its temperature*. Now wl is its mass in grains, and therefore wls is its whole thermal capacity; and if we divide the preceding expression by this, we find

$$e^2/(Jl^2 s\sigma)$$

for the rate per second at which it commences to rise in temperature at the instant when the battery is applied. If we call e/l the electromotive force per foot, we may enunciate the result thus:

The rate at which a linear conductor of uniform metallic substance commences rising in temperature at the instant when an electric current commences passing through it, is equal to the square of the electromotive force per unit of length divided by the continued product of Joule's equivalent into the specific heat of the substance, into the specific resistance of the substance.

Let us suppose, for example, that the conductor in question is copper of best electric conductivity. Its specific resistance will be about 7×10^6, and its specific heat about ·1. The value we must use for Joule's equivalent will be 32·2 times the number 1390, which Joule found for the mechanical value in foot-grains of the thermal unit Centigrade, since the absolute unit of force, being that force which acting on a grain of matter during a second of time generates 1 foot per second of velocity, is 1/32·2 of the weight of a grain in middle latitudes of Great Britain. Thus we find

$$J = 44758.$$

---

* As soon as it has risen sensibly in temperature it will begin to give out heat by conduction, or by conduction and radiation, to the surrounding matter; and the rate at which it will go on rising in temperature will be the rate expressed by the formula in the text (with the true specific resistance, &c., for each temperature), diminished by the rate of loss to the surrounding matter.

Hence the expression for the rate in degrees Cent. per second, at which the temperature begins rising in a copper conductor, is

$$(e/l)^2/(313 \times 10^8).$$

I have found the electromotive force of a single cell of Daniell's to be about $2{\cdot}3 \times 10^6$ British absolute units*; and if we suppose $1/n$ of this to go to each foot of the conductor in question, we shall have

$$(e/l)^2 = (2{\cdot}3^2 \times 10^{12}/n^2) = (5{\cdot}29 \times 10^{12}/n^2);$$

and therefore the expression for the rate of heating becomes

$$1{\cdot}69 \times (100/n^2).$$

Now, by using a sufficiently large single cell, we may make the electromotive force, E, between S and T', be as little short as we please of the whole electromotive force of the cell. We might then, in testing by equality, with a standard and a tested conductor each three inches or so long, and using a single cell, have nearly as much as half the electromotive force of one cell acting per quarter foot of these conductors, or two cells per foot. Hence if either is of best conductive copper, its temperature would commence rising at the rate of $4 \times 169°$ or $676°$ Cent. per second. It would be almost impossible to work with so high a heating effect as this. But if we use only $\frac{1}{10}$th of the supposed electromotive force, that is to say $\frac{1}{5}$th of a cell per foot of the copper conductor, the rate of heating will be reduced to $\frac{1}{100}$, that is to say, will be $6{\cdot}76°$ per second. By using only very brief battery applications, it would be possible to work with so high a rate of heating as that, without having the results much vitiated by it. But $\frac{1}{50}$ of a cell per foot will give only $\cdot 0676°$ of heating effect per second, and will be quite a sufficient battery power to use in most cases. In the case we have supposed, for instance, of conductors only three inches long, the electromotive force on each would then be about $\frac{1}{200}$ of the electromotive force of the cell. What we denoted above by e and f in equations (7) would therefore each have this value. Hence, by equation (4), we see that the effect of a difference of $\frac{1}{1000}$ between SS' and TT' would be to give $q - p$ the value $\frac{1}{400000}$ of the electromotive force of a single cell. Now one of the light mirror† galvanometers, which I commonly use, reflecting the

---

* *Proceedings of the Royal Society*, February 1860.

† The mirror is a circle of thin "microscope glass" about three-eights of an inch in diameter, silvered in the ordinary manner; and a small piece of flat file

image of a gas or paraffine lamp to a scale 25 inches distant, would, if made with a coil of 50 yards of copper wire of moderate quality, weighing 5 grains per foot, give a deflection of half a division of $\frac{1}{40}$ of an inch on this scale, with an electromotive force of $\frac{1}{400000}$ of a single cell*. Hence by using such a galvanometer, and primary and secondary conductors of sufficient resistances to fulfil the condition of doing away with sensible error from imperfect connexions in the manner explained above, but yet of resistances either less than or not many times greater than the resistance of the galvanometer coil, it is easy to test to $\frac{1}{1000}$ the resistance of a copper wire or bar not more than 3 inches long. The current we have found to be sufficient for this object would only produce a heating effect of ·14°C. in two seconds, which, with good apparatus, is more than enough of time, as I shall show presently. The influence of this heating effect may be regarded as nearly insensible, since even as much as ·2° only alters the resistance of copper by about $\frac{3}{4000}$.

In all measurements of electric resistance, whatever degree of galvanic power is used, a spring "make and break" key† ought to be placed in one of the battery electrodes, so that the current may never flow except as long as the operator wills to keep it flowing, and presses the key. I introduce a second similar spring key in one of the galvanometer electrodes (that is between either Q or P

steel of equal length, attached to its back by lac varnish, constitutes the "needle" of the galvanometer. The whole weight of mirror and needle amounts to from 1 to 1½ grains. It is suspended inside the galvanometer coil by single silk fibre about ⅓ inch long. It is necessary to try many mirrors thus prepared, each with its magnet attached, before one is found giving a good enough image. I am much indebted to Mr White, optician, Glasgow, for the skill and patience which he has applied to the very troublesome processes involved.

* In this state of sensibility the needle is under Glasgow horizontal magnetic force of the earth alone; and, with its mirror, it makes a vibration one way in about ·7 of a second. In many uses of my form of mirror galvanometer, both for telegraphic and for experimental purposes, I find it convenient to make its indications still more rapid, though, of course, less sensitive, by increasing the directive force by means of fixed steel magnets. On the other hand, I use fixed steel magnets to diminish the earth's directing force and make the needle more sensitive, when very high sensibility is wanted; but this would be inconvenient for the application described in the text, because effects of thermo-electric action would be made too prominent.

† Morse's original telegraph key, which instrument-makers have "improved" into the in every respect worse form in which it is now commonly made—a massive contact-lever urged by a spring.

and the galvanometer coil), so arranged that the pressure of the operator's finger on a little block of vulcanite attached to either spring shall first make the contact of the first spring (completing the battery circuit), and when pushed a little further, shall make the contact of the second spring and complete the galvanometer circuit. The test for the balance of resistances will then be that not the slightest motion of the needle is observable as a consequence of this action on the part of the operator. The sensibility of the arrangement is doubled by a convenient reverser in the galvanometer circuit, by which the current, if any, may be reversed easily by the operator while keeping the two connexions made by full pressure on the double spring key just described. Another convenient reverser should be introduced into the battery circuit, to eliminate effects of thermo-electric action if sensible.

It may often happen, unless the galvanometer is at an inconveniently great distance from the conductors tested, that its needle will be directly affected to a sensible extent by the main testing-current; but with the arrangement I have proposed the observer tests whether or not this is the case by pressing the double spring-key to only its middle position (battery contact alone made), and watching whether or not the needle moves perceptibly. If it does not move perceptibly, he has nothing more to do than immediately to press the double key home, to test the balance of resistances. If the needle does move when the key is pressed to its middle position, he may, when in other respects allowable, keep the current flowing by holding the key in its middle position till the needle comes to rest, or at least till it shows the point towards which its oscillations converge, and then press home to test the balance of resistances. When the very highest accuracy is aimed at, or when, for any reason (as, for instance, extreme shortness in the standard or tested conductor), only the shortest possible duration of current is allowable, the position of the galvanometer, with reference to the battery and the other portions of circuit, must be so arranged that its needle may show no sensible deflection when the key is pressed to the middle position. Ignorant or inadvertent operators are probably often led into considerable mistakes in their measurements of resistance by confounding deflections due to direct electro-magnetic influence of battery, battery electrodes, or standard, tested, or testing-conductors, on the needle of the galvanometer, with the proper influence of a current through its own coil,—a confusion which can only be resolved by making or breaking the galvanometer circuit while the battery circuit is kept made, for which there is no provision in the ordinary plans of Wheatstone's balance. We may, however, suppose that most experimenters will be sufficiently upon their guard against error from such a source. But there is another and a much more important advantage in the double-break arrangement which I now propose. Electro-magnetic inductions will generally be sensible* in some or in all of the different branches of the compound circuit, and cannot, except in very special cases, be exactly balanced as regards electro-motive force between P and Q with the arrangement which makes an exact balance of resistances. Hence, at the moment when the battery contact is made, there must generally be an electromotive impulse between Q and P, which will drive a current through the galvanometer coil, and make an embarrassing deflection of the needle if the galvanometer circuit is complete at that instant (as it is in the common plans of Wheatstone's balance), and will require the observer to wait until the needle comes to rest, or until he can tell precisely to what point its oscillations converge, the current being kept flowing all the time, before he can discover whether the balance of resistances has been attained or not. This absolutely precludes very refined testing, since, whether by the heating and consequent augmentation of resistance of some part of the balanced branches, or by thermo-electric reactions consequent on heating and cooling effects at junctions of dissimilar metals when the branches of the balance are not all of one homogeneous metal, or last, though not least, by the eye losing the precise position where the galvanometer needle or indicating image rested, it is not possible to use the full sensibility of the galvanometer for testing a zero if its needle is allowed to receive such a shock in the course of the weighing. Embarrassment from this source is completely done away with by using the double spring key described above, and giving time, from its first to its second contact, to allow the electro-magnetic

---

* I make them as little sensible as possible in my coiled testing-conductors, and in sets of coiled standards of resistance, by either doubling each coil or each branch of each coil on itself, or by reversing the lathe at regular intervals in winding on any single coil on a bobbin,—a plan which has also the advantage of rendering the direct electro-magnetic action of any coil so wound very small or quite insensible on any galvanometer needle in its neighbourhood.

induction to subside. An extremely small fraction of a second is enough in almost all cases; and the operator may therefore generally press the key home almost as sharply as he will or can. But when there is a large "electrodynamic capacity*" in any part of the balance-circuit, as, for example, when the coil of a powerful electro-magnet with soft iron† core is the conductor whose resistance is tested, it may be necessary to keep the key in its middle position for a few seconds before pressing it home, to avoid obtaining what might be falsely taken for an indication of too great a resistance to conduction (or "frictional" resistance, as I have elsewhere called it‡), being a true indication of resistance or reaction of inertia to the commencement of the current in the electro-magnetically-loaded branch§. In such cases it is impossible, either by electrodynamic balance or in any other way, to obtain a measurement of resistance without keeping the battery applied for the few seconds required to produce sensibly its final strength of current undiminished by inductive reaction, over and above the time required to get an indication from the galvanometer. But, as already remarked, in all ordinary cases, the

---

* This term I first introduced in a communication "On Transient Electric Currents" (*Phil. Mag.* June 1853), to designate what for any electric current through a given conductor is *identical* in meaning with the "simple-mass equivalent" in the motion of Attwood's machine as ordinarily treated. A rule for calculating the electrodynamic capacity is given in that communication; also the rule, with an example, in Nichol's *Cyclopædia*, article "Magnetism—Dynamical Relations of."

† Giving a resistance to the commencing, to the ceasing, or to any other variation in the strength of an electric current (precisely analogous to the effect of inertia on a current of common fluid),—which it seems quite certain must be owing to true inertia (not of what we should at present regard as the electric fluid or matter itself flowing through the conductor, but) of motions accompanying the current, chiefly rotatory with axes coinciding with the lines of magnetic force in the iron, air, and other matter in the neighbourhood of the conductor, and continuing unchanged as long as the current is kept unchanged. See Nichol's *Cyclopædia*, article "Magnetism—Dynamical Relations of," edition 1860 [*Electrostatics and Magnetism*, pp. 446—8]; also *Proceedings of the Royal Society*, June 1856; or *Phil. Mag.* Vol. XIII. 1857. [*Baltimore Lectures*, Appendix F.]

‡ "Dynamical Theory of Heat, Part VI, Thermo-electric Currents," *Transactions of the Royal Society of Edinburgh*, 1854; and *Phil. Mag.* 1856. [*Math. and Phys. Papers*, Vol. I. pp. 232—291.]

§ It is probable that a Wheatstone's balance, perfectly adjusted for equilibrium of resistances to conduction, and used with the galvanometer circuit constantly made, so as to show the whole effect of the inductive impulse, may afford the best means for making accurate metrical investigations on electro-magnetic induction and especially for determining "electrodynamic capacities" in absolute measure.

inductive reaction becomes insensible after a very small fraction of a second, and the operator may press the double key home to its second contact almost as sharply as he pleases. With such a galvanometer as I have described, he need not hold it down for more than 7 of a second (the time of the simple vibration of the needle*) to test the balance of resistances. The order of procedure will therefore generally be this:—The operator will first strike the key sharply, allowing it to rise again instantly, adjust resistances in the balance-circuit according to the indication of the galvanometer; strike the key sharply again, readjust resistances; and so on, until the balance is nearly attained. He will go on repeating the process, but holding the key down rather longer each time. At the last he will press the key gently down, hold it pressed firmly for something less than a second of time, and let it rise again; and if the spot of light reflected from the mirror of the galvanometer does not move sensibly, the resistances are as accurately balanced as he can get them.

* The mirror galvanometers commonly used in Germany have all much longer periods (ten or twenty times as long in many cases) for the vibration of their needles, and want proportionately longer contacts to obtain full advantage of their sensibility,—in each case a contact during a time equal to that of the vibration of the needle one way being required for this purpose.

187. On the Electromotive Force induced in the Earth's Crust by Variations of Terrestrial Magnetism.

[From *Phil. Trans.* Vol. CLII. pt. ii. 1862, pp. 637, 638; being a note on Balfour Stewart's paper on "The Nature of the Forces concerned in producing the greater Magnetic Disturbances."]

THE evidence from observation adduced in the preceding paper tending to show that some "earth-currents" which have been actually observed have been the electro-magnetically induced effects of variations of terrestrial magnetism, appears to be a very important contribution to the discovery of the complete theory of these most interesting and perplexing phenomena. It necessarily, however, suggests the question, Is the electromotive force induced by variations of terrestrial magnetism comparable in amount with that which is found in observations on earth-currents? There is scarcely occasion at present for working out a complete mathematical theory of the currents induced in the earth's crust by any fully specified magnetic variations. This can easily be done as soon as observation supplies data enough to make it useful. In the mean time a very rough theoretical estimate of the absolute amount of electromotive force induced by such magnetic variations as we know to exist, is sufficient to render it certain that this electro-magnetic induction does very sensibly influence the observed phenomena of earth-currents. For if there be, as we know there is every day, a gradual variation of terrestrial magnetism over a large portion of the earth's surface, amounting to a minute of angle in the direction of the dipping-needle, or to some thousandths, or not much less than one thousandth of force during several hours, this change of magnetism must induce in a length of a few hundred miles in the earth's crust an electromotive force, which we readily see must be comparable with that which would be induced in a linear conductor of the same length if carried across the lines of magnetic force at the rate of a minute of arc (or a nautical mile) in several hours. In two articles communicated eleven years ago to the *Philosophical Magazine**, I explained the principles on which such an electromotive force as this is to be compared with familiar standards, as, for instance, that of an element of Daniell's battery. Thus a horizontal conductor, 1,400,000 feet long (or about 270 British statute miles), carried at the rate of 600 feet (or about one-tenth of a minute of arc) per hour in a horizontal direction perpendicular to its own length across the British Isles or neighbouring Atlantic ocean (where the vertical magnetic force averages about 10 British absolute units of magnetic force), would experience an induced electromotive force amounting to $\frac{600}{3600} \times 10 \times 1,400,000$, or about 2,300,000 British absolute units of electromotive force. But, as I showed in the second of those articles, the electromotive force of a single cell of Daniell's is about 2,500,000 British absolute units. Hence the induced electro-magnetic force in question is about equal to that of a single cell. Some such electromotive force as this, therefore, must be induced in a length of 270 miles of the earth's crust, by the ordinary diurnal variations of terrestrial magnetism; and the much more rapid variations in magnetic storms must produce much greater electromotive forces, which we may conceive may not unfrequently be as much as that of ten or twenty cells, and sometimes may amount to 100 cells or more. Just such amounts of electromotive force were those which I actually observed in the Atlantic cable, as the following extract from the *Encyclopædia Metropolitana*, article "Telegraph, electric," shows.

"In the failure of the Atlantic Cable in September 1858, the portion terminating at Valencia came to give nearly the same indications as an insulated conductor about 270 miles long, laid out westward, and connected with a copper plate sunk at a little less than that distance in the Atlantic. In these circumstances the writer found that from 1 to 9 or 10 twentieths of the electromotive force of two Daniell's elements was generally sufficient to balance the earth-current; not unfrequently 14 or 15 were required; sometimes, although rarely, 20, or the full electromotive

---

* See 1851, Second half year. "On the Mechanical Theory of Electrolysis," and "Applications of the Principle of Mechanical Effect to the Measurement of Electromotive Forces, and of Galvanic Resistances, in Absolute Units." [*Math. and Phys. Papers*, Vol. I. pp. 472—503.]

force of two Daniell's elements, was insufficient; and once or twice in the course of the month of September, earth-currents were received so strong that five or six Daniell's elements would have been required to balance them."

It seems therefore quite certain that the ordinary every-day earth-currents in that locality must be very sensibly influenced by electro-magnetic induction from the ordinary diurnal variations of terrestrial magnetism; but it is also quite certain that they are only in part due to this cause, and that some more powerful, but as yet unknown, agency is at work to produce them. For although I found that a day seldom, if ever, passes without the direction of the current changing several times, yet there was no relation between the times of such changes and the solar hours. I conclude with the following additional extract from the same article, expressing views regarding earth-currents which I think will be found to agree with the extensive and careful observations of Mr C. V. Walker, which have been published since it was written, although they seem quite at variance with the theory which has recently been advocated by Prof. Lamont and Dr Lloyd, that earth-currents, however they are themselves generated, do directly produce the magnetic variations.

"Earth-currents are certainly related to the irregular variations of terrestrial magnetism, since they are always found unusually strong during brilliant displays of aurora borealis; for it has long been known that, on these occasions, the magnetic disturbances are unusually strong. Being related to the variations of terrestrial magnetism, it is probable that the earth-currents also will be found to have daily periods; but, in the mean time, we only know that, while the diurnal variation in terrestrial magnetism is observable in general every day, and is only on rare occasions overborne by irregular disturbances, the earth-currents vary each day from hour to hour, like the wind, under some overpowering non-periodic influence, and can only show daily periodicity in residual averages derived from lengthened series of observations. It is probable that careful synchronous observations of auroras, earth-currents, and variations of terrestrial magnetism, will lead to a discovery of the primary influence, whether in the earth, or terrestrial atmosphere, or surrounding interplanetary air, which causes these phenomena."

188. EFFECT OF UNIFORMLY IMPERFECT INSULATION ON SIGNALS THROUGH SUBMARINE CABLES.

[From *Phil. Trans.* Vol. CLII. pt. ii. [read Jan. 14, 1863], pp. 1011—1017.]

[Appendix to F. Jenkin's paper on "Experimental Researches on the Transmission of Electric Signals through Submarine Cables." Not reprinted.]

189. ON THE RESULT OF REDUCTIONS OF CURVES OBTAINED FROM THE SELF-RECORDING ELECTROMETER AT KEW.

[From *British Association Report*, 1863, p. 27. Abstract.]

THE author said, that all the photographs up to last March had been reduced to numbers, and the monthly averages taken. Each month shows a maximum in the morning, sometimes from 7 to 9 A.M., and another in the evening, from 8 to 10 P.M. There are pretty decided indications of an afternoon maximum, and another in the small hours after midnight, but the irregularities are too great to allow any conclusion to be drawn from a mere inspection of the monthly averages. He intended to calculate three terms, if not more, of the harmonic series for each month, and thus be able to judge whether the observations show any consistence in a third term (which alone would give four maxima and four minima), or a first term (which alone would give one maximum and one minimum in the twenty-four hours). There is a very decided winter maximum and summer minimum on the daily averages. That for January is more than double of that for July. This part of the subject will also require much labour to work it out. In the reductions hitherto made he had included negatives with positives, and all the sums have been "algebraic" (*i.e.* with the negative terms subtracted). Very important results with reference to meteorology will, no doubt, be obtained by examining the negative indications separately; and, again, by taking daily and monthly averages of the *fine-weather readings* alone. This part of the subject he had not been able to attack at all yet. Nor had he yet been able to go through a comparison of the amounts of effect with wind in different quarters.

190. On Electrically impelled and Electrically
controlled Clocks.

[From *Glasg. Phil. Soc. Proc.* Vol. VI. [read Jan. 24, 1866], pp. 61—64.]

Professor William Thomson exhibited and explained an
improved arrangement of magnets for the electro-magnets and
coil used in the Jones's system of controlling clocks electrically.
The object aimed at in the system is to force the time of vibration
of the controlled pendulum to agree with that of the controlling
clock, and to interfere as little as possible with the extent of
range through which the controlled pendulum vibrates.   To
produce such an effect the proper place for the controlling force
to act is as near as possible to the ends of the range through
which the pendulum swings.   A force acting on a pendulum
when it is near or passing through its middle position tends
either to stop it or to make it vibrate more widely.   Accordingly,
the middle position is the proper place for the impelling force to
act, by which the pendulum is kept vibrating; and this is what
is put in practice in the dead-beat escapement, also in Bain's
electrical pendulum.   Bain's arrangement of coil and magnets,
which was introduced by him for the purpose of keeping a
pendulum, and through it the works of a clock, vibrating by the
energy of an electric current, proved to be a very convenient
way of applying electro-magnetic force to a pendulum.   It was
accordingly adopted by Mr Jones for the very different purpose
contemplated in his admirable invention, which was to regulate a
clock kept going by its own weights, and compel it to agree in
its rate with the distant controlling clock.   In Bain's arrangement two bar-magnets are used, fixed in a horizontal line, with
their north poles together.   The pendulum carries at its lower
end a coil or bobbin of silk-covered copper wire, with a wide
aperture or core, through the centre of which the line of the
magnets passes.   The pendulum in vibrating carries this coil

from one side of the double north pole to the other in every vibration, and the mutual action between the double pole and alternating electric currents through the coil gives rise to horizontal forces alternately in opposite directions, by which the pendulum is kept vibrating, while resisted by the air, friction in the clock-work, &c. The distant south poles may be regarded as practically inoperative, because of their comparatively great distance from the coil; but the small practically insensible forces which they produce are, however, at each instant opposed to the true resulting effect produced by the central north poles. Whichever way the current flows, this arrangement gives the maximum of force to the pendulum at the middle of its arc; but it gives enough of force, when the pendulum is near the ends of its range, to produce the desired controlling power, with very moderate and cheap electric force, as is proved by the success of Mr Jones's plan, even without the improvements of arrangement which have been introduced in Glasgow. The first of these improvements consists in drawing out the magnets, and fixing them so that their north poles shall be just reached, or barely covered by the coil, as the pendulum swings to either end of its range. The effect of this is to increase the electro-magnetic force where it is wanted—that is, at the ends of the range—and to diminish it in the middle, where it is not only not wanted, but has sometimes been found in practice very detrimental. It must happen occasionally that by some accident, whether the telegraph wires be carried away by a storm, or some temporary stoppage or failure occurs in the controlling-clock-contacts or battery, the controlling current is interrupted for a time. During such an interruption the controlled clocks are simply left each to go at its own rate as an ordinary clock, and in the course of a few hours any one of them may gain or lose several seconds. When the current is re-established, they ought all immediately to submit again to the control, and go on each with the few seconds gain or loss it may have made while it was left to itself. But if the pendulum of any one of them chances to be vibrating on the whole against, or considerably more *against* than *with*, the re-established alternating electro-magnetic force, it may be stopped because of the greatness of this force on the pendulum in or near its middle position; and after a time (several minutes in general) it might be started again in the proper direction by the timed impulses which it constantly receives. (This curious phenomenon was illustrated by experiments performed before the Society.) Accordingly, after accidental stoppages of the current, some of the controlled clocks, during the first few months of the Glasgow experiment, were more than once found going, but several minutes slow. After the magnets were drawn out, according to the theory explained above, no such derangement has ever been experienced; and the greatest derangement that a stoppage of the current has caused in any one of the controlled clocks seems to have been two seconds, four seconds, or six seconds. Such a derangement can mislead no one, as the University of Glasgow requires, as the sole condition under which it will supply electric time-signals in any case, that a galvanometer showing (by the lost deflection on the sixtieth second) the precise moment of the beginning of each minute shall be exhibited with every clock under control which has a seconds' hand; and no one who is so exact as to look to seconds will omit to verify by the galvanometer the time-signal he takes.

A further improvement consists in as it were bending the bar-magnets round into the horse-shoe shape, so as to bring their south poles from the positions in the axis of the coil where they do damage, mitigated only by the greatness of their distances from the field of action, to proper positions close outside the coil at each end of its swing, where they act forcibly at close quarters in aid of the useful action of the north poles when the coil is at either end of its range, and yet mitigate very effectively its detrimental action on the coil in the middle part of its swing. One of the many new clocks now in the course of construction by Messrs Mitchell, to be controlled from Professor Grant's mean time clock in the University Observatory, had been so nearly completed, with the proper form of coil and disposition of horse-shoe magnets on the new plan as to allow the action to be exhibited to the Society; but Professor Thomson had only had it to try for an hour before the meeting of the Society. He hoped in the course of a few weeks to make experiments on it which would allow him to report to the Society—

1. The number of seconds per day of error in its uncontrolled rate from which a definitely specified amount of electric current would hold it controlled.

2. The greatest number of seconds per day of error which could be checked in it by any current incapable of stopping it by catching it the wrong way.

This improvement is even more important for the two seconds' pendulum of a turret clock than for the ordinary seconds' pendulum. After nearly a year of experiments and trials of coils and magnets more and more powerful, the 2½ cwt. pendulum of St George's Church clock had, by a proper disposition of merely bar-magnets, been brought under control in the same circuit as the other clocks, all of which have seconds' pendulums, without the necessity for a special wire and modified system of currents, as specified by Mr Jones in his patent. It may be confidently expected that the new improvement now brought forward will render the control of other heavy two seconds' pendulums a much easier matter (that is to say, much more stringent with the same amount of electric coil); and it is to be hoped that the example first set in Scotland by the University of Glasgow, in putting the tower clock under control, will be followed not only in St George's Church clock, but very soon in every public clock in Glasgow, Paisley, Greenock, and Ayr, and even in Edinburgh, notwithstanding the time gun.

191. On a new Form of the Dynamic Method for Measuring the Magnetic Dip.

[From *Manchester Phil. Soc. Proc.* Vol. VI. 1867 [read April 16, 1867], pp. 157, 158; *Manchester Phil. Soc. Mem.* Vol. III. 1868, pp. 291, 292.]

Seven years ago an apparatus was constructed for the natural philosophy class of the University of Glasgow for illustrating the induction of electric currents by the motion of a conductor across the lines of terrestrial magnetic force. This instrument consisted of a large circular coil of many turns of fine copper wire, made to rotate by wheelwork about an axis, which can be set to positions inclined at all angles to the vertical. A fixed circle, parallel to the plane containing these positions, measured the angles between them. The ends of the coil were connected with fixed electrodes, so adjusted as to reverse the connexions every time the plane of the coil passes through the position perpendicular to that plane. When in use, the instrument should be set as nearly as may be in the magnetic meridian. The fixed electrodes being joined to the two ends of a coil of a delicate galvanometer, a large deflection is observed when the axis of rotation forms any considerable angle with the line of magnetic dip. On first trying the instrument I perceived that its sensibility was such as to promise an extremely sensitive means for measuring the dip. Accordingly, soon after I had a small and more portable instrument constructed for this special purpose; but up to this time I had not given it any sufficient trial. On the occasion of a recent visit, Dr Joule assisted at some experiments with this instrument. The results have convinced us both that it will be quite practicable to improve it so that it may serve for a determination of the dip within a minute of angle. I hope, accordingly, before long to be able to communicate some decisive results to the Society, and to describe a convenient instrument which may be practically useful for the observation of this element.

192. MODIFICATION OF WHEATSTONE'S BRIDGE TO FIND THE
RESISTANCE OF A GALVANOMETER-COIL FROM A SINGLE
DEFLECTION OF ITS OWN NEEDLE.

[From *Roy. Soc. Proc.* Vol. XIX. 1871 [read Jan. 19, 1871], p. 253;
Phil. Mag. Vol. XLI. 1871 (Supplement), pp. 537, 538.]

IN any useful arrangement in which a galvanometer or electrometer and a galvanic element or battery are connected, through
whatever trains or network of conductors, let the galvanometer
and battery be interchanged. Another arrangement is obtained
which will probably be useful for a very different, although re-
ciprocally related object. Hence, as soon as I learned from
Mr Mance his admirable method of measuring the internal
resistance of a galvanic element (that described in the first of
his two preceding papers), it occurred to me that the reciprocal
arrangement would afford a means of finding the resistance of a
galvanometer-coil, from a single deflection of its own needle, by a
galvanic element of unknown resistance. The resulting method
proves to be of such extreme simplicity that it would be incredible
that it had not occurred to any one before, were it not that I fail
to find any trace of it published in books or papers; and that
personal inquiries of the best informed electricians of this country
have shown that, in this country at least, it is a novelty. It
consists simply in making the galvanometer-coil one of the four
conductors of a Wheatstone's bridge, and adjusting, as usual, to
get the zero of current when the bridge contact is made, with
only this difference, that the test of the zero is not by a galvanometer in the bridge showing no deflection, but by the galvanometer itself, the resistance of whose coil is to be measured,
showing an unchanged deflection. Neither diagram nor further
explanation is necessary to make this understood to any one who
knows Wheatstone's bridge.

193. On a Constant Form of Daniell's Battery.

[From *Roy. Soc. Proc.* Vol. xix. 1871 [read Jan. 19, 1871], pp. 253—259 ;
Phil. Mag. Vol. xli. 1871 (Supplement), pp. 538—543; *Nature*, Vol. iii.
March 2, 1871, pp. 350, 351.]

GRAHAM'S discovery of the extreme slowness with which one
liquid diffuses into another, and Fick's mathematical theory of
diffusion, cannot fail to suggest that diffusion alone, without intervention of a porous cell or membrane, might be advantageously
used for keeping the two liquids of a Daniell's battery separate.
Hitherto, however, no galvanic element without some form of
porous cell, membrane, or other porous solid for separator, has
been found satisfactory in practice.

The first idea of dispensing with a porous cell, and keeping
the two liquids separate by gravity, is due to Mr C. F. Varley, who
proposed to put the copper-plate in the bottom of a jar, resting
on it a saturated solution of sulphate of copper, resting on this a
less dense solution of sulphate of zinc, and immersed in the sulphate
of zinc the metal zinc-plate fixed near the top of the jar. But
he tells me that batteries on this plan, called "gravity-batteries,"
were carefully tried in the late Electric and International Telegraph
Company's establishments, and found wanting in economy. The
waste of zinc and of sulphate of copper was found to be more in
them than in the ordinary porous-cell batteries. Daniell's batteries
without porous cells have also been tried in France, and found
unsatisfactory on account of the too free access of sulphate of
copper to the zinc, which they permit. Still, Graham's and Fick's
measurements leave no room to doubt but that the access of
sulphate of copper to the zinc would be much less rapid if by true
diffusion alone, than it cannot but be in any form of porous-cell
battery with vertical plates of copper and zinc opposed to one
another, as are the ordinary telegraphic Daniell's batteries which
Mr Varley finds superior to his own "gravity-battery." The

comparative failure of the latter, therefore, must have arisen from mixing by currents of the liquids. All that seems necessary, therefore, to make the gravity-battery much superior instead of somewhat inferior to the porous-cell battery, is to secure that the lower part of the liquid shall always remain denser than the upper part. In seeking how to realize this condition, it first occurred to me to take advantage of the fact that saturated solution of sulphate of zinc is much denser than saturated solution of sulphate of copper. It seems* that, at temperature 15° C., saturated aqueous solution of sulphate of copper is of 1·186 sp. gr, and contains in every 100 parts of water 33·1 parts of the crystalline salt; and that at 15° the saturated solution of sulphate of zinc is of sp. gr. 1·44, and contains in every 100 parts of water 140·5 parts of sulphate of zinc, both results being from Michel and Krafft's experiments†. Hence I made an element with the zinc below; next it saturated solution of sulphate of zinc, gradually diminishing to half strength through a few centimetres upwards; saturated sulphate of copper resting on this; and the copper-plate fixed above in the sulphate-of-copper solution. In the beginning, and for some time after, it is clear that the sulphate of copper can have no access to the zinc otherwise than by true diffusion. I have found this anticipation thoroughly realized in trials continued for several weeks; but the ultimate fate of such a battery is that the sulphate of zinc must penetrate through the whole liquid, and then it will be impossible to keep sulphate of copper separate in the upper part, because saturated solution of sulphate of zinc certainly becomes denser on the introduction of sulphate of copper to it. To escape this chaotic termination I have introduced a siphon of glass with a piece of cotton-wick along its length inside it, so placed as to draw off liquor very gradually from a level somewhat nearer the copper than the zinc; and a glass funnel, also provided with a core of cotton-wick, by which water semisaturated with sulphate of zinc may be continually introduced at a somewhat lower level. A galvanic element thus arranged will undoubtedly continue remarkably constant for many months; but it has one defect, which prevents me from expecting permanence for years. The zinc being below, must sooner or later, according

* Storer's *Dictionary of Solubilities of Chemical Substances.* Cambridge, Massachusetts: Sever and Francis, 1864.

† *Ann. Ch. et Phys.* (3) Vol. xli. pp. 478, 482: 1854.

to the less or greater vertical dimensions of the cell, become covered with precipitated copper from the sulphate of copper which finds its way (however slowly) to the zinc. On the other hand, if the zinc be above, the greater part of the deposited copper falls off incoherently from the zinc through the liquid to the copper below, where it does no mischief, provided always that the zinc be not amalgamated,—a most important condition for permanent batteries, pointed out to me many years ago by Mr Varley. Placing the zinc above has also the great practical advantage that, even when after a long time it becomes so much coated with metallic copper as to seriously injure the electrical effect, it may be removed, cleaned, and replaced without otherwise disturbing the cell; whereas if the zinc be below, it cannot be cleaned without emptying the cell and mixing the solutions, which will entail a renewal of fresh separate solutions in setting up the cell again. I have therefore planned the following form of element, which cannot but last until the zinc is eaten away so much as to fall to pieces, and which must, I think, as long as it lasts, have a very satisfactory degree of constancy.

The cell is of glass, in order that the condition of the solutions and metals which it contains may be easily seen at any time. It is simply a cylindrical or rectangular jar with a flat bottom. It need not be more than 10 centimetres deep; but it may be much deeper, with advantage in respect to permanence and ease of management, when very small internal resistance is not desired. A disk of thin sheet copper is laid at its bottom. A properly shaped mass of zinc is supported in the upper part of the jar. A glass tube (which for brevity will be called the charging-tube) of a centimetre or more internal diameter, ending in a wide saucer or funnel above, passes through the centre of the zinc, and is supported so as to rest with its lower open end about a centimetre above the copper. A glass siphon with cotton-wick core is placed so as to draw liquid gradually from a level about a centimetre and a half above the copper. The jar is then filled with semisaturated sulphate-of-zinc solution. A copper wire or stout ribbon of copper coated with india-rubber or gutta-percha passes vertically down through the liquid to the copper-plate below, to which it is riveted or soldered to secure metallic communication. Another suitable electrode is kept in metallic communication with the zinc above. To put the cell in action, fragments of sulphate of copper, small enough to fall down through the charging-tube, are placed in the funnel above. In the course of a very short time the whole liquid below the lower end of the charging-tube becomes saturated with sulphate of copper, and the cell is ready for use. It may be kept always ready by occasionally (once a week for instance) pouring in enough of fresh water, or of water quarter saturated with sulphate of zinc at the top of the cell, to replace the liquid drawn off by the siphon from near the bottom. A cover may be advantageously added above, to prevent evaporation. When the cell is much used, so that zinc enough is dissolved, the liquid added above may be pure water; or if large internal resistance is not objected to, the liquid added may be pure water, whether the cell has been much used or not; but after any interval, during which the battery has not been much in use, the liquid added ought to be quarter saturated, or even stronger solution of sulphate of zinc, when it is desired to keep down the internal resistance. It is probable that one or more specific-gravity beads kept constantly floating between top and bottom of the heterogeneous fluid will be found a useful adjunct, to guide in judging whether to fill up with pure water or with sulphate-of-zinc solution. They may be kept in a place convenient for observation by caging them in a vertical glass tube perforated sufficiently to secure equal density in the horizontal layers of liquid, to be tested by the floaters.

An extemporized cell on this plan was exhibited to the Royal Society, and its resistance (measured as an illustration of Mance's method, described in the first of his two previous communications) was found to be ·29 of an Ohm (that is to say, 290,000,000 centimetres per second). The copper and zinc plates of this cell, being circular, were about 30 centimetres in diameter, and the distance between them was about 7·5 centimetres. A Grove's cell, of such dimensions that forty in series would give an excellent electric light, was also measured for resistance, and found to be ·19 of an Ohm. Its intensity was found to be 1·8 times that of the new cell, which is the usual ratio of Grove's to Daniell's; hence seventy-two of the new cells would have the intensity of forty of Grove's. But the resistance of the seventy-two in series would be 209 Ohms, as against 76 Ohms of the forty Grove's; hence, to get as powerful an electric light, threefold surface, or else diminished resistance by diminished distance of the plates, would be required. How much the resistance may be diminished by diminishing the distance rather than increasing the surface, it is impossible to deduce from experiments hitherto made.

Two or three cells, such as the one shown to the Royal Society, will be amply sufficient to drive a large ordinary turret-clock without a weight; and the expense of maintaining them will be very small in comparison with that of winding the clock. The prime cost of the heavy wheel-work will be avoided by the intro- duction of a comparatively inexpensive electro-magnetic engine. For electric bells, and all telegraphic testing and signalling on shore, the new form of battery will probably be found easier of management, less expensive, and more trustworthy than any of the forms of battery hitherto used. For use at sea, it is probable that the sawdust Daniell's, first introduced on board the 'Agamemnon' in 1858, and ever since that time very much used both at sea and on shore, will still probably be found the most convenient form; but the new form is certainly better for all ordinary shore uses.

The accompanying drawing represents a design suitable for the electric light, or other purposes, for which an interior resistance

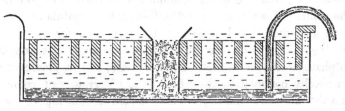

not exceeding $\frac{2}{10}$ of an Ohm is desired. The zinc is in the form of a grating, to prevent the lodgment of bubbles of hydrogen gas,

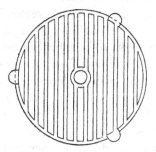

which I find constantly, but very slowly, gathering upon the zincs of the cells I have tried, although the solutions used have no free

acid, unless such as may come from the ordinary commercial
sulphate-of-copper and commercial sulphate-of-zinc crystals which
were used.

POSTSCRIPT.   February 2, 1871.

The principle which I have adopted for keeping the sulphate
of copper from the zinc is to allow it no access to the zinc except
by true diffusion.   This principle would be violated if the whole
mass of the liquid contiguous to the zinc is moved toward the
zinc.   Such a motion actually takes place in the second form of
element (that which is represented in the drawing, and which is
undoubtedly the better form of the two) every time crystals
of sulphate of copper are dropped into the charging-tube.   As the
crystals dissolve, the liquid again sinks, but not through the whole
range through which it rose when the crystals were immersed.
It sinks further as the sulphate of copper is electrically precipitated
on the copper plate below in course of working the battery.
Neglecting the volume of the metallic copper, we may say, with
little error, that the whole residual rise is that corresponding to
the volume of water of crystallization of the crystals which have
been introduced and used.   It becomes, therefore, a question
whether it may not become a valuable economy to use anhydrous
sulphate of copper instead of the crystals; but at present we are
practically confined to the "blue vitriol" crystals of commerce,
and therefore the quantity of water added at the top of the cell
from time to time must be, on the whole, at least equal to the
quantity of water of crystallization introduced below by the
crystals.   Unless a cover is added to prevent evaporation, the
quantity of water added above must exceed the water of crystallization introduced below by at least enough to supply what has
evaporated.   There ought to be a further excess, because a downward movement of the liquid from the zinc to the level from
which the siphon draws is very desirable to retard the diffusion
of sulphate of copper upwards to the zinc.   Lastly, this downward
movement is also of great value to carry away the sulphate of
zinc as it is generated in the use of the battery.   The quantity
of water added above ought to be regulated so as to keep the
liquid in contact with the zinc a little less than half saturated
with sulphate of zinc, as it seems, from the observations of various

experimenters, that the resistance of water semisaturated with sulphate of zinc is considerably less than that of a saturated solution. A still more serious inconvenience than a somewhat increased resistance has been pointed out to me by Mr Varley as a consequence of allowing sulphate of zinc to accumulate in the battery. Sulphate of zinc crystallizes over the lip of the jar, and forms pendents like icicles outside, which act as capillary siphons, and carry off liquid. Mr Varley tells me that this curious phenomenon is not unfrequently observed in telegraph-batteries, and sometimes goes so far as to empty a cell and throw it altogether out of action. Even without this extreme result, the crystallization of zinc about the mouth of the jar is very inconvenient and deleterious. It is of course altogether avoided by the plan I now propose.

In conclusion, then, the siphon-extractor must be arranged to carry off all the water of crystallization of the sulphate of copper decomposed in the use of the cell, and enough of water besides to carry away as much sulphate of zinc as is formed in the use of the battery. Probably the most convenient mode of working the system in practice will be to use a glass capillary siphon, drawing quickly enough to carry off in a few hours as much water as is poured in each time at the top; and to place, as shown in the drawing, the discharging end of the siphon so as to limit the discharge to a level somewhat above the upper level of the zinc grating. It will no doubt be found convenient in practice to add measured amounts of sulphate of copper by the charging-tube each time, and at the same time to pour in a measured amount of water, with or without a small quantity of sulphate of zinc in solution.

As 100 parts by weight of sulphate of copper crystals contain, as nearly as may be, 36 parts of water, it may probably answer very well to put in, for every kilogramme of sulphate of copper, half a kilogramme of water. Experience (with the aid of specific-gravity beads) will no doubt render it very easy, by a perfectly methodical action involving very little labour, to keep the battery in good and constant action, according to the circumstances of each case.

When, as in laboratory work, or in arrangements for lecture-illustrations, there may be long intervals of time during which the battery is not used, it will be convenient to cease adding sulphate of copper when there is no immediate prospect of action being required, and to cease pouring in water when little or no colour of sulphate of copper is seen in the solution below. The battery is then in a state in which it may be left untouched for months or years. All that will be necessary to set it in action again will be to fill it up with water to replace what has evaporated in the interval, and stir the liquid in the upper part of the jar slightly, until the upper specific-gravity bead is floated to near the top by sulphate of zinc, and then to place a measured amount of sulphate of copper in the funnel at the top of the charging-tube.

194. A Hint to Electricians [on the necessity of Scientific Standardising].

[From *Nature*, Vol. III. Jan. 26, 1871, pp. 248, 249.]

Mr Mance's method for measuring the internal resistance of a single galvanic element or battery, communicated to the Royal Society at its meeting of last week, and the modifications of Wheatstone's bridge suggested by myself for finding the resistance of a galvanometer coil from the deflection of its own needle, supply desiderata in respect to easy and rapid measurement, which have been long *felt* by telegraph electricians and *needed* by other scientific investigators and by teachers of science. Year after year the latter, in their arrangement of batteries, electrodes, and galvanometers, have darkly and wastefully followed the method which from workmen we learn to call rule of thumb; while the former, with admirable scientific art, measure every element with which they are concerned, in absolute measure. How many physical professors are there in Europe or America who could tell (in millions of centimetres per second) the resistance of any one of the galvanometers, induction coils, or galvanic elements which they are daily using? How many of them, in ordering an electro-magnet, require of the maker that the specific resistance of the copper shall not exceed 16,000 (gramme-centi-metre-seconds)? How many times have eight Grove cells been set up to produce a degree of electro-magnetic effect which four would have given, had the professor exacted of the instrument-maker the fulfilment of a simple and inexpensive scientific condition, as submarine telegraph companies have done in their specifications of cables? If every possessor of an electro-magnet were to cut a metre off its coil, weigh the piece, measure its resistance, and send the result to *Nature*, and if every maker of Ruhmkorff coils would do the like for every coil of copper wire designed for his instruments, a startling average might be shown. And what of the items? I venture to say that (provided the instruments of the great makers are not excluded) specific resistance above 30,000 would not be a singular case. I could tell something of galvanometers of 1869, comparable only to submarine cables of 1857. I refrain:—but let makers of galvanometers, Ruhmkorff coils, and electro-magnets beware; surely *Nature* will find them out if they do not reform before 1872.

195. Note on Mr Gore's Paper on Electrotorsion.

[From *Phil. Trans.* Vol. CLXIV. 1874, pp. 560—562*.]

In Section 5, " General cause of the torsions," the phenomena are attributed to the combined influence of ordinary magnetic polarity and the magnetic condition of iron at right angles to that. To see precisely how this combined influence produced the results discovered by Mr Gore, we have only to look to Joule's discovery of the effects of magnetism on the dimensions of iron and steel bars, and of the musical sounds consequently produced in an electromagnet every time battery-contact is made or broken. This great discovery was first described in public on the occasion of a *conversazione* held at the Royal Victoria Gallery of Manchester, on February 16th, 1842. A printed account of it is to be found in the eighth volume of Sturgeon's *Annals of Electricity*, p. 219, and in the *Philosophical Magazine*, 1847, first half year. The following are the chief results obtained by Joule :—

1. When a wire or bar of iron (or steel) is alternately subjected to, and left free from, longitudinal magnetizing force, it alternately becomes longer and shorter.

2. In the same circumstances its volume remains sensibly unaltered; and therefore it experiences lateral shrinking to an extent equal to half the extension in length†.

3. Joule verified the lateral shrinking by passing a current through an insulated wire along the axis of a piece of iron gas-pipe‡, 1 yard long, $\frac{3}{16}$ of an inch in bore, and $\frac{3}{16}$ of an inch in

---

* Not read before the Society, but ordered to be printed.

† It is understood, of course, that the shrinking is reckoned in proportion to the transverse diameter, and the extension in proportion to the length, as is usual in the geometry of strains and in the theory of elasticity.

‡ The bends of the insulated wire outside the gas-pipe in Joule's experiment complicate the circumstances somewhat by superimposing upon the circular poleless magnetization, which a single straight wire along the axis of the pipe would pro-

thickness, and found, as he anticipated, that the *length* of the gas-pipe became *diminished* when the current was instituted, and *increased* when the current was stopped.

4. Residual magnetism leaves residual changes of dimension in iron and steel of the same signs as those exhibited when magnetizing force is first applied or afterwards re-applied.

5. Longitudinal pull*, if sufficiently intense, reduces to zero the magnetic extensions and contractions; and if more intense still, puts the metal into such a state that opposite strains are produced by it. An iron or steel wire stretched vertically by a small weight becomes elongated by magnetization, but if kept stretched by a constant sufficiently heavy weight it is shortened by magnetization†.

Now the passage of a current along a straight iron or steel wire of circular section gives rise to poleless magnetization in circles perpendicular to the length of the wire and with their centres in its axis. Let γ be the strength of the current through the wire, reckoned of course in absolute units. If the wire be infinitely long, the resulting field of force (whether the wire be of iron or of any other metal) is fully specified by saying that the lines of

duce, magnetization in which there is northern polarity along one semicylinder, and southern polarity along the other semicylinder of the outer boundary of the iron pipe, and fainter opposite polarities on the inner cylindrical surface. But if the wire had been continued straight for several inches outside the pipe at each end, and then carried away to the battery without ever being brought near the gas-pipe externally, it is clear that effects in the same direction, though of slightly less magnitude (by an almost infinitesimal difference), would have been observed.

* Rankine's nomenclature regarding stresses and strains (which is consistent with Huyghens' celebrated *ut tensio sic vis*) ought to be carefully followed. It is therefore necessary to introduce two nouns, pull and thrust, common enough in familiar language, but not hitherto much used in the theory of elasticity, to express longitudinal forces in the directions which would elongate or shorten the bar or wire. With reference to a stretched wire we ought to talk of the pull along the wire, and ought not to use the word strain or tension to express a stretching force. The only objection to the word pull is that some people might consider it too familiar; but surely it is not a valid objection to the mathematician or philosopher that a word, the use of which enables him to avoid ambiguity in scientific statements, is already understood by non-scientific people. According to Rankine's nomenclature we must confine the word strain to a change of dimension or figure caused by stress; thus the longitudinal strain of a wire or of a beam experiencing a pull or thrust is the (positive or negative) elongation produced by the force.

† Hence the "Young's modulus" of iron or steel is increased by longitudinal magnetization.

force are circles in planes perpendicular to the axis and having their centres in this line, and that the intensity of the force is

$$(2\gamma/a^2)\, r \text{ for points in the substance of the wire,}$$

and $\qquad\qquad 2\gamma/r$ for external points,

where a denotes the radius of the wire, and r denotes distance from its axis. (The same description, as is now well known— thanks to the beautiful illustrations and diagrams of iron-filings by which Faraday showed it in the Royal Institution—is approxi- mately applicable to the field of force in the neighbourhood of a straight portion of wire conveying a current, provided no other part of the wire is near.) Hence the intensity of the magnetiza- tion of the substance is equal to

$$\mu\,(2\gamma/a^2)\, r,$$

where μ denotes the "magnetic susceptibility"* of the substance. Let now a uniform magnetizing force X be applied along the whole length of the wire. This, combined with the force due to the current through the wire, gives at any point of the substance a resultant force equal to $\sqrt{\{(4\gamma^2 r^2/a^4) + X^2\}}$ in a direction inclined to the length of the wire at the angle whose tangent is $2\gamma r/a^2 X$. The lines of force in the resultant field are therefore spirals. The wire being supposed infinitely long, the magnetization will still be poleless†, and will be everywhere in the direction of the resultant force; and its intensity will be equal to the resultant force multiplied by the magnetic susceptibility. The extension of the substance along the spiral lines of magnetization, and its shrinkage along the orthogonal spirals, to be anticipated from Joule's old results, give rise to Gore's phenomena of electro- torsion.

Although we thus see Gore's "electrotorsion" as a geometrical consequence of the earlier discovery of Joule, we must, never- theless, regard Mr Gore's investigation as having led to an independent discovery of a remarkably interesting character, enhanced by the well-designed and necessarily laborious working out of varied details described in his paper.

---

* Thomson's reprint of *Electrostatics and Magnetism*, § 610. 3.

† The free polarity in the actual experiments due to the finiteness of the iron bar or wire and of the magnetizing helix reduces somewhat the magnitude of the effects, but does not alter their general character.

It is difficult to conceive any physical investigation, except Faraday's magnetic rotation of the plane of polarization of light, more important towards a physical theory of magnetism than Joule's result (No. 5) above. It suggests an interesting extension of Mr Gore's investigation. Let the wire rod or tube experimented upon be stretched by a heavy weight, and at the same time subjected to a constant twisting-couple of sufficient magnitude, then, no doubt, the torsions as well as the elongations observed under varying magnetic influences will be the reverse of those discovered by Mr Gore. The investigation ought, of course, to be varied by applying couple alone and longitudinal pull alone.

196. ELECTROLYTIC CONDUCTION IN SOLIDS. FIRST EXAMPLE, HOT GLASS.

[From *Roy. Soc. Proc.* Vol. XXIII. 1875 [read June 10, 1875], pp. 463, 464.]

MANY years ago I projected an experiment to test the voltaic relations between different metals with glass substituted for the electrolytic liquid of an ordinary simple voltaic cell, and with so high a temperature that the glass would have conducting-power sufficient to allow induction through it to rule the difference of potentials between the two metals. Imperfect instrumental arrangements, and want of knowledge of the temperature at which glass would have sufficient conductivity to give satisfactory results, have hitherto prevented me from carrying out the proposed investigation. The quadrant electrometer has supplied the first of these deficiencies, and Mr Perry's recent experiments* on the conductivity of glass at different temperatures the second. The investigation has now been resumed; and in a preliminary experiment I have already obtained a very decided result.

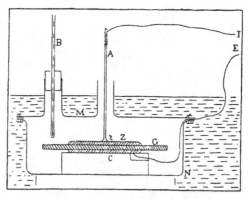

The drawing shows the arrangement adopted. *MN* is a brass case immersed in an oil-bath. A copper plate, *C*, of 5 centims.

* See *Proc. Roy. Soc.* Vol. XXIII. 1875, p. 468.

diameter, lies in the case on a block of wood; it is kept metallically connected with the outside case, E, of the electrometer. A
flint-glass plate, G, which is found to insulate very well at ordinary
temperatures, is laid upon C. A zinc plate, Z, lies on the glass,
and is connected with the insulated electrode, I, of the electrometer, by means of a wire attached to the end of a stout metallic
stem, Az, passing through the centre of an open vertical tube
reaching above the level of the oil. The glass was heated
gradually, and was usually kept between 100° and 120° C., the
temperature being measured by a thermometer, B.

Even below 50° C. there is a decided result, but shown less
rapidly than at higher temperatures. If the glass is kept at 50° C.
for some time, and I, after having been metallically connected
with C, is left insulated, it soon becomes sensibly charged; and
the charge increases till it is approximately equal to that acquired
when zinc and copper plates in a liquid electrolyte are metallically
connected with I and E respectively. With the hot glass, as with
the liquid electrolyte, the charge given by the zinc to the insulated
electrode of the electrometer is negative. The charge ultimately
reached when the temperature is 50° C. is not exceeded at higher
temperatures; but, as said above, when the zinc is connected
with the copper and then insulated, the charge increases towards
its ultimate value much more rapidly at higher temperatures
than at lower.

At temperatures between 100° and 120° C. there is a sensible
diminution of the ultimate charge after the zinc has been kept
for a short time connected with the copper and then insulated.
There is also a slow diminution of the ultimate, or, as we may
now call it, the temporarily static charge when the zinc plate is
left insulated for several hours in connexion with I.

If a small quantity of either negative or positive electricity
be given to I (always in metallic connexion with the zinc), the
temporarily static state is reached at about the same rate as the
zero would be reached by conduction through the hot glass
(according to Mr Perry's experiments, communicated to the
present Meeting) were the plates both of copper or both of zinc.

After the experiment the surfaces of the copper and zinc
plates in contact with the glass were found to be thickly oxidized.
The glass plate was quite cloudy after the experiment, and a

repetition of the experiment increased its cloudiness. This plate is the flattened bottom of a flint-glass electrometer-jar.

Three smoother glass plates, tried since, show as yet no signs of decomposition. At first they only became "exhausted" (in their power to produce the normal charge in zinc and copper) after the plates had been connected for nearly a day, the glass being at from 100° to 120° C.; but after a time, although they still gave the normal charge at the beginning of the morning's experiments, the charge fell to zero quite rapidly (that is, in about an hour), even when the zinc was kept insulated.

Keeping for a length of time the zinc charged negatively (so as to give to I a greater negative charge than that which it would have in the "temporarily static" condition of the copper, hot glass, and zinc) seemed to have no effect in restoring the normal electrolytic condition; but I propose to pursue this trial further, especially with longer time for the restorative electrification.

I propose also to return to similar experiments which I made many years ago on the electric relations of copper, ice, and zinc.

197. ELECTRICAL MEASUREMENT

[*Lecture before the Section of Mechanics at the Conferences held in connection with the Special Loan Collection of Scientific Apparatus at the South Kensington Museum, May* 17, 1876.]

[Reprinted in *Popular Lectures and Addresses*, Vol. I. pp. 423—454.]

198. RAPPORT SUR LES MACHINES MAGNÉTO-ÉLECTRIQUES GRAMME.

[Rapports présentées à l'Exposition de Philadelphie. From *Journ. de Phys.*
Vol. VI. Aug. 1877, pp. 240—242.]

PLUSIEURS spécimens des machines bien connues de Gramme sans aimant d'acier forment cette collection ; quelques-uns des appareils exposés ont fonctionné et produit de la lumière électrique.

Les particularités fondamentales de la machine Gramme méritent les plus vifs éloges ; la disposition excellente et très-originale des bobines mobiles enroulées sur un anneau de fer doux, entre les pôles d'un aimant, permet d'obtenir non-seulement une force électromotrice et une résistance presque parfaitement uniformes, mais aussi, je crois, une grande économie électrodynamique.

Les machines Gramme de cette collection sont intéressantes aussi comme exemple de la méthode introduite pour la première fois, je crois, par Siemens, et consistant à alimenter le magnétisme inducteur par le courant induit. Si l'on considère une telle machine avec son circuit complet, et que son fer soit dépourvu de tout magnétisme rémanent, si de plus elle n'est influencée par aucune action magnétique exterieure, pas même celle de la Terre, on peut la faire tourner avec une vitesse quelconque, aussi grande que l'on voudra, sans produire aucun courant électrique ni aucune aimantation. A moins que la vitesse ne dépasse une certaine limite, ce repos électrique et cette neutralité magnétique constituent un état d'équilibre stable ; mais, si la vitesse dépasse une certaine limite, qui dans la machine de Gramme, ou dans la machine de Siemens, est bien inférieure aux vitesses qu'on donne pratiquement à l'appareil, l'état de repos électrique et de neutralité magnétique constitue un équilibre instable. Une aimantation infiniment petite du fer produite sous l'influence de la moindre action magnétique extérieure, ou un courant d'une faiblesse extrême développé dans le circuit par une force électromotrice quelconque (par exemple celle qui peut être due à des différences de température entre les soudures des différents métaux du circuit) déterminerait immédiatement un courant dans une direction ou dans l'autre, et l'intensité augmenterait jusqu'à atteindre très-rapidement une limite dépendant de la vitesse de rotation.

La force électromotrice induite par le mouvement peut s'exprimer par la formule suivante :

$$N(M + CY),$$

dans laquelle N désigne le nombre de tours dans l'unité de temps, M une quantité dépendant de l'aimantation du fer, C une constante dépendant de la forme et de la dimension des bobines, et Y l'intensité du courant qui traverse actuellement le circuit. Si R est la résistance de ce circuit, nous devrons donc avoir

$$Y = \frac{N(M + CY)}{R}.$$

Supposons maintenant que l'aimantation du fer dépende uniquement de l'action magnétique du courant ; en négligeant le magnétisme rémanent, nous avons

$$M = C_1 IY,$$

où C_1 est une constante qui dépend de la machine, I la capacité inductrice magnétique du fer, ou une moyenne de la capacité inductrice magnétique efficace, quand le courant est assez intense (ce qui sera généralement le cas) pour rendre la capacité inductrice inégale dans les différents points. On doit, en général, considérer I comme une quantité variant avec Y, toujours positive, constante (soit I_0) pour de très-petites valeurs de Y, et qui enfin varie comme $1/Y$ quand Y est infiniment grand. Nous voyons ainsi que, pour de très-petites valeurs de Y, il ne peut y avoir équilibre que si $Y = 0$, et l'équilibre est instable, à moins que $N < R/(C_1 I_0 + C)$. Quand $N > R/(C_1 I_0 + C)$, le courant part d'une valeur très-petite, dans un sens quelconque, et croît jusqu'à ce que I devienne assez petit pour rendre $N = R/(C_1 I + C)$*.

* [Cf. *infra*, foot of p. 437 ; also the general graphical procedure as developed by J. and E. Hopkinson, *Phil. Trans.* 1886, or *Collected Papers* of J. Hopkinson, Vol. I. p. 84.]

S'il arrivait que l'on eût $N > R/C$, le courant augmenterait sans limite et les bobines fondraient inévitablement, quelque méthode de refroidissement que l'on puisse employer pour les maintenir froides; mais, pour les vitesses extrêmes que peut atteindre la machine dans la pratique, on a certainement $N < R/C$, et l'intensité du courant est pratiquement limitée par une réduction modérée de I vers son minimum qui est nul. La loi suivant laquelle l'intensité du courant augmente avec la vitesse de rotation de la machine est un sujet important pour des recherches scientifiques.

En outre de ces relations théoriques très-intéressantes, les machines Gramme exposées ont des applications pratiques à l'électro-métallurgie, à l'éclairage des manufactures, des phares et des navires, et serviront peut-être aussi, dans l'avenir, à l'éclairage des villes et des maisons d'habitation.

199. THE EFFECTS OF STRESS ON THE MAGNETIZATION
OF IRON, COBALT, AND NICKEL.

[From *Roy. Instit. Proc.* Vol. VIII. 1879 [May 10, 1878], pp. 591—593*.]

THOUGH, as discovered by Faraday, every substance has a
susceptibility for inductive magnetization, the three metals, iron,
nickel, cobalt, stand out so prominently from among the other
known chemical elements, that they only are commonly regarded
as *the* magnetic metals, and the magnetism of all other substances
is so feeble as to be comparatively almost imperceptible.

The magnetization of each of the three magnetic metals is
greatly affected by mechanical stress. From the beginnings of
magnetic science it must have been known that the magnetism
of iron and steel is disturbed, sometimes lost or much diminished,
by blows, striking the metal with a hammer, or letting it fall on
hard ground. Gilbert, nearly three hundred years ago, showed
that bars of soft iron held in the direction of the dipping needle
and struck violently by a hammer acquire much more magnetism,
and again more reverse magnetism when inverted in that line,
than when placed in those positions gently without shock of any
kind. An ordinary fireside poker shows these effects splendidly,
with no other apparatus than a little pocket-compass or a sewing
needle, magnetized and slung horizontally, hanging by a fine silk
thread. If habitually the poker rests upright the upper end will
be found a true north pole, the lower a true south when first
tested by the needle. Holding the poker with exceeding gentle-
ness, invert it:—The end that was down, though now up, is still a
true south pole, and repels the north end (or true south pole) of the
movable suspended needle. A gentle tap with the hand on the
poker now produces a surprising result. Instantly it yields to
the terrestrial influence, and its upper end, becoming a true north
pole, attracts the northern end of the suspended needle. Even

* [For the complete experimental investigation, see *Phil. Trans.* May 23, 1878;
reprinted in *Math. and Phys. Papers*, Vol. II. pp. 358—395 and pp. 403—407.]

more surprising is the slightness of the agitation which suffices
to shake the retained magnetism of a former position out of a
soft iron wire, and let it take the magnetization due to the
position in which it is held. A superstitious person would say—
that is animal magnetism!—when he sees an iron wire becoming
a notably effective magnet when held vertically and rubbed gently
from end to end between finger and thumb.

Changes of magnetism produced by mechanical agitation are
shown to a much greater degree in thin bars than in thick ones;
and when the diameter exceeds a quarter of the length they are
hardly sensible. Hence when the "Flinders bar" is applied to
compensate the error produced in a ship's compass by change of
magnetic latitude, its length ought not to be more than six or
seven times its diameter; and for the same reason long iron rods
or stanchions in the neighbourhood of a compass are very detrimental to its trustworthiness. Half a hundredweight of iron in
the shape of rails or awning stanchions, too often to be found very
near a compass, is more dangerous than tons or hundreds of
tons in the shape of heavy steam steering gear, or of armoured
turrets in an ironclad.

A piece of iron left in the Royal Institution by Faraday, with
a label in his own handwriting to the effect that it had been
between three hundred and four hundred years fixed in a vertical
position in the stonework of the Oxford Cathedral, having been
given to him by Dr Buckland in May, 1835, was exhibited and
tested. It was found to have its upper end a true north pole.
It was inverted before the audience, and instantly that end became
a true south pole and the other a true north pole. Thus nearly
four hundred years in one position had done nothing to *fix* the
magnetism. In its inverted position it was hammered violently
on each end by a wooden mallet: this increased the magnetism
somewhat, but did not *fix* it. The bar was inverted again, and
then, in its first position, its original upper end, now up again,
became again a true north pole.

The stoutness of the bar (that is to say, the greatness of the
proportions of its breadth and thickness to its length) were such,
that if of modern iron, it probably would not have behaved as it
did; but probably also it may have been superior in "softness" to
the ordinary run of modern bar iron.

Bars of nickel and cobalt, unique and splendid specimens, for which the speaker was indebted to the celebrated metallurgical chemist, Mr Wharton, of Philadelphia, were exhibited, and found to show effects of concussion quite as do bars of iron of different qualities.

An altogether new effect of stress was discovered about ten years ago by Villari, according to which longitudinal pull augments the temporary induced magnetism of soft iron bars or wires when the magnetizing force is less than a certain critical value; and diminishes it when the magnetizing force exceeds that value; and augments the residual magnetism when the magnetizing force, whether it has been great or small, has been removed.

The speaker had measured approximately the Villari critical value, and found it to be about twenty-four times the vertical component of the terrestrial magnetic force (or about 10 c.g.s. units). The maximum effect in the way of augmentation by pull he had found with about six times the Glasgow vertical force. He had found for bars of nickel and cobalt opposite effects to those of Villari for soft iron, and had found a maximum value, with a certain degree of magnetizing force, and evidence making it probable that a critical magnetizing force would be found for each of these metals also, such that the magnetization would be *increased* by pull when the magnetizing force exceeds it.

The speaker had found corresponding effects of *transverse pull* in soft iron, and had found them to be correspondingly opposite to those discovered by Villari for longitudinal pull. The transverse pull was produced by water pressure in the interior of a gun-barrel applied by a piston and lever at one end. Thus a pressure of about 1000 lbs. per square inch, applied and removed at pleasure, gave effects on the magnetism induced in the vertical gun-barrel by the vertical component of the terrestrial magnetic force, and, again, by an electric current through a coil of insulated copper wire round the gun-barrel, which were witnessed by the audience. When the force magnetizing the gun-barrel was any-thing less than about sixty times the Glasgow value of the vertical component of the terrestrial force, the magnetization was found to be less with the pressure on than off. When the magnet-izing force exceeded that critical value, the magnetization was *greater* with the pressure on than off. The residual (retained) magnetism was always less with the pressure on than off (after ten or a dozen "*ons*" and "offs" of the pressure to shake out as much of the magnetization as was so loosely held as to be shaken out by this agitation).

It is remarkable that the critical amount of the magnetizing force in respect to effect of transverse pull is more than double that of the Villari effect of longitudinal pull. Thus for intermediate amounts of force (say forces between 10 and 25 c.g.s. units), both longitudinal pull and transverse pull diminish the induced magnetization. Hence it is to be inferred that equal pull in all directions would diminish, and equal positive pressure in all directions would increase, the magnetization under the influence of force between these critical values, and through some range above and below them; and not improbably for all amounts, however large or small, of the magnetizing force; but further experiment is necessary to answer this question.

A most interesting further inquiry in connection with this subject is to find if æolotropic stress (pressure unequal in different directions), beyond the limits of elasticity, leaves in iron, nickel, or cobalt a permanent æolotropic difference of magnetic susceptibility in different directions analogous to that discovered thirty years ago by Tyndall in the diamagnetic quality of soft, imperfectly elastic material, such as fresh bread. Special difficulties prevented the speaker from obtaining any results thirty years ago, when he tried to discover corresponding effects in iron; but the investigation is not hopeless, and he intends to resume it.

200. ON THE EFFECT OF TORSION ON THE ELECTRICAL CONDUCTIVITY OF METALS.

[From *Phys. Soc. Proc.* Vol. III. [May 25, 1878, title only], p. 4 ; *Nature*, Vol. XVIII. June 13, 1878, pp. 180, 181.]

THE lecturer pointed out that torsion of a metal tube within its limits of elasticity produces æolotropic stress, of which the mutually perpendicular lines of maximum extension and maximum contraction are spirals, each very nearly at 45 deg. to the length of the tube. From the author's early experiments—described in his paper on the Electrodynamic Qualities of Metals, published in the *Transactions* of the Royal Society for 1856—showing a diminution of electric conductivity by pulling force in metallic wires, and Mr Tomlinson's recent confirmations and extensions of those results, it is to be expected that the conductivity of the substance will be less in the direction of extension and greater in the direction of contraction in the stressed substance than the conductivity, equal in all directions, of the substance when free from stress. Hence, if an electric current be maintained along a tube, torsion would cause it to flow in spiral stream lines, with spirality of opposite name to that of the twist. The whole flow may be resolved into two components, one right along the tube, the other round it. The latter would—like the current through a galvanometer coil—deflect a needle hung in the interior of the tube, with its axis perpendicular to the tube when undisturbed; or it would magnetize a bar or wire of soft iron placed within the tube. The current itself would, except near the ends of the tube, produce no external effect directly, but either of those appliances may be used to give an external indication. Since the last meeting of the Physical Society, when the lecturer raised the question of the spiral electric stream lines in a twisted tube, experiments had been made for him by Mr Macfarlane in the physical laboratory of the University of Glasgow on the last-mentioned plan, and on the former plan by Mr J. T. Bottomley in the physical laboratory of King's College, London, by kind permission of Professor Adams, and with the valuable assistance of his staff. Mr Macfarlane, using a small mirror magnetometer suspended externally in the neighbourhood of one end of an iron wire placed within a brass tube, found that when the spiral was right-handed the end of the wire next that end of the tube by which the current enters becomes a true north pole. Mr Bottomley, with the cell and suspended mirror and needle of an ordinary dead beat mirror galvanometer, supported by an independent support within a brass tube along which a current is maintained, found that the true north pole of the needle is moved towards the end of the tube by which the current enters. Thus, both Mr Macfarlane's and Mr Bottomley's observations confirm the anticipation that the electric conductivity is least in the direction of greatest extension, and greatest in the direction of greatest contraction of metal.

The apparatus by which Mr Bottomley had made his experiments was exhibited to the meeting. It included a mode of balancing the effect on the internal needle by placing a circular portion of the main circuit at a proper distance from it, the centre and the plane of the circle being in and perpendicular to the axis of the tube. From a measurement of the distance from the centre of the circle to the needle, when the balance is obtained, the ratio of the maximum to the minimum conductivity can be calculated.

201. PARLIAMENTARY EVIDENCE REGARDING ELECTRIC LIGHT AND POWER. [Abstract.]

[From *Nature*, Vol. xx. May 29, 1879, pp. 110, 111.]

SIR WILLIAM THOMSON gave some valuable evidence on Friday before the Select Committee engaged in considering the subject of the electric light. He said that whereas one-horse-power of energy would only produce 12-candle gas light, it might produce 2,400-candle electric light. "The upshot of the experiments made at the factory of Messrs Siemens, at Woolwich, and at the natural philosophy class of the University of Edinburgh, was that, allowing the practical estimate of one-horse-power applied in driving the engine, it had produced 1,200 candles of actual visible electric light, half the gross energy going to produce the light while the other half was lost in heating the machine and the wires. As the electric light was such an economical producer he anticipated that it had a great and immediate future before it. He believed before long it would be used in every case where a fixed light was required, whether in large rooms or small ones—even in passages and staircases of private dwellings. There was immense promise in the actual work carried out by practical men in the present day. There was a prodigiously greater economy in the transmission of mechanical force into energy in the case of the electric light than in the case of gas. With regard to regulators for the electric light, he had seen one the previous day—the Siemens regulator—which gave a steady, pure, and quiet light. The electric light was especially adapted for being placed high where it illuminated a wide area. It might be put upon an iron pole raised 60 feet high, or the old French plan of swinging a lamp on a wire from one side of the street to the other might be followed with advantage. Such a plan would be useful in doing away with the necessity for opal globes, which destroyed a large quantity of the illuminating quality of the light. Indeed, he was surprised that these globes had ever been used, wasting as they did 50 or 60 per cent. of the illuminating power. He considered that the advantages of using the electric light within buildings would be very great, because of the small effect it would have when compared with gas in heating and vitiating the atmosphere. In the case of electricity, the waves of light only became converted into sensible heat, not in the air, but on the ceiling or walls and floor of the room after they had done their work. With regard to the subdivision of the light, according to practical experiments, if the same amount of energy that was used in producing one large light was employed in producing ten feebler lights, none of those lights gave one-tenth of the amount of illumination of the one large concentrated light. Still there was nothing mathematically impossible in the matter, and it was quite possible that a plan of subdivision might be found by which the ten feebler lights would give a sum of illumination equal to that of the one larger light. He considered that the electric light as now developed was fit for use in large rooms. He was also of opinion that a great deal of natural energy which was now lost might be advantageously applied in the future to lighting and manufactures. There was a deal of energy in waterfalls. In the future, no doubt, such falls as the Falls of Niagara would be extensively used—indeed, he believed the Falls of Niagara would in the future be used for the production of light and mechanical power over a large area of North America. The electricity produced by them might be advantageously conducted for hundreds of miles, and the manufactories of whole towns might be set in motion by it. Powerful copper conductors would have to be used—conductors of a tubular form with water flowing through them to keep them cool. There would be no limit to the application of the electricity as a motive power; it might do all the work that could be done by steam-engines of the most powerful description. It seemed to him that legislation, in the interests of the nation and in the interests of mankind, should remove as far as possible all obstacles such as those arising from vested interests, and should encourage inventors to the utmost. As to the use of electricity by means of the Falls of Niagara, his idea was to drive dynamic engines by water power in the neighbourhood of the Falls, and then to have conductors to transmit the force to the places where illumination or the development of mechanical power was wanted. There would be no danger of terrible effects being brought about accidentally by the use of such a terrific power, because the currents employed would be continuous and not alternating." This may be called a fanatical view of the electric light.

202. On a Method of Measuring Contact Electricity.

[From *British Association Report*,.1880, pp. 494—496; *Nature*, Vol. XXIII.
April 14, 1881, pp. 567, 568.]

IN my reprint of Papers on 'Electrostatics and Magnetism,'
§ 400 (of date January 1862), I described briefly this method, in
connection with a new physical principle, for exhibiting contact
electricity by means of copper and zinc quadrants substituted for
the uniform brass quadrants of my quadrant electrometer. I had
used the same method, but with movable discs for the contact
electricity, after the method of Volta, and my own quadrant electrometer substituted for the gold-leaf electroscope by which Volta
himself obtained his electric indications, in an extended series of
experiments which I made in the years 1859–61.

I was on the point of transmitting to the Royal Society a
paper which I had written describing these experiments, and
which I still have in manuscript, when I found a paper by Hankel
in Poggendorff's *Annalen* for January 1862, in which results
altogether in accordance with my own were given, and I withheld
my paper till I might be able not merely to describe a new
method, but, if possible, add something to the available information
regarding the properties of matter to be found in Hankel's paper.
I have made many experiments from time to time since 1861 by
the same method, but have obtained results merely confirmatory
of what had been published by Pfaff in 1820 or 1821, showing the
phenomena of contact electricity to be independent of the surrounding gas, and agreeing in the main with the numerical values
of the contact differences of different metals which Hankel had
published; and I have therefore hitherto published nothing except
the slight statements regarding contact electricity which appear
in my 'Electrostatics and Magnetism.' As interest has been
recently revived in the subject of contact electricity, the following

description of my method may possibly prove useful to experimenters. The same method has been used to very good effect, but with a Bohnenberger electroscope instead of my quadrant electrometer, in researches on contact electricity by Monsieur H. Pellat, described in the *Journal de Physique* for May 1880.

The apparatus used in these experiments was designed to secure the following conditions:—To support two circular discs of metal about four inches in diameter in such a way that the opposing surfaces should be exactly parallel to each other and approximately horizontal; and that the distance between them might be varied at pleasure from a shortest distance of about $\frac{1}{80}$ of an inch to about a quarter or half an inch. The lower plate, which was the insulated one, was fixed in a glass stem rising from the centre of a cast-iron sole plate. The upper plate was suspended by a chain to the lower end of a brass rod sliding through a steadying socket in the upper part of the case. A stout brass flange fixed to the lower end of this rod bears three screws, one of which, S, is shown in the drawing [omitted], by which the upper plate can be adjusted to parallelism to the lower plate. The other apparatus used consisted of a quadrant electrometer and a gravity Daniell's cell of the form which I described in *Proc. R. S.* 1871 (pp. 253—259) with a divider by which any integral number of per cents. from 0 to 100 of the electromotive force of the cell could be established between any two mutually insulated homogeneous metals in the apparatus.

Connections.

The insulated plate was connected by a brass wire passing through the case of the contact instrument to the electrode of the insulated pair of quadrants. The upper plate was connected to the metal case of the Volta condenser and to the metal case of the electrometer, one pair of quadrants of which were also connected to the case. One of the terminals of the divider, which connected the poles of the cell, was connected to the case of the electrometer, and to the other terminal was attached one of the contact wires, which was a length of insulated copper wire having soldered to its outer end a short piece of platinum. The other contact surface was a similar short piece of platinum fixed to the insulated electrode of the electrometer. Hence it will be seen that contact between the two plates was effected by putting the divider at zero and bringing into contact the two pieces of platinum wire.

Order of Experiment.

The sliding piece of the divider was put to zero, and contact made and broken and the upper plate raised, when the deflection of the spot of light was observed. These operations were repeated, with the sliding piece at different numbers on the divider scale, until one was found at which the make-break and separation caused no perceptible deflection. The number thus found on the divider scale was the number of per cents. which was equal to the contact electric difference of the Volta condenser.

203. ON THE EFFECT OF MOISTENING WITH WATER THE OP-POSED METALLIC SURFACES IN A VOLTA-CONDENSER, AND OF SUBSTITUTING A WATER ARC FOR A METALLIC ARC IN THE DETERMINING CONTACT.

[From *Edinb. Roy. Soc. Proc.* Vol. XI. [read Feb. 21, 1881], p. 135, title only.]

204. THE STORAGE OF ELECTRIC ENERGY.

[From *Nature*, Vol. XXIV. June 16, 1881, pp. 137, 156, 157.]

I AM continuing my experiments on the Faure accumulator with every-day increasing interest. I find M. Reynier's statement, that a Faure accumulator, weighing 75 kilograms (165 lbs.) can store and give out again energy to the extent of an hour's work of one-horse-power (2,000,000 foot-pounds) amply confirmed. I have not yet succeeded in making the complete measurements necessary to say exactly what proportion of the energy used in the charging is lost in the process of charging and discharging. If the processes are pushed on too fast there is necessarily a great loss of energy, just as there is in driving a small steam-engine so fast that energy is wasted by "wire-drawing" of the steam through the steam pipes and ports. If the processes are carried on too slowly there is inevitably some loss through local action, the spongy lead becoming oxidised, and the peroxide losing some of its oxygen viciously, that is to say, without doing the proper proportion of electric work in the circuit. I have seen enough however to make me feel very confident that in any mode of working the accumulator not uselessly slow, the loss from local action will be very small. I think it most probable that at rates of working which would be perfectly convenient for the ordinary use of fixed accumulators in connection with electric lighting and electric transmission of power for driving machinery, large and small, the loss of energy in charging the accumulator and taking out the charge again for use will be less than 10 per cent. of the whole that is spent in charging the accumulator : but to realise such dynamical economy as this prime cost in lead must not be stinted. I have quite ascertained that accumulators amounting in weight to three-quarters of a ton will suffice to work for six hours from one charge, doing work during the six hours at the uniform rate of one-horse-power, and with very high economy. I think it probable that the economy will be so high that as

much as 90 per cent. of the energy spent in the charge will be
given out in the circuit external to the accumulator. When, as
in the proposed application to driving tramcars, economy of weight
is very important, much less perfect economy of energy must be
looked for. Thus, though an eighth of a ton of accumulators
would work very economically for six hours at one-sixth of a
horse-power, it would work much less economically for one hour
at one horse-power; but not so uneconomically as to be practically
fatal to the proposed use. It seems indeed very probable that a
tramcar arranged to take in, say $7\frac{1}{2}$ cwt. of freshly-charged
accumulators, on leaving headquarters for an hour's run, may be
driven more economically by the electric energy operating through
a dynamo-electric machine than by horses. The question of
economy between accumulators carried in the tramcar, as in
M. Faure's proposal, and electricity transmitted by an insulated
conductor, as in the electric railway at present being tried at
Berlin by the Messrs Siemens, is one that can only be practically
settled by experience. In circumstances in which the insulated
conductor can be laid, Messrs Siemens' plan will undoubtedly be
the most economical, as it will save the carriage of the weight
of the accumulators. But there are many cases in which the
insulated conductor is impracticable, and in which M. Faure's
plan may prove useful. Whether it be the electric railway or
the lead-driven tramcar, there is one feature of peculiar scientific
interest belonging to electrodynamic propulsion of road carriages.
Whatever work is done by gravity on the carriage going down
hill will be laid up in store ready to assist afterwards in drawing
the carriage up the hill, provided electric accumulators be used,
whether at a fixed driving station or in the carriage itself.

[Letters to the *Times*, June 6, 9, 11.]

The marvellous "box of electricity" described in a letter to
you. which was published in the *Times* of May 16, has been
subjected to a variety of trials and measurements in my laboratory
for now three weeks, and I think it may interest your readers to
learn that the results show your correspondent to have been by
no means too enthusiastic as to its great practical value. I am
continuing my experiments to learn the behaviour of the Faure
battery in varied circumstances, and to do what I can towards
finding the best way of arranging it for the different kinds of

service to which it is to be applied. At the request of the Conseil d'Administration of the Société de la Force et la Lumiere, I have gladly undertaken this work, because the subject is one in which I feel intensely interested, seeing in it a realisation of the most ardently and unceasingly felt scientific aspiration of my life—an aspiration which I scarcely dared to expect or to hope to live to see realised.

The problem of converting energy into a preservable and storable form, and of laying it up in store conveniently for allowing it to be used at any time when wanted, is one of the most interesting and important in the whole range of science. It is solved on a small scale in winding up a watch, in drawing a bow, in compressing air into the receiver of an air-gun or of a Whitehead torpedo, in winding up the weights of a clock or other machine driven by weights, and in pumping up water to a height by a windmill (or otherwise, as in Sir William Armstrong's hydraulic accumulator) for the purpose of using it afterwards to do work by a waterwheel or water pressure on a piston. It is solved on a large scale by the application of burning fuel to smelt zinc, to be afterwards used to give electric light or to drive an electro-magnetic engine by becoming, as it were, unsmelted in a voltaic battery. Ever since Joule, forty years ago, founded the thermodynamic theory of the voltaic battery and the electro-magnetic engine, the idea of applying the engine to work the battery backwards and thus restore the chemical energy to the materials so that they may again act voltaically, and again and again, has been familiar in science. But with all ordinary forms of voltaic battery the realisation of the idea to any purpose seemed hopelessly distant. By Planté's admirable discovery of the lead and peroxide of lead voltaic battery, alluded to by your correspondent, an important advance towards the desired object was made twenty years ago ; and now by M. Faure's improvement practical fruition is attained.

The "million of foot pounds" kept in the box during its seventy-two hours' journey from Paris to Glasgow was no exaggeration. One of the four cells, after being discharged, was recharged again by my own laboratory battery, and then left to itself absolutely undisturbed for ten days. After that it yielded to me 260,000 foot pounds (or a little more than a quarter of a million). This not only confirms M. Reynier's measurements, on the faith of which your correspondent's statement was made; it seems further to show that the waste of the stored energy by time is not great, and that for days or weeks, at all events, it may not be of practical moment. This, however, is a question which can only be answered by careful observations and measurements carried on for a much longer time than I have hitherto had for investigating the Faure battery. I have already ascertained enough regarding its qualities to make it quite certain that it solves the problem of storing electric energy in a manner and on a scale useful for many important practical applications. It has already had in this country one interesting application, of the smallest in respect to dynamical energy used, but not of the smallest in respect to beneficence, of all that may be expected of it. A few days ago my colleague, Prof. George Buchanan, carried away from my laboratory one of the lead cells (weighing about 18 lbs.) in his carriage, and by it ignited the thick platinum wire of a galvanic *écraseur* and bloodlessly removed a nævoid tumour from the tongue of a young boy in about a minute of time. The operation would have occupied over ten minutes if performed by the ordinary chain *écraseur*, as it must have been had the Faure cell not been available, because in the circumstances the surgical electrician, with his paraphernalia of voltaic battery to be set up beforehand, would not have been practically admissible.

The largest useful application waiting just now for the Faure battery—and it is to be hoped that the very *minimum* of time will be allowed to pass till the battery is supplied for this application—is to do for the electric light what a water cistern in a house does for an inconstant water supply. A little battery of seven of the boxes described by your correspondent suffices to give the incandescence in Swan or Edison lights to the extent of 100 candles for six hours, without any perceptible diminution of brilliancy. Thus, instead of needing a gas engine or steam engine to be kept at work as long as the light is wanted, with the liability of the light failing at any moment through the slipping of a belt—an accident of too frequent occurrence—or any other breakdown or stoppage of the machinery, and instead of the wasteful inactivity during the hours of day or night when the light is not required, the engine may be kept going all day and stopped at night, or it may be kept going day and night, which will undoubtedly be the most economical plan when the electric light comes into general enough use. The Faure accumulator, always kept charged from the engine by the house supply wire, with a proper automatic stop to check the supply when the accumulator is full, will be always ready at any hour of the day or night to give whatever light is required. Precisely the same advantages in respect of force will be gained by the accumulator when the electric town supply is, as it surely will be before many years pass, regularly used for turning lathes and other machinery in workshops and sewing-machines in private houses.

Another very important application of the accumulator is for the electric lighting of steamships. A dynamo-electric machine of very moderate magnitude and expense, driven by a belt from a drum on the main shaft, working through the twenty-four hours, will keep a Faure accumulator full; and thus, notwithstanding irregularities of the speed of the engine at sea or occasional stoppages, the supply of electricity will always be ready to feed Swan or Edison lamps in the engine-room and cabins, or arc lights for mast-head and red and green side lamps, with more certainty and regularity than have yet been achieved in the gas supply for any house on *terra firma*.

I must apologise for trespassing so largely on your space My apology is that the subject is exciting great interest among the public, and that even so slight an instalment of information and suggestions as I venture to offer in this letter may be acceptable to some of your readers.

Your leading article in the *Times* of yesterday, on the storage of electricity, alludes to my having spoken of Niagara as the natural and proper chief motor for the whole of the North American Continent. I value the allusion too much to let it pass without pointing out that the credit of originating the idea and teaching how it is to be practically realised by the electric transmission of energy is due to Mr C. W. Siemens, who spoke first, I believe, on the subject in his presidential address to the Iron and Steel Institute in March, 1877. I myself spoke on the subject in support of Mr Siemens's views at the Institution of Civil Engineers a year later. In May, 1879, in answer to questions put to me by the Select Committee of the House of Commons on Electric Lighting, I gave an estimate of the quantity of copper conductor that would be suitable for the economical transmission of power by electricity to any stated distance*; and, taking Niagara as example, I pointed out that, under practically realisable conditions of intensity, a copper wire of half an inch diameter would suffice to take 26,250 horse-power from water-wheels driven by the Fall, and (losing only 20 per cent. on the way) to yield 21,000 horse-power at a distance of 300 British statute miles; the prime cost of the copper amounting to 60,000 $l.$, or less than 3 $l.$ per horse-power actually yielded at the distant station.

If you do me the honour to publish a letter which I wrote to you yesterday regarding the electric transmission of energy it will be seen that I thoroughly sympathise with Prof. Osborne Reynolds in his aspirations for the utilisation of Niagara as a motor, but that neither Mr Siemens nor I agree with him in the conclusion which he asserts in his letter to you, published in the *Times* of to-day, that electricity has been tried and found wanting as a means for attaining such objects. The *transmission* of power was not the subject of my letter to you published in the *Times* of the 9th inst., and Prof. Reynolds' disappointment with M. Faure's practical realisation of electric *storage*, because it does not provide a method of *porterage* superior to conduction through a wire, is like being disappointed with an invention of improvements in water cans and water reservoirs because the best that can be done in the way of movable water cans and fixed water reservoirs will never let the water-carrier supersede water-pipes wherever water-pipes can be laid.

The $1\frac{1}{2}$ oz. of coal cited by Prof. Osborne Reynolds as containing a million of foot-pounds stored in it is no analogy to the Faure accumulator containing the same amount of energy. The accumulator can be re-charged with energy when it is exhausted, and the fresh store drawn upon when needed, without losing more than 10 or 15 per cent. with arrangements suited for practical purposes. If coal could be unburned—that is to say, if carbon could be extracted from carbonic acid by any economic process of chemical or electric action, as it is in nature by the growth of plants drawing on sunlight for the requisite energy—the result would be analogous to what is done in Faure's accumulator.

* [Cf. *supra*, p. 420.]

205. On some Uses of Faure's Accumulator in connection with Lighting by Electricity.

[From *British Association Report*, 1881, p. 526.]

The largest use of Faure's accumulator in electric lighting was to allow steam or other motive power and dynamos to work economically all day, or throughout the twenty-four hours where the circumstances were such as to render this economical, and storing up energy to be drawn upon when the light was required. There was also a very valuable use of the accumulator in its application as an adjunct to the dynamo, regulating the light-giving current and storing up an irregular surplus in such a manner that stoppage of the engine would not stop the light, but only reduce it slightly, and that there would always be a good residue of two or three hours' supply of full lighting power, or a supply for eight or ten hours of light for a diminished number of lamps. The speaker showed an automatic instrument which he had designed and constructed to break and make the circuit between the Faure battery and the dynamo, so as automatically to fulfil the conditions described in the paper. This instrument also guarded the coils of the dynamo from damage, and the accumulator battery from loss, by the current flowing back, if at any moment the electro-motive force of the dynamo flagged so much as to be overpowered by the battery.

206. On the Economy of Metal in Conductors
of Electricity.

[From *British Association Report*, 1881, pp. 526—528; *Lum. Élec.*
Vol. v. Oct. 12, 1881, pp. 65, 66.]

THE most economical size of the copper conductor for the
electric transmission of energy, whether for the electric light or
for the performance of mechanical work, would be found by comparing the annual interest of the money value of the copper with
the money value of the energy lost in it annually in the heat
generated in it by the electric current. The money value of a
stated amount of energy had not yet begun to appear in the City
price lists. If £10 were taken as the par value of a horse-power
night and day for a year, and allowing for the actual value being
greater or less (it might be very much greater or very much less)
according to circumstances, it was easy to estimate the right
quantity of metal to be put into the conductor to convey a current
of any stated strength, such as the ordinary strength of current
for the powerful arc light, or the tenfold strength current (of
240 webers) which he (Sir William Thomson) had referred to in
his address as practically suitable for delivering 21,000 horse-power
of Niagara at 300 miles from the fall.

He remarked that (contrary to a very prevalent impression
and belief) the gauge to be chosen for the conductor does not
depend on the length of it through which the energy is to be
transmitted. It depends solely on the strength of the current
to be used, supposing the cost of the metal and of a unit of
energy to be determined.

Let A be the sectional area of the conductor; s the specific
resistance (according to bulk) of the metal; and c the strength
of the current used. The energy converted into heat and so lost,
per second per centimetre, is sc^2/A ergs.

Let p be the proportion of the whole time during which, in the course of a year, this current is kept flowing. There being $31\frac{1}{2}$ million seconds in a year, the loss of energy per annum is

$$31 \cdot 5 \times 10^6 psc^2/A \text{ ergs} \quad\text{......................(1).}$$

The cost of this, if E be the cost of an erg, is

$$31 \cdot 5 \times 10^6 psc^2 E/A \quad\text{........................(2).}$$

Let V be the money value of the metal per cubic centimetre The cost of possessing it, per centimetre of length of the wire, at 5 per cent. per annum, is

$$VA/20 \quad\text{.............................(3).}$$

Hence the whole annual cost, by interest on the value of the metal, and by loss of energy in it, is

$$\tfrac{1}{20}VA + 31 \cdot 5 \times 10^6 psc^2 EA^{-1} \quad\text{.................(4).}$$

The amount of A to make this a minimum (which is also that which makes the two constituents of the loss equal) is as follows :

$$A = \sqrt{(31 \cdot 5 \times 10^6 psc^2 E/\tfrac{1}{20}V)}$$
$$= c\sqrt{(63 \times 10^7 ps E/V)} \quad\text{.....................(5).}$$

Taking £70 per ton as the price of copper of high conductivity (known as 'conductivity copper' in the metal market), we have £·00007 as the price of a gramme. Multiplying this by 8·9 (the specific gravity of copper), we find, as the price of a cubic centimetre,

$$V = \text{£·00062} \quad\text{..........................(6),}$$

and the assumption of £10 as the par value of one horse-power day and night for 365 days gives, as the price of an erg,

$$\text{£10}/(31\tfrac{1}{2} \times 10^6 \times 74 \times 10^6) = (23 \times 10^{15})^{-1} \text{ of £1} \quad\text{...(7).}$$

Supposing the actual price to be at the rate of $e \times$ £10 per year for the horse-power, we have

$$E = \frac{e}{23 \times 10^{15}} \text{ of £1} \quad\text{....................(8).}$$

Lastly, for the specific resistance of copper, we have

$$s = 1640 \quad\text{.............................(9).}$$

Using (8) and (9) in (5), we find

$$A = c\sqrt{\frac{63 \times 10^7 \times 1640 \times pe}{23 \times 10^{15} \times \cdot 00062}} = c\sqrt{\frac{pe}{13 \cdot 8}} \quad\text{......(10).}$$

Suppose, for example, $p = \cdot 5$ (that is, electric work through the conductor for twelve hours of every day of the year to be provided for), and $e = 1$. These suppositions correspond fairly well to ordinary electric transmission of energy in towns for light, according to present arrangements. We have

$$A = c \sqrt{\frac{1}{27 \cdot 6}} = \frac{c}{5 \cdot 25} \fallingdotseq \cdot 19 \times c.$$

That is to say, the sectional area of the wire in centimetres ought to be about a fiftieth of the strength of the current in webers*. Thus, for a powerful arc-light current of $2 \cdot 1$ webers, the sectional area of the leading wire should be $\cdot 4$ of a square centimetre, and therefore its diameter (if it is a solid round wire) should be $\cdot 71$ of a centimetre.

If we take $e = 1/27 \cdot 6$, which corresponds to £1,900 a year as the cost of 5,250 horse-power (see Presidential Address, Section A†), and if we take $p = 1$, that is reckon for continued night and day electric work through the conductor, we have

$$A = \frac{c}{\sqrt{381}} \fallingdotseq \frac{c}{19 \cdot 5} :$$

and if $c = 24$, $A = 1 \cdot 24$, which makes the diameter $1 \cdot 26$ centimetres, or half an inch (as stated in the Presidential Address). But even at Niagara it is not probable that the cost of an erg can be as small as $\frac{1}{28}$ of what we have taken as the par value for England; and probably therefore a larger diameter for the wire than $\frac{1}{2}$ inch will be better economy if so large a current as 240 webers is to be conducted by it.

---

* [This takes the weber to be the same as the modern unit, the ampère, which is one-tenth of the c. G. s. unit. The number $2 \cdot 1$ in the next line should thus be corrected to 21.]

† [Reprinted in *Popular Lectures and Addresses*, Vol II. pp. 433—450. The practical problem is of course entirely altered by the modern use of alternating currents of high voltage, transformed down before use at the end of the cable.]

207. On the Proper Proportions of Resistance in the Working Coils, the Electro-magnets and the External Circuits of Dynamos.

[From *British Association Report*, 1881, pp. 528—531; *Compt. Rend.* Vol. xCIII. 1881, pp. 474—479; *Nature*, Vol. XXIV. Sept. 29, 1881, pp. 526, 527; *Lum. Élec.* Vol. IV. Sept. 24, 1881, pp. 385—387.]

For the electro-magnet;

Let L be the length of the wire,

> B „ bulk of the whole space occupied by wire and insulation,
>
> n „ ratio of this whole space to the bulk of the copper alone (that is, let B/n be the bulk of the copper),
>
> A „ the sectional area of wire and insulator,
>
> R „ the resistance of the wire.

For the working coil, let the corresponding quantities be L', B', n', A', R'. Lastly, let s be the specific resistance of the copper. We have

$$B = AL, \qquad R = ns\, L/A = ns\, B/A^2.$$

Hence
$$A = \sqrt{(ns\, B/R)} = K/\sqrt{R} \quad \dots\dots\dots\dots\dots(1),$$

and similarly,
$$A' = \sqrt{(n's'\, B'/R')} = K'/\sqrt{R'} \dots\dots\dots\dots\dots(2),$$

where K and K' denote constants.

Now, let c be the current through the magnet coil, and c' that through the working coil, and let v be the velocity of any chosen point of the working coil. Denoting by p the average electromotive force between the two ends of the working coil, we have

$$p = Icv/AA' \dots\dots\dots\dots\dots\dots\dots(3);$$

where I is the quantity depending on the forms, magnitudes, and relative positions of B and B', and on the magnetic susceptibility of iron, diminishing as the susceptibility diminishes with increased strength of current, or with any change of R and R' which gives increase of magnetising force.

In the single-circuit dynamo (that is, the ordinary dynamo) c' is equal to c, but not so in the shunt-dynamo. In each, the whole electric activity (that is, the rate of doing work) is pc'; or, by (3),

$$Icc'v/AA' \dots\dots\dots\dots\dots(4),$$

or, by (1) and (2),

$$I\sqrt{(RR')}\,cc'v/KK' \dots\dots\dots\dots(5).$$

Of this whole work, the proportions which go to waste in heating the coils and to work in the external circuit are

$$Rc^2 + R'c'^2 \dots\dots\dots \text{ waste } \dots\dots\dots(6),$$

$$\frac{I\sqrt{(RR')}\,cc'v}{KK'} - (Rc^2 + R'c'^2)\dots\text{useful work}\dots\dots(7).$$

By making v sufficiently great, the ratio of (6) to (7) (waste to useful work) may be made as small as we please. Our question is, how ought R and R' to be proportioned to make the ratio of waste to work a minimum, with any given speed? or, which comes to the same thing, to make the speed required for a given ratio of work to waste a minimum? To answer it, let r be the ratio of the whole work to the waste. We have, by (5) and (6),

$$r = \frac{I\sqrt{(RR')}\,cc'}{Rc^2 + R'c'^2}\,\frac{v}{KK'} \dots\dots\dots\dots(8).$$

For the single-circuit dynamo we have $c = c'$, and (8) becomes

$$r = \frac{I\sqrt{(RR')}}{R+R'}\,\frac{v}{KK'}, \text{ or } r = \frac{I\sqrt{\{R(S-R)\}}\,v}{SKK'} \dots(9),(10),$$

where $$S = R + R' \dots\dots\dots\dots\dots(11).$$

Suppose now S to be given, and suppose for a moment I to be constant. The problem of making r a maximum with v given, or v a minimum with r given, requires simply that $R(S-R)$ be a maximum; which it is when $R = \frac{1}{2}S$, that is, when the resistances in the working coil and the electro-magnet are equal. But in reality I is not constant; it diminishes with increase of the magnetising force. As it generally depends chiefly on the soft iron of the electro-magnet, and comparatively but little on the soft iron of the moving armature, or on iron magnetised by the current through the moving coils, it will generally be the case that I will, *cœteris paribus*, be diminished by increasing R and diminishing R'. Hence the maximum of r/v is shown by (10) to require R' to be somewhat greater than $\frac{1}{2}S$: how much greater we cannot find from the formula, without knowing the law of the variation of I.

Experience and natural selection seem to have led in most of the ordinary dynamos, as now made, to the resistance in the electro-magnet being somewhat less than the resistance in the working coil, which is in accordance with the preceding theory.

Whether the useful work of the dynamo be light-giving, or power, or heating, or electro-metallurgy, we may, for simplicity, reckon it in any possible case by referring to the convenient standard case of a current through a conductor of given resistance E connecting the working terminals of the dynamo. This conductor, in accordance with general usage, I call the 'external circuit,' which is an abbreviation for the part of the whole circuit which is external to the dynamo. In the case of the single-circuit dynamo, the current in the external circle is equal to that through the working coil and electro-magnet, or c of our notation. Hence, by Ohm's law,

$$c = p/(E + R + R') \quad\dots\dots\dots\dots\dots\dots(12),$$

or, by (3), (1), and (2),

$$c = c\,\frac{I\sqrt{(RR')}\,v}{KK'(E + R + R')}\quad\dots\dots\dots\dots(13).$$

Hence either

$$c = 0, \quad\text{or}\quad I = KK'(E + R + R')/\sqrt{(RR')}\,v\dots(14),(15).$$

The case of $c = 0$ is that in which

$$v < KK'(E + R + R')/I_0\sqrt{(RR')}\quad\dots\dots\dots\dots(16),$$

where I_0 denotes the value of I for $c = 0$. To understand it, remember we are supposing no residual magnetism. For any speed subject to (16), the dynamo produces no current. When this limit is exceeded the electric equilibrium in the circuit becomes unstable; an infinitesimal current started in either direction rises rapidly in strength, till it is limited by equation (15), through the diminution of I, which it produces*. Thus, regarding I as a function of c, we have in (15) the equation mathematically expressing the strength of the current maintained by the dynamo when its regular action is reached. Using (15) in (10), we find

$$r = (E + S)/S \dots\dots\dots\dots\dots(17),$$

which we all knew forty years ago from Joule.

In the shunt-dynamo the whole current, c', of the working coil branches into two streams, c through the electro-magnet, and $c' - c$ through the external circuit, whose strengths are inversely as the resistances of their channels. Still calling the resistance of the external circuit E, we therefore have

$$cR = (c' - c)\, E, \text{ which gives } c = \frac{E}{R + E}\, c' \quad \dots\dots(18).$$

Hence, by Joule's original law, the expenditures of work per unit of time in the three channels are respectively

$$\left.\begin{array}{l} R'c'^2 \quad \dots\dots\dots\text{working coil} \\[2mm] R\left(\dfrac{E}{R + E}\right)^2 c'^2 \dots\text{electro-magnet} \\[2mm] E\left(\dfrac{R}{R + E}\right)^2 c'^2 \dots\text{external circuit} \end{array}\right\} \dots\dots\dots\dots(19).$$

Hence, denoting as above by r the ratio of the whole work to the heat developed in the external circuit, we have

$$r = \left\{ R' + R\left(\frac{E}{R + E}\right)^2 + E\left(\frac{R}{R + E}\right)^2 \right\} \Big/ E\left(\frac{R}{R + E}\right)^2 \dots(20),$$

whence

$$\left.\begin{array}{l} R^2 r = R'\,(R + E)^2\, E^{-1} + R\,(R + E) \\[2mm] \quad = R'\,R^2\, E^{-1} + (R + R')\, E + R\,(2R' + R) \end{array}\right\} \dots\dots\dots(21).$$

Suppose now R and R' given, and E to be found to make r a minimum. The solution is

$$E^2 = R'R^2/(R + R') \dots\dots\dots\dots\dots(22),$$

and this makes

$$r = 2\sqrt{\frac{R'\,(R + R')}{R^2}} + \frac{2R' + R}{R} \quad \dots\dots\dots(23).$$

---

* [Cf. *supra*, p. 412.]

Put now

$$R'/R = e \dots\dots\dots\dots\dots\dots(24),$$

(22) and (23) become

$$E^2 = RR'/(1 + e)\dots \quad \dots\dots\dots\dots\dots(25)$$

and $$r = 1 + 2 \sqrt{\{e\,(1 + e)\}} + 2e \ \dots\dots\dots\dots(26).$$

For good economy r must be but little greater than unity; hence e must be very small, and therefore *approximately*

$$E = \sqrt{(RR')}, \text{ and } r = 1 + 2 \sqrt{e}\dots\dots\dots\dots(27).$$

For example, suppose the resistance of the electro-magnet to be 400 times the resistance of the working coil—that is $e = \frac{1}{400}$— and we have, approximately,

$$E = 20R', \text{ and } r = 1 + \tfrac{1}{10}.$$

That is to say, the resistance in the external circuit is 20 times the resistance of the working coil, and the useful work in the external circuit is approximately $\frac{10}{11}$ of that lost in heating the wire in the dynamo.

208. On the Illuminating Powers of Incandescent Vacuum
Lamps with Measured Potentials and Measured
Currents. By Sir W. Thomson and J. T. Bottomley.

[From *British Association Report*, 1881, pp. 559—561.]

The electromotive force used in these experiments was derived
from Faure secondary batteries, kindly supplied for the purpose
by the Société la Force et la Lumière in their London office.

Two galvanometers were used simultaneously, one called the
potential galvanometer for measuring the difference of potentials
between the two terminals of the lamp, the other (called the
current galvanometer) for measuring the whole strength of the
current through the lamp.

The potential galvanometer had for its coil several thousand
metres of No. 50 (B. W. G.) silk-covered wire (of which the copper
weighs about $\frac{1}{30}$ gramme per metre, and therefore has resistance
of about 3 ohms per metre). Its electrodes were applied direct
on the platinum terminals of the lamp.

The current galvanometer had for its coil a single circle of
about 10 centimetres diameter, of thick wire placed in the direct
circuit of the lamp, by means of electrodes kept close together at a
sufficient distance from the galvanometer to ensure no sensible
action on the needle except from the circle itself. The directive force
on the needle which was produced by a large semi-circular horse-
shoe magnet of small sectional area was about $2\frac{1}{2}$ c.g.s., or 15
times the earth's horizontal magnetic force in London. This
arrangement would have been better for the potential galvanometer also than the plan actually used for it, which need not be
described here. The scale of each galvanometer was graduated
according to the natural tangent of the angle of deflection, so
that the strength of the current was simply proportional to the
number read on the scale in each case.

Three lamps were used, Nos. II and III of a larger size than No. I. The experiment was continued with higher and higher potentials on each lamp till its carbon broke.

The illuminating power was measured in the simplest and easiest way (which is also the most accurate and trustworthy), by letting the standard light and the lamp to be measured shed their lights nearly in the same direction on a white ground (a piece of white paper was used); and comparing the shadows of a suitable object (a pencil was used); and varying the distance of the standard light from the white ground till the illuminations of the two shadows were judged equal. The standard used was a regulation 'standard candle,' burning 120 grains of wax in the hour. The burning was not actually tested by weighing; but it was no doubt very nearly right; nearly enough for our purpose, which was an approximate determination of the illuminating powers of each lamp through a wide range of electric power applied to it. The following results were obtained [tables omitted].

Some of the irregularities of the results in the preceding tables are very interesting and important, as showing the effect of the blackening of the glass by volatilization of the carbon when too high electric power came to be applied.

The durability of the lamp at any particular power must be tested by months' experience before the proper intensity for economy can be determined.

209. [GALVANOMETERS FOR] THE MEASUREMENT OF
ELECTRIC CURRENTS AND POTENTIALS.

[From *Glasg. Phil. Soc. Proc.* Vol. xv. 1884 [read Jan. 9, 1884], pp. 96—101.]

IN this communication, after referring to the electrometer as
being, when available, in one form or other, the most proper
instrument for measuring differences of electric potential (inasmuch as it disturbs the difference of potential to be measured not
at all, because the insulation of the insulated part or parts may
be made practically perfect), it was shown that the functions of
an electrometer can, for many practical purposes, in a thoroughly
satisfactory manner, be performed by means of a galvanometer of
high resistance. Thus, galvanometers of from 200 or 300 to
30,000 ohms resistance can be very conveniently used for measuring
differences of potentials of from 1 to 600 volts, provided that in
each case the coil is not heated by the current produced in it so
much as to cause more than allowable error by augmentation of
its resistance. To obviate the need for a temperature correction
on account of difference of atmospheric temperature it is necessary,
for fairly satisfactory accuracy in ordinary practice, to have the
coils of potential galvanometers made of German silver wire
instead of copper. In cases, however, in which the sensibility
obtainable by copper is desired, the galvanometer coils may be
made of copper wire, but the proper temperature correction,
amounting approximately to ·39 per cent. per degree centigrade,
must be carefully applied. But there is no need for the sensibility
obtainable by copper coils in almost any of the practical applications of electricity except telegraphy; and for general use in
scientific laboratories, or in electrical factories, or in connection
with electric light installations, the potential galvanometer ought
to have its coils of German silver wire, because the increase of
resistance for this alloy is only about 04 per cent. per degree

centigrade, and is less than for any other known and practically available metal.

The author showed a new form of potential galvanometer, with four German silver coils acting on the two needles of an astatic combination hung by a single silk fibre. The astatic combination has for most practical applications great advantage over the single needle galvanometer, in respect to liability to disturbance by magnets in its neighbourhood. The new form of astatic needle galvanometer, whether for current or potential, may be used even in close proximity to a dynamo machine without being sensibly disturbed. In this instrument the two needles of the astatic combination are controlled by two equal magnets, or combinations of magnets, symmetrically arranged, adjustable for zero, and adjustable in respect to power. The method for adjustment to any desired sensibility, for the potential galvanometer, through a wider range than can be conveniently given by the magnetic controller, is by the use of adjustable resistances placed in series with the coils of the galvanometer. A very convenient and ready appliance of this kind constituted part of a complete instrument shown to the Society and designated by the author a long-range potential galvanometer. After many years of trials and anxious thought, the author had come to prefer, for practical purposes, the use of a standard Daniell cell to any other means hitherto realised for standardising a potential galvanometer. Accordingly he has made, to be used in connection with the new galvanometer, a convenient form of a standard Daniell cell.

A specimen of the cell was shown to the Society. It consists of a shallow rectangular tray of sheet copper, which may be about 15 centimetres square, having vertical sides about 5 centimetres high. The vertical sides are, for electrical reasons, enamelled or painted, to prevent contact between them and the liquid of the cell. Within this tray is placed a zinc plate about 3 mm. thick, and 14 centimetres square, supported firmly in notches of four wooden blocks resting securely in the four corners of the tray. In the centre of the zinc plate is a hole about one centimetre diameter, through which the stem of a filler, long enough to reach the bottom of the tray, passes, and which is used to pour into the bottom of the tray a measured quantity of a saturated solution of sulphate of copper. The tube of the filler is very fine, and the capacity of the wide part is sufficient to hold the whole amount of copper solution, which is poured quickly into the filler, and flows slowly out of it in the course of about a minute, so as to spread slowly over the bottom of the cell. The wide part of the filler is flat, so as to let it rest stably on the top of the zinc plate. To set up in action this cell, a solution of sulphate of zinc of 1·02 specific gravity is poured into the tray so as to fill it as high up as the zinc plate, and by means of the filler a concentrated solution of sulphate of copper, sufficient in quantity to make a layer of about half a centimetre deep, is poured in along the bottom of the tray.

The instrument, without any resistance added to its coil, is of good sensibility for a single cell. In this condition its coefficient is determined. Then a very simple calculation, or a table of numbers, tells the proper quantity of resistance to be added in order to give exactly the right sensibility with single divisions or round numbers of divisions, corresponding to single volts or round numbers of volts, according to convenience, in the special application for which the instrument is wanted. The user can always, with great ease, and with very little expenditure of time, standardise his own instrument in this manner by aid of a standard cell and standard solutions which are supplied along with the instrument.

The pointer by which the deflections are indicated consists of a light tube of aluminium with one end shaped to a fine edge, by which the deflection is read on a circular scale, divided and numbered, as in the author's graded galvanometers, according to tangents of angles of deflection. As in those previous instruments, the scale is of paper pasted on a horizontal mirror, by aid of which error from parallax is avoided with great ease in taking the readings. To the other end of this pointer is attached about a centimetre of fine platinum wire, turned downwards so as to dip about a quarter of a centimetre below the surface of a large flat dish of heavy paraffin oil, with which a little olive oil may be mixed when more of viscous resistance than is given by the paraffin is desired. The author has found that by this means a most satisfactory dead-beat effect is obtained, without any sensible error in respect to the position of equilibrium, whether for zero or for the deflection produced by a current. A properly shaped metallic pipette bottle, by which, when the galvanometer is to be moved at any time, the oil may be drawn off from the oil-vessel with great ease, carried away without risk of losing any of it, and replaced in the vessel in the galvanometer, is supplied with the instrument. To make the galvanometer itself portable, the frame-work carrying the steel needles and pointer, which is hung by a silk fibre when the instrument is in use, is let down by aid of a screw so as to hang upon a brass pin, passing through a ring which forms part of the framework. In this condition the instru-ment may be turned upside down or carried about in the roughest way, without the possibility of breaking the silk fibre. To set the instrument in action all that is necessary is to place it in position and level it, and by the screw pull up the upper end of the fibre to its proper position.

A current galvanometer, with portable silk fibre suspension, oil-vessel to give the dead-beat quality, astatic needles acted on by two circular coils in the main circuit surrounding the two needles respectively, and two magnetic systems of controlling magnets acting symmetrically on the two needles, was shown to the Society. This instrument is in all respects exceedingly con-venient for the practical measurement of currents, except for the comparative difficulty of standardising from time to time, necessary on account of the inconstancy of the steel magnets. This may be done electrolytically, with sufficient accuracy for most practical purposes, by weighing the deposit of copper from a solution of sulphate of copper, in the manner described at the end of Chap. VII. of Mr Andrew Gray's book, *Absolute Measurements in Electricity and Magnetism* (Macmillan, 1884). The process is, however, somewhat troublesome, and takes a good deal of time. The standardising may be done with much greater ease by comparison with a standard current meter on the principle of Weber's electro-dynamometer. A very valuable and convenient instrument of this class, Siemens' well-known electro-dynamometer, with its novel and ingenious method for measuring torque by means of a spiral spring acting on the moveable coil of arc, which is suspended by a silk thread passing down along the axis of the spring from a fixed point above, and has its two ends dipping in cups of mercury below in the same vertical with the bearing thread, was shown to the Society.

Among the instruments shown to the Society was also a

Siemens Watt-meter, which is the same as his electro-dynamometer except that the fixed coil, instead of being of a small number of turns of thick copper wire in the main circuit, is of a very large number of turns of very fine German silver wire, connected by its two ends to two points of the main circuit, between which the measurement in Watts, of the electrical activity of the current, is to be made. The moveable coil is, by means of its cups of mercury, joined in series with the main circuit. The instrument is an exceedingly convenient and valuable realisation of an idea which has no doubt occurred to many electricians*, but which, so far as the author knew, was first published by Prof. John Perry in the *Journal of the Society of Arts* for April 15, 1881, as embodied in an instrument designed by Prof. Ayrton and himself.

The author remarked upon the great want which existed of a name for the unit of conductivity. He repeated a suggestion which he had made in his lecture on "Electrical Units of Measurement," delivered before the Institute of Civil Engineers in London, on 3rd May 1883, that the name "mho" might be adopted for the unit of conductivity corresponding to the resistance of one ohm. He showed to the Society a new instrument which he had designed, and which he termed a mho-meter. It illustrated the great advantages in using the conductivity method in many electric measurements.

This instrument is simply an astatic-needle current galvano-meter, with the steel controlling-magnets replaced by coils of very fine German silver or copper wire, placed perpendicularly to the galvanometer coils proper. The two ends of the controlling coils are put in metallic connection with two points of a conductor, between which the conductivity is to be determined, the conductivity to be measured being exceedingly great in comparison with that of the controlling coils. The galvanometer coils proper are connected, so that the whole current through them passes through the conductor whose conductivity is to be measured, except the small part of it which goes through the controlling coils. With these connections it is clear that the conductivity to be tested is

---

* Prof. Perry, *Journal of the Society of Arts* for 15th April 1881 ; Sir William Thomson and Profs. Ayrton and Perry, *British Association Report*, York, 1881, and *Journal of Society of Telegraph Engineers and Electricians*, Vol. XI. May 1882.

equal to an absolute constant multiplied into the conductivity of
the controlling coils and into the tangent of the angle of deflection.
One main object of the instrument is to give a ready measurement
of the conductivity of dynamo armatures and electric light mains,
or of specimens of copper submitted for tests of their specific
conductivity. For this application the controlling coils are made
of copper, so that no temperature correction may be required
(provided the temperature of the galvanometer is approximately
enough the same as that of the conductor tested), and the
instrument thus made may be called a "copper conductivity
meter."

When the object is to measure electric light currents, the
controlling coils are made of German silver wire, and are fixed in
series with an adjustable resistance, also of fine German silver
wire, so wound in a convenient position in the base of the instrument as to exert no electro-magnetic force on the needles. In
this instrument the conductivity of the controlling coils, and of the
resistance wire in series with them, varies so little with the practical
variations of temperature that the conductivity to be measured is,
for most practical purposes, given with amply sufficient accuracy
by the tangent of the deflection multiplied by an absolutely
constant coefficient. The added resistance may be adjusted to
make this coefficient such that one division corresponds to a mho,
or a decamho, or a hectamho, or to any other value, within a wide
range, which may be found convenient. It may, for example, be
adjusted to give deflections at the rate of one division per lamp,
and for this application the instrument may conveniently be called
a "lamp meter."

The author has such an instrument now in use on the lighting
system of his house, in which varying numbers of from 1 to 40
Swan lamps, some of the old type, 42 volts, and some of the new
type, but of only 84 volts (instead of the 110, which of course
would be preferable if it were not for the old lamps, which he has
had in use for two years, and which he wishes to keep in use for
as many years longer as they will last). It is beautiful to see the
pointer moving up two divisions when a pair of the old lamps in
series, and by one division when one lamp of the new type is
lighted, and falling similarly when any number of the lamps are
extinguished; and it is very useful to be able, merely by looking

at the instrument, to tell how many lamps are lighted at any time. The inequality of the lamps gives scarcely an uncertainty of one lamp in the whole number when 30 or 40 are incandesced.

The potential galvanometer, which is also always kept on the mains, shows considerable inequalities (checked always, in the course of a few seconds, by the author's potential regulator, with its pair of mutually-geared horizontal cog-wheels dipping on one side or the other, according as the potential is too high or too low, into a hollow centrifugal cylinder of oil rotating rapidly round them), but the pointer of the lamp meter remains absolutely unmoved (or rather, with absolutely no perceptible motion) throughout these inequalities.

210. On Constant Gravitational Instruments for
Measuring Electric Currents and Potentials.

[From *British Association Report*, 1885, pp. 905, 906; *Nature*,
Vol. XXXII. Oct. 1, 1885, p. 535.]

THESE instruments, the author stated, were parts of two series
of electric measuring instruments, for current and potential, which
he was now working out. In the two current instruments—the
milliamperemeter and the hecto-amperemeter—the mode of
effecting the measurement was founded on Faraday's law, according
to which a ferro-magnetic mass placed in a variable magnetic field
experiences forces tending to make it move from places of weaker
to places of stronger force. The essential parts of the milliamperemeter are shown in the sketch diagram, fig. 1. It consists
of an electro-magnetic coil, fixed with its axis vertical, and a little
cylindrical mass of soft iron hung from one end of a light balanced
lever so as to be free to move up and down in a circular arc,
deviating but little in its middle and at its two ends from the
axis of the coil.

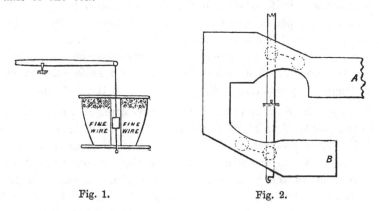

Fig. 1.   Fig. 2.

The measurement is given by the deflections indicated on a
scale by the end of the balanced lever, when a weight of known
amount is hung on the ring below the iron mass. To screen the

iron from the effects of the earth's magnetism the coil is enclosed in an iron box.

In the hecto-amperemeter the variable magnetic field is obtained by a suitable disposition of the metallic conductor conveying the current to be measured. The conductor may be taken as consisting of two thick copper plates, shaped each according to the sketch, fig. 2, supported in a vertical position parallel to each other, say one centimetre apart, and metallically connected at the place indicated by B. At A is fixed a suitable electrode. The course of the current is therefore from A to B, and from B across to and through the other plate to the part of it corresponding to A, which forms the other electrode. In this way, two similarly varying magnetic fields are produced, and the balanced lever, capable of motion in a plane situated midway between the plates, carries two masses of iron, one in each field. In other respects, the instrument is similar to the milliamperemeter.

The electrometer consists of an air condenser with one of its plates capable of a to-and-fro motion so as to vary the capacity of the condenser.

The fixed brass plates are supported so as to be accurately parallel to each other and in metallic connection, while they are thoroughly insulated from the case of the instrument. The movable plate is of aluminium, and is supported in a vertical position on a knife edge; the plane of its motion being parallel to the fixed plates and situated midway between them. The upper end of this movable plate has a fine prolongation which serves as a pointer for indicating the deflections on the scale of the instrument, and at its lower end is fixed a knife edge having its length perpendicular to the plane in which the plate moves.

When the fixed and movable plates are connected respectively to two points of an electric circuit between which there exists a difference of potential, the movable plate tends to move so as to augment the electrostatic capacity of the instrument, and the magnitude of the force concerned in any measurement is proportional to the square of the difference of potential by which it is produced. In the use of the instrument this force of attraction is balanced by the horizontal component of a weight of any convenient amount hung on the knife edge at the bottom of the movable plate.

211. On a Method of Multiplying Potential from a
Hundred to several Thousand Volts.

[From *British Association Report*, 1885, p. 907.]

The method described by the author was to arrange in series
a number *n* of condensers, where *n* is the number indicating the
required multiplication. A terminal is connected to the junction
between each pair of adjacent condensers. This series of $n+1$
terminals is conveniently placed so that by a suitable mechanism
a pair of movable electrodes, between which a known difference
of potentials exists, may be brought successively and repeatedly,
at short intervals of time, into contact with each pair of adjacent
terminals in the series, moving always in the same direction
along them. In this way the difference of potentials established
between the two end terminals of the series of condensers is *n*
times the known difference of potentials between the movable
electrodes.

212. Discussion on "Electrolysis" at the British
Association.

[From *Nature*, Vol. xxxiii. Nov 5, 1885, p. 20, Abstract;
British Association Report, 1885, pp. 723—772.]

Sir W. Thomson referred, in his remarks on Prof. Lodge's
paper, to a matter of importance in electro-plating—viz. the
selection which takes place in the electrolysis of solutions containing several salts, as, for instance, in the electrolysis of copper
sulphate containing ferrous sulphate, which, when decomposed
by a strong current gives a deposit containing impurities, whereas
a slower decomposition yields a very pure deposit. Sir W.
Thomson spoke also of the necessity for the careful investigation
of those cases in which the formation of deposits between the
electrodes had been observed, and it would be important to know
whether deposits could be formed in the line of conduction
without a nucleus at all. Such matters are of importance to
physiology, indicating a possible danger in the passing of long
continued currents through the human body.

213. On a Double Chain of Electrical Measuring Instru-
ments to Measure Currents from the Millionth of a
Milliampere to a Thousand Amperes, and to Measure
Potentials up to Forty Thousand Volts.

[From *Glasg. Phil. Soc. Proc.* Vol. xviii. [April 20, 1887], pp. 249—256;
La Lumière Électrique, Vol. xxiv. June 4, 1887, pp. 476—479.]

A FUNDAMENTAL requisite of a measuring instrument is that
its application to make a measurement shall not alter the
magnitude of the thing measured. When this condition is not
fulfilled (as is essentially the case with an electric measuring
instrument not kept permanently in or on the electric circuit or
system to which it is applied), it is the magnitude as influenced
or modified by the measuring instrument which is actually
measured, and the measurement is to be interpreted on this
understanding whatever may be the circumstances. Suppose,
for example, the thing to be measured is the diameter of a fine
wire, or of the carbon filament for an Edison-Swan lamp, or of a
hair, or of a silk fibre. If the measurement is made by a microscope, with the proper optical apparatus for measurement, the
thing measured is absolutely unaltered by the measuring appliance.
But if the measurement is made by an ordinary screw gauge, or
by any other mechanical fitting appliance however gentle, it is
impossible to avoid some diminution of the diameter to be
measured by the pressure of the measuring appliance, which
introduces some slight uncertainty even in the measurement of
steel or copper wire, and very considerable uncertainty when a
filament of softer material is to be measured.

The nearest approach in electric measuring instruments to the
fulfilment of this condition, of not altering the magnitude of the
thing measured, is attained by the electrometer when applied to
measure differences of potential between different points of a wire,
or metallic mass of any shape, in which electricity is kept flowing

by a battery or dynamo or other electro-motive apparatus. The insulation of any practical electrometer is so nearly perfect that the conduction of electricity through the instrument does not sensibly diminish the difference of potentials of the points touched by the electrodes. In this respect, therefore, the electrometer would be ideally perfect: but, alas, it is only for potentials of more than 400 or 500 volts that the electrometer in any shape has been made a convenient and fairly-accurate standard measuring instrument for ordinary practical use. For less than 400 volts the practical solution in connection with electric lighting is afforded by a current-measuring instrument with a known resistance in its circuit; though for many differential measurements, and particularly for measurements of the insulation of submarine cables, and determinations of the insulating quality of insulators for many practical purposes the quadrant electrometer is found useful as a differential measuring instrument.

The quadrant electrometer in its most sensitive adjustment indicates about $\frac{1}{100}$ of a volt, and with modified adjustments (heterostatic and idiostatic) can be used for measuring up to 300 or 400 volts. It is described in detail in the "Report on Electrometers and Electrostatic Measurements" published in the British Association volume for 1867 (Report of the Committee on Electrical Standards), and reprinted as Article XX. of my *Collected papers on Electrostatics and Magnetism*, so I need say nothing more of it at present. The Electrostatic Voltmeter exhibited and shown in action this evening, and represented in the annexed drawing is an idiostatic standard instrument for measurement of from 400 volts to 10,000 volts.

It consists of an air condenser with one of its plates capable of a to-and-fro motion so as to vary the capacity of the condenser. The fixed brass plates are supported so as to be accurately parallel to each other and in metallic connection, while they are thoroughly insulated from the case of the instrument. The movable plate is of aluminium, and is supported in a vertical position on a knife edge; the plane of its motion being parallel to the fixed plates and situated midway between them. The upper end of this movable plate has a fine prolongation which serves as a pointer for indicating the deflections on the scale of the instrument, and at its lower end is fixed a knife edge having its length perpendicular to the plane in which the plate moves.

There are two pairs of terminals, one pair for the fixed plates and the other pair for the movable plate, and each terminal is insulated from the case of the instrument. Of the pair on the left-hand side of the case, the terminal towards the back of the instrument is in metallic connection with the fixed brass plates, while that towards the front (which may be called the working terminal) is simply an insulated brass pin. A glass U-tube is suspended between these two terminals, and contains a safety-arc of finest copper wire connecting them. The terminal toward

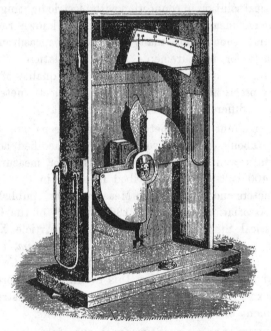

the back of the instrument on the right-hand side is in metallic connection, through the V-groove support, with the movable plate; in other respects the pair of terminals on the right are similar to the pair on the left.

In order to save time in taking readings an arrangement is provided for checking the oscillations of the movable plate, and stops are placed to limit its range and prevent damage to the pointer.

When the fixed and movable plates are connected respectively to two points of an electric circuit, between which there exists a difference of potential, the movable plate tends to move so as to augment the electrostatic capacity of the instrument, and the magnitude of the force concerned in any case is proportional to the square of the difference of potential by which it is produced. In the use of the instrument this force of attraction is balanced by the horizontal component of a weight of any convenient amount hung on the knife edge at the bottom of the movable plate.

The scale is graduated from 0 to 60, and the divisions represent equal differences of potential—the actual magnitude of the difference per division being dependent upon the weight in use at the time. A set of three weights is sent with each instrument, of respectively 32·5, 97·5, and 390 milligrammes, providing for three grades of measurement in the proportion of 1 : 2 : 4. Thus the instrument shows one division per 50 volts with the link (the lightest weight) alone on; one division per 100 volts with the medium weight hanging on the link, and one division per 200 volts with all three weights on.

At from 8,000 to 10,000 volts there is some liability of a spark passing between the movable plate and one or other of the four fixed quadrant plates between which it moves, hence the highest potential for which the instrument can be used is about 10,000 volts. For higher potentials my long-range electrometer [*Collected papers on Electrostatics and Magnetism*, §§ 383, 384] is, so far as I know, the only standard electrometer which has hitherto been practically used. But with the experience which I have had of gravity instruments for electric balances in general, and particularly of the electrostatic voltmeter, I am now convinced that an electric balance for weighing the direct attractive force between a fixed plate and a movable disc will be the best form of standard electrometer for all potentials exceeding 8,000 or 10,000 volts. I hope before the close of next session to be able to show to the Philosophical Society a convenient instrument of this kind, adapted to measure from 10,000 to 40,000 volts and possibly even to 70,000 or 80,000 volts; although I cannot anticipate for measurement of such high electric potentials any great practical demand, as far as the future of electric technology can be conjectured at present.

The electrometer is available with equal convenience for the measurement of electric potentials in circuits of direct current or of alternate current. In the latter application the result is quite

definitely the square root of the time average value of the square
of the difference of potentials between the points touched by its
electrodes. A good deal is now being done for electric lighting
and the electric transmission of power by means of circuits at high
potentials of from 700 to 2,000 and 3,000 volts; and for such
purposes the electrostatic voltmeter, described and illustrated
above, is conveniently available. The method of finding potentials
by measurement of current through a known resistance is also
available even for these very high potentials, but the electrometer
is preferable because of the cumbrousness and expensiveness of the
resistance coils required by the current-measuring method when
the potential is more than 600 or 700 volts.

For potentials of from 500 volts downwards the method of
measuring current through resistance is thoroughly convenient and
practical. Thus we are led to the consideration of the current-
measuring branch of our "Two-branch" chain of measuring
instruments.

Beginning with the feeblest currents, I may refer to the astatic
mirror galvanometer of the form introduced by me about twenty-
eight years ago, and now much used for laboratory and telegraph
testing. This instrument is capable of being adjusted to measure
currents as low as the fifty millionth of a milliampere with a coil
of from 5,000 to 10,000 ohms. It has, besides, the great advantage
that its sensibility can be easily varied through a wide range. Thus
an instrument, which, when in its most sensitive state, will measure
the fifty millionth of a milliampere can be easily arranged to
measure the thousandth of a milliampere. Professor S. P. Langley
of the Allegheny Observatory, U.S., recently used in his radiation
experiments one of these instruments which he had specially made,
and which included several important improvements of his own,
such as the use of a very long suspending fibre and a very perfect
mirror. He finds an instrument of 20 ohms resistance capable of
being used with perfect accuracy for the measurement of the two
millionth of a milliampere. In ordinary telegraph-testing these
galvanometers are commonly adjusted to measure currents down
to about the millionth of a milliampere, and for special tests, as,
for instance, the measurement of the insulation of short lengths
of core for submarine cable, sensibilities as high as those specified
above are often used.

The current-measuring instrument, commonly called by the name of galvanometer, whether it be the sine galvanometer, the tangent galvanometer, or any of the different varieties of mirror galvanometers in use, is essentially magneto-static. That is to say, it is an instrument in which a controlling magnet is used to balance the electro-magnetic force produced on a magnetised steel needle by a conductor or coil through which flows the current to be measured. In the sine galvanometer and tangent galvanometer, as originally used, and as still largely used for many important electrical measurements, the controlling magnet is the earth; but it is only in a locality far from dynamos and from wires carrying continuous* currents for electric lighting, that the terrestrial magnetic field suffices for a sine or a tangent galvanometer when any approach to accuracy is required; and thus in telegraph offices, workshops, and engine-rooms it is generally desirable, if not absolutely necessary, to use a much stronger magnetic field than that of the earth for controlling the needle of a galvanometer. But whether the controlling magnet be the earth or a steel magnet its force is essentially inconstant; and, therefore, a magneto-static galvanometer of any kind requires some means for determining from time to time the coefficient by which the absolute value of the current measured can be calculated from its indications, or a means of freshly adjusting the field to give any convenient absolute value to the readings on the scale of the instrument.

Sixteen or seventeen years ago I introduced a form of tangent galvanometer which came to be called the "paddle wheel galvanometer" from the appearance of the two coils being somewhat like the paddles of a paddle steamer. In this instrument two coils, adjustable to any desired equal distance on the two sides of the centre, act on a small needle or group of small needles hung by a single silk fibre, and carry a pointer of about 8 centimetres length, which shows deflections on a scale of unequal divisions proportional to the tangents of the deflections. Ordinarily the needle is acted upon by the terrestrial magnetic force alone, but sometimes in laboratory experiments a controlling magnet is placed either to augment or to diminish the controlling force on the needle. This instrument, though capable of considerable accuracy

---

* The neighbourhood of a conductor carrying an alternate current for electric lighting, however strong, would not disturb the terrestrial magnetic field for a sine galvanometer or tangent galvanometer in its neighbourhood.

for absolute measurement in a locality free from local magnetic
disturbance, was replaced by my graded galvanometers*, which
have been found much more convenient for general purposes and
have been somewhat largely used since they were brought out five
years ago. I have never, however, been quite satisfied with them,
and I have been incessantly occupied in the attempt to produce
an adjustable magneto-static tangent galvanometer which shall be
both more convenient for ordinary use, and more susceptible of
high accuracy for scientific investigation than the graded galvanometers. I am not able this evening to show more than a half-
finished attempt to realize this object, but I feel that I am now
nearly touching it, and I hope at the commencement of next
session to be able to place before you a working instrument of the
magneto-static class, which shall be thoroughly convenient for
showing the number of lamps a-light at any time in an ordinary
dwelling-house lighted by Edison-Swan lamps with direct current,
and which shall also be available as a scientific measuring instrument to measure currents of from one to sixty amperes with a
proportionate accuracy of one quarter per cent. in the best part of
its scale.

I have also made great efforts during the last six years to
produce satisfactory constant-standard instruments for measuring
electric currents, and I have from time to time placed before you
results in which the object was to some degree realized, but, as
I have always told you, not satisfactorily. This time last year
I placed before you and explained instruments depending on
Faraday's discovery of the tendency of a globe, or cube, or short
bar of soft iron, to move from places of weaker to places of stronger
force in a magnetic field†. Two of those instruments which you
saw, and which were then nearly completed, are indicating the
potential of the current by which we are lighted this evening.
You see when I increase the potential between the terminals of
the instrument (by taking out lamps in multiple-arc between the
electric light mains), how the indicator shows—now an augmentation of $2\frac{1}{2}$ per cent. which makes the lights considerably brighter;
now again a diminution of 1 per cent.; and again the previous
potential, and the lights as they were. Since you first saw them
these two instruments have been incessantly at work on the

* Patent No. 5,668, of 1881, 26th December.
† Patent No. 11,106 [Provisional Specification], 9th August 1884.

electric light wires in my house and laboratory, and have done good work all this time. I may say in passing they have allowed me to use much higher incandescence without risk to the lamps than would have been practicable without a trustworthy potential indicator, and personal, if not automatic, attention to regulate the potential according to its indications.

Though, however, they have so far been practically useful to myself, I am not satisfied with the instruments, because they involve the use of soft iron; the retentiveness of which is always a serious trouble that I have only been able to keep within bounds, not completely to eliminate, by the use of the current reversor which you see in connection with one of the instruments before you. I have, therefore, returned to the original discovery by Ampère of the mutual force between movable and fixed portions of an electric circuit, first utilized by Weber in his "Electro-dynamometer," to obtain a constant-standard instrument for measuring electric currents without any of the trouble and residual inaccuracy entailed by the use of soft iron. After many trials, I have succeeded in improving an electro-dynamic balance towards which I made a great many trials five years ago; and within the last three months I have succeeded in my laboratory in making accurate measurements of currents of from 20 milli-amperes to 200 amperes, by means of four instruments, on the general plan of the rudimentary centi-ampere balance before you, in which a single movable coil is repelled and attracted upwards by two fixed coils, one below it and the other above it, and the total of the electro-magnetic force is balanced by a weight hung on a knife edge attached to the balance*. At the commencement of next session I hope to show you some of these instruments in a state fit for practical work. In the meantime, I can only thank you for the patience with which you have listened to me this evening, and for the kindness with which now, as on previous occasions, you have allowed me to bring half-done work before you, and to tell you of what I hoped to do, but had not yet done, in the way of producing practically useful instruments.

* Patent No. 2,028 [Provisional Specification], 21st April 1883.

214. NEW ELECTRIC BALANCES.

[From *British Association Report*, 1887, pp. 582, 583; *Electrician*, Vol. xx. Dec. 16, 1887, pp. 130, 131; *Nature*, Vol. xxxvi. Sept. 29, 1887, p. 522.]

THESE balances are founded on the mutual forces, discovered by Ampère, between the fixed and movable portions of an electric circuit. The mutually-influencing portions are usually circular rings. Circular coils or rings are fixed, with their planes horizontal, to the ends of the beam of a balance, and are each acted on by two horizontal fixed rings placed one above and the other below the movable ring. Six grades of instrument are made, named centi-ampere, deci-ampere, ampere, deca-ampere, hecto-ampere, and kilo-ampere balance. The range of each balance is about 25. Thus, the centi-ampere balance will measure currents of from 2 to 50 centi-amperes, while the kilo-ampere balance will measure currents of from 100 to 2500 amperes. Since the indications of the instrument depend on the mutual forces between two parts of an electric circuit of permanent form and relative position, they are not subject to the changes with time which are so troublesome in instruments the constants of which depend on the strength of permanent magnets.

The most important novelty in these balances is the connexion between the movable and the fixed parts of the circuit. The beam of the balance is suspended by two flat ligaments made up of fine copper wires placed side by side. These ligaments serve instead of knife-edges for the balance, and at the same time allow the current to pass into and out from the movable coils. The number of wires in each ligament varies from 20 in the centi-ampere to 900 in the kilo-ampere balance. The diameter of the wire is about $\frac{1}{10}$ of a millimetre, and each centimetre breadth of the ligament contains about 100 wires.

The electric forces produced by the current are balanced by means of weights which can be moved along a graduated scale by means of a self-relieving pendant. Two scales are provided—one a scale of equal divisions, the other a scale the numbers on which are double the square roots of the numbers on the scale of equal divisions. The square-root scale allows the current to be read off directly to a sufficient degree of accuracy for most purposes. When high accuracy is required, the fine scale of equal divisions may be used, and the exact value of the current obtained from a table of doubled square roots supplied with the instrument.

An engine-room voltmeter on a similar plan was described. It consists of a coil fixed to the end of a balance arm (suspended as above described) and acted on by one fixed coil placed below it. The distance apart of the two coils is indicated by means of a magnifying lever, and serves to indicate the difference of potential between the leads to which the instrument is connected. The coils of the instrument are of copper wire, and an external platinoid resistance of considerably greater amount is joined in circuit with it. The electrical forces are balanced by means of a weight placed in a trough fixed to the front of the movable coil and weights suited to the temperatures 15°, 20°, 25°, 30° C., as indicated by a thermometer with its bulb in the centre of the coil, are provided.

Two other instruments were described—namely, a marine voltmeter suitable for measuring the potential of an electric circuit on board ship at sea, and a magneto-static current-meter suitable for a lamp-counter.

In the marine voltmeter an oblate spheroid of soft iron is suspended in the centre of, and with its equatorial plane inclined at about 40° to, the axis of a small coil of fine wire, by means of a stretched platinoid wire. When a current is passed through the coil, the oblate of soft iron tends to set its equatorial plane parallel to the axis of the coil, and this tendency is resisted by the rigidity of the suspension wire.

The lamp-counter is a tangent galvanometer with special provision for preventing damage to its silk fibre suspension, and for allowing the constant to be readily varied by the user to suit the lamps on his circuit.

215. On the Application of the Centi-ampere or the Deci-ampere Balance for the Measurement of the E.M.F. of a Single Cell.

[From *British Association Report*, 1887, pp. 610, 611; *Electrician*, Vol. xx. 1888, pp. 130, 131, 272; *Nature*, Vol. xxxvi. Sept. 29, 1887, p. 522; with additions, *Phil. Mag.* Vol. xxiv. Dec. 1887, pp. 514—516 and Vol. xxv. Feb. 1888, p. 164.]

The method described in this paper for the determination, in absolute measure, of the electromotive forces of voltaic cells, consists in the use of one of my standard ampere-balances instead of the tangent-galvanometer in the method given in the following statement, which I quote from Kohlrausch's *Physical Measurements*, pp. 223, 224, 230:—"The only method applicable to inconstant elements, of which the electromotive force varies with the current-strength, is to bring the current to zero by opposing an equal electromotive force. Poggendorff's method, which is very convenient, as it involves no measurement of internal resistance, requires the use of a galvanoscope, *G*, a galvanometer, *T*, and a rheostat, *R*, and, in addition, that of an auxiliary battery, *S*, of constant electromotive force, greater than either of those which are to be compared. The arrangement of the experiment is shown in the figure. In the left division of the circuit are the galvanoscope *G*, and the electromotive force *E* to be measured; in the right, the auxiliary battery *S* and the galvanometer *T*. *E* and *S* are so placed that their similar poles are turned towards each other. In the middle part of the circuit, which is common to both batteries, is the rheostat *R*.

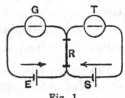

Fig. 1.

"As much rheostat resistance *W* must now be intercalcated as will cause the current in *EG* to vanish, and the current-strength *J* in *T* must then be observed....If *J* = current-strength in *T*, the electromotive force of the battery *E* is *E = WJ*."

The deci-ampere balance, or, when a sufficient number of battery-cells is available, the centi-ampere balance, answers well for the current measurements here required. An arrangement of the circuit which is convenient for most purposes is shown in the diagram (fig. 2); but it may be remarked that the reversing-keys there shown may be replaced by ordinary make-break keys. Referring to the diagram, a battery of a sufficient number of cells is joined in circuit through a reversing-key with a rheostat, a deci-ampere balance, and a standard resistance. The poles of the cell to be tested are connected in circuit with a key and a sensitive mirror-galvanometer to the two ends of the standard

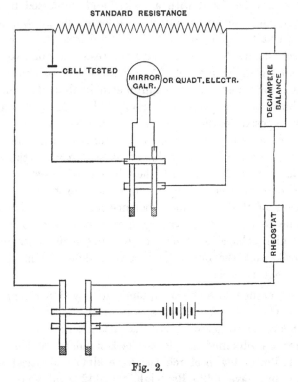

Fig. 2.

resistance in such a way that both the battery and the cell to be tested tend to send a current in the same direction through that resistance. Care should be taken that the circuit of the cell to be tested is well insulated, and that both it and the standard resistance are free from other electromotive force. When, as in the case of Clark-standard cells, the cell is incapable of maintaining a current, a high-resistance galvanometer or an additional resistance should be included in its circuit. If very great sensibility is not required, a quadrant-electrometer may in such cases be substituted for the galvanometer. The standard resistance must be of such a form that no sensible error is introduced through heating by the passage of the current. A good plan is to wind well-insulated platinoid wire in one layer on the outside of a brass or copper cylindrical vessel which can be filled with water. The temperature of the water, when it is nearly the same as that of the air outside will be very approximately the temperature of the coil. For still greater accuracy the cylinder may be fitted with a jacket and immersed in a vessel of water, and appliances introduced for changing the water in each part and keeping account of its temperature. For use with the deci-ampere balance, a platinoid resistance of two ohms is sufficient for any single cell. A resistance of two ohms, made of insulated platinoid wire one millimetre in diameter, and wound on a brass tube capable of holding half a litre of water and simply exposed to the air outside, will carry a current of one ampere for an hour without changing its resistance more than a tenth per cent. The water should be stirred when the readings are taken, and, if necessary, the change of resistance can be approximately allowed for by taking its temperature. To measure by means of the deci-ampere balance an electromotive force of from one to two volts, a battery of two small secondary cells or four Daniell cells, a resistance of two ohms such as has just been described, with the other appliances as indicated in the figure, are all that is necessary.

The method has been applied in my laboratory by Mr Thomas Gray for the measurement of the electromotive forces of standard and other cells, and has been found very convenient. The results obtained for four Clark-standard cells set up by Mr J. T. Bottomley in March last were almost identical with one another, and gave 1·439 Rayleigh, or 1·442 legal, volts at 11° C. The variation of the electromotive force of these cells with temperature has not yet been determined; but assuming the average value obtained for this variation by Lord Rayleigh, namely a fall of ·077 per cent. per degree centigrade rise of temperature, and correcting to 15° C., we obtain 1·4346 Rayleigh volts at that temperature. This result is interesting as showing a difference of less than $\frac{1}{30}$th per cent. from that obtained by Lord Rayleigh for similar cells, which was 1·435 at 15° C.

216. On a New Composite Electric Balance.

[From *Glasg. Phil. Soc. Proc.* Vol. XIX. [read Feb. 4, 1888], pp. 273, 274.]

THIS instrument has been designed for the purpose of providing, in one piece of apparatus, the means of measuring (1) the
difference of potential between two points of an electric circuit,
as, for instance, the difference of potential between the supply-
conductors of an electric-light installation; (2) the current flowing
in such a circuit; and (3) the rate of working in the circuit.
The instrument thus forms a combined voltmeter, ampère-meter,
and watt-meter. The general form of the instrument is shown
in Fig. 1. In this figure, *a* and *b* are two coils of silk-
covered copper wire fixed one above the other, with their planes
horizontal, on a slab of slate, *S*. Two coils, *c* and *d*, of similar
wire, are made up in the form of anchor rings, and fixed to the
ends of a balance beam, *B*, which is suspended by two flat ligaments, *e* and *f*, of fine copper wires, one of which is shown at *e*,
in such a position that one of the coils fixed to the ends of the
beam is suspended mid-way between the coils, *a* and *b*, with its
plane parallel to the planes of these coils, and with its centre in
the line joining the centres of them. Other two coils, *g* and *h*,
capable of carrying strong currents, are fixed to the sole-plate, *S*,
in positions which, relatively to *d*, are similar to those occupied
by *a* and *b* relatively to *c*. When the instrument is to be used
for the measurement of continuous currents, the coils *g* and *h* are
made of several turns of thick copper ribbon, capable of carrying
currents up to a maximum strength of five hundred ampères.
When it is to be used for the measurement of alternating currents
these coils are made of two or three turns of a stranded copper
conductor. Each wire of the stranded conductor is covered with
silk, so as to insulate it from the others, and, in order as far as
possible to annul the effect of induction in causing the current

K V.   30

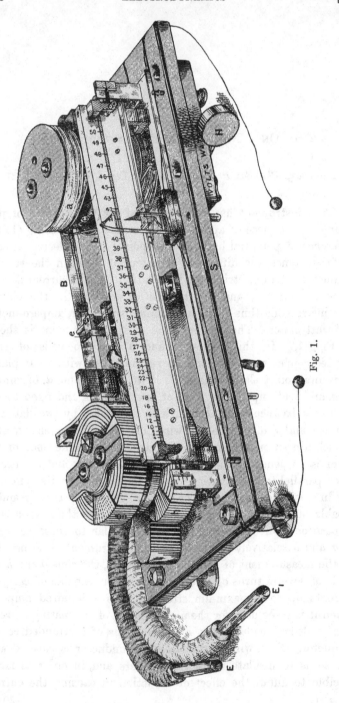

Fig. 1.

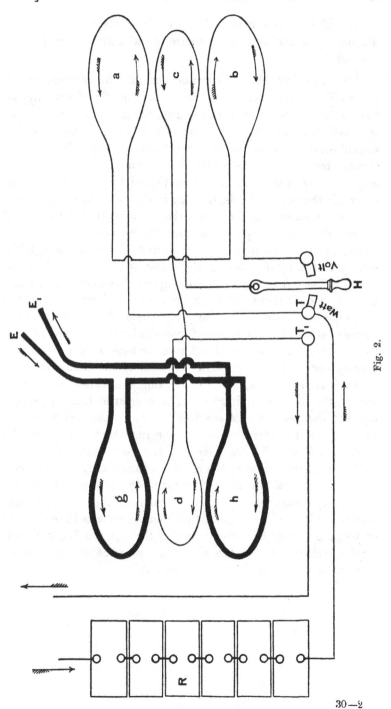

Fig. 2.

to be different at different distances from the axis of the conductor, the strand is given one turn of twist for each turn round the coil.

The arrangement of the connections in the instrument will be readily understood by reference to Fig. 2. First, suppose the instrument to be used for the measurement of potentials, that is to say, as a voltmeter. It is connected to the circuit through a suitable resistance, R, wound anti-inductively, through which the current passes to the terminal, T, from which the course of the current through the coils to T_1 is indicated by the arrows in the diagram; the switch handle, H, being in this case turned to "Volt." For the measurement of ampères the switch is turned to "Watt," a measured current is passed through the suspended coils of the balance, and the current to be measured is passed through the coils, g and h, by introducing the electrodes, E and E_1, into the circuit. The current through the suspended coils may sometimes be measured by means of the instrument itself arranged for the measurement of volts. This may be done by first measuring the current which the difference of potential between the supply-conductors of an electrical installation, or between the poles of a battery, causes to flow through the coils of the instrument and its external resistance, and then turning the switch to "Watt," and, at the same time, introducing a resistance into the circuit equal to the resistance of the fixed coils. When the balance is used as a watt-meter the switch is turned to "Watt," and the terminals, T and T_1, are joined to the two supply-conductors, while the current through the circuit is passed through the coils, g and h. When the rate of working in an alternate-current circuit is measured by such a balance, the anti-inductive resistance, R, must be so great that there is no sensible difference of phase between the currents flowing through the fine wire coils of the instrument and the electromotive force on the supply-conductors to which they are connected.

217. ELECTROMETRIC DETERMINATION OF " v." By Sir WILLIAM
THOMSON, Prof. W. E. AYRTON, and Prof. J. PERRY.

[From *Brit. Assoc. Report*, 1888, p. 616 [title only]; *Electrician*, Vol. XXI.
1888, p. 681 ; *Lum. Élec.* Vol. XXX. Oct. 13, 1888, p. 79.]

THE standard measuring instruments used were an absolute
electrometer and a centiampere balance, with a resistance coil
for measurement of potential. Their indications of potential in
electrostatic measure and in electro-magnetic measure were compared by aid of two electrostatic voltmeters (A and B), which
had been carefully compared in Glasgow, and which were used as
intermediaries. The absolute electrostatic value of the scale of
one of the intermediaries (A) was determined in London by
Messrs Ayrton and Perry, using the absolute electrometer. They
consider that as neither of the two instruments is capable of very
minute accuracy, their determination might be three-quarters per
cent. wrong, but they believe that it may be quite trusted to be
within one per cent. of the truth. The voltage of the other
electrostatic voltmeter (B) was determined from a potential of
about 80 volts between the two ends of a resistance of 600 ohms,
with a current of about 133 milliamperes through it, very
accurately measured by a centiampere balance. The 80 volts
multiplied 16 times by the method of condensers in series
described by Sir William Thomson to Section A at Aberdeen in
1885 gave the voltage of the B electrostatic voltmeter. As this
instrument is not capable of very minute accuracy, the determination may have been wrong half per cent., but it may be
trusted as probably not three-quarters per cent. wrong.

The comparison of these measurements then gave the voltage
corresponding to electric potential reckoned in electrostatic units.
The final result is that the electrostatic C.G.S. unit of potential is
equal to 292 volts—that is to say, 292×10^8 C.G.S. units of electromotive force in electro-magnetic measure ; in other words, "v" is

292×10^8 centimetres per second. This result may, we believe, be trusted to as within $1\frac{3}{4}$ per cent. of the truth. It therefore seems that "v" cannot be greater than 297×10^8 nor less than 287×10^8. We present it merely as an approximate result*, and look forward to measuring this very important physical constant within one-quarter per cent. by the purely electrometric method carried out by an absolute electrometer, and intermediary electrostatic voltmeters, capable of greater accuracy than the instruments which we have used hitherto.

* [The modern value is of course 300×10^8: cf. *infra*, p. 473.]

218. Electrostatic Measurement.

[From *Roy. Inst. Proc.* Vol. xii. 1889 [Feb. 8, 1889], pp. 561, 562 ;
Nature, Vol. xxxix. March 14, 1889, pp. 465, 466.]

A FUNDAMENTAL requisite of a measuring instrument is that its application to make a measurement shall not alter the magnitude of the thing measured. When this condition is not fulfilled (as is essentially the case with an electric measuring instrument not kept permanently in or on the electric circuit or system to which it is applied), it is the magnitude as influenced or modified by the measuring instrument which is actually measured, and the measurement is to be interpreted on this understanding, whatever may be the circumstances.

The nearest approach in electric measuring instruments to the fulfilment of this condition, of not altering the magnitude of the thing measured, is attained by the electrometer when applied to measure differences of potential between different points of a wire or metallic mass of any shape, in which electricity is kept flowing by a battery or dynamo or other electromotive apparatus. The insulation of any practical electrometer is so nearly perfect that the conduction of electricity through the instrument does not sensibly diminish the difference of potentials of the points touched by the electrodes, and the consumption of energy is therefore practically *nil*. In this respect, therefore, the quadrant electrometer would be ideally perfect : but it is only available for potentials of a few volts, and in its most sensitive adjustment indicates about $\frac{1}{100}$ of a volt. The lecturer has therefore designed for ordinary use in connection with electric lighting and the other practical applications of electric energy, a series of instruments which will measure by electrostatic force potentials of from 40 volts to 50,000 volts. The construction of the various types of this series was fully explained.

The standardisation of these instruments up to 200 or 300 volts is made exceedingly easy, by aid of his centiampere balance and continuous rheostat, with a voltaic battery of any kind, primary or secondary, capable of giving a fairly steady current of $\frac{1}{10}$ of an ampere through it and the platinoid resistance in series with it. The accuracy of the electromagnetic standardisation, within the range of the direct application of this method, is quite within $\frac{1}{20}$ per cent. A method of multiplication by aid of condensers, which was explained, gives an accuracy quite within $\frac{1}{5}$ per cent. for the measurement in volts up to 2000 or 3000 volts; and with not much less accuracy, by aid of an intermediate electrometer, up to 50,000 volts.

He also explained, and illustrated by a drawing, an absolute electrometer which he had constructed for the purpose of measuring "v," the number of electrostatic units of potential or electromotive force in the electromagnetic unit of potential. This number "v" is essentially a velocity, and experiments have proved it to be so nearly equal to the velocity of light that from all the direct observations hitherto made we cannot tell whether it is a little greater than, or a little less than, or absolutely equal to, the velocity of light.

The determination was made by comparing the electromagnetic with the electrostatic value, in C.G.S. units, as given by the balance, for a potential of 10,000 volts: but hitherto he has not been able to make sure of the absolute accuracy of the electrostatic balance to closer than $\frac{1}{2}$ per cent.

The results of a great number of measurements which had been made in the Physical Laboratory of the University of Glasgow during the previous two months gave the required number, "v," within $\frac{1}{2}$ per cent. of 300,000 kilometres per second; the velocity of light is known to be within $\frac{1}{4}$ per cent. of 300,000 kilometres per second. Results of previous observers for determining "v" had almost absolutely proved at least as close an agreement with the 300,000,000 metres. He expressed his obligations to his assistants and students in the Physical Laboratory of Glasgow University, Messrs Meikle, Shields, Sutherland, and Carver, who worked with the greatest perseverance and accuracy, in the laborious and often irksome observations by which he had attempted to determine "v" by the direct electrometer method, as exactly as, or more exactly than, it has been determined by other observers and other methods.

Note added March 14th, 1889.

The measurement of "v" by Sir William Thomson and Profs. Ayrton and Perry, communicated to the British Association at Bath, was too small (292) on account of the accidental omission of a correction regarding the effective area of the attracted disk in the absolute electrometer. When this correction is applied their result is brought up to 298, which exactly agrees with Profs. Ayrton and Perry's previous determination by another method, in Japan. Prof. J. J. Thomson's result is 296·3. It is understood that Rowland has found 299. The result of Sir William Thomson's latest observations, founded wholly on the comparison of electrometric and electro-magnetic determinations of potential in absolute measure, is 30·1 legal ohms, or 30·04 Rayleigh ohms. Assuming, as is now highly probable, that the Rayleigh ohm is considerably nearer than the legal ohm to the true ohm, the result for "v" is 300,400,000 metres per second. Sir William does not consider that this result can be trusted as demonstrating the truth within $\frac{1}{3}$ per cent.

219. ON THE SECURITY AGAINST DISTURBANCE OF SHIPS'
COMPASSES BY ELECTRIC LIGHTING APPLIANCES.

[From *Inst. Elec. Engrs. Journ.* Vol. XVIII. 1890 [May 23, 1889], pp 567—571,
579 ; *Lum. Élec.* Vol. XXXIII. Aug. 10, 1889, pp. 290, 291.]

THE danger to be avoided is sufficiently explained in the
following short statement by Mr William Bottomley, which
appeared in the *Nautical Magazine* for December, 1885 :—

"The following example of a case which might occur in any
large ship, will show the amount of error which may be produced on the compass" [by the electric lighting apparatus] "unless
precautions are taken to guard against it.

"Suppose a main lead from the engine-room to the fore part
of the ship, to light up 100 lamps, is brought along the centre
of the ship. It may be at a distance of 10 metres, or 33 feet,
from the standard compass, and will run almost underneath it.
If we suppose that each lamp takes one ampere of current
there will be a current of 100 amperes altogether in this lead.
Now, the effect" [of an infinitely long straight current] "on the
compass" [above it] "at a distance D in centimetres is given
by the formula $F = 2 \times \frac{1}{10} C/HD$, where C is the current in amperes
and H is the horizontal magnetic force. In this case we have
$C = 100$ amperes and $D = 1,000$ centimetres. Therefore

$$F = 20/1,000H = 0.02/H.$$

At Glasgow the horizontal force may be taken as 0·15 in C.G.S.
units, therefore the effect on the compass will be ·02/·15 = 1/7·5.
This will be expressed in degrees by multiplying by 57·3, the
number of degrees in the radian, or angle subtended at the
centre of a circle by an arc equal in length to the radius.
Therefore, the amount of error produced by such a current on
the compass will be 57·3/7·5 = 7·6 degrees.

"The foregoing refers to a single wire and a continuous current machine, but if an alternate current machine is employed no effect will be produced on the compass even when the ship's side is used for the return. When a continuous current machine is used, the danger of producing an error on the compass can be avoided by using two wires close to one another, but these wires should be well insulated from the ship's side. If in any way one of the wires is brought in contact at two points of its length with the iron of the ship there may be no change observable in the lighting, but the current may produce as much error on the compass as it would if there was only a single wire.

"The following points should therefore be attended to in all cases of lighting ships by electricity :—

"First.—With continuous current machines two wires, well insulated, should always be employed.

"Second.—The insulation of the wires should be tested periodically; if any connection with the iron of the ship is found, the fault should at once be corrected.

"Third.—When an alternate current machine is used, a single wire may be employed and the iron of the ship used to complete the current without producing any effect on the compass.

"What makes this question of the greatest importance is that the error may be produced without ever being detected by the officers of the ship. On board ship the errors of the compass are usually determined during the day, in the morning and afternoon, but the electric light is only used at night. The captain may therefore carefully determine his errors every day, and set his course quite correctly; but at night, when the electric light is turned on, the ship may be going several degrees off her proper course, although she is being correctly steered by the compass.

"In connection with the lighting of ships with electricity, there is another point which should also be attended to—that is, the position of the dynamo. If it is placed near an iron bulkhead, the upper end of which is near the compass, the bulkhead may become magnetised by induction so powerfully that it will produce a considerable error on the compass."

The subject was also referred to in Mr Bottomley's paper on *The Magnetism of Ships and the Mariner's Compass*, read before the Society of Arts, January 28, 1886, and published in the Journal of the Society for February 5, 1886. In the discussion which followed, and in which Captain Creak, of the Admiralty Compass Department, Mr Alexander Siemens, and Dr Hopkinson took part, it appeared that in three ships, lighted on the single-wire system with direct currents, small but not unimportant errors in the compass, due to the lighting currents, had been actually observed. Since that time several cases have been reported to me of large passenger ships, lighted with direct currents on the one-wire system, in which as much as 4° or 5° of error on the compass has been produced by the electric lighting. In the latest of these cases, a few weeks ago, an error of 4° on the North course was found when the light was put on. The light was put on and off several times with the ship's head North, and every time the same error was produced.

The precautions for security which I have to suggest are—

1.　The use of the two-wire method exclusively (unless, which is now rarely the case, alternate currents are used).

2.　The most simple and convenient test for faults of insulation capable of disturbing any of the compasses on board is a lamp set up in the neighbourhood of the dynamo, with one end permanently connected with the ship's iron, and a switch for readily putting its other terminal in connection with either of the dynamo mains at any time. The switch should occasionally be moved each way by the engineer in charge, and if either motion lights the lamp to any visible degree, a defect of insulation on the corresponding main is proved, and ought to be immediately corrected. But unless the lamp is lighted to full brilliance, the fault is certainly not so great as to sensibly disturb a compass. Full brilliance proves only one fault; and there must be more than one such fault before any error of practical importance can be produced in any of the compasses.

3.　Care that there is not magnetic "leakage" from the dynamo (as practical men, guided by Faraday's ideas, theory, and language, have now taught the scientific world to call it) enough to produce any compass-disturbance of practical moment. Capt. Creak, speaking at the beginning of 1886, in the Society of Arts discussion previously referred to, said that in one ship the direct compass-disturbance produced by the generating machine was "felt through a distance of 55 ft.," and across iron bulkheads, and that it was perceptible also in other ships of the Royal Navy electrically lighted on the two-wire system.

My impression is that the improved dynamos now made have much less of magnetic leakage than those made prior to 1886, but we still want information as to their disturbing magnetic effect at such distances as have to be considered in connection with the compass question.

4. To ascertain that there is no perceptible compass-disturbance, or if there is any to test its amount, the compass should be observed while the current through the dynamo is started and stopped, either by starting and stopping the dynamo itself, or by making and breaking the circuit of the field magnets. This should always be done before the electric light installation is taken over from the contractors. It is best and most easily done when the ship is in dock, or lying steadily at anchor. On no account ought it to be delayed, in a new ship, till she goes out for compass adjustment. A determination of the amounts of the disturbance, if any, for all courses of the ship can be made by aid of my deflector without moving the ship. But a sufficient practical test may be made by first observing the effect of starting and stopping the current, on the compass as it stands; then adjusting a small magnet placed on, or supported a little above, the glass of the bowl, to deflect the compass about 45° first on one side and then on the other side of its undisturbed position, and in each case observing the effect of starting and stopping the electric current. This effect ought not to be as much as 2° in any of the three cases.

5. A small electric lamp, with its two electrodes insulated and twisted together in the usual manner, may safely (and with very great advantage in most cases of electrically lighted ships) be used to light the compass. The effect, if any perceptible, of its current on the compass ought to be tested in the manner described in No. 4.

After a practical discussion, the President, Sir W. Thomson, continued as follows:—

With reference to Major Cardew's very important remark,

that precaution must be used in attempting to light compasses by an electric lamp, I may say that it is easy to shape the filament so that its magnetic moment, with the current through it, shall be insufficient to produce any sensible disturbance on the compass, however the lamp is placed outside the bowl, for convenience of lighting; and this with a lamp amply powerful enough to light the compass.

The proper arrangement of filament for a binnacle-lamp is to shape it like a hair-pin, with its two sides not more than half a centimetre or a centimetre apart. Suppose, for example, a 50-volt 6-candle lamp (which is a more than amply sufficient light to steer by), the filament would be about 6 centimetres long, giving area 3 square centimetres, and magnetic moment, when excited by $\frac{2}{3}$ of an ampere through it, ·2 c.g.s. The maximum magnetic force of this at a distance of 10 cm. (and it could hardly be placed nearer, even for the smallest yacht compass!) is $2 \times ·2 \times 10^{-3}$, or 1/2500, which could not disturb the compass by more than about a tenth of a degree in these latitudes.

It is, I think, quite certain that, with any practical arrange-ment of the wiring, an alternating current cannot demagnetise a compass to a partial extent. Regarding the magnitude of the disturbance produced by the one-wire system with direct current, I may say that, although Mr Bottomley's illustration was a rough-and-ready example of an extreme case, you have only to vary the figures. Take 150 amperes instead of 100 amperes, or take 60 amperes instead of 100 amperes, and take 20 feet instead of 30 feet; vary it about, and instead of an infinitely long wire, which is convenient for calculation, take any actual length of wire concerned in any particular case, and the well-known formulas will show you that the effect on the compass is practically very considerable. But I must say that theoretical calculations of this kind are mere examples of what are possibilities. If the calculation of such an example as that shown in Mr Bottomley's calculation gave only 2° or 1° for the greatest possible disturbance, then we might rest contented that in no practical circumstance would it be very serious. All we can do by theoretical examples of that kind is to let us know before we go into iron and steel and compasses and ships, before we go out of the laboratory or the workshop—to let us know what can be expected as a possible disturbance. If we know that the greatest possible disturbance is insensible, we may be satisfied; but if we know that the disturbance can be considerable, then experience alone can tell us whether we may neglect the thing in any particular case, or agree to neglect it in general, or not. Now I must say, as a practical matter, that in the first place it is better to avoid a disease altogether than to let a disease be produced and then to find a remedy for it; and in the next place I would say, with reference to proposed remedies, the doctor's bill for curing the disease is liable to be much more expensive than adopting the arrangement in the beginning by which the disease can be prevented, and *the cure is essentially imperfect at best, after all that can be done short of almost complete re-wiring.*

By a troublesome and expensive shunting or doubling of wires in a part of the ship, you may annul the disturbance on one particular compass, but then there is another compass and *another* compass, all three incessantly used in the navigation of the ship. It is practically not possible to arrange the mains on the one-wire system so that there is no sensible error on one or other of the compasses. I did not care to mention the names of ships or Companies, but I know many cases in which there are errors of 3°, 4°, and 5°, undoubtedly due to the electric lighting. Now I would remark that it is not at all satisfactory to have a changing error in the steering compass, although the standard compass may be unaffected. The ship is steered by the steering compass. An officer in another part of the ship looks frequently at the standard compass, and if he finds the course of the ship is wrong he passes an order or a caution to the steersman; but it is exceedingly inconvenient if the officer in command, having known that his steering compass *was* all right at a certain time, should at some uncertain time—when the saloon is lighted up for No. 2 passengers' dinner, for example!—find the ship off her course 2°. He does not know whether it is careless steering or an error in the compass, and it may take ten minutes to find out which it is that has caused the ship to go off 2°. That is an intolerable state of things. Anything that introduces errors at all adds to the complication, great enough and perplexing enough as it is, that already exists, whether with the officers, the watch, or the men steering, and should if possible be avoided; and if the electric lighting of the ship could not be done otherwise than by methods which introduce errors of from 2° to 5° at uncertain times in different parts of the ship, it would be a serious question whether the electric lighting should not be given up altogether, or the captain and officers of the ship should make up their minds to pay careful attention to it and look out for the changes. To depend upon the steward sending a message that he is going to light up a cabin or part of the ship would be a very inconvenient state of things.

It seems to me that if the one-wire system is to be used at all, it ought to be obligatory to use only alternate current with it. If direct currents are used, the two-wire system alone ought to be admitted on board ship. Electricians must not suppose that if sailors do not complain there is nothing to complain of. In the first place, sailors do not always know that they have suffered from the error. Many a man has been steering for several hours, and has never imagined that his compass had been disturbed owing to the lighting of the ship. Even a thoroughly careful man may not have discovered that it was the lighting that has caused some disturbance which he may have noticed in his reckoning.

I have occupied your time too long, but I would just, in conclusion, beg the Institution to consider, as far as the influence of its members is concerned (and I hope my friends will forgive me for being so urgent), whether it would not be better to adopt the two-wire system universally in ship lighting.

220. NOTE ON THE DIRECTION OF THE LONGITUDINAL CURRENT IN IRON AND NICKEL WIRES INDUCED BY TWIST WHEN UNDER LONGITUDINAL MAGNETIZING FORCE.

[From *Phil. Mag.* Vol. XXIX. Jan. 1890 [dated Dec. 21, 1889], pp. 132, 133. Cf. *supra*, pp. 418, 419.]

To avoid circumlocutions suppose the iron or nickel wire to be vertical, and the magnetizing current to be in the opposite direction to that of the motions of the hands of a watch held with its face up.   The undisturbed magnetization is downwards*. Now suppose a right-handed twist to be given to the wire.   Its elongational spiral is right-handed, and its contractional spiral is left-handed.   If the substance is iron, the lines of magnetization become left-handed spirals; if nickel, right-handed.   Now a downward current, in the downwardly magnetized wire, would, by the superposition of circular magnetization in the direction opposite to that of the hands of a watch, cause the lines of magnetization to become left-handed spirals.   Hence the sudden right-handed twist induces in iron a current upwards, in nickel a current downwards.   Thus we have the following simple specification for the

---

* Much of circumlocution is avoided, and of clearness gained, throughout dynamics and physics, by introducing the substantive noun *ward* (as has been done by my brother Prof. James Thomson in his lectures on Engineering, and in lithographed sheets put into the hands of his students, in the University of Glasgow) to signify *line and direction in a line*;—that which is represented ordinarily by a barbed arrow.   Taking advantage of this usage I now define the ward of magnetization as the ward in which the magnetizing force urges a portion of the ideal northern magnetic matter or northern polarity.   By northern polarity, I mean polarity of the same kind as that of the earth's northern hemisphere.   It is that which is marked *blue* by Sir George Airy to distinguish it from southern magnetic matter or southern polarity, which he marked *red*.   According to a usage condemned 300 years ago by Gilbert, but not yet quite dead, English instrument-makers still sometimes mark with an N the true south pole, and with an S the true north pole, of their steel bar-magnets.   All confusion due to this unhappy mode of marking magnets is done away with by Sir George Airy's red and blue.

K. V   31

directions of the induced longitudinal currents in the two sub-stances, without reference to "up" or "down."

From any point, *P*, on the surface of the wire, draw same-wards parallels to the current in the nearest part of the magnetizing solenoid, and to the direction of the induced longitudinal current. Draw a helix through *P* making an acute angle with each of these lines. This helix is of same name as the elongational helix for iron, and as the contractional helix for nickel.

221. On Electrostatic Stress.

[From *Nature*, Vol. XLI. Feb. 13, 1890, p. 358, Abstract; *Edinb. Roy. Soc. Proc.* Vol. XVII. [Jan. 20, 1890, title only], p. 412.]

A COMPLETE dynamical illustration of electro-dynamic action may be had in an elastic solid, homogeneous in so far as rigidity is concerned, permeated with pores of unalterable size containing liquid. These pores may be in part in communication with each other, and in part closed by elastic partitions. These cases cor-respond to conductors and non-conductors respectively. Electro-static stress depends on the curvature and extension of the partitions. The law of capacity in the model is identical with that in conductors.

222. On an Accidental Illustration of the Effective Ohmic Resistance to a Transient Electric Current through a Steel Bar.

[From *Edinb. Roy. Soc. Proc.* Vol. XVII. [read March 17, 1890], pp. 157—167. Reprinted in *Math. and Phys. Papers*, Vol. III. art. xci. pp. 473—483.]

223. On the Time-integral of a Transient Electro-
magnetically Induced Current.

[From *Phil. Mag.* Vol. xxix. March, 1890, pp. 276—280.]

It has hitherto been generally supposed that, in ordinary
apparatus for electromagnetic induction, with or without soft
iron, the oppositely directed transient currents, in the secondary
circuit, induced by startings and stoppings of current in the
primary circuit, have equal time-integrals.

I have recently perceived [been wrongly led to imagine] that
this may be far from being practically the case by the following
considerations. The starting and stopping of the current in the
primary circuit was, in Faraday's original discovery of this kind
of electromagnetic induction (Exp. Res. Series I. Nov. 1831), and
is generally in elementary illustrative experiments, produced by
making and breaking a circuit consisting of a voltaic element or
battery and the inductor-wire. In this arrangement the starting
of the inductor-current is generally much less sudden than the
stopping. Hence a thicker shell of the secondary wire (or portion
inwards from the outward boundary) is utilized for conducting
the secondary current, on the make, than on the break*. Hence
the effective ohmic resistance in the secondary circuit is less to
the current induced by the make than to the current induced by
the break ; and the time-integral of the former current is correspondingly greater than the time-integral of the latter.

Faraday in his first experiment found the current induced
by the make to be greater than that induced by the break, but
he explained it by the running down of his voltaic battery during
the time the current was passing through the primary, in consequence of which the magnitude of the current stopped on the
break was smaller than that of the current instituted on the

---

* [In reality the whole cross sectional area of the secondary conductor is
utilized, equably in all its parts, in conducting the secondary current. See Postscript of February 23. (W. T.)]

make. This was undoubtedly a *vera causa*, and probably one of considerable potency, considering that Faraday had then no Daniell's battery and had no storage-cells to serve him in his work. Another *vera causa* is the heating of the circuit, which, even with a battery of constant E.M.F., may render the current started very considerably greater than the current stopped in ordinary experiments. Faraday, in his original experiments*, had only magnetization of steel wires to discover the induced currents by, and to test their magnitude; and he had no galvanometer in the primary circuit. If he had had a ballistic galvanometer in his secondary circuit, and any suitable galvanometer for steady currents in his primary circuit, he might possibly have found† that the time-integral (as shown by the ballistic galvanometer) of the primary current exceeded that of the secondary current by a greater difference than could be accounted for by the current being suddenly started and the current suddenly stopped in the primary. So far as I know no one, from 1831 till now, has made any experimental examination of the question suggested in Faraday's Exp. Res. Series I. 16; and his idea that the two currents are equal has been generally accepted‡. I have therefore asked Mr Tanakadaté to make some experiments on the subject in my laboratory. He immediately obtained results§ seeming to demonstrate a considerable excess in the time-integral (as shown by a ballistic galvanometer) of the secondary current on the make above that on the break. A magneto-static galvanometer in the primary circuit showed the current before the break to be always a little less than immediately after the make. This difference was due to heating of the primary circuit, because a potential galvanometer applied to the two terminals of the voltaic element used, which was a large storage-cell, showed no sensible drop of potential during the flow

---

* Exp. Researches, Series I. 1831.

† [No. On the contrary, he would have found that his first idea, of perfect equality of the two currents, was perfectly true! February 23. (W. T.)]

‡ See Maxwell's *Electricity and Magnetism* (1873), Vol. II. § 537, p. 171.

§ But these results we find are quite untrustworthy because of the susceptibility of the steel needle of the galvanometer to magnetic induction, which with the currents produced through its coil by the induced currents due to the make and break in the primary circuit may *largely alter its effective magnetism*. It is in fact well known that ballistic galvanometers with steel needles give very erratic results if they are used in attempting to find time-integrals of very intense transient currents of very short duration.

of the current in the primary circuit. It was, however, insufficient to account for the large differences found between the ballistic deflexions produced by the induced currents on the make and on the break. A rapid succession of makes and breaks has given large irregular permanent deflexions of the ballistic galvanometer, sometimes in one direction, sometimes in the other, which we have found to be chiefly due to magnetic susceptibility of the steel needle of the ballistic galvanometer: and we find that this cause has probably vitiated the observations on the effects of single makes and breaks. I have therefore arranged to have experiments continued, with a coil of very fine wire bifilarly suspended, instead of the steel needle of the ballistic galvanometer; a steady current through the fine wire being maintained by an independent voltaic battery.

Hitherto the make and break have been performed by hand, dipping a wire into and lifting it out of a cup of mercury.

Various well-known methods may be used to render either the break or the make so gradual that we may be sure of the induced current running practically *full-bore* through the secondary. On the other hand, very sudden breaks may be effected by separating two little balls or other convex pieces of copper by the blow of a hammer. The experiment may be varied by short-circuiting the ends of the inductor and allowing the current from the battery to continue flowing through electrodes of sufficient resistance not to allow an injuriously great amount of current to flow. These electrodes, between the voltaic element and the ends of the inductor-wire, may have large self-inductance given to them by coiling them round a closed magnetic circuit of soft iron. The starting of the current in the inductor-wire will thus be rendered much more sudden than the stopping; and the induced current in the secondary will no doubt be found stronger on the stoppage than on the start. Mr Tanakadaté is continuing the experiments with these modifications in view.

The mathematical foundation of the common opinion that the time-integrals of the induced currents on the make and break are equal is as follows*.

---

* [*February* 23.—Worked out more perfectly it shows, as follows, that the common opinion is correct!

Imagine the whole secondary conductor divided into infinitely small filaments of cross-section $d\Omega$: and let ξ be the current-density, at time t, in any one of these.

Let β and γ be the currents at time t in the primary and in the secondary circuits respectively; J the self-inductance of the secondary circuit; and M the mutual inductance of the two. We have

$$R\gamma = -M\frac{d\beta}{dt} - J\frac{d\gamma}{dt}.$$

Hence if $\gamma = 0$ when $t = 0$, *and if we suppose R to be constant,*

$$\int_0^t \gamma\,dt = \frac{M}{R}(\beta_0 - \beta) - \frac{J}{R}\gamma.$$

Now let T be any value of t so large that β has become sensibly constant, and γ has subsided to zero. We have

$$\int_0^T \gamma\,dt = \frac{M}{R}(\beta_0 - \beta_T).$$

This shows that the time-integral of the induced current in the secondary circuit would depend solely on the difference of values of the current in the primary at the beginning and end of the time included in the reckoning, and would be quite independent of suddenness, *if the effective ohmic resistance were constant.* But this supposition is not true; and that it is very effectively untrue for copper wires of a millim. diameter or more, and times of change in the primary less than $\frac{1}{300}$ of a second, we see readily by looking to the diffusional curve and the time-number, $\frac{1}{20}$ of a second for curve 10, corresponding to the diffusivity of copper for electric currents (which is 131 square centimetres per second) given and explained in my short article on a Five-fold Analogy, in the

Instead of γ in the text take $\xi d\Omega$, and instead of R take $l/(\sigma d\Omega)$; σ denoting the specific conductivity of the material, and l the length of the circuit. We have

$$l\sigma^{-1}\xi = -M\frac{d\beta}{dt} - J\frac{d\xi d\Omega}{dt} - S,$$

where S denotes the sum of effects due to the risings and fallings of current in the different parts of the secondary conductor. Their time-integral from 0 to T is essentially zero; as is also that of the infinitesimal middle term of the second member. Hence we have

$$l\sigma^{-1}\int_0^T dt\,\xi = M(\beta - \beta_0);$$

which shows that the time-integral of the current, $\xi d\Omega$, *in each filament of the secondary conductor,* is exactly equal to that which is calculated according to the ordinary elementary theory! The whole details of the fallacy in the text are now clear!]

British Association Report for Manchester, 1888 (to be found also in the Electrical Journals, and *Nature*).

[POSTSCRIPT, *February* 23, 1890.—The thermal analogy, which is very simple for the case of electric currents in parallel straight lines, has, as soon as I have considered it, shown me the fallacy pervading the text; and has led me to make the corrections in the insertion and footnotes enclosed in brackets [ ], all of which are of the date of this postscript.   Preliminary experiments with the suspended coil instead of the steel magnet, in a ballistic galvanometer now nearly completed by Mr Tanakadaté, have already disproved the *large* differences which I expected between the time-integrals of the secondary currents on the break, and on the make; and as far as they have yet gone are consistent with the *perfect equality* which I now find proved by theory.]

224.   ON THE SUBMARINE CABLE PROBLEM, WITH ELECTRO-MAGNETIC INDUCTION.

[From *Nature*, Vol. XLII. July 17, 1890, pp. 287, 288, Abstract; *Edinb. Roy. Soc. Proc.* Vol. XVII. [title only], p. 420.   Cf. *infra*, p. 489.]

SIR W. THOMSON read a paper on the submarine cable problem, with electromagnetic induction   The solution of the problem with intermittent or alternating currents of period so long that the distribution of current over a given cross-section of the core is uniform, is already well known.   Sir W. Thomson extends the solution, through all intermediate stages, to the case in which the period is so short that the current is confined to an exceedingly thin surface-layer of the core.   He has worked out the conditions which obtain with a core and sheath of any forms.   The thickness of the layer depends only, other things being equal, upon the period of alternation—the law being that given by Fourier for the penetration of the annual and diurnal heat-waves into the earth's crust.   The distribution of density throughout the layer depends upon the form and relative position of the core and the sheath.

225. ON AN ILLUSTRATION OF CONTACT ELECTRICITY PRE-
SENTED BY THE MULTICELLULAR ELECTROMETER.

[From *Brit. Assoc. Report*, 1890, p. 728; *Nature*, Vol. XLII. Oct. 9, 1890,
p. 577; *Lum. Élec.* Vol. XXXVIII. Dec. 6, 1890.]

IN the multicellular electrometer, the force between the
aluminium needles and the brass cells is modified by the 'contact-
electricity' difference between polished brass and polished
aluminium. In the trial instruments made up to about a year
ago, the result was scarcely perceptible; probably because care
had not in them been bestowed to give high polish to the metallic
surfaces.

In the instrument as now made, differences of from two-
tenths to three-tenths of a volt are found; averaging about ¼ of
a volt.

The force by which ¼ of a volt of difference of potential could
be shown, on a difference of 100 volts, bears to the force by which
the same difference could be shown with the two metals in metallic
connection, the ratio of $(100 + \frac{1}{4})^2 - 100^2$ to $(\frac{1}{4})^2$ or of 800 to 1.
Thus the use of the multicellular electrometer gives a new and
very interesting direct proof of Volta's contact electricity.

Some careful observations in Major Cardew's new standardising
laboratory of the Board of Trade, made by Mr Rennie at frequent
intervals during the last six weeks, have given doubled differences
of from ·65 to ·5, seeming to show a slight tendency to decrease
with time.

226. ON ALTERNATE CURRENTS IN PARALLEL CONDUCTORS OF
HOMOGENEOUS OR HETEROGENEOUS SUBSTANCE.

[From *Brit. Assoc. Report*, 1890, pp. 732—736; *Nature*, Vol. XLII.
Oct. 9, 1890, p. 577.]

THIS Paper consists of a description of some of the results of
a full mathematical investigation of the subject, which I hope
to communicate to the *Philosophical Magazine* for an early
number* :

1. Two or more straight parallel conductors, supposed for
simplicity to be infinitely long, have alternating currents maintained in them by an alternate-current dynamo or other electromotive agent applied to one set of their ends at so great a
distance from the portion investigated that in it the currents
are not sensibly deviated from parallel straight lines. The
other sets of ends may, indifferently in respect to our present
problem, be either all connected together without resistance, or
through resistances, or through electro-motive agents. All that
we are concerned with at present is, that the conductors we
consider form closed circuits, or one closed circuit, and that
therefore the total quantities of electricity per unit of time at
any instant traversing the normal sections in opposite directions
are equal.

2. We suppose the period of the alternation to be very great
in comparison with the time taken by light to traverse a distance
equal to the greatest diameter of cross-section of our whole
group of conductors. This supposition is implied in the previous
assumptions of parallel rectilinearity of the electric stream lines,
and of equality of the quantities of electricity traversing, in
opposite directions, the several areas of a normal section.

3. We farther suppose that the length of our conductors

---

* [For more detailed treatment, cf. Presidential Address on "Ether Electricity
and Ponderable Matter," Jan. 10, 1889, reprinted in *Math. and Phys. Papers*,
Vol. III. pp. 484—513.]

and their effective ohmic resistances are so moderate* that the quantities of electricity deposited on and removed from their boundaries to supply the electrostatic forces along the conductors required for producing the alternations of the currents, are negligible in comparison with the total quantity flowing in either direction in the half-period. This supposition excludes important practical problems of telegraphy and telephony, the problem of long submarine cables, for instance; but it includes the problem of electric lighting by alternating currents transmitted at high tension through considerable distances; as, for example, from Deptford to London.

4. The general investigation includes as readily any number of separate circuits of parallel conductors as a single circuit, but, for simplicity in describing results, I suppose our system of conductors to be so joined at their ends as to constitute a single simple circuit of two parallel conductors†. It may be either two parallel conductors or one conductor, one of which may or may not surround the other, as shown in Figs. 1 and 2, representing cross-sections. Each conductor may be single, as in Figs. 1 and 2, or either may be multiple parallels.

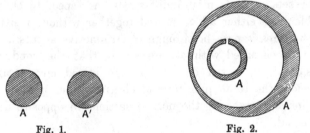

Fig. 1.　　　　　　　　　　　　Fig. 2.

* The circumstances in which this condition is fulfilled may be usefully illustrated by considering the important practical cases of submarine cables, and of metallic circuits of two parallel wires insulated at a distance anything less than a few hundred times their diameter. For all these cases the numeric expressing the electrostatic capacity of either conductor per unit of its length (the other supposed for the moment to be at zero potential) is between 2 and 0·1, and for our present rough comparison may be regarded as moderate in comparison with unity. On this supposition the condition of the text requires for fulfilment that the mean proportional between the velocity which expresses in electro-magnetic measure the resistance of one of the conductors and the velocity of a body travelling the length of the conductor, in a time equal to half the period of alternation, shall be exceedingly small in comparison with the velocity of light.

† Cf. "Anti-Effective Copper in Parallel or in Coiled Conductors for Alternate Currents," *infra* [p. 496].

5. We suppose each conductor to be homogeneous in substance, and in cross-section from end to end, but not necessarily homogeneous in different parts of the cross-section. Thus the two conductors, or the different parts of either, may be of different metals, or either conductor or any part of either conductor may consist of two metals (as iron and copper, or iron and lead) laid parallel and soldered together.

6. We shall call A and A' the cross-sectional areas or groups of areas of the two conductors respectively of the other. All the different portions of A are connected metallically at their two ends, and are thus all of them at one potential at one end and another potential at the other end; and similarly for A'. The homogeneousness of the material and of the cross-sections along the length of the conductors and the uniformity of the total currents assumed in § 3, implies that all the different parts of A in one cross-sectional plane are at one potential, even though A consist of mutually isolated parts, or A' consist of isolated parts. If, as in Figs. 1 and 2, all the parts of A are in mutual metallic connection, and all the parts of A' are in mutual metallic connection, this would entail uniformity of potential through A, and uniformity of potential through A', even without the limitation of our subject laid down in § 3.

7. The following are some of the most noteworthy results of the full mathematical treatment of the subject:

I. When the period of alternation is large in comparison with 400 times the square of the greatest thickness or diameter of any of the conductors, multiplied by its magnetic permeability, and divided by its electric resistivity, the current intensity is distributed through each conductor inversely as the electric resistivity; the phase of alternation of the current is the same as the phase of the electro-motive force; and the current across every infinitesimal area of the cross-section is calculated, according to the electro-motive force at each instant, by simple application of Ohm's law.

II. When the period is very small in comparison with 400 times the square of the smallest thickness, or diameter of any of the conductors, multiplied by its magnetic permeability and divided by its electric resistivity, the current is confined to an exceedingly thin surface-stratum of the conductors. The thickness of this stratum is directly as the square root of the quotient of resistivity divided by magnetic permeability, of the substance in different parts of the surface. The total quantity of the current per unit breadth of the surface is independent of the material, and, except in such cases as those referred to at the end of Case II below, varies in each cross-section in simple proportion to the electric surface density of the static electrification induced by the electro-motive force applied between the extremities for maintaining the current. The distribution of this electric density is similar in all cross-sections, but its absolute magnitude at corresponding points of the cross-section varies along the length of the conductor in simple proportion to the difference of electric potential between A and A', and is zero at one end, in the particular case in which the conductors are connected through zero resistance at one end, while the electro-motive force is applied by an alternate current dynamo at the other end. On the other hand, the surface distribution of electric current is uniform throughout the whole length of the conductors, and it is only its distribution in different parts of the cross-section that varies as the electric density.

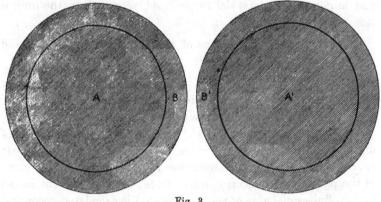

Fig. 3.

The proportionality of surface intensity of the current to electric density, asserted above, fails clearly in any case in which the circumstances are such that the distance we must travel along the surface to find a sensible difference in electric density is not very great in comparison with the thickness of the current-stratum. Such a case is represented in Fig. 3, which is drawn to scale for alternate currents of period $\frac{1}{80}$th of a second in round

rods of copper of six centimetres diameter. The spaces between
the outer circular boundaries and the inner fine circles indicate
what I have called the mhoic thickness*, being 0·714 of a centimetre for copper of resistivity 1611 square centimetres per second.
The full solution for such a case as that represented in Fig. 3
belongs to the large class of cases intermediate between Cases I
and II, and could only be arrived at by a kind of transcendent
mathematics not hitherto worked. But, without working it out,
it is easy to see how the time-maximum intensity of the current
will diminish inwards from the surface, and will be, at any point of
either of the inner fine circles, about one-half or one-third of what
it is at the nearest point of the boundary surface; and that at
points in the surface, distant from BB' by one-half or one or two
times the mhoic thickness, the current intensity will be much
smaller than it is at B and B'.

III.   In Case I the heat generated per unit of time, per unit
of volume, in different parts of the conductors, is inversely as
the electric resistivity of the substance, and directly as the square
of the total strength of current, at any instant. In Case II the
time-average of the heat generated per unit of time, per unit of
area of the current stratum, is as the time-average of the square
of the quantity of current per unit breadth, multiplied by the
square root of the product of the electric resistivity into the
magnetic permeability.

IV.   Example of Case III : Let the conductor A be a thin flat
bar, as shown in the diagram (Fig. 4), A' being a tube surrounding
A, or another flat bar like A, or a
conductor of any form whatever,
provided only that its shortest distance from A is a considerable
multiple of the breadth of A. The
thickness of A must be sufficiently
great to satisfy the condition of
Case II, and its breadth must be a

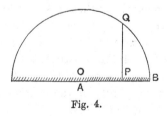

Fig. 4.

large multiple of its thickness. (For copper carrying alternating
currents of frequency 80 periods per second, these conditions
will be practically fulfilled by a flat bar of 4 cms. thickness
and 30 or 40 cms. breadth.) The current in it is chiefly con-

---

* *Collected Papers*, Vol. III. Art. cii, § 35.

fined to two strata extending to small distances inwards from its two sides. (For copper and frequency 80 periods per second, the time-maximum of intensity of the current at the surface will be about ϵ^2, or 7·4, times what it is at a distance 1·43 cms. in from the surface.) The quantity of current per unit breadth, or, as we may call it for brevity, the surface-density of the current in each stratum, is determined by the well-known solution of the problem of finding the surface-electric-density of an electrified ellipsoid of conductive material undisturbed by any other electrified body. The case we have to consider is that of an ellipsoid, whose longest diameter is infinite, medium diameter the breadth of our flat conductor, and least diameter infinitely small. In this case the electric density varies inversely as $\sqrt{(OB^2 - OP^2)}$. The graphic construction in the drawing shows $PQ = \sqrt{(OB^2 - OP^2)}$, and we conclude that the time-maximum of the surface-density of the current varies inversely as PQ. The infinity, which in the electric problem we find for electric density of the ideal conductor, is obviated for the electric current problem by the proper consideration of the rectangular corners or the rounded edge (as the case may be) of our copper bar, which, though exceedingly interesting, is not included in the present communication. Suffice it to say that there will be no infinities, even if the corners be true mathematical angles.

V. Example of Cases I and II: Let A consist of three circular wires, C, L, and I, of copper, lead, and iron respectively. In Case I the quantities of the whole current they will carry and the quantities of heat generated per unit of time in them will be inversely as their resistivities In Case II, if the centres of the three circular cross-sections form an equilateral triangle, the quantities of heat generated in them will be directly as the square roots of the resistivities for C and L; and for Case I would be as the square root of the product of the resistivity into the magnetic permeability, if the magnetic permeability were constant and the viscous or frictional resistance to change of magnetism nothing for the iron in the actual circumstances. This last supposition is probably true approximately with a permeability of $\frac{1}{80}$ for iron or steel, according to Lord Rayleigh, if the current is so small that the greatest magnetising force acting on the iron is less than ·1 C.G.S.

VI. The dependence of the total quantity carried on extent

of surface, according to the electrostatic problem described in
Case II, justifies Snow Harris, and proves that those who condemned him out of Ohm's law were wrong, in respect to his advising tubes or broad plates for lightning conductors; but does not
justify him in bringing them down in the interior of a ship (even
through the powder magazine) instead of across the deck and
down its sides, or from the masts along the rigging and down the
sides to the water. The non-dependence of the total quantities
of current on the material, whether iron or non-magnetic metals,
seems quite in accordance with Dr Oliver Lodge's experiments
and doctrines regarding "alternative path" and lightning conductors. The case of alternate currents is, of course, not exactly
that of lightning discharges; but from it, by Fourier's methods,
we infer the main conclusions of Cases II and V, whether the discharges be oscillatory or non-oscillatory, provided only that it be
as sudden as we have reason to believe lightning discharges are.

227. On Anti-effective Copper in Parallel Conductors or in Coiled Conductors for Alternate Currents.

[From *Brit. Assoc. Report*, 1890, pp. 736—740; *Nature*, Vol. XLII. Oct. 9, 1890, pp. 577, 578; *Lum. Élec.* Vol. XXXVIII. pp. 317—321.]

1. It is known that by making the conductors of a circuit too thick we do not get the advantage of the whole conductivity of the metal—copper, let us say—for alternate currents. When the conductor is too thick, we have in part of it comparatively ineffective copper present; but, so far as I know, it has generally been supposed that the thicker the conductor the greater will be its whole effective conductance, and that thickening it too much can never do worse than add comparatively ineffective copper to that which is most effective in conveying the current. It might, however, be expected that we could get a positive augmentation of the effective ohmic resistance, because we know that the presence of copper in the neighbourhood of a circuit carrying alternate currents causes a virtual increase of the apparent ohmic resistance of the circuit in virtue of the heat generated by the currents induced in it. May it not be that anti-effective influence, such as is thus produced by copper not forming part of the circuit, can be produced by copper actually in the circuit, if the conductor be too thick? Examining the question mathematically, I find that it must be answered in the affirmative, and that great augmentation of the effective ohmic resistance is actually produced if the conductor be too thick; especially in coils consisting of several layers of wire laid over one another in series around a cylindric or flat core, as in various forms of transformer.

2. Fig. 1 may be imagined to represent the secondary coil of a transformer consisting of solid square copper wire in three layers. For simplicity we suppose the axial length to be infinitely great, and straight; but the uniformity which this involves, and a close practical application to its simplicity, is realised in that

excellent form of transformer which consists of a toroidal iron core completely covered by primary and secondary wires laid on toroidal surfaces. To simplify the mathematical work, I suppose the whole thickness of the three layers to be small in comparison with the greatest radius of curvature of the circular or flat cylindric surface on which the wire is wound, but if it is not so the solution is easily obtained, for the case of circular cylinders, in terms of the Fourier-Bessel functions. It is of no consequence for our present question what there be inside of coil No 3, and, if we please, we may imagine there to be nothing but air; the drawing, however, indicates an iron core and a space which might be occupied by the primary coil, if a transformer is the subject; or our coil *AAAA* may be the primary coil of a transformer with secondary coil and

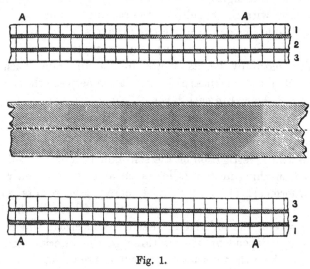

Fig. 1.

core inside it, and the alternate current maintained in it by an external electromotive agent acting in an arc between its ends outside. Our present results are applicable to all these varieties of cases indifferently, all that is essential being that the total quantity of current be given at each instant, and be uniform throughout the whole length of the coiled conductor.

3. This last condition is secured by perfectness of insulation between all contiguous turns of the coil, unless we were considering so enormously long a coil that the quantity of electricity required for the essential changes of static electrification would be sensible

as constituting drafts from, or contributions to, the current in
the coil. The consideration of static electrification, involved in
the maintenance of alternate currents through a coil such as that
represented in Fig. 1, is exceedingly curious and interesting; but
we do not enter on it at present at all, as in all practical cases
the quantities concerned are quite infinitesimal in comparison
with the whole quantity flowing in one direction or the other in
the half period.

4. In the drawing the section of the wires is represented as
square; but this is not essential, and in practice a flat rectangular
ribbon would, no doubt, for some dimensions of coils, be preferable.
I assume the thickness of the insulation between the successive
squares or rectangles in each layer to be infinitely small in
comparison with the breadth of the rectangle; but the thickness
of the insulation between successive layers, which is a matter of
indifference to my calculations, may be anything; and would, in
practice, naturally be, as shown in the diagram, considerably
greater than the thickness of the insulation between the contiguous
portions of the coil in each layer.

5. The full mathematical work which I hope to communicate
to the *Philosophical Magazine* for publication in an early number
includes an investigation of the self-induction of the coil with or
without anything in its interior (such as core or primary wire of
a transformer); but at present I merely give results, so far as
effective ohmic resistance, or generation of heat in the interior
of the wire of the coil $AAAA$ itself, is concerned, which, as said
above, is independent of everything in the interior, and of the
mode in which the alternating current is produced, provided only
that the total amount of electricity crossing the section of the
wire per unit of time be given at each instant.

6. As a preliminary to facilitate the expression of these
results, it is convenient first to give a general statement of the
solution of the problem of laminar diffusion of a simple harmonic
variation, applied to the case of electric currents in a homogeneous
conductor. Let the periodically varying magnetic force in the air
or other insulating material in the neighbourhood of so small a
portion, S, of the surface of a conductor that we may regard it as
plane, be given. Resolve this magnetic force into two components,
one, perpendicular to S, which we may neglect, as it has no influence

in connection with the currents we are to consider, the other, parallel to S, which we shall call the effective component and denote by Y. Through any point O, of S, draw three rectangular lines OX, OY, OZ, of which OY and OZ are in S, and OY is parallel to the direction of the effective magnetic force component Y. Let now the value of Y at time t be

$$Y = M \cos \frac{2\pi t}{T},$$

where M denotes a constant, and T the period of the alternation. The varying magnetic force Y, to whatever cause it may be due, implies currents parallel to OZ in the conductor, expressed by the formula for γ, the current intensity at distance x from the plane S, provided T be small enough to fulfil the condition stated below :—

$$\gamma = \frac{M}{\lambda \sqrt{2}} \epsilon^{-\frac{2\pi x}{\lambda}} \cos \left(\frac{2\pi t}{T} - \frac{2\pi x}{\lambda} + \frac{1}{4}\pi \right);$$

where λ denotes what we may call the wave-length of the disturbance, and is given in terms of T, the period of the disturbance, and ρ and Π the resistivity and magnetic permeability of the substance, by the following formula :—

$$\lambda = \sqrt{\frac{T\rho}{\Pi}}.$$

For copper we have $\Pi = 1$, and $\rho = 1611$ square centimetres per second; and thus for 80 periods per second $\lambda = 4\cdot49$, or, say, $4\frac{1}{2}$ centimetres. In order that the formula for γ may be approximately true it is necessary, in the first place, that λ must be small in comparison with the distance we must travel in any direction in the surface of S before finding any deviation of it from the tangent plane through O comparable with λ. Secondly, for a very good approximation, λ must be so small that we may be able to travel inwards in any direction from O, through a space equal to at least twice λ, without coming to any other part of the bounding surface of the conductor. If, for example, the surface be a flat plate, this condition requires that the thickness be more than twice λ. But (because $\epsilon^{-\pi}$ is less than $\frac{1}{23}$) the formula gives a very fair approximation requiring for a half the thickness of the plate inwards from S no greater correction than about 4 per cent., even if the thickness of our plate be no greater than λ. When the thickness of the plate is less than 2λ, we may consider waves of electric current as travelling inwards from its two sides, and being both sensible at the middle of the plate; and a complete solution of the problem is readily found by the method of images. But a direct analytical investigation, by which the proper conditions of relation to varying magnetic force on the two sides of the plate are fulfilled, is the most convenient way of fully solving the problem, and it is thus that the results given below have been obtained.

7.   The smallness of the insulating space between the successive turns in each layer of our coil $AAAA$, and the equality of the whole current through them all, prevent any surface disturbance from being produced at the contiguous faces, and allow the problem to be treated as if, instead of a row of squares or rectangles, we had a continuous plate forming each stratum. The smallness of the thickness of this plate in comparison with the radius of the cylindric surface to which it is bent allows, as said above, the mathematical treatment for an infinite plate bounded by two parallel planes to be used without practical error. I have thus found an expression for the intensity of the current at any point in the metal of any one of the layers of a coil of one, two, three, or more layers; and have deduced from it an expression for the quantity of heat generated per unit of time, at any instant, per unit breadth in any one of the layers. I need not at present quote the former expression; the latter is as follows:—With q to denote the dynamical value of the time-average of the heat generated per unit of time at different instants of the period, per unit breadth and unit length in layer No. i, from the outside of the coil, c^2 the time-average of the square of the total current per unit breadth, ξ the distance from the surface and a the thickness of the layer,

$$q = \frac{2\pi\xi}{\lambda}\,\Theta c^2,$$

where

$$\Theta = \frac{\epsilon^{2\theta} + 2\sin 2\theta - \epsilon^{-2\theta}}{\epsilon^{2\theta} - 2\cos 2\theta + \epsilon^{-2\theta}} + 2i\,(i-1)\,\frac{\epsilon^{\theta} - 2\sin\theta - \epsilon^{-\theta}}{\epsilon^{\theta} + 2\cos\theta + \epsilon^{-\theta}}$$

and $$\theta = \frac{2\pi a}{\lambda}.$$

8.   The numerical results shown in the table have been calculated, and the accompanying graphic representation (Fig. 2) drawn for me by Mr Magnus Maclean.

9. We see from the tables and curves that each curve has a minimum distance from the line of abscissas, and that each comes to an horizontal asymptote, parallel to the line of abscissas, for $\theta = \infty$. By looking at the formula we see that there is, in fact, an

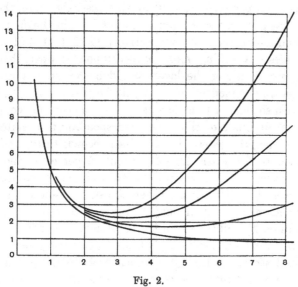

Fig. 2.

Table of Values of Θ.

$\dfrac{16\theta}{\pi}$	$i=1$	$i=2$	$i=3$	$i=4$
1	5·113	5·118	5·127	5·141
2	2·553	2·592	2·669	2·786
4	1·316	1·634	2·270	3·224
6	·9854	1·997	4·019	7·053
8	·9173	2·993	7·143	13·37
10	·9452	4·062	10·30	19·65
12	·9822	4·899	12·73	24·48
14	1·000	5·276	13·83	26·66
16	1·002	5·362	14·08	27·16
∞		5·000	13·00	25·00

infinite succession of minimums and maximums in the expression for Θ; but it is only the first minimum and following maximum that occur within the range of variation of Θ, which we regard as sensible. In the case of $i = 1$ the formula gives $\theta = \frac{1}{2}\pi$ for the first minimum. The curves show for the cases of $i = 2$, 3, 4, respectively the first minimum at $16\theta/\pi = 4\frac{1}{2}$, 3, and 2·6 respectively. The thickness which corresponds to $\theta = \pi$ is the half-wave length of the electric disturbance, which, as we have seen, is for copper 2·244 centimetres when the frequency of the alternations is 80 periods per second; and for this case, therefore, the thicknesses that give minimum generation of heat in the first, second, third, and fourth layers are respectively 11·22, 6·31, 4·21, and 3·65 millimetres. Anything more of continuous copper than these thicknesses in any of the layers would be not merely ineffective or comparatively ineffective, but would be positively anti-effective. Even with so small a thickness as 2·8 millimetres, for copper and frequency 80, line 2 of the table (corresponding to a sixteenth of the wave length) shows, in the first, second, third, and fourth layers, losses of 0·3 per cent., 2 per cent., 5 per cent., and 10 per cent. in excess of that due to the true ohmic resistance of the copper were it all effective. When the size chosen for the transformer and the amount of output required of it are such that a thickness of $2\frac{1}{2}$ millimetres in the direction perpendicular to the layers is insufficient, a remedy is to be had by using braided wire, or twisted strand, with slight insulation of varnish or whitewash, crushed or rolled into rectangular or square form of the desired thickness and breadth. A very slight resistance between the different wires thus crushed together would suffice to cause the current to run nearly enough full bore to do away with any sensible loss from the cause which forms the subject of this communication.

228. On a Method of Determining in Absolute Measure
the Magnetic Susceptibility of Diamagnetic and
Feebly Magnetic Solids.

[From *Brit. Assoc. Report*, 1890, pp. 745, 746; *Nature*, Vol. XLII. Oct. 9, 1890,
p. 578; *Lum. Élec.* Vol. XXXVIII. Dec. 27, 1890, p. 611.]

THE communication was suggested from two directions in
which the subject had been treated—(1) Professor Rücker's
investigations of the magnetic susceptibility of basaltic rocks, to
which he was led in the interpretation of the results of the great
magnetic surveys made by himself in conjunction with Dr Thorpe,
by which remarkable disturbances due to magnetisation of the
rocks and mountains were found; (2) Quincke's determinations
of the magnetic susceptibility of liquids. The method proposed
by the author consisted in measuring the mechanical force experienced by a properly shaped portion of the substance investigated,
placed with different parts of it in portions of magnetic field between
which there was a large difference of the magnetic force. A
cylindrical or rectangular or prismatic shape, terminated by planes
perpendicular to its length, was the form chosen; the component magnetic force in the direction of its length was equal to
$\frac{1}{2}\mu(R^2 - R'^2)A$; where μ denoted the magnetic susceptibility,
R, R' the magnetic force in the portions of the field occupied by
its two ends, and A the area of its cross-section. For bodies of
very feeble susceptibility the best arrangement of field was that
originally adopted by Faraday, and pushed so far recently by
Professor Ewing, in the way of giving exceedingly intense fields.
One end of the prism, or plate, or wire was in the air between
flat ends and conical magnetic portions; the other might be in a
place practically out of the field, or, if the portion of the substance
given were exceedingly small, it might be in the field, but in a
place of much less force than in the centre of the field. The
measurement of the magnetic force of the field was easily made

by known methods; best by measuring the force experienced by
a short element of wire carrying a measured current. This
portion of wire should be placed in the positions occupied by the
two ends of the plate or wire of the substance, first in one position
and then in the other. But when the second position was in a
place of sensibly known force, the single measurement with the
element of the wire in the first position sufficed.

229. A New Electric Meter. The Multicellular Volt-meter. An Engine-room Voltmeter. An Ampere Gauge. A New Form of Voltapile, useful in Standardising Operations.

[From *Nature*, Vol. XLII. Sept. 25, 1890, p. 534, Abstract; *Brit. Assoc. Report*, 1890, p. 956 [title only].]

THE subject which first occupied Sir William's attention was
the new electric meter which he has recently brought out. This
apparatus is yet in the experimental stage. Perhaps Sir William
will be able to do something towards cheapening the design. An
example of the meter was shown in operation on the platform.
In the discussion which followed, Prof. Fleming made some
pertinent remarks on the effect of rough and smooth surfaces.
The multicellular voltmeter and the engine-room voltmeter
described by the author had previously been brought before the
public through the medium of technical literature. A new form
of voltapile, also described, was an instrument which was intended
for standardising operations.

230. On Electrostatic Screening by Gratings, Nets, or Perforated Sheets of Conducting Material.

[From *Roy. Soc. Proc.* Vol. xlix. 1891 [April 9, 1891], pp. 405—418.]

1.   Maxwell, in his "Theory of a Grating of Parallel Wires" (*Electricity and Magnetism*, Arts. 203—205*, and Plate XIII), gives a very valuable and interesting two-dimensional investigation of electrostatic screening, and a most instructive diagram of "Lines of Force near a Grating," which powerfully helps to understand and extend the theory, and to acquit it of an accusation wrongly made against it in the last two sentences of Art. 205.   It is only on the supposition of the grate-bars being circular cylinders that the investigation is less than rigorous : and that supposition nowhere enters into the investigation ; it merely appears in the word "radius," in the first line of the last sentence but one of Art. 204, and it is contradicted in lines 3 and 4, and by the rest of the sentence, and by the next sentence.   (See § 6 below.)

2.   The conclusion, "$\alpha = -0.11a$," in the last sentence of Art. 205, condemned as "evidently erroneous," is quite correct, and very interesting.   It shows that a corrugated metal plate agreeing with the equipotential surface $c = \frac{1}{2}a$, exceeds in electrostatic capacity a plane metal surface through the poles of the diagram (Plate XIII, reproduced in § 9 below), with the surroundings described in Art. 204, and supplies the datum requisite for finding the exact amount of the excess.   The reason for the greatness of the excess clearly is that the surface $c = \frac{1}{2}a$, which just touches the plane through the poles of the diagram midway between the poles, is everywhere nearer than this plane to the other plate of the condenser.   (See § 7 below.)

---

* In formula (7) of Art. 204, delete λ ; in Art. 204, delete 2 in last line of p. 250 (Edition 1873) ; and delete 2 in lines 6 and 16 from foot of the page.   [Cf. also H. Lamb, 'On the Reflection and Transmission of Electric Waves by a Metallic Grating,' *Proc. Lond. Math. Soc.* xxix. 1898.]

3.   For $c = a/6$ we have, by (11) of Art. 205, $\alpha = 0$, and the corresponding equipotential, partially shown in Maxwell's diagram, is a set of curves concave towards $z = -\infty$, and asymptotic to the lines $x = (i \pm \frac{1}{4})\, a$, i denoting any integer. (See §§ 10 to 13 below.) For every value of c less than $a/6$, the equipotential is a row of ovals; and the grating formed by constructing these ovals in metal has less electrostatic capacity in the circumstances described in Art. 205 than a plane through the poles or the ovals (this being no doubt what is meant by "a plane......in the same position" as the grating).

4.   For every value of c exceeding $a/6$ the equipotential, instead of being the boundary of a grating, is a continuous corrugated surface, and its electrostatic capacity exceeds that of the plane through the poles.

5.   Begin now afresh, and let it be required to find the electric force in the air on either side of an infinite row of parallel bars at equal consecutive distances, a, each uniformly charged with electricity.   Let ρa be the quantity per unit length on each bar, so that ρ would be the surface density, if the same quantity were uniformly distributed over the plane of the bars.   Taking O in one of the bars, OX perpendicular to the bars, and OZ perpendicular to their plane, we find (by Fourier's method) for the z-component of force at any point (x, z) for which z is positive,

$$Z = 4\pi\rho\,(\tfrac{1}{2} + \epsilon^{-mz}\cos mx + \epsilon^{-2mz}\cos 2mx + \&c.)......(1),$$

where $$m = 2\pi/a \quad ....(2).$$

Summing this we find

$$Z = 2\pi\rho\,\frac{\epsilon^{mz} - \epsilon^{-mz}}{\epsilon^{mz} - 2\cos mx + \epsilon^{-mz}} \quad ....(3).$$

This has equal positive and negative values for equal positive and negative values of z, and it therefore gives the value of the z-force, not only for positive, but also for negative, values of z. Taking now $-\int Z dz$, with constant assigned to make the integral zero for $z = \pm D$, we find

$$V = \rho a\left(\log\frac{1}{\epsilon^{mz} - 2\cos mx + \epsilon^{-mz}} + mD\right) ....(4)$$

as the potential due to the grating, and two parallel planes at equal distances, D, on its two sides, each uniformly electrified with half the quantity of electricity of opposite sign to that on the grating.

6. If now we construct in metal, C, any one complete equipotential surface, V_0, of this system, and electrify it with the same quantity of electricity as that which we gave originally to the infinite row of infinitely thin bars; and if we place metal planes, B, B', at the two places of zero-potential ($z = \pm D$), we have an insulated conductor at potential V_0, between two planes, B, B', at zero potential, and at distance $2D$ asunder, on each of which the electric density is $\frac{1}{2}\rho$. For brevity, I shall denote the insulated conductor by I.

Its electrostatic capacity per unit area of its medial plane (the plane of the original infinitely thin bars) is ρ/V_0.

7. This conductor, I, is symmetrical on each side of its medial plane, and consists either of an infinite number of isolated parallel bars, each surrounding one of the original infinitely thin bars, or of a plate symmetrically corrugated on its two sides, with maximum and minimum thicknesses respectively at the places of the infinitely thin bars, and the lines midway between them. For the case of isolated bars, let $2c$ be the diameter of each, in the medial plane. Then, to find V_0, we must put $x = \pm c$ and $z = 0$, in (4). Thus we find

$$V_0 = \rho \left\{ 2\pi D - a \log \left(4 \sin^2 \frac{\pi c}{a} \right) \right\} \quad \ldots\ldots\ldots\ldots(5).$$

Hence the electrostatic capacity of I in the circumstances is

$$\left\{ 2\pi D - a \log \left(4 \sin^2 \frac{\pi c}{a} \right) \right\}^{-1} \quad \ldots\ldots\ldots\ldots(6),$$

which is greater or less than $1/(2\pi D)$, the electrostatic capacity that it would have if reduced to its medial plane, according as $c >$ or $< \frac{1}{6}a$. The conductor I, to be a grating, implies $c < \frac{1}{2}a$, or $\sin^2 \pi c/a < 1$, and therefore requires that

$$V_0 > 2\pi\rho \left(D - \frac{a}{2\pi} \log 4 \right) = 2\pi\rho \left(D - \cdot 22a \right) \quad \ldots\ldots(7).$$

When V_0 exceeds this critical value, the conductor I is the continuous plate corrugated on each side, which was described in § 7. The critical value corresponds to an intermediate case of a plate so deeply furrowed on each side as to be just cut through by its two surfaces crossing at right angles; and (7) shows that the electrostatic capacity of the conductor I so constituted is equal to that of a plane sheet of thickness

$$2a \log [2^2]/(2\pi), \text{ or } \cdot 44a \quad\text{................}(8),$$

insulated midway between the two earth plates B, B', at the same distance asunder as they had with I between them.

8. By (4), (5), and (7), we have for the equation of the surface constituting the two sides of I in this critical case,

$$\epsilon^{mz} - 2 \cos mx + \epsilon^{-mz} = 4 \quad\text{................}(9).$$

Taking double the positive value of z which this gives when $x = 0$, we find

$$2a \log [(1 + \sqrt{2})^2]/(2\pi), \text{ or } \cdot 562a \quad\text{............}(10)$$

as the maximum thickness of I. This is $\log(1 + \sqrt{2})^2/\log 2^2$, or $1\cdot273$, times the amount shown in (8) for the thickness of the plane-sided plate of equal electrostatic capacity; which is just such a relation as is expected before calculation!

9. If $\phi(z, x)$ denote what V becomes when in place of mD we substitute $- mz$ in (4), we have the potential due to a uniform

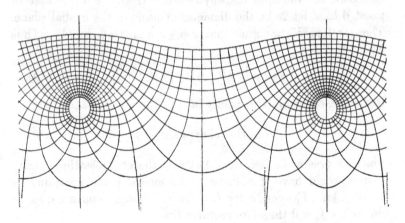

electrical force ρam, or $2\pi\rho$, added to the z-component, of the force due to the grating with its given charge of ρa quantity per unit length of each bar; and it is the equipotentials and lines of force of this system that are represented in Maxwell's diagram of Plate XIII, reproduced here. In it the resultant force for infinitely large positive values of z is parallel to OZ, and of constant value $4\pi\rho$; and it is zero for infinitely large negative values of z. The approximation to these values is very close, at only so moderate a distance as a on either side of the grating.

10. Choosing, in the system of § 6, any one of the multiple-oval equipotentials around the infinitely thin bars, let c be the distance from the infinitely thin primary bar within it, at which it is cut by the plane of the primary bars. By putting, in the expression for $\phi(z, x)$, $z = 0$, and $x = c$, we find

$$\phi(0, c) = \rho a \log \left(4 \sin^2 \frac{\pi c}{a}\right)^{-1} \quad \dots\dots\dots\dots(11)$$

as the potential at the surface of each of these chosen ovals. Construct now each of these ovals in metal, and let the supposed uniform force, $2\pi\rho$, be produced by uniform electrification of density $-\rho$, on a metal plane, B, at any great distance, b, on the positive side of the grating. We thus construct a grating of thick bars of oval-shaped cross section which, when electrified with the same quantity of electricity as that which we gave initially to the infinitely thin bars, and subjected to the influence of the equal quantity of negative electricity on B, has $\phi(z, x)$ for potential through external space from $B(z = b)$, to infinite distance on the other side of the grating ($z = -\infty$), and has for potential through all the portions of space within the surfaces of the grate-bars the constant value expressed by (11). In this system the potential, for positive values of z great in comparison with a, is, by (4) with $-mz$ instead of $+mD$,

$$\phi(z, x) \fallingdotseq -4\pi\rho z \quad \dots\dots\dots\dots\dots(12).$$

The difference of potentials between B and the grating is, by (6) and (5),

$$\phi(0, c) - \phi(b, x) \fallingdotseq 4\pi\rho \left\{b + \frac{a}{4\pi} \log \left(4 \sin^2 \frac{\pi c}{a}\right)^{-1}\right\} \dots(13).$$

Hence the electrostatic capacity of the mutually insulated system, B, and the grating of oval-shaped bars is equal to the capacity of a pair of parallel planes, B and a plane at a distance beyond the plane of the primitive infinitely thin bars equal to

$$-\frac{a}{4\pi} \log \left(4 \sin^2 \frac{\pi c}{a}\right)^{-1} \quad \dots\dots\dots\dots(14).$$

11. If in (4) we put $-nz$ in place of $+mD$, we have the potential of a system in which besides the electricity of the

primary bars there is distant electricity such as all in all to give at great enough distances on the two sides of the primitive bars uniform fields of z-force respectively equal to

$$\rho\,(m+n),\ \text{for}\ z \leqq + \infty\ ;\ \text{and}\ \rho\,(m-n),\ \text{for}\ z \leqq -\infty \ ...(15).$$

If, in (4) with $-nz$ instead of $+mD$, we put

$$\epsilon^{-V/\rho a} = C ....(16),$$

we find, as the equation of the equipotential surfaces,

$$-2\cos mx + \epsilon^{mz} + \epsilon^{-mz} = C\epsilon^{-nz} \ ....(17).$$

By taking $n=0$, or $n=m$, we fall back on the cases of §§ 5—8, and §§ 9, 10, respectively.

12. To find an approximate equation for the equipotentials at distances around primitive bars small in comparison with a, the distance from bar to bar, let x and z be so small that we may neglect all powers of mx, mz, and nz, above the square, which implies that C is small of the same order as $(mx)^2$ and $(mz)^2$, (17) becomes

$$x^2 + \left(\frac{z + \frac{1}{2}nr^2}{1 + \frac{1}{4}n^2r^2}\right)^2 = r^2\,(1 + \frac{1}{4}n^2r^2) \ ....(18),$$

where r^2 denotes $m^{-2}C$. This shows that, to the degree of approximation in which we neglect cubes and higher powers of mx, mz, nz, the equipotential is a row of elliptic cylinders of eccentricity $nr/\sqrt{2}$, with their greater diametral planes perpendicular to the plane of the primitive bars. When $n=0$, the equipotential is a row of circular cylinders having the primitive bars for their axes; and this is true to the higher approximation in which we need only neglect powers above the cubes of mx and mz, as we see by going back to equation (17), with $n=0$.

13. The conclusions of § 12 are useful for detailed investigation of the screening effect of plane gratings of circular or elliptic, straight parallel bars electrified with given quantities of electricity and placed with their planes perpendicular to the lines of force in a uniform field of force, and to corresponding problems in which potentials are given, as in Maxwell's §§ 203—205.

14. Instead of a single row of parallel equidistant infinitely thin bars in one plane, let us take for primitives two or more such rows, parallel or not parallel, all in one plane or not in one plane. We may thus form an endless variety of force-systems

available for illustrating or helping to solve problems which may occur. Towards the several problems of electric screening we find important contributions by considering in two parallel planes rows of primitive lines parallel to one another for one case and perpendicular to one another for another case. The consideration of three rows of primitive lines in one plane, dividing it into equal and similar triangles alternately oriented in opposite directions, leads to a complete theory of electrostatic screening by a triangular lattice of metallic wire or ribbon. The fundamental potential formula for this system obtained by summation of expressions, each given by an application of (4) to one of the three rows, is

$$V = - \log \left(\epsilon^{lz} - 2 \cos lp + \epsilon^{-lz} \right)^{\varpi a} \left(\epsilon^{mz} - 2 \cos mq + \epsilon^{-mz} \right)^{\rho b}$$
$$\left(\epsilon^{nz} - 2 \cos nr + \epsilon^{-nz} \right)^{\sigma c} + 2\pi \left(\varpi + \rho + \sigma \right) D \ldots (19);$$

where a, b, c denote the intervals between the successive lines of the three systems; ϖa, ρb, σc, the quantities of electricity per unit length of bar in the three systems; p, q, r, z, special coordinates of the point for which (19) expresses the potential, viz., z, its distance from the plane of the primitive bars, and p, q, r its distances from three planes drawn perpendicular to this plane through a bar of each of the three systems; D the value of $\pm z$ for planes on the two sides of the net for which the potential is zero; and, lastly,

$$l = 2\pi/a, \ m = 2\pi/b, \ n = 2\pi/c \ \ldots\ldots\ldots\ldots(20).$$

For the present, however, we may confine ourselves to the case of two rows of primitive lines dividing a plane into squares and charged, both rows, with equal quantities, $\frac{1}{2}\rho a$, of electricity per unit length. The potential formula, a particular case of (19), is

$$V = \tfrac{1}{2}\rho a \left[- \log \left(\epsilon^{mz} - 2 \cos mx + \epsilon^{-mz} \right)\left(\epsilon^{mz} - 2 \cos my + \epsilon^{-mz} \right) + 2mD \right]$$
$$\ldots\ldots\ldots(21).$$

15. The consideration of the equipotentials of this surface is very interesting. The equipotential lines in the plane of the primitive bars are given by the equation

$$16 \left(\sin \frac{\pi x}{a} \sin \frac{\pi y}{a} \right)^2 = \epsilon^{-2V/\rho a + 2mD} \ldots\ldots(22).$$

16. Considerations quite analogous to those of §§ 6, 7, 8, and again the other considerations analogous to those of §§ 9—13, are, after the full explanations there given, easily completed so as to

formulate a full theory of electrostatic capacity and electrostatic screening for square nets of wire exposed to electric action giving uniform fields of force at distances on each or on one side of the plane of the net considerable in comparison with a, the side of each square.

17. In what follows we shall for brevity call any thin sheet, whether plane or not plane, which answers to the description contained in the title of this paper, a *perforated sheet* or a *perforated surface*; understanding that its radii of curvature are everywhere large in comparison with its thickness. The diameters of holes must be large in comparison with the thickness in order that the approximations which we use below may be valid. We shall call the electric density of a perforated sheet the total quantity of electricity with which it is electrified, reckoned per unit area of continuous surface approximately agreeing with it, and passing through the middle of cage bars, bosses, &c. This continuous surface I shall call the medial surface, or sometimes, for brevity, *the medial*.

18. In what precedes we have virtually a complete investigation of the screening effect of a homogeneous plane perforated sheet against the electric force of a uniform field with lines of force perpendicular to the plane. Let it now be required to find the screening effect of a non-plane perforated sheet against a uniform field of electrostatic force, and of a perforated sheet S, plane or not plane, against the electrostatic force of any given electrified bodies.

19. Let ϕ be the potential of the given electrified bodies at any point (x, y, z) of the space occupied by S, and let ρ be the unknown electric density of S at (x, y, z), under the influence of those bodies. To make the problem of finding ρ determinate, we might suppose either the total quantity of electricity on S, or the potential at which its metal is kept, to be given. We shall take the latter supposition, and call the given potential K.

20. Let ϕ denote the potential which would be produced by the electricity of S if it were spread continuously over the medial with electric density equal to ρ at (x, y, z); and let

$$\phi + \mu\rho \quad\quad\dots\dots\dots\dots\dots(23)$$

denote the potential in the metal of S, due to the actual distribution of electricity on its surface.

21. To understand the meaning of this notation (μ), consider a large area A around (x, y, z), so large that its border is very distant from (x, y, z) in comparison with the thickness of the sheet, and with the diameters of its apertures, but not so large as to deviate sensibly from the tangent plane at (x, y, z). Let the electricity of all the surface of S beyond A be changed from the imagined continuous distribution to the actual distribution on the surface of the perforated metal. This change will make no sensible difference in the potential at (x, y, z). Next, let the imagined continuous distribution of uniform electric density ρ, over the continuous area A, be changed to the actual distribution of the same quantity over the surface of the perforated metal of the porous sheet A. The augmentation of potential at (x, y, z) produced by this charge is what we denote by $\mu\rho$, where μ is a coefficient depending on the shapes and magnitudes of the perforations, that is to say, on the complex surface of the perforated metal. It would be zero if there were no perforations, and we shall see that the greater it is the less is the screening efficiency. We shall therefore call μ the electric permeability, and μ^{-1} the electric screening efficiency of the perforated sheet. The sheet is homogeneous as to permeability or screening efficiency if μ has the same value for all parts of it, but we need not assume this to be the case; on the contrary, we shall suppose μ to be any known function of (x, y, z). In §§ 5—16 we have the explanations necessary for determining μ in the various cases of gratings and nets there described. For similarly perforated surfaces, the values of μ are as the linear dimensions of a perforation or of the bars or bosses of the structures.

22. The equation of electric equilibrium is

$$\phi + \mu\rho = K \text{ (a constant)} \quad \dots\dots\dots\dots(24),$$

when S, being insulated and electrified, is not under the influence of any other electrified matter.

It is

$$\phi + \mu\rho = K - V \quad \dots\dots\dots\dots(25),$$

when S is under the influence of any given electrified bodies producing a given potential, V, at (x, y, z).

23. As a first example, going back to (24), let μ be such that ϕ shall be constant. This makes, if we denote by k a constant,

$$\mu = k\phi/\rho \quad \dots\dots\dots\dots(26),$$

(k being a constant), which means that the screening efficiency is, in different places of S, inversely proportional to the electric density at similarly situated places of a continuous electrified conductor of the same shape as S. Let, for instance, S be an ellipsoid; then, if the sizes of the perforations be inversely proportional to the perpendicular from the centre to the tangent plane, (26) is satisfied. Generally, to fulfil this condition, the net must be finer in the more convex and more projecting parts, and coarser in the flatter and less projecting parts.

24. If any perforated conductor or cage, S, fulfilling the condition of § 23, be electrified and insulated away from the disturbing influence of other conductors, or electrified bodies, the charge distributes itself so as to have in every part the same quantity per unit area of the medial, as a smooth continuous metallic surface agreeing with the medial and electrified with the same total quantity. When the medial is a closed surface, the electricity on the perforated surface does not confine itself to the parts of it outside the medial: on the contrary, when the apertures are very wide in comparison with diameters of cage-bars, bosses, &c., the electricity distributes itself almost equally on the parts of the complex surface inside and outside the medial.

25. Seeing that the electric density (as defined in § 17) is the same for a perforated surface fulfilling the condition of § 23 as for the medial constructed in continuous metal, we naturally ask the question, what then is the difference between the two cases, if any, besides the fact of the electricity being equally but very unequably distributed over the outer and inner portions of the complex surface in one case, and equably over the outside of the smooth medial in the other? There is a very important and interesting difference. The electrostatic capacity of the perforated conductor, S, is less, in the ratio of 1 to $1+k$, than that of the medial constructed in continuous metal; as we see by (23) and (26).

26. As a sub-example, suppose S to be a spherical surface. If homogeneously perforated, it will fulfil the condition of § 23: and if its screening efficiency is the same as that of a grating of parallel bars (circular cross section of diameter $2r$; distance from centre to centre a), we have, by (5) of § 7, when $\pi c/a$ is very small,

$$\mu = 2a \log (a/2\pi c) \quad \dots\dots\dots\dots\dots\dots(27).$$

Now, S being spherical, if R denotes its radius, we have (§ 20)

$$\phi = 4\pi R\rho \ldots\ldots\ldots\ldots\ldots\ldots(28).$$

Hence, by (26) and (27),

$$k = \frac{a}{2\pi R}\log\frac{a}{2\pi c} = \frac{1}{N}\log\frac{a}{2\pi c} \ldots\ldots\ldots\ldots(29),$$

where N denotes the number of bars in the equatorial belt of the cage of § 27 below.

27.   To illustrate a realisation of § 26, let a spherical cage be made up of a narrow equatorial belt of approximately straight parallel bars of diameter $2c$, and distance from middle of one bar to middle of next, a; completed by polar caps (nearly hemispheres) of thin metal perforated so as to have everywhere the same effective electric screening efficiency $\{2a\log(a/2\pi c)\}^{-1}$.

Suppose, for instance, the bars to be of "No. 18 gauge" ($2c = 0\cdot122$ cm.) and $a = 5$ cm.   We have

$$\log(a/2\pi c) = \log 13 = 2\cdot57.$$

Hence, for this case, and any other in which the ratio a/c is the same, we have, by (27) and (29),

$$\mu = 5\cdot14a, \quad k = 0\cdot409a/R \ldots\ldots\ldots(30), (31).$$

Thus, if $a = 5$ cm., and $R = 50$ cm., $k = 0\cdot0409 = \frac{1}{24}$; and (§ 25) the electrostatic capacity of the spherical cage is $\frac{24}{25}$ of that of a simply continuous spherical surface of the same magnitude.

28.   Let now an electrified metal globe, or globe of insulating material uniformly electrified, G, be insulated concentrically within S. It may be of any magnitude, large or small, provided only that the interval between the two surfaces be at least two or three times the diameter of the largest of the perforations of S. Let S be connected with the earth, and let Q denote the quantity of (positive) electricity with which G is electrified, and Q' the quantity of the opposite electricity which it induces on S. The potential in the metal of S due to Q' is, by (23),

$$-\left(\frac{Q'}{R} + \mu\frac{Q'}{4\pi R^2}\right)\ldots\ldots\ldots\ldots\ldots(32).$$

This, added to Q/R, the potential due to G, must be zero, and therefore

$$Q = Q'(1 + \mu/4\pi R) \ldots\ldots\ldots\ldots(33),$$

or, by (26), $$Q = Q'(1 + k)\ldots\ldots\ldots\ldots\ldots\ldots(34).$$

Hence, in the particular case of § 27 (31),

$$Q = Q' \left(1 + 0 \cdot 409a/R\right) \quad \ldots\ldots\ldots\ldots(35);$$

and when $R = 10a$, we find $Q - Q' \doteqdot \frac{1}{25}Q$, and conclude that the effect of S, earthed, with G electrified and insulated within it, is just 4 per cent. of the effect of G unscreened.

29. If S is connected with the earth, and supported at a height above the earth equal to at least six or eight times its diameter, the quantity of electricity (positive in fine weather) induced on it will be $1/(1 + k)$ of that which would be induced on a simply continuous metal globe of the same size. Hence the potential at any point of the air within S at not less distance inwards than $2a$ will be $k/(1 + k)$ of the undisturbed atmospheric potential at the same height above the ground, or 4 per cent. in our particular case. This is quite in accordance with the imperfectness of the screening effect against atmospheric electricity found by Roiti* within earthed wire cages, supported at a considerable height above the ground, by a bracket attached to the top of a wall of a building in Florence, tested by a water-dropper with its nozzle inside the cage connected by an insulated wire with a quadrant electrometer in the buildings.

30. The problem of finding the distribution of electricity on a spherical cage, of equal electric permeability, μ, in all parts of its surface, formulated in (25) of § 22, is easily solved by aid of spherical harmonics. Confining ourselves for brevity to the case of external influencing bodies, let their potential at any point, P, within S be

$$V = - \Sigma S_i r^i / R^i \quad \ldots\ldots\ldots\ldots\ldots(36),$$

where S_i denotes a given spherical surface-harmonic of order i, and r the distance of P from the centre of S. And ρ_i, denoting an unknown surface-harmonic of order i, let

$$\rho = \Sigma \rho_i \quad \ldots\ldots\ldots\ldots\ldots(37)$$

be the harmonic expression for ρ, the required electric density. Going back to § 20 for the definition of ϕ, we find, by the elements of spherical harmonics,

$$\phi = \Sigma \frac{4\pi \rho_i}{2i + 1} \frac{r^i}{R^{i-1}} \quad \ldots\ldots\ldots\ldots(38).$$

* "Osservazioni Continue della Elettricità Atmosferica" (*Pubblicazioni del R. Istituto di Studi Superiori in Firenze*), Florence, 1884.

Hence, by (25),

$$\rho_0 = \frac{K + S_0}{4\pi R + \mu}, \qquad \rho_i = \frac{(2i+1)S_i}{4\pi R}\left\{1 + \frac{(2i+1)\mu}{4\pi R}\right\}^{-1} \dots (39), (40),$$

and

$$\phi = K + \Sigma \left\{1 + \frac{(2i+1)\mu}{4\pi R}\right\}^{-1} \frac{S_i r^i}{R^i} \qquad \dots\dots\dots(41).$$

In (39) we have virtually the same result as in (33). The approximation on which we are founding in §§ 17—29 is valid in (40) and (41) only for values of i small in comparison $2\pi R/a$: but, as in virtue of greatness of the logarithm for the case formulated in (27), μ may be great in comparison with a; and therefore the denominator of (40) need not be only infinitesimally greater than unity, and may be any numeric however great.

31. Taking $S_1 r = x$, $S_2 = 0$, $S_3 = 0 \dots$, we see by (41) that if an insulated unelectrified spherical cage be brought into a uniform field of electric force, X (that of atmospheric electricity, for example, at any height above the ground exceeding five or six diameters of the cage), the force within the cage is

$$X - X\left\{1 + \frac{3\mu}{4\pi R}\right\}^{-1} \qquad \dots\dots\dots\dots\dots(42),$$

or, according to (27), and (29),

$$X - X\left\{1 + \frac{3a}{2\pi R}\log\frac{a}{2\pi c}\right\}^{-1}, \text{ or } X - X\left\{1 + \frac{3}{N}\log\frac{a}{2\pi c}\right\}^{-1} \dots (43).$$

This result is also applicable to a hemispherical screen of radius R, simply placed on the ground. For the particular proportions of § 27, it makes the force under the hemispherical cage $\frac{1}{9}$ of the undisturbed force outside. A cage of ordinary gardener's (anti-rabbit) hexagonal wire-net (of $5\frac{1}{2}$ cm. from parallel to parallel) cannot be very different from this. If, instead of the radius being 50 cm. it be 200, but the cage still of the same net, the force inside would be only 3 per cent. of the undisturbed force outside.

32. In every case the force at any distance from the perforated surface, on either side of it, more than the diameter of a perforation, is, as is easily proved by Fourier's methods, very nearly the same as if the electricity were spread equably over the medial surface, with the same quantity per unit area of the medial as the grating has in each part of it. Hence, in the case of § 31, the force is

uniform throughout the interior of the cage, except within distances
from the net of two or three times the aperture.   Hence a second
screen, similar but slightly smaller, placed inside the first will
reduce the force farther in the same ratio; so that, if eX denote
the force inside the single screen, the force inside the inner screen
when there are two will be e^2X, provided the distance between
the two is nowhere less than the diameter of the perforation.
Thus, with screens such as those in the last particular case of
§ 31, the force inside the inner screen would be only 9/10,000 of
the undisturbed force far enough outside the outer.   The two
screens, if placed close together, so as to narrow the apertures
as much as possible, would have little more than double the
screening efficiency of either singly, as we may judge from (27)
of § 26, and from (21) of § 14.   The principle that, to duplicate a
screen with best advantage, the two screens should be placed, not
in one surface but in two, with not less distance between them
than the diameter of their apertures, is not only theoretically
interesting, but is of great practical importance in the screening
of electrometers against disturbing electric force.

33.   Questions analogous to those of §§ 26 —32, but for circular
cylindric (mouse-mill) cages of equidistant parallel bars, instead of
the spherical or hemispherical cage which we have been considering,
are readily answered by the simpler work corresponding to that of
§ 30 (with $\sin i\theta$ and $\cos i\theta$ instead of spherical harmonics).   But
it deserves more complete synthetic investigation, not limited by
the approximational conditions of §§ 21, 22, if for no other reason,
because of Hertz's mouse-mill.   This must, however, be reserved
for a future communication.   Meantime, it is worth saying that
sudden variations of electric current, or alternating electric currents,
distribute themselves between different straight parallel conductors in the same proportion as static electrification is distributed
in corresponding electrostatic arrangements, whenever the suddenness, or the frequency, is sufficient to cause the impedance by
mutual induction of the separate parallel conductors (and therefore, *a fortiori*, the impedance by self-induction of each) to be
very large in comparison with ohmic resistance.   Hence Hertz's
mouse-mill screening follows (though by utterly different physical
action), simply the electrostatic law, except in any case in which
his wave-length is less than a considerable multiple of the diameter
of his mouse-mill.

231. ELECTRIC AND MAGNETIC SCREENING.

[From *Roy. Inst. Proc.* Vol. XIII. 1893 [April 10, 1891], pp. 345—355.]

THERE are five kinds of screening against electric and magnetic influences, which are quite distinct in our primary knowledge of them, but which must all be seen in connected relation with one another when we know more of electricity than we know at present:—I. Electrostatic screening; II. Magnetostatic screening; III. _Variational screening against electromotive force; IV. Variational screening against magnetomotive force; V. Fire-screens and window-blinds or shutters.

I.

Electrostatic screening is of fundamental significance throughout electric theory. It has also an important place in the history of Natural Philosophy, inasmuch as consideration of it led Faraday from Snow Harris's crudely approximate but most interestingly suggestive doctrine of non-influence of unopposed parts and action in parallel straight lines between the mutually visible parts of mutually attracting conductors, to his own splendid theory of inductive attraction transmitted along curved lines of force by specific action in and of the medium intervening between the conductors.

A continuous metallic surface completely separating enclosed air from the air surrounding it acts as a perfect screen against all electrostatic influence between electrified bodies in the portions of air so separated. This proposition, which had been established as a theorem of the mathematical theory of electricity by Green, in the ninth article of his now celebrated essay*, was admirably illustrated by Faraday, by the observations which he made inside

---

* See pp. 14 and 48 of the reprint edited by Ferrers.

the wooden cube covered all around with wire netting and bands of tinfoil, which he insulated within this lecture-room*: "I went into the cube and lived in it; and, using lighted candles, electrometers, and all other tests of electrical states, I could not find the least influence upon them, or indication of anything particular given by them, though all the time the outside of the cube was powerfully charged, and large sparks and brushes were darting off from every point of its outer surface."

The doctrine of electric images is slightly alluded to, and an illustrative experiment performed, showing the fixing of an electric image. The electroscope used for the experiments is an electrified pith ball, suspended by a varnished double-silk fibre of about 9 or

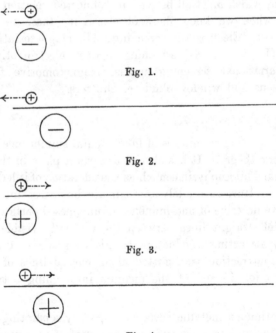

Fig. 1.

Fig. 2.

Fig. 3.

Fig. 4.

10 feet long. Figs. 1—4 represent experimental illustrations, in which the pith ball, positively electrified, experiences a force due to electrified bodies, optically screened from it by a thin sheet of tin-plate. In Figs. 1 and 2 the pith ball is attracted round a corner by a stick of rubbed sealing-wax, and in Figs. 3 and 4

* Experimental Researches, 1173—1174.

repelled round a corner by a stick of rubbed glass. In Fig. 2 the sealing-wax *seems* to repel the pith ball, and in Fig. 4 rubbed glass *seems* to attract it. This experiment constituted a very palpable illustration of Faraday's induction in curved lines of force.

In the present lecture some experimental illustrations were given of electrostatic screening by incomplete plane sheets and curved surfaces of continuous metal, and of imperfectly conducting material, such as paper, slate, wood, and a sheet of vulcanite, moist or dry, window glass at ordinary temperatures in air of ordinary moisture, and by perforated metal screens and screens of network, or gratings of parallel bars.

The fixing of an electric image is shown in two experiments: (1) the image of a stick of sealing-wax in a thin plane sheet of vulcanite, moistened, warmed, and dried under the electric influence by the application and removal of a spirit-lamp flame; (2) the glass jar of a quadrant electrometer with a rubbed stick of sealing-wax held projecting into it, while the outer surface is moistened, warmed, and dried by the application and removal of a ring of flame produced by cotton wick wrapped on an iron ring and moistened in alcohol.

Fig. 5 is copied from a diagram of Clerk Maxwell's to illustrate screening by a plane grating of parallel bars of approximately circular cross section, with distance from centre to centre twelve times the diameter of each bar*. It represents the lines of force due to equal quantities of opposite electricities on the grating itself, and a parallel plane of continuous metal (not shown in the diagram) at a distance from the grating of not less than one and a half times the distance from bar to bar. The shading shows the lines of force for the same circumstances, but with oval bars instead of the small circular bars of Maxwell's grating. It is interesting to see how every line of force ends in a bar of the grating, none straying to an infinite distance beyond it, which is necessarily the case when the quantities of electricity on the grating and on the continuous plane are equal and opposite. If an insulated electrified body, with electricity of the same name as that of the grating, for example, is brought up from below,

---

* *Electricity and Magnetism*, Vol. I. Art. 203, Plate xiii. [The diagram is on p. 508 *supra*.]

it experiences no electric force differing sensibly from that which would be produced by its own inductive effect on the grating, till it is within a less distance from the grating than the distance from bar to bar, when it experiences repulsion or attraction, according as it is under a bar of the grating or under the middle of a space between two bars. If there be a parallel metal plane below the grating, kept at the same potential as the grating, it takes no sensible proportion of the electricity from the grating, and experiences no sensible force when its distance from the grating exceeds a limit depending on the ratio of the diameter of each bar to the distance from bar to bar. The mathematical theory of this action was partially given by Maxwell*, and yesterday I communicated an extension of it to the Royal Society.

II.

Magnetostatic screening by soft iron would follow the same law as electrostatic screening, if the magnetic susceptibility of

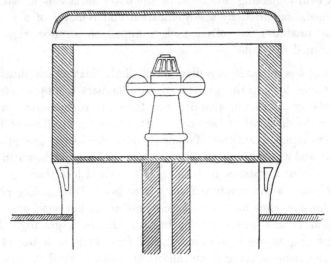

Fig. 6.

the iron were infinitely great. It is not great enough to even approximately fulfil this condition in any practical case. The nearest approach to fulfilment is presented when we have a thick iron shell completely enclosing a hollow space, but the thickness

* Arts. 203—205.

must be a considerable proportion of the smallest diameter, not less than $\frac{1}{10}$, perhaps, for iron of ordinary magnetic susceptibility to produce so much of screening effect that the magnetic force in the interior should be anything less than 5 per cent. of the force at a distance outside, when the shell is placed in a uniform magnetic field. The accompanying diagram, Fig. 6, representing the conning-tower of H.M.S. '*Orlando*,' and the position of the compass within it, has been kindly sent to me by Captain Creak, R.N., for this lecture, by permission of the Controller of the Navy. It gives an interesting illustration of magnetic screening effect by the case of a belt of iron, 1 foot thick, 5 feet high, and 10 feet in internal diameter, with roof and floor of comparatively thin iron. Captain Creak informs me that the average horizontal component of the magnetic directing force on the compass in the centre of this conning-tower is only about one-fifth of that of the undisturbed terrestrial magnetism.

An evil practice, against which careful theoretical and practical warnings were published two or three years ago*, and which is now nearly, though, I believe, not at this moment quite thoroughly, stopped, of what is called single wiring in the electric lighting of ships, has been fallaciously defended by various bad reasons, among them an erroneous argument that the ship's iron produced a sufficient screening effect against disturbance of the ship's com-- passes, by the electric light currents, when that plan of wiring is adopted. The argument would be good for a ship 50 feet broad and 30 feet deep, if the deck and hull were of iron 3 feet thick. As it is, mathematical calculation shows that the screening effect is quite small in comparison with what the disturbance of the compass would be if the ship and her decks were all of wood. Actual observation, on ships electrically lighted on the single wire system by some of the best electrical engineers in the world. has shown, in many cases, disturbance of the compass of from 3 degrees to 7 degrees, produced by throwing off and on the groups of lights in various parts of the ship, which are thrown on and off habitually in the evenings and nights, in ordinary and necessary practice of sea-going passenger ships. When the facts become known to shipowners, single wiring will never again be admitted at sea unless the alternating current system of electric lighting is again adopted. But, although this system was largely

* See *The Electrician*, Vol, xxiii. p. 87 [*supra*, p. 474].

used when electric lighting was first introduced into ships, the economy and other advantages of the direct-current system are so great that no one would think of using the alternate system for the trivial economy, *if any economy there is,* in the single wire, as compared with the double insulated wire system.

An interesting illustration of a case in which iron, of any thickness, however great, produces *no screening effect* on an electric current, steady or alternating, is shown by the accompanying diagram [omitted], which represents in section an electric current along the axis of a circular iron tube, completely surrounding it. Whether the tube be long or short, it exercises no screening effect whatever. A single circular iron ring, supported in the air, with its plane perpendicular to the length of a straight conductor conveying an electric current, produces absolutely no disturbance of the circular endless lines of magnetic force which surround the wire ; neither does any piece of iron, wholly bounded by a surface of revolution, with a straight conductor conveying electricity along its axis*.

A screen of imperfectly conducting material is as thorough in its action, when time enough is allowed it, as is a similar screen of metal. But if it be tried against rapidly varying electrostatic force, its action lags. On account of this lagging, it is easily seen that the screening effect against periodic variations of electrostatic force will be less and less, the greater the frequency of the variation. This is readily illustrated by means of various forms of idiostatic electrometers. Thus, for example, a piece of paper supported on metal in metallic communication with the movable disc of an attracted disc electrometer annuls the attraction (or renders it quite insensible) a few seconds of time after a difference of potential is established and kept constant between the attracted disc and the opposed metal plate, if the paper and the air surrounding it are in the ordinary hygrometric conditions of our climate. But if the instrument is applied to measure a rapidly alternating difference of potential, with equal differences on the two sides of zero, it gives very little less than the same average force as that found when the paper is removed and all other circumstances kept the same. Probably, with ordinary clean

---

* [The remainder of this Lecture appears also in *Proc. Roy. Soc.* Vol. XLIX. April 9, 1891, from which it is reprinted in *Baltimore Lectures,* Appendix K, pp. 681—7.]

white paper in ordinary hygrometric conditions, a frequency of alternation of from 50 to 100 per second will more than suffice to render the screening influence of the paper insensible. And a much less frequency will suffice if the atmosphere surrounding the paper is artificially dried. Up to a frequency of millions per second, we may safely say that, the greater the frequency, the more perfect is the annulment of screening by the paper; and this statement holds also if the paper be thoroughly blackened on both sides with ink, although possibly in this condition a greater frequency than 50 to 100 per second might be required for practical annulment of the screening.

Now, suppose, instead of attractive force between the two bodies separated by the screen, as our test of electrification, that we have as test a faint spark, after the manner of Hertz. Let two well insulated metal balls, *A*, *B*, be placed very nearly in contact, and two much larger balls, *E*, *F*, placed beside them, with the shortest distance between *E*, *F* sufficient to prevent sparking, and with the lines joining the centres of the two pairs parallel. Let a rapidly alternating difference of potential be produced between *E* and *F*, varying, not abruptly, but according, we may suppose, to the simple harmonic law. Two sparks in every period will be observed between *A* and *B*. The interposition of a large paper screen between *E*, *F*, on one side, and *A*, *B*, on the other, in ordinary hygrometric conditions, will absolutely stop these sparks, if the frequency be less than, perhaps, 4 or 5 per second. With a frequency of 50 or more, a clean white paper screen will make no perceptible difference. If the paper be thoroughly blackened with ink on both sides, a frequency of something more than 50 per second may be necessary; but some moderate frequency of a few hundreds per second will, no doubt, suffice to practically annul the effect of the interposition of the screen. With frequencies up to 1000 million per second, as in some of Hertz's experiments, screens such as our blackened paper are still perfectly transparent, but if we raise the frequency to 500 million million, the influence to be transmitted is light, and the blackened paper becomes an almost perfect screen.

Screening against a varying magnetic force follows an opposite law to screening against varying electrostatic force. For the present I pass over the case of iron and other bodies possessing magnetic susceptibility, and consider only materials devoid of magnetic susceptibility, but possessing more or less of electric conductivity. However perfect the electric conductivity of the screen may be, it has no screening efficiency against a steady magnetic force. But if the magnetic force varies, currents are induced in the material of the screen which tend to diminish the magnetic force in the air on the remote side from the varying magnet. For simplicity, we shall suppose the variations to follow the simple harmonic law. The greater the electric conductivity of the material, the greater is the screening effect for the same frequency of alternation; and, the greater the frequency, the greater is the screening effect for the same material. If the screen be of copper, of specific resistance 1640 sq. cm. per second (or electric diffusivity 130 sq. cm. per second), and with frequency 80 per second, what I have called the "mhoic effective thickness"* is 0.71 of a cm.; and the range of current intensity at depth $n \times 0.71$ cm. from the surface of the screen next the exciting magnet is ϵ^{-n} of its value at the surface.

Thus (as $\epsilon^3 = 20.09$) the range of current intensity at depth 2.13 cm. is $\frac{1}{20}$ of its surface value. Hence we may expect that a sufficiently large plate of copper of $2\frac{1}{4}$ cm. thick will be a little less than perfect in its screening action against an alternating magnetic force of frequency 80 per second.

Lord Rayleigh, in his "Acoustical Observations," †after referring to Maxwell's statement, that a perfectly conducting sheet acts as a barrier to magnetic force‡, describes an experiment in which the interposition of a large and stout plate of copper between two coils renders inaudible a sound which, without the copper screen, is heard by a telephone in circuit with one of the coils excited by electromagnetic induction from the other coil, in which an intermittent current, with sudden, sharp variations of strength, is produced by a "microphone clock" and a voltaic battery. Larmor, in his paper on "Electromagnetic Induction in Conducting Sheets and Solid Bodies"§ makes the following very interesting statement:—"If we have a sheet of conducting matter in the neighbourhood of a magnetic system, the effect of a disturbance of that system will be to induce currents in the sheet of such kind as

---

* *Collected Papers*, Vol. III. Art. cii. § 35.

† *Phil. Mag.* 1882, first half-year.

‡ *Electricity and Magnetism*, § 665.

§ *Phil. Mag.* 1884, first half-year.

will tend to prevent any change in the conformation of the tubes [lines] of force cutting through the sheet. This follows from Lenz's law, which itself has been shown by Helmholtz and Thomson to be a direct consequence of the conservation of energy. But if the arrangement of the tubes [lines of force] *in* the conductor is unaltered, the field on the other side of the conductor into which they pass (supposed isolated from the outside spaces by the conductor) will be unaltered. Hence, if the disturbance is of an alternating character, with a period small enough to make it go through a cycle of changes before the currents decay sensibly, we shall have the conductor acting as a screen.

"Further, we shall also find, on the same principle, that a rapidly rotating conductor sheet screens the space inside it from all magnetic action which is not symmetrical round the axis of rotation."

Mr Willoughby Smith's experiments on "Volta-electric induction," which he described in his inaugural address to the Society of Telegraph Engineers of November 1883, afforded good illustration of this kind of action with copper, zinc, tin, and lead, screens, and with different degrees of frequency of alternation. His results with iron are also very interesting: they showed, as might be expected, comparatively little augmentation of screening effect with augmentation of frequency. This is just what is to be expected from the fact that a broad enough and long enough iron plate exercises a large magneto-static screening influence; which with a thick enough plate, will be so nearly complete that comparatively little is left for augmentation of the screening influence by alternations of greater and greater frequency.

A copper shell closed around an alternating magnet produces a screening effect which on the principle stated above we may reckon to be little short of perfection if the thickness be $2\frac{1}{4}$ cm. or more, and the frequency of alternation 80 per second.

Suppose now the alternation of the magnetic force to be produced by the rotation of a magnet M about any axis. First, to find the effect of the rotation, imagine the magnet to be represented by ideal magnetic matter. Let (after the manner of Gauss in his treatment of the secular perturbations of the solar system) the ideal magnetic matter be uniformly distributed over the circles described by its different points. For brevity call I the ideal

ELECTRODYNAMICS [231

magnet symmetrical round the axis, which is thus constituted. The magnetic force throughout the space around the rotating magnet will be the same as that due to I, compounded with an alternating force of which the component at any point in the direction of any fixed line varies from zero in the two opposite directions in each period of the rotation. If the copper shell is thick enough, and the angular velocity of the rotation great enough, the alternating component is almost annulled for external space, and only the steady force due to I is allowed to act in the space outside the copper shell.

Consider now, in the space outside the copper shell, a point P rotating with the magnet M. It will experience a force simply equal to that due to M when there is no rotation, and, when M and P rotate together, P will experience a force gradually altering as the speed of rotation increases, until, when the speed becomes sufficiently great, it becomes sensibly the same as the force due to the symmetrical magnet I. Now superimpose upon the whole system of the magnet, and the point P, and the copper shell, a rotation equal and opposite to that of M and P. The statement just made with reference to the magnetic force at P remains unaltered, and we have now a fixed magnet M and a point P at rest, with reference to it, while the copper shell rotates round the axis around which we first supposed M to rotate.

A little piece of apparatus, constructed to illustrate the result experimentally, was submitted to the Royal Institution and shown in action. The copper shell is a cylindric drum, 1·25 cm. thick, closed at its two ends with circular discs 1 cm. thick. The magnet is supported on the inner end of a stiff wire passing through the centre of a perforated fixed shaft which passes through a hole in one end of the drum, and serves as one of the bearings; the other bearing is a rotating pivot fixed to the outside of the other end of the drum. The accompanying sections*, drawn to a scale of three-fourths full size, explain the arrangement sufficiently. A magnetic needle outside, deflected by the fixed magnet when the drum is at rest, shows a great diminution of the deflection when the drum is set to rotate. If the (triple compound) magnet inside is reversed, by means of the central wire and cross bar outside, shown in the diagram, the magnetometer outside is greatly

* [For diagram, see *Baltimore Lectures*, p. 686.]

affected while the copper shell is at rest; but scarcely affected perceptibly while the copper shell is rotating rapidly.

When the copper shell is a figure of revolution, the magnetic force at any point of the space outside or inside is steady, whatever be the speed of rotation; but if the shell be not a figure of revolution, the steady force in the external space observable when the shell is at rest becomes the resultant of the force due to a fixed magnet intermediate between M and I compounded with an alternating force with amplitude of alternation increasing to a maximum, and ultimately diminishing to zero, as the angular velocity is increased without limit.

If M be symmetrical, with reference to its northern and southern polarity, on the two sides of a plane through the axis of rotation, I becomes a null magnet, the ideal magnetic matter in every circle of which it is constituted being annulled by equal quantities of positive and negative magnetic matter being laid on it. Thus, when the rotation is sufficiently rapid, the magnetic force is annulled throughout the space external to the shell. The transition from the steady force of M to the final annulment of force, when the copper shell is symmetrical round its axis of rotation, is, through a steadily diminishing force, without alternations. When the shell is not symmetrical round its axis of rotation, the transition to zero is accompanied with alternations as described above.

When M is not symmetrical on the two sides of a plane through the axis of rotation, I is not null; and the condition approximated to through external space with increasing speed of rotation is the force due to I, which is an ideal magnet symmetrical round the axis of rotation.

A very interesting simple experimental illustration of screening against magnetic force may be shown by a rotating disc with a fixed magnet held close to it on one side. A bar magnet held with its magnetic axis bisected perpendicularly by a plane through the axis of rotation would, by sufficiently rapid rotation, have its magnetic force almost perfectly annulled at points in the air as near as may be to it, on the other side of the disc, if the diameter of the disc exceeds considerably the length of the magnet. The magnetic force in the air close to the disc, on the side next to the magnet, will be everywhere parallel to the surface of the disc.

232. ON A NEW FORM OF AIR LEYDEN, WITH APPLICATION TO THE MEASUREMENT OF SMALL ELECTROSTATIC CAPACITIES.

[From *Roy. Soc. Proc.* Vol. LII. 1893 [June 2, 1892], pp 6—10 ; *Nature*, Vol. XLVI. [June 30, 1892], pp. 212, 213; *Lum. Élec.* Vol. XLV. July 16, 1892, pp. 139—141.]

IN the title of this paper as originally offered for communication "*Air Condenser*" stood in place of "*Air Leyden*," but it was accompanied by a request to the Secretaries to help me to a better designation than "Air Condenser" (with its ambiguous suggestion of an apparatus for condensing air), and I was happily answered by Lord Rayleigh with a proposal to use the word "leyden" to denote a generalised Leyden jar, which I have gladly adopted.

The apparatus to be described affords, in conjunction with a suitable electrometer, a convenient means of quickly measuring small electrostatic capacities, such as those of short lengths of cable.

The instrument is formed by two mutually insulated metallic pieces, which we shall call *A* and *B*, constituting the two systems of an air condenser, or, as we shall now call it, an air leyden. The systems are composed of parallel plates, each set bound together by four long metal bolts. The two extreme plates of set *A* are circles of much thicker metal than the rest, which are all squares of thin sheet brass. The set *B* are all squares, the bottom one of which is of much thicker metal than the others, and the plates of this system are one less in number than the plates of system *A*. The four bolts binding together the plates of each system pass through well-fitted holes in the corners of the squares ; and the distance from plate to plate of the same set is regulated by annular distance pieces which are carefully made to fit the bolt, and are made exactly the same in all respects. Each system is bound firmly together by screwing home nuts on the ends of the bolts, and thus the parallelism and rigidity of the entire set is secured.

The two systems are made up together, so that every plate of *B* is between two plates of *A*, and every plate of *A*, except the two end ones, which only present one face to those of the opposite set, is between two plates of *B*. When the instrument is set up for use, the system *B* rests by means of the well-known "hole, slot, and plane arrangement*," engraved on the under side of its bottom plate, on three upwardly projecting glass columns which

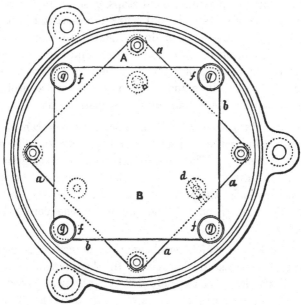

Fig. 1.

are attached to three metal screws working through the sole plate of system *A*. These screws can be raised or lowered at pleasure, and by means of a gauge the plates of system *B* can be adjusted to exactly midway between, and parallel to, the plates of system *A*. The complete leyden stands upon three vulcanite feet attached to the lower side of the sole plate of system *A*.

In order that the instrument may not be injured in carriage, an arrangement, described as follows, is provided by which system

---

* Thomson and Tait's *Natural Philosophy*, § 198, example 3.

B can be lifted from off the three glass columns and firmly clamped to the top and bottom plates of system A.

The bolts fixing the corners of the plates of system B are made long enough to pass through wide conical holes cut in the top and bottom plates of system A, and the nuts at the top end of the bolts are also conical in form, while conical nuts are also fixed to their lower ends below the base plate of system A. Thumbscrew nuts, f, are placed upon the upper ends of the bolts after they pass through the holes in the top plate of system A.

When the instrument is set up ready for use these thumb-screws are turned up against fixed stops, g, so as to be well clear of the top plate of system A; but when the instrument is packed for carriage they are screwed down against the plate until the conical nuts mentioned above are drawn up into the conical holes in the top and bottom plates of system A; system B is thus raised off the glass pillars, and the two systems are securely locked together so as to prevent damage to the instrument.

A dust-tight cylindrical metal case, h, which can be easily taken off for inspection, covers the two systems and fits on to a flange on system A. The whole instrument, as said above, rests on three vulcanite legs attached to the base plate of system A; and two terminals are provided, one, i, on the base of system A, and the other, j, on the end of one of the corner bolts of system B.

The air leyden which has been thus described is used as a standard of electrostatic capacity. In the instrument actually exhibited to the Society there are twenty-two plates of the system B, twenty-three of the system A, and therefore forty-four octagonal air spaces between the two sets of plates. The thickness of each of these air spaces is approximately 0·301 of a centimetre. The side of each square is 10·13 cm., and there-fore the area of each octagonal air space is 85·1 sq. cm. The capacity of the whole leyden is therefore approximately $44 \times 85 \cdot 1/(4\pi \times 0 \cdot 301)$, or 990 cm. in electrostatic measure; or $1 \cdot 1 \times 10^{-18}$ c.g.s., electromagnetic measure; or $1 \cdot 1 \times 10^{-9}$ farads, or $1 \cdot 1 \times 10^{-3}$ microfarads. This is only an approximate estimate founded on a not minutely accurate measurement of dimensions, and not corrected for the addition of capacity, due to the edges and projecting angles of the squares and the metal cover.

I hope to have the capacity determined with great accuracy by comparison with Mr Glazebrook's standards in Cambridge.

To explain its use in connexion with an idiostatic electrometer for the direct measurement of the capacity of any insulated conductor, I shall suppose, for example, this insulated conductor to be the insulated wire of a short length of submarine cable core, or of telephone, or telegraph, or electric light cable, sunk under water, except a projecting portion to allow external connexion to be made with the insulated wire.

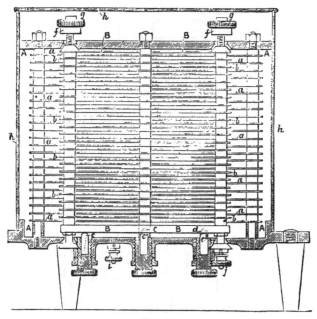

Fig. 2.

The electrometer which I find most convenient is my "multicellular voltmeter," rendered practically dead-beat by a disc under oil hung on the lower end of the long stem carrying the electric "needles" (or movable plates). In the multicellular voltmeter used in the experimental illustration before the Royal Society, the index shows its readings on a vertical cylindric surface, which for electric light stations is more convenient than the horizontal scale of the multicellular voltmeters hitherto in use; but for the measurement of electrostatic capacity the older horizontal scale instrument is as convenient as the new form.

To give a convenient primary electrification for the measurement, a voltaic battery, VV' (fig. 3), of about 150 or 200 elements, of each of which the liquid is a drop of water held up by capillary attraction between a zinc and copper plate about 1 mm. asunder. An ordinary electric machine, or even a stick of rubbed sealing-wax may, however, be used, but not with the same facility for giving the amount of electrification desired as the voltaic battery.

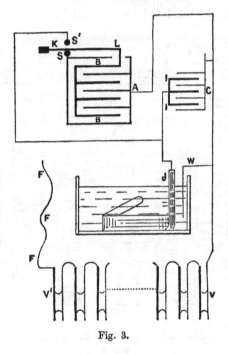

Fig. 3.

One end of the voltaic battery is kept joined metallically to a wire, W, dipping in the water in which the cable is submerged, and with the case C of the multicellular, and with the case and plates A of the leyden, and with a fixed stud, S, forming part of the operating key to be described later. The other end of the voltaic battery is connected to a flexible insulated wire, FFF, used for giving the primary electrification to the insulated wire J of the cable, and the insulated cells II of the multicellular kept metallically connected with J. The insulated plates, B, of the leyden are connected to a spring, KL, of the operating key referred to above, which, when left to itself, presses down on the metal stud S, and which is very perfectly insulated when lifted from contact with S by a finger applied to the insulating handle K. A second well insulated stud, S', is kept in metallic connection with J and I (the insulated wire of the cable and the insulated cells of the multicellular).

To make a measurement the flexible wire F is brought by hand to touch momentarily on a wire connected with the stud S', and immediately after that a reading of the electrometer is taken and watched for a minute or two to test either that there is no sensible loss by imperfect insulation of the cable and the insulated cells of the multicellular, or that the loss is not sufficiently rapid to vitiate the measurement. When the operator is satisfied with this he records his reading of the electrometer, presses up the handle K of the key, and so disconnects the plates B of the leyden from S and A, and connects them with S', J, I. Fifteen or twenty seconds of time suffices to take the thus diminished reading of the multicellular, and the measurement is complete.

The capacity of the cable is then found by the analogy:— As the excess of the first reading of the electrometer above the second is to the second, so is the capacity of the leyden to the capacity of the cable.

The preceding statement describes the arrangement which is most convenient when the capacity of the cable exceeds the capacity of the leyden. The plan which is most convenient in the other case, that is to say, when the capacity of the cable is less than that of the leyden, is had by interchanging B and J throughout the description. In this case, a charge given to the leyden is divided between it and the cable. The capacity of the cable is then found by the analogy:—As the second reading of the electrometer is to the excess of the first above the second, so is the capacity of the leyden to the capacity of the cable.

A small correction is readily made with sufficient accuracy, for the varying capacity of the electrometer, according to the different positions of the movable plates, corresponding to the different readings, by aid of a table of corrections determined by special measurements for capacity of the multicellular.

233. Instructions for the Observation of Atmospheric Electricity for the use of the Arctic Expedition of 1875.

[From the *Manual of the Natural History, Geology, and Physics of Greenland*, prepared for the use of the Arctic Expedition of 1875, under direction of the Arctic Committee of the Royal Society. London, 1875, 8vo. pp. 20—24.]

The instrument to be used is the portable electrometer described in Sir Wm. Thomson's reprint of "Papers on Electrostatics and Magnetism," §§ 368—378*. Full directions for keeping the instrument in order, preparing it for use, and using it to make observations of atmospheric electricity, are to be found in sections 372—376; these are summarized in the following short practical rules:—

I. The instrument having been received from the maker with the inner surface of the glass, and all the metallic surfaces within, clean and free from dust or fibres, and the pumice dry. To prepare it for use:

(1) Remove from the top the cover carrying the pumice. Drop upon the pumice a small quantity of the prepared sulphuric acid supplied with the instrument, distributing it as well as may be over the whole surface of the stone. There ought not to be so much acid as to show almost any visible appearance of moisture when once it has soaked into the pumice. Replace the cover without delay, and screw it firmly in its proper position, and then leave the instrument for half an hour or an hour, or any longer time that may be convenient to allow the inner surface of the glass to be well dried through the drying effect of the acidulated pumice on the air within.

(2) Turn the micrometer screw till the reading is 2,000. (There are 100 divisions on the circle which turns with the screw

* A copy of this book has been sent by the author for the use of the officer or officers to whom the observations of atmospheric electricity are committed.

on the top outside, and the numbers on the vertical scale inside
show full turns of the screw. Thus each division on the vertical
scale inside corresponds to 100 divisions on the circle; and 20
on the vertical scale is read "2,000.") Introduce the charging rod
and give a charge of negative electricity by means of the small
electrophorus which accompanies the instrument. When enough
has been given to bring the hair a little below the middle of the
space between the black dots, give no more charge; but remove
the charging rod and close the aperture immediately. If now the
hair is still seen a little below the middle of the space between
the black dots, turn the screw head in such a direction as to raise
the attracting disc, and so diminish the attraction till the hair is
exactly midway between the dots.. Watch the instrument for a
few minutes, and if the hair is seen to rise, as it generally will
(because of the electricity which has been given, spreading over
the inner surface of the glass), turn the micrometer screw in the
direction to lower the attracting plate, so as to keep the hair midway
between the dots.

(3) The insulation will generally improve for several hours,
and sometimes for several days, after the instrument is first
charged. The instrument may be considered to be in a satisfactory state if the earth reading does not diminish by more than
30 divisions per 24 hours. If the maker has been fortunate with
respect to the quality of the substance of the glass jar, the earth
reading may not sink by more than 30 divisions per week, when
the pumice is sufficiently moistened with strong and pure sulphuric
acid. Recharge with negative electricity occasionally so as to
keep the earth reading between 1,000 and 2,000.

II. To keep the instrument in order. Watch the pumice
carefully, looking at it every day. If it begins to look moist,
remove the cover, take out the screws holding the lead cup,
remove the pumice and dry it on a shovel over the galley fire.
When cool put prepared sulphuric acid on it, replace it in the
instrument, and re-electrify according to No. I.

*Never leave the pumice unwatched, in the instrument, for as
long as a week.* WHEN THE INSTRUMENT IS TO BE OUT OF USE
FOR A WEEK OR LONGER TAKE THE PUMICE OUT OF IT.

III. To use the portable electrometer for observing atmospheric electricity:

(1)   The place of observation, if on board ship, must be as far removed from spars and rigging as possible. In a sailing ship or rigged steamer the best position for the electrometer generally is over the weather quarter when under way, or anywhere a few feet above the tafferel when at anchor. On shore or on the ice a position not less than 20 yards from any prominent object (such as a hut or a rock or mass of ice or ship), standing up to any considerable height above the general level, should be chosen. Whether on board ship or in an open boat or on shore or on the ice, the electrometer may be held by the observer in his left hand while he is making an observation; but a fixed stand, when conveniently to be had, is to be preferred, unless in the case of making observations from an open boat.

(2)   To make an observation in ordinary circumstances the observer stands upright and holds or places the electrometer in a position about five feet above the ground (or place on which he stands), so as to bring the hair and two black dots about level with his eye. The umbrella of the principal electrode being *down* to begin with (and so keeping metallic connection between the principal electrode and the metallic case of the instrument), the observer commences by taking an "earth reading*." The steel wire, with a match stuck on its point, being in position on the principal electrode, the match is then lighted, the umbrella lifted, and the micrometer screw turned so as to keep the hair in the middle between the black dots. After the umbrella has been up and the match lighted for 20 seconds or half a minute, a reading may be taken and recorded, called an "air reading." A single such reading constitutes a valuable observation. But a series of readings taken at intervals of a quarter of a minute, or half a minute, or at moments of maximum or minimum electrification during the course of two or three minutes, the match burning all the time, is preferable. In conclusion, remove the match if it is not all burned away, lower the umbrella home, and take an earth reading.

(3)   The electric potential of the air at the point of the burning match is found by subtracting the earth reading from the air reading at any instant. When the air reading is less than the earth reading the air potential is negative, and is to be recorded as the difference between the earth reading and the air

---

* *Electrostatics and Magnetism*, § 375.

reading with the sign — prefixed. The earth reading may be generally taken as the mean between the initial and final earth readings. But the actual earth readings and air readings ought all to be recorded carefully, and the full record kept.

(4) Note and record the wind at the time of each observation, also the character of the weather.

IV. Observations to be made :

(1) At the commencement of the Expedition, in the course of the northward voyage, observations of atmospheric electricity ought to be taken regularly three or four times a day; also occasionally during the night to give the observer some practice in the use of a lantern for reading the divisions on the circle and of the vertical scale.

(2) When stationary in winter quarters observations should be made three times a day at intervals of six hours; for example, at 8 a.m., 2 p.m., and 8 p.m., or at 7.30 a.m., 1.30 p.m., and 7.30 p.m. Whatever times are most convenient may be chosen provided they be separated by intervals of six hours.

(3) It is very desirable that hourly observations should be made, if only for a few days, in winter and in summer. If possible arrangements to do so at least for six consecutive days in winter, and for six consecutive days in summer should be made. The results will be very interesting as showing whether there is a diurnal or semi-diurnal period in either the Arctic winter or summer, as we know there is at every time of year in places outside the Arctic circle.

(4) Make occasionally special observations when there is anything peculiar in the weather, especially with reference to wind.

V. Special precautions :

(1) In the Arctic climate more care may be necessary than in ordinary climates as to earth connections. Therefore put a piece of metal on the stand on which the electrometer is placed during an observation on board ship, and keep this in metallic communication with the ship's coppers or lightning conductors. If the electrometer is held in the hand with or without a glove, a fine wire ought to be tied round the brass projection which carries the lens, or otherwise attached to the outer case of the electrometer, and by this wire sufficient connection maintained

with the earth during an observation. The connection will probably be sufficient if a short length of the wire is laid on the ice and the observer stands on it. Enough, however, is not yet known as to electric conductivity of ice: and to make sure it *may* be necessary to have a wire or chain let down to the water through a hole in the ice, and metallic connection kept up by a fine wire between this and the electrometer case during an observation.

(2) The observer's cap (particularly if of fur) and his woollen clothing, and even his hair if not completely covered by his cap, will be apt in the Arctic climate to become electrified by the slightest friction, and so to give false results when the object to be observed is atmospheric electricity. A tin foil cover for cap and arms, kept in metallic communication by a fine wire with the hand or hands applied to the case of the electrometer or to the micrometer screw head, should therefore be used by the observer (and assistant, if he has an assistant to carry lanthorn, or for any other purpose), unless he has made sure that there is no sensible disturbance from those causes, without the precaution.

VI. Instruments, stores, and appliances for observation of atmospheric electricity sent with the Expedition:

1. Two portable electrometers, Nos. 35 and 36, each with one steel wire for carrying match, one charging rod, and one electrophorus for charging the jar.

2. Six spare steel wires (three to go with each instrument).

3. Supply of matches ready made. (The slower the match burns, the better. If those supplied burn too fast, steep them in water and dry them again.)

4. White blotting paper and nitrate of lead to make more matches when wanted. (Moisten the paper with weak solution of nitrate of lead, and roll into matches with thin paste made with a very little nitrate of lead in the water.)

5. Six spare pumices (three for each electrometer); india rubber bands to secure pumice in lead case.

6. Eight small stoppered bottles of prepared sulphuric acid (four for each electrometer).

7. Tin foil and fine wire.

ELECTRICAL CONTRIBUTIONS TO ENGINEERING SOCIETIES

THE MALTA AND ALEXANDRIA SUBMARINE TELEGRAPH CABLE.

[From DISCUSSION ON H. C. FORDE'S PAPER, *Inst. Cir. Engrs. Proc.* Vol. XXI. [May 13 and 20, 1862], pp. 535, 536.]

PROFESSOR W. THOMSON considered, that with a cable of eleven hundred knots, having the same sectional area and dimensions as the Malta and Alexandria cable, a telegraphic speed of ten words per minute ought to be obtained. That estimate was based upon actual experience of the working of the Atlantic Telegraph. Morse's alphabet had been used, and a telegraphic speed of two and a half words per minute, fully spelt and without abbreviations, was attained. According to a well-established law, the speed of working through half the Atlantic cable would have been four times the speed actually attained through the whole length. From the preceding data he had no doubt that in the case of the thirteen hundred knots, which formed the entire length of the Malta and Alexandria cable, a speed of twelve fully-spelt words per minute might be obtained with instruments adapted not only for high speed, but also for convenience and certainty. With reference to the battery power which would be required, he believed, that the smallest battery power was sufficient to produce the speed he had mentioned, and that ten cells would be quite sufficient.

The theoretical result from the law of the squares, using the data derived from the working of the Atlantic Telegraph cable, and taking into account the greater lateral dimensions of the

Malta and Alexandria line, would considerably exceed twelve words per minute *.

It was to be hoped, that the forebodings which had been expressed, as regarded the permanency of the cable, might not be realized, and either that repairs might not be required, or that if required they might be successfully executed. It would, however, be extremely unwise to lose sight of any such possible contingency, and in designing future submarine telegraph cables, every effort should be made to insure the permanent protection of the insulating medium. From the rapid advance of telegraphic science and experience, he hoped that this important and difficult problem would be solved, and that a submarine telegraph cable would be designed, constructed, and laid, with the same prospect of success and permanency as a bridge, or a railway.

* This calculation has since been made, and the result gives fifteen and a half words per minute for 1,330 knots of cable.—W. T.

[TESTS OF TELEGRAPHIC APPARATUS.]

I.   On a New Form of Joule's Tangent Galvanometer.
II. On the Measurement of Electrostatic Capacity.
III. Tests of Battery.
IV. Tray Battery for the Siphon Recorder.

[From Original Communications, in the *Journal of the Society of Telegraph Engineers*, Vol. I. 1872, pp. 392—406.]

I.   On a New Form of Joule's Tangent Galvanometer.

NEARLY thirty years ago Joule showed how to make a good tangent galvanometer*. Yet, up to the present time, the tangent galvanometer described in the text-books and made by the chief instrument makers has for needle a heavy steel bar two or three inches long, supported by an agate capsule on a steel point; and it is nowhere properly placed except in a historical museum. It is certainly useless in the laboratory or the electrician's testing room.

Joule's improvement, subsequent to the meeting of the British Association at Cork in 1843, consists in substituting for the point and agate capsule, suspension by a single silk fibre ; for the great steel bar, a very light needle about a quarter inch long, or, at the most, half an inch ; or two or more such needles cemented parallel to one another on an exceedingly light cradle: and for the bar itself as pointer, an exceedingly light glass rod or tube cemented to the needle or cradle, pointing degrees of deflection on a graduated horizontal circle. The cradle ought to be the very lightest that can bear the needles, and the glass pointer the lightest that can bear its own weight without drooping to an inconvenient extent. By fulfilling these conditions, and particularly by making for the

---

* Not before the Meeting of the British Association at Cork, in 1843, certainly; not long after, probably. See, in the Association's Report of that year, abstract of his paper " On a Galvanometer."

pointer a glass tube about six inches long, and so light that its ends droop about a quarter inch each below the level of its middle, and by making the floor and glass roof of the case protecting it from currents of air as close together as they can conveniently be to allow the pointer freedom to turn without touching them, Joule long ago produced an instrument with the distinguishing merit of my more recent "dead-beat mirror." When the current through the coil is suddenly made, or reversed or broken, the indicator moves direct to its fresh position without shooting past it, and reaches it sensibly in, at the most, two or three seconds of time. I do not think it an exaggeration to say that, with an instrument having these qualities, as many accurate observations can be made in ten minutes as can be made in an hour with the ordinary instruments which oscillate most tediously before settling sufficiently to show the position of equilibrium. With an instrument constructed for the University of Glasgow, about twenty-five years ago (by Mr Dancer of Manchester, under the direction of Joule himself), in which the circle is six inches diameter, and is divided into degrees and halves of a degree, I find that by using reversals, and taking the means of the four readings of the two ends of the needle in the two positions, I can readily obtain results, trustworthy to within 2′ or 3′, of the true value of the single deflection.

The new form of Joule's instrument, now before the Society, has several modifications, which I have introduced chiefly to render it more convenient for the practical electrician, but partly, also, to save time in measuring the resistances and intensities of galvanic elements in the scientific laboratory. These modifications are:

No. 1. Instead of a single fixed coil, two coils, supported on geometrical slides, so as to be movable to different distances from the needle, in the direction perpendicular to the magnetic meridian on the two sides.

No. 2. Simple appliance to produce change from connection of the two coils in series, to connection in double arc, when one of them is brought nearer to the needle than a certain position on its slide.

No. 3. Two long coils, with a great number of convolutions, of fine wire, so as to give a resistance immensely greater than that of any single element, or even battery, to the measurement of which the instrument is to be frequently applied.

No. 4. Copper wire of high conductivity, a condition which, so far as I am aware, has never been attended to in any tangent galvanometer hitherto made, except a few by White, including the instrument now before the Society.

No. 5. Division of the scale, to show equal differences of tangents of the angles, from zero, instead of equal angular differences.

No. 6. Simplified details of suspension and adjustment, including an exceedingly-convenient mode devised by Mr White for lifting the cradle, and supporting it firmly, without breaking the fibre, so as to render the instrument readily portable.

The object of No. 1 is to render the instrument readily applicable to test the resistances and intensities of batteries of from 2 to 100 cells. The object of No. 2 is to give sufficient sensibility for satisfactorily measuring the intensity of a single Daniell's cell. The object of No. 3 is to give the simplicity of the method by electrometer, to testing the resistance and intensity of a battery by galvanometer. The object of No. 4 is obvious, and it is worthy to be noticed by the Society of Telegraph Engineers, that it has hitherto been so lamentably neglected. The object of No. 5 is to avoid the necessity of using a table of natural tangents, and to save time when numerous tests have to be made.

The instrument shown this evening to the Society is merely a rough first trial towards the construction of a galvanometer fulfilling the conditions I have described. I hope at a future meeting to be able to show an instrument with the constructive details improved in many important particulars. Such an instrument, ordered by Mr Saunders for the Eastern Telegraph Company, is now being made by Mr White, and a sufficient description, if not the instrument itself, will I hope soon be placed before the Society. Among other improvements it will have a plain mirror of silvered glass, instead of white paper, for the floor of the case, in the immediate neighbourhood of the graduated arc. This, according to a well-known old German plan, will greatly promote accuracy in reading the deflections, by facilitating the avoidance of parallactic error.

II. On the Measurement of Electrostatic Capacity.

The ordinary methods of measuring electrostatic capacity, in which a galvanometer is used to measure the electro-magnetic impulse, on the same dynamical principle as that adopted by Robins in his ballistic pendulum, are liable to very serious objections:

No. 1. It is difficult to secure that the whole duration of the impulse shall be a sufficiently small fraction of the needle's period.

No. 2. The resistance of the air or of other fluid employed to damp the vibrations of the needle, directly affects the observed throws, to amounts which cannot easily, if at all, be determined with accuracy.

No. 3. On account of (1) and (2) it is necessary for accuracy that the period be long, and the resistance against the motions of the needle so small that the diminution of amplitude in successive swings shall be small. An instrument fulfilling these conditions is exceedingly inconvenient for practical use, because it is difficult, or at least tedious, to get the needle brought to rest, as it must be before taking a discharge upon it. Such an instrument is indeed rather suitable for the physical laboratory than for a cable-station, a ship, or a telegraph engineer's testing room.

No. 4. The intensity of the electro-magnetic action during the instant of the discharge is liable to alter the magnetization of the galvanometer needle (which is sometimes a fatal objection to the method, whether a single discharge or contrary discharges through a differential galvanometer be used).

The difficulty No. 2 has been met in practice hitherto, most commonly, I believe, by using a differential galvanometer to reduce the observed throw to zero, or to a small proportion of that due singly to one or other of the two quantities compared. But the use of a differential galvanometer aggravates curiously, and sometimes inconveniently, the difficulty No. 1. Unless the period of the needle is much longer than it is in any of the galvanometers ordinarily used, the contrary discharges through the two coils must be very accurately simultaneous, to avoid giving undue influence to the greater of the two discharges, whichever of the two be the first made.

Whether the differential galvanometer be used, or the method by single discharges, the period of the needle must be many times greater than any that would be convenient for the practical electrician, if either method is to be applied to measure the capacity of lengths of submarine cables exceeding one or two hundred miles.

De Sauty's beautiful method (Clark and Sabine, *Electrical Tables and Formulæ*, Edition 1871, page 62), though nearly free from all the difficulties described above, when applied to short lengths of submarine cable, or to condensers or other ordinary laboratory apparatus, is not applicable to great lengths of submarine cable on account of the manner in which the currents concerned in it are influenced by the inductive retardation.

Two methods, one or other of which I have generally preferred, even for laboratory work, are quite free from all the difficulties referred to above, and are equally applicable to several thousand miles of submarine cable, and to leyden jars, or ordinary electrostatic apparatus of even smaller capacity. The first of these methods was described at the meeting of the British Association in Glasgow, in 1855, and a short statement of a method virtually agreeing with the second was communicated to the British Association Committee on electrical measurements, four or five years ago, but no full practical description of either has been hitherto published.

My first method consists of a purely electrostatic arrangement analogous to the method for currents commonly known as Wheatstone's bridge. Four condensers are arranged in two series of two each. The two first plates (a, A) of the two series are connected together. So are the two last plates (a', A'). An electrometer* or galvanometer is employed to test differences of potential between the intermediate connected plates (b, b') of one series, and the connected plates (B, B') of the other series, after a difference of potentials has been suddenly established between the first plates and the last plates of the series. If the difference of potentials is found zero, we conclude that

$$c : c' :: C : C',$$

* It was for this application that sixteen years ago I first thought of the "divided ring electrometer," of which the modern quadrant electrometer is a species.

where c and c' denote the capacities of the two condensers in one series, and C and C' those of the other two in the same order, in the other series. When a galvanometer is employed to test the difference of potentials, its circuit must not be made until time enough has been allowed for the establishment of electrostatic equilibrium among the condensers. The time required for this is a very small fraction of a second, in ordinary laboratory work, and in applications of the method to short lengths of submarine cable, or to the testing of condensers of hundreds of microfarads, provided there is no resistance of more than a few ohms in any of the connections concerned. On the other hand, in testing considerable lengths of submarine cable (or in experiments in which,

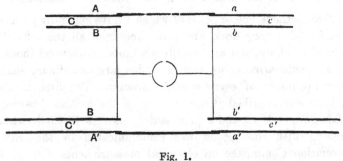

Fig. 1.

as with Varley's "artificial line," condensers of great capacity are connected through great resistances) two, three, or more seconds may be necessary before the zero test can be taken. To produce the zero, one or more of the four capacities c, c', C, C' must be altered by a proper method of adjustment. When an electrometer is used, and when the insulation is very good, the zero may be obtained by this adjustment, without a fresh electric charge; but, after having obtained the zero, it will, of course, be right to discharge the apparatus and re-test with fresh charge. On the other hand, when a galvanometer is used, the condensers must be thoroughly discharged, then re-charged and left for time enough to secure electric equilibrium before testing again the desired equality of potentials by the (quasi-bridge) discharge through the galvanometer: and so after repeated trials the adjustment of capacities giving zero effect on the galvanometer, and therefore securing the proportion $c : c' :: C : C'$, is found. The details of this method admit of much variation, according to the nature and character of the investigation to which it is applied, and of the available instruments. One way of carrying it out, adapted to very small capacities, is fully described in a paper entitled "Measurement of Specific Inductive Capacity of Dielectrics," by John Gibson, M.A., and Thomas Barclay, M.A.—*Transactions of the Royal Society*, Feb., 1871.

My second method is still analogous to the Wheatstone bridge, but differs from the first in this, that one only of the two connected series consists of condensers, and that the other is a line of current through resistance. The two condensers (*c, c'*) to be compared are put in series, that is to say, one plate of the condenser *c*, which will be called its second plate, is put in connection with one plate of condenser *c'*, which will be called its first plate. The first plate of condenser *c* will be called *a*, the connected second plate of *c* and

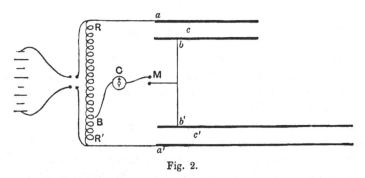

Fig. 2.

first plate of *c'* will be called *b*, and the second plate of *c'* will be called *a'*. Join *a* and *a'* in metallic connection through a wire *aBa'* of not less than several thousand ohms resistance, and let *B*, a point in this line of conduction, be put in connection with one terminal of an electrometer or galvanometer, the other terminal of which is to be occasionally connected with *b*, by a make and break key, *M*. To commence, make and break contact at *M* several times, not too rapidly, and observe the effect on the indicator whether of electrometer or of galvanometer. Then with contact at *M* broken, establish a difference of potentials between *a* and *a'* by means of a battery, and after time has been allowed for electric equilibrium between *c* and *c'*, make contact at *M*, and keep it made for a sufficient time. Then break contact at *M* and reverse the battery electrodes between *a* and *a'*. Then after sufficient time for re-establishment of electric equilibrium between *c* and *c'*, make contact at M. The ratio of the resistances R, R', between a and B, and between B and a' must be adjusted by varying one or both of them until the effect of making contact at M is the same with successive reverse applications of the battery to a and a'. (This effect will be zero, generally, in laboratory and factory tests, but, because of earth-currents, not exactly so in a submerged cable.) When the adjustment has been satisfactorily made, we conclude that

$$R' : R :: c : c'.$$

This method is applicable, notwithstanding earth-currents in moderation, to measure the resistance of a submerged cable of two or three thousand miles length, with only a single microfarad as standard for capacity, and a battery of not more than 100 cells to charge it.

III. Tests of Battery.

To measure the intensity and resistance of a cell or series of cells, use a galvanometer with a very long coil, and add resistance to it to make up a total resistance many times greater than the greatest internal battery resistance to be measured. A tangent galvanometer of the new form made for this purpose, with arc divided according to tangents, so that the readings on its scale are simply proportional to the currents, is very convenient. The resistance of its coils in series is about 4000 ohms, and by varying the distance of the coils from the needle it may be applied without any other change of adjustment to any battery of from one to fifty cells. A mirror galvanometer with strong controlling magnet or magnets to give very quick action to the needle is also very convenient. Its sensitiveness will still be sufficient to require the addition of from one hundred to four hundred thousand ohms resistance to its circuit. The proper degree of sensitiveness ought to be produced by controlling magnets and by adding resistance; without in any case applying a shunt to its coil. The galvanometer should be kept in one place, and convenient leading wires used as electrodes from whatever cell or battery is tested. The suspending fibre, whether the mirror or the tangent galvanometer be used, must be so fine as not sensibly to disturb the zero. To make an observation, provide a suitable resistance coil, which for brevity will be called the shunt. This coil should have thick flexible

electrodes convenient for direct application to the two terminals
of the cell or battery to be tested. Its resistance may be anything
from about one half, to two or three times the resistance of the
battery to be tested. Its length ought never to be less than two
metres for testing a single cell; or twice as many metres as there
are cells in the series to be tested. If its resistance is to be
anything lower than one ohm it may be conveniently made by
a strand of two, three, or more wires twisted together. These
wires may be silk-covered or cotton-covered, copper or German
silver. Or bare wire of either metal may be used: and if so the
strand must be taped, or otherwise sufficiently insulated, to
securely separate the different turns in the coil from metallic
contact with one another. If too short a length is used the
heating action of the cell is too great; and therefore although
two metres may be allowed as a minimum, a considerably greater
length is preferable. When the length is only two metres per cell
tested, the current ought not to be allowed to flow through the
shunt for more than two or three seconds at a time, or it may
become too much heated*.

To make an observation:—First observe the deflection (D) produced by the battery through the simple circuit of the galvanometer,
and added resistance if any. Then apply the shunt direct to the
terminals of the battery, making sure of good contacts; and observe
the reduced deflection D'. Let S be the resistance of the shunt,
and B the resistance of the battery, we have

$$B = S(D - D')/D' \dagger.$$

If $D - D'$ is too small a number of divisions to give a good result,
a shunt of less resistance must be used. If, on the other hand,

* If the current is kept flowing through a copper wire, with an electromotive
force amounting to that of one good Daniell's cell per n metres, the temperature of
the wire would rise at the rate of $15/n^2$ degrees centigrade per second, were no heat
allowed to leave it.

† The [corrected] rigorous formula is as follows:

$$B = S \frac{D - D'}{D'(G + S)/G - DS/G},$$

where G denotes the resistance of the galvanometer circuit—that is to say, the
resistance of the galvanometer coil, together with the added resistance if there is
any. When G is very great in comparison with S, $G/(G + S)$ is very nearly unity,
so that the numerator becomes $D - D'$, nearly enough for practical purposes; and
S/G being a very small fraction the denominator becomes nearly enough D'; thus
we have the simple approximate formula in the text.

D' is too small, a shunt of greater resistance must be used. When a battery consisting of any number (n) of cells in series is thus tested, B/n is the mean internal resistance per cell. When a single cell is tested, B is simply the resistance of that cell. The intensity of the cell is inferred from the single deflection D, by noting what a single cell of good quality would give when tested in the same way. Thus, if V denote the deflection which would be produced by the electromotive force of one volt, a single good Daniell's cell would give a deflection equal to $1 \cdot 07 \times V$. By applying once for all a good Daniell's cell and dividing the observed deflection by $1 \cdot 07$, we have a fair approximation to the value of V for the particular adjustment of galvanometer used. Of course, if the adjustment is changed, the value of V must be re-determined. Supposing, then, V to be known for the adjustment actually used at any time, we have $E = D/V$, where E denotes the intensity of the electromotive force, in volts, of the cell or battery tested in the process described above. If E/n is found to be much less than $1 \cdot 07$ the battery is defective in intensity. Let E be the electromotive force of the battery in volts, B its internal resistance, and R the internal resistance of the electro-magnetic coils through which its circuit is completed when at work, each reckoned in ohms. The quantity of electricity circulating per second will be $E/(B + R)$ reckoned in terms of a unit equal to the quantity required to charge a condenser of one farad to a potential of one volt. The quantity of zinc dissolved in each cell amounts therefore to $\frac{1}{3000} \times E/(B + R)$ grammes per second, or $1 \cdot 2 \times E/(B + R)$ grammes per hour, and the quantity of sulphate of copper used is $4 \cdot 75 \times E/(B + R)$ grammes per hour.

For example, suppose twenty of the ordinary recorder trays in good condition to be employed. The electromotive force will be about 21 volts, and the whole internal resistance about 3 ohms. The coils of the electro-magnet, if in series, will give a resistance of about 14 ohms. Putting then $E = 21$, $B = 3$, $R = 14$, we have $4 \cdot 75 \times 21 \div 17$, that is $5 \cdot 9$ grammes, for the consumption of copper in each cell per hour, or 100 grammes* in seventeen hours.

The efficiency of the whole battery, and of each cell of it

---

* This is as nearly as may be at the rate of four ounces in 19 hours, or a pound in 76 hours. An ounce avoirdupois is $28 \cdot 3$ grammes. 1 lb. is 453 grammes.

separately, when actually at work on the recorder electro-magnet, is to be tested as follows, without in any way interrupting or disturbing the use of the instrument for signalling:—Carry electrodes from the galvanometer first to the two terminals of the electro-magnet, and observe the deflection D. Then place them successively upon the two terminals of each cell, and observe the deflections in each case, D_1, D_2, etc. It will be found that the sum of these is equal to D. At any time when the instrument is not required for signalling, the intensity and the resistance of each cell may be determined, without the use of a separate shunt, by the following process which will be found very easy and convenient in practice:—Keep the electrodes of the whole battery on their proper places on the terminals of the electro-magnet coil, and find D as above. Then as above find D_1. Then without moving the galvanometer electrodes from the terminals of the first cell, break the electro-magnet circuit by the proper make and break switch. Let D_1' be the deflection then observed. Proceed similarly for each of the other cells, and let D_2, D_2', D_3, D_3', etc., be the observed deflections. Then if B_1, B_2, etc., denote the resistance of the several cells, and E_1, E_2, etc., their intensities, we have

$$E_1 = 1\cdot07 \times \frac{D_1'}{C}, \qquad E_2 = 1\cdot07 \times \frac{D_2'}{C}, \quad \text{etc.}$$

$$B_1 = R\frac{E_1 - D_1}{D}, \qquad B_2 = R\frac{E_2 - D_2}{D}, \quad \text{etc.}$$

where R denotes the resistance of the electro-magnet coil. In the recorder as now made, R will be about 16 ohms when the two coils are in series, and about 4 ohms when they are connected in double arc. Let C be the deflection given by a good Daniell's cell direct through the galvanometer. The actual efficiency of the whole battery, as working on the electro-magnet reckoned in volts, will be $1\cdot07 \times D/C$; and the efficiency of the several separate cells, in terms of the same unit, will be $1\cdot07 \times D_1/C$, $1\cdot07 \times D_2/C$, etc.

In observing the deflection for each cell, be careful to note its direction. There might be a fair degree of deflection from any one cell of a battery in series, but yet its direction might be opposite to that corresponding to positive efficiency. This would be the case, for example, if, in a battery of n cells in good order as to intensity, the resistance of any one cell exceeds one nth part of the sum of all the resistances, internal and external, in the circuit. When this is found to be the case for any cell, either short circuit it immediately, or refresh it by putting in crystals of sulphate of copper.

The cells in actual use are to be tested, in the manner just described, as frequently as is necessary to make sure of their efficiency. Until a cell has been a fortnight at work, it will, generally speaking, not show any failure in efficiency except through want of sulphate of copper.

Supposing 14,000 grammes to be the weight of water in each tray, the dissolution of 840 grammes of zinc will raise the specific gravity from 1·1 to 1·3. This would result from the consumption of 3400 grammes (or 120 ounces) of sulphate of copper. Keep account of the quantities of sulphate of copper put into each tray, from time to time; also occasionally, before putting in the fresh sulphate of copper, test the specific gravity by a hydrometer. Before the specific gravity of the stratum of liquid level with the zinc grating reaches 1·35, draw out some of the liquor, and pour fresh water into the space over the zinc.

IV. Tray Battery for the Siphon Recorder.

The battery consists of square wooden trays lined with lead, and zinc gratings resting in them on wooden or stone-ware props. In the lower edge of each tray, on the outside, a groove is cut to facilitate pouring in fresh water over the zincs, during the use of the battery. A stout copper wire is soldered to the lead lip of each tray, to facilitate making connection to electrodes when required. To allow the deposit of copper to be readily removed, a narrow slip of sheet copper is soldered to the lead near the middle of the bottom of each tray, and all the rest of the lead both over the bottom and on the slant sides is carefully coated with paraffin, bees-wax, or other non-conducting material, such as varnish of any convenient kind; and a piece of dutch metal paper, with its metallic side up, and with the slip of copper passing through a hole in it to its upper side, is pasted flat down over the varnished bottom. The slip is then bent and sprung so as to press with its end firmly on the metallic surface.

The same object may be attained by the following alternative process, with less trouble, in case of setting up afresh old lead trays, which, from want of the precaution, have been ruffled by a deposit of copper adhering to them too firmly to be removed. The whole of the lead, bottom and slant sides, of the tray, must be thickly varnished, except a small area of a quarter of an inch diameter at the centre, which must be cleaned so as to present a fresh metallic surface, to make contact with a sheet of copper placed in the bottom of the tray, and to receive enough of deposited copper to establish a permanent metallic connection between the lead and the sheet of copper. To make sure of an initial contact between the copper and the lead, a burred hole is made in the centre of the sheet of copper (by a round-headed punch, the copper being placed on a piece of soft wood) before it is put into the tray. The under surface of the copper plate must be also carefully varnished except the burr projecting downwards. In laying the sheet of copper in the bottom of the lead tray, it must be so placed that the burr presses on the clean portion of the lead. When properly placed, secure it by wax at its corners and edges so as to prevent it from shifting, and to keep it lying flat in case of any part of it showing a tendency to curl up.

Each zinc is to be protected by a square of parchment paper bent round below it, and folded neatly at the corners and fixed with sealing wax*. Care must be taken that the edge of the paper be generally $\frac{3}{4}$ inch (and in no place less than a $\frac{1}{4}$ inch) above the upper level of the bars of the zinc grating. It must be bound firmly to the zinc, by twine passing under the parchment paper, and tied over the zinc above; also by a long piece of twine several times round the square. Each lead tray should have a stout copper wire soldered to it, projecting about three inches from one corner.

To support a pile of trays take four blocks of wood each four or five inches square in horizontal dimensions and of any convenient height: and place them in positions to bear the four corners of a tray. The pile must be so placed as to give ready access to each

* Press the paper against the zinc between finger and thumb on each side of the corner and draw the bight or bend of the paper diagonally away from the corner; then fold the bight round the vertical corner of the zinc, and press it against the flat zinc surface on one side or other of the corner. Secure with sealing wax in the bight, and where one side of it is pressed against the paper on the vertical zinc surface. Then tie it round carefully with cord in the manner described in the text.

of its sides. Put a piece of thick sheet gutta-percha, six inches square, on the top of each of the wooden squares, and then lay down the first tray upon them, seeing that it is properly levelled. Put four hard wood or stone blocks, each $1\frac{1}{2}$ inch cube, in the corners of the tray, and put one of the zinc gratings resting with its four corners on these props. Put a quarter-saturated solution of sulphate of zinc (specific gravity about 1·1) into the tray, pouring in first between the lead and the parchment paper, and afterwards filling up to the level of the top of the zinc grating, by pouring some of the solution directly on the zinc over the paper. See that the top corners of the zinc, and the bottom corners of the tray to rest upon it, are all properly tinned, and clean and dry. Place a lead tray resting with its four corners on the upper pro-jecting corners of the zinc. Place four wooden or stone props in the corners of this second tray, put a second zinc upon them, and fill with solution as before. Observe in placing the trays, to turn them with their grooved edges on the same side of the pile, and *that* the side most convenient for pouring in fresh water. Proceed thus until a pile of from six to ten trays, one over the other, is made and filled with liquid. Solder a stout copper wire for electrode to one of the corners of the top zinc. In the same way make as many piles as are required. Leave a space of about one foot breadth between each pile and its neighbour. Connect these piles in series, the top zinc of one pile to the lowest lead tray of the next one.

The crystals of sulphate of copper to be used should be broken into small pieces, the largest not more than the size of a pea, and weighed out in quantities of an ounce each. To put the battery in action drop in four ounces to each cell; one ounce separately on each side, distributing it as equally as may be, along the space between the wooden props; and immediately after doing so, short-circuit the cell, and keep it on short-circuit till required for use. Within a short time (ten minutes or a quarter of an hour) the battery will be ready to act at full power. Take care not to drop in any crystals so near the corners as to fall against the wooden or stone props.

From time to time put in more sulphate of copper; always regularly four ounces to each tray, one ounce separately along each side, as in first charging the battery: but never put in a fresh supply of the crystals until the previous supply is nearly all used up. If at any time the battery is to be out of use for several days, short-circuit it until all the sulphate of copper is used up. The short-circuit should be broken when there is no sulphate of copper remaining in any of the trays. From time to time draw off by a siphon from a point a little below the lowest level of the zinc, enough of the liquor to sink the surface by about a quarter of an inch; and then fill up to the proper height by pouring in fresh water by a funnel to the space above the grating bars. The specific gravity of the liquor drawn off each time should be tested by a hydrometer, and the quantity drawn off should be regulated so as to keep the specific gravity of the liquor in the cell at about from 1·15 to 1·35.

Some Recent Improvements in Dynamo-Electric Apparatus [and Transmission of Power thereby].

[From Discussion of Higgs' and Brittle's Paper in *Inst. Civ. Engrs. Proc.* Vol. LII. 1878 [Jan. 22, 1878], pp. 81—83.]

Sir William Thomson said Dr Siemens had just brought forward an idea which appeared to him to be quite new, and to be of great practical importance; namely, that the electric light, if there were a sufficient number of lamps, could be produced with equal economy at a great and at a small distance. By way of illustration he would suppose the cost of a central station dispensing by four hundred machines the electric current to four hundred lamps, each at a distance of 1 mile. Taking those four hundred miles of wire and putting them in a line, having four hundred engines in series, and putting the four hundred lamps at short distances from one another, without any change of circumstances, the same effect would be produced at 400 miles as at 1 mile. The question of the heat developed in the wire was, as Dr Siemens had remarked, the fundamental question with reference to the quantity of metal required to communicate the effect to a distance. It appeared to him that the most practical way of producing the result would be to put the wire in the shape of a copper tube. Having a copper tube, with a moderate amount of copper in its sectional area, and a current of water flowing through it, with occasional places to let it off, and places to allow water to be admitted for the purpose of cooling, there would be, without any injury to the insulation, a power of carrying off heat practically unlimited. He believed that with an exceedingly moderate amount of copper it would be possible to carry the electric energy for one hundred, or two hundred, or one thousand electric lights, to a distance of several hundred miles. The economical and engineering moral of the theory appeared to be, that towns henceforth would be lighted by coal burned at the pit's mouth, where it was cheapest. The carriage expense of electricity was nothing, while that of coal was sometimes the greater part of its cost. The dross at the pit's mouth (which was formerly wasted) could be used for working dynamo engines of the most economical kind; and in that way he had no doubt that the illumination of great towns would be reduced to a small fraction of the present expense. Nothing could exceed the practical importance of the fact to which attention had been called; that no addition was required to the quantity of copper to develope the electric light at a distance. The same remarks would apply to the transmission of power. Dr Siemens had mentioned to him in conversation that the power of the Falls of Niagara might be transmitted electrically to a distance. The idea seemed as fantastic as that of the telephone or the phonograph might have seemed thirteen months ago; but what was chimerical then was an accomplished fact now. He thought it might be expected that, before long, towns would be illuminated at night by an electric light produced by the burning of coal at the pit's mouth, or by a distant waterfall. The power transmissible by the machines was not simply sufficient for working sewing machines and turning lathes, but by putting together a sufficient number any amount of H.P. might be developed. Taking the cost of the machines required to develope one thousand H.P., he believed it would be found comparable with the cost of a thousand H.P. engine; and he need not point out the vast economy to be obtained by the use of such a fall as that of Niagara, or the employment of waste coal at the pit's mouth.

As to the transmissibility of different kinds of light through clear or foggy air, that was a matter for experimental investigation. Theoretically it was an exceedingly interesting question. The whole quantity of energy in the vibrations must be the same, therefore it might be expected that the absorption would be the same, whether the light proceeded from a small surface, as in the electric light, or from a much larger surface, as in oil lamps of lighthouses. But the mode of absorption of the vibrations according as the same colour was produced from one or other of the two sources was a question that mathematicians could not solve, and it could only be decided by experiment. Perhaps Mr Douglass could give some further information derived from experiments as to the relative transmissibility of light of the same colour from a larger area less intense, or from a smaller area more intense.

On New Standard and Inspectional Electric Measuring Instruments *.

[From *Soc. Teleg. Engrs. Journ.* Vol. XVII. [May 24, 1888], pp. 540—556, and 566, 567.]

THE general principles and many of the details of the instruments placed on the table for the inspection of the meeting are already well known, having been described in the electrical journals. It has been considered, however, that there are members of this Society who might like to see some of the instruments themselves, and to have a few explanations, which I may be able to give personally, regarding details that have not hitherto been published; and that is my excuse for bringing before the Society a subject of which so large a part has already been published and is known to members of the Society.

All the instruments which you see on this table fall under the designation of standard electric measuring instruments, and all except one depend upon Ampère's discovery of the mutual action between fixed and movable parts of an electric circuit. The one instrument which I except is that which I hold in my hand—a portable marine voltmeter or milli-ampere meter. This instrument, for instance, is a milli-ampere meter as it is now before you, but with the addition of resistances wound on the tubular guard surrounding the stretched wire (which I shall describe presently) it becomes a voltmeter. Resistances amounting to about 930 ohms of platinoid wire wound upon these cylinders render the instrument a voltmeter, capable of measuring through a range of from 90 to 120 volts.

In enumerating the standard instruments I said that with one exception those to be explained this evening are founded upon Ampère's discovery, and all come under the designation of standard electric measuring instruments. But there is another instrument on the table, which I shall at this time

* [See also Nos. 209, 210, 213, 214, 216 *supra*.]

merely mention—a magneto-static current meter. This type of
instrument I use as an adjunct to the series of standard instruments because it is exceedingly useful in extending the scale
above and below the range of any one of that series. Two of the
magneto-static species with one of the standard instruments give
the means of measuring currents through the whole range from
one milli-ampere up to several hundred amperes.

I shall briefly describe the marine voltmeter before passing
on to the other standard instruments. It consists essentially of
an oblate of soft iron suspended in the centre of a solenoid. I am
sorry that I am not able easily to open up this instrument in
order that you may see the interior, but I may tell you that
I have a more recent instrument fulfilling exactly the general
description I shall give of this type, but with somewhat more
convenient mechanical details, in which, by simply taking out a
screw, I can remove the cap and show you the working parts.
The resultant action between the electro-magnetic force due to
the solenoid, and the oblate of soft iron with its equator oblique
to the lines of force, is a couple; and that couple is balanced by
the torsion of a stretched platinoid wire on which the oblate is
suspended, the platinoid wire being kept stretched to about one-
third its breaking weight by two springy arrangements at its two
ends. The sketch on the board represents the type of an instrument which is being made in Glasgow. In *that* type the bobbin
is divided for convenience, the ground plan showing two halves
of the bobbin, which can be put together with great ease; between
them there is a hollow cut out, sufficiently wide to let a fine
platinoid wire pass down through the centre, bearing in the axis
of the solenoids an oblate of soft iron. The instrument before
you is constructed with a single bobbin, and has on that account
complicated details which I need not now speak of, because
I believe that I shall abandon them altogether. I feel confident
that I shall not adhere to the type of single bobbin, considering
the much less simplicity that it involves in the more delicate and
vital part of the instrument, viz., the suspension of the oblate and
the index-needle, which in the new form will be lighter, and
therefore better adapted for working at sea, and less liable to
damage from any very rough usage that it might experience in
carriage on shore. To explain the principle of the instrument,
I may remind you that, as Faraday long ago showed, a long bar

of soft iron tends to place its length along the lines of force in a uniform field of magnetic force; an oval or a flat disc tends to place its equatorial plane along the lines of force. I had to choose between an oval and an oblate, and I found that the oblate was better—that a larger magnetic moment was available in the circumstances by choosing an oblate than by choosing an oval. In the oblate used the equatorial diameter is rather less than a centimetre, the polar diameter rather less than half of that amount. The tendency of the oblate to place its equatorial plane in the direction of the lines of force of the solenoid, is the force due to the electro-magnetic action, which is balanced mechanically by torsion in the use of the instrument. The principles upon which I was led to the choice of this configuration of iron depended upon the circumstance that when we do not use excessively high force in the magnetic field, so that the magnetisation of the iron shall be very far short of saturation, a figure of a short character, in which there is no one diameter very much more than twice as long as any other diameter, is much less disturbed by magnetic retentiveness than an elongated figure. In the instruments of this class the electro-magnetic force is very nearly in exact proportion to the *square* of the strength of the current, which is rigorously the law of the Ampère-force used in, and in that respect they bear very much resemblance to, the gravity-balance instruments, by which the highest possible accuracy is to be attained. On the other hand, in instruments of the class to which Ayrton and Perry's ammeter belongs, the force is more nearly in simple proportion to the strength of the current, owing to iron being used nearly in a state of saturation. These details will no doubt be interesting to some members of the Society, for all are scientific enough to know that consideration of details is of the very essence of success in practical work. The elements of electrical and mechanical science—rudimentary drawings of levers, springs, solenoids, and magnets—are valuable, but without the application of physical science in the construction of electrical instruments —without consideration of the qualities, both electrical and mechanical, of the matter involved in the design of the details— the instrument or piece of apparatus will not answer its purpose, and it is really the working out of such details that is the life-work of anyone who is engaged in practical science.

I wish to point out one difference between this instrument and the type now being made. In this species of instrument the torque* due to the electro-magnetic action is balanced by the elastic torque, or the torsional couple, as it is more commonly called, in this twisted wire. In the instrument before you we have a latent zero, and to verify the zero I must undo the torsion by turning the torsion-head accordingly. The range of the instrument is essentially limited. As it stands just now it is made to read from 90 volts to 120 volts. Indeed, it is impossible to make this a very satisfactory *long-range* instrument, except by graduated resistance-coils. The needle is not allowed to come to the zero at all; and my reason for that is that it was only so that I was able to eliminate practical error from change of zero in the copper and platinum wires which I used until lately, when I adopted platinoid. The elasticity of fine platinoid wire has proved so good that, after a year's use of the instrument continuously in the electric light circuit of my house, I found, on testing it by an absolute standard, that the latent zero had not changed more than one-fifth per cent. in its effect on the indication of strength of current. I found, in fact, no sensible variation whatever when I took the twist out and verified the zero, afterwards replacing the twist. We all know that the permanence of the elasticity of this piece of wire may be considered as quite perfect; so that now or a hundred years hence, or even as many million years hence as would give the earth a very different gravitational force, this wire will remain constant. So that in one respect this portable marine voltmeter is a more constant instrument for all time and for every place than any gravitational instrument can be. At the present time gravity differs notably in different parts of the earth, being one-half per cent. less at the equator than at the poles. But for all practical purposes you must feel that gravity is the most constant force we have to deal with; although, looking at the elasticity of metals, and looking forward a sufficient number of millions of years, it is something interesting to know that the elasticity of a piece of

---

* The word "torque," introduced by my brother, Professor Jas. Thomson, often saves a great deal of awkwardness, and is exceedingly convenient when we have to speak, not of a couple of forces, but of a system of elastic or other forces balanceable by a Poinsot couple.

metal is a more perennial standard than gravity itself. Either elasticity or gravity may therefore be considered as practically quite perfect, provided the material of the spring does not alter in any way by becoming rusted or by evaporation going on, but retains its qualities unchanged from age to age. The marine voltmeter is really an accurate and standard instrument, while it is perfectly portable and bears the roughest usage. The platinoid-wire spring has been kept stretched for a considerable time, and heated to the temperature of boiling water, and after such treatment—or *ageing*, as we may call it—the instrument preserves its constancy; and we find that turning it upside down, knocking it roughly down on its side, or carrying it in a cab through a rough street, does not alter the indication at all. But if even still more rough usage, or long-continued rough usage, or accident, should alter the indication, we have a ready means of adjusting the zero, even in this form now before you (and still more easily in the improved form), and so can absolutely reinstate it as a standard instrument at any moment. The two things that require constancy to render this a perennial instrument are the oblate and the platinoid wire. These are constants that we regard as practically absolute, so that from generation to generation that little instrument will really be a standard current meter.

This instrument has not yet been much used at sea. I have not pushed it forward, because I have been anxious to satisfy myself as to some of the mechanical details; but an instrument exactly like the one before you has been in use on board one of the State Line steamships (the "State of Nevada"), and has now made five or six voyages across the Atlantic, between Glasgow and New York. In the first four voyages it saved breakdowns, from over-incandescence, estimated at about 30 lamps, and from that time forward the consumption of lamps has been very much less than formerly.

The other standard instruments now before you are all founded, as I said, on the mutual forces between the movable and fixed parts of an electric circuit. I will take these instruments in order, beginning with the one adapted to the measurement of the smallest current. The useful range of the first instrument of the series—the centi-ampere* balance—is from 1 to 50 centi-amperes. The next instrument is the deci-ampere balance, a specimen of which is now before you. The centi-ampere and the deci-ampere balances are quite similar in appearance, the difference being entirely in the number of turns and the thickness of the wire in the coils. The instrument now before you may be taken as representing the deci-ampere or the centi-ampere balance, but the type which is made now differs somewhat from the type before you. The type before you would not be improved by the difference if it were to be used solely for currents in one direction; but if it were to be used for alternate currents, the copper bobbins on which, as you see, the wire is wound, and the thick double-sheet copper constituting the sole-plate and frame of the instrument, would, by the currents induced in them, interfere very seriously with the indications. Accordingly, for the instruments to be used for the measurement of alternate currents, the type, while it is exactly the same in principle as that instrument, is like another of the instruments on the table before you—the deka-ampere balance—in which we have a slate base. Passing over the ampere meter for a moment, the range of the deka-ampere meter is from 2 to 100 amperes. This is a type of which a considerable number were made for Sir Coutts Lindsay and Company for the Grosvenor Gallery installation, at the request of Mr Ferranti. The deci-ampere meters and the deka-ampere meters of this type have been made for the purpose of being available for alternate as well as for direct currents, and therefore they have a slate base, and slate cores and slate ends for the coils, instead of copper; but in other respects the arrangements are the same as in the instruments shown.

I will not weary the Society by going too much into details, and pass over without further reference the principle of these instruments, because that, I think, you all understand already. In each type of this series of balance instruments, except the hekto-ampere and kilo-ampere balances, the movable part of the circuit consists of two coils, one carried at each end of a horizontal balance-beam, and the fixed part consists

---

* The French nomenclature is, whatever classical critics may say of it, exceedingly convenient and valuable—the Latin numerals for the sub-multiples, the Greek numerals for the multiples—milli-ampere (the thousandth of an ampere), the centi-ampere, deka-ampere, hekto-ampere, and kilo-ampere.

of four fixed coils placed one above and one below each movable coil.

Look at this instrument. The movable coil here at one end of the balance-beam is attracted down by the coil below it and repelled down by the coil above it, while at the same time this movable coil at the other end of the balance-beam is attracted up by the coil above it and repelled up by the coil below it. We have thus two *movable* and four *fixed* coils in this type of instrument, and the advantages so obtained cannot be had with anything simpler. We might take away the two upper fixed coils, and leave only the two lower coils, but a slight change in the height of the axis of suspension, as might be due to a slight stretching of the suspension ligament, would introduce a sensible error in the use of the instrument. On the other hand, we have a place of *minimum* force for the middle position of each movable coil between its two fixed coils, and by choosing the sighted position of the index to correspond very nearly to the place of minimum force we have a condition of things suited to give the highest degree of accuracy in the results. I may just give you an illustration of the practical accuracy and handiness attained by this arrangement. As a practical matter, my ambition has been that boxes passing from the workshop of my instrument maker, Mr White, should be labelled "Glass—without care; any side up." That is the ideal, but we are very far yet from having attained it; and although any of these instruments *when packed* may be turned upside down, somehow or other there are a number of mischances that do happen, and which cannot well be guarded against. One of these instruments, carefully packed in a box marked "Glass—with care," sent through the Belgian Custom House, arrived broken, and it was returned to Glasgow with the balance-arm bent and other serious injury, which showed that the damage was not due to the breakdown of the suspension ligament or anything of that kind, but that the box containing the instrument had probably simply been let fall upon a stone floor in the Custom House, and been subjected to carelessly rough usage. The arms were simply re-straightened, the balance re-suspended, and everything replaced by the instrument maker. It was then sent up to my laboratory and tested, with the result that it was found to agree to within one-twentieth per cent. with its indications before it met with the accident giving that very disintegrating rough usage. Now that kind of accuracy can only be attained by working in the neighbourhood of the minimum or maximum.

Here are a few particulars as to the resistances, number of turns, and so on, in the coils.

In the centi-ampere balance, of which the range is from 1 to 50 centi-amperes, each movable coil contains 440 turns of No. 30 copper wire, resistance 19·6 ohms, say 20 ohms (it depends on the temperature, of course); each of the four fixed coils contains 1,295 turns of No. 26 copper wire, resistance 30 ohms; total resistance of the instrument, 161 ohms. That is a large resistance, but to have a standard instrument for measuring such small currents I do not see any way of doing it without a large resistance; and, after all, it does not take a great number of cells to give the current, and in most places where the instrument is used there is sufficient power to give the electric potential for the useful current.

The deci-ampere balance, with a range of from 1 to 50 deci-amperes, has a total resistance of only 2 ohms. The two movable coils have each 64 turns of No. 17 copper wire, resistance ·18 ohm, and the four fixed coils each 166 turns of No. 17 copper wire, resistance ·428 ohm in each; total resistance of the instrument, 2·072, or say 2·0 ohms. The deci-ampere balance has a great advantage over the centi-ampere balance; for, although founded on exactly the same principles, there is in the former a far smaller proportion of space occupied by insulation, and a much larger proportion of space occupied by copper, in the vital part of the instrument; and thus we have a much more than proportionately small insulation, if we calculate the proportion of the dimensions on the idea of the different instruments being exactly similar. The different instruments are essentially not similar, because of the insulation being necessarily so much thicker in proportion with fine wire than with thick wire.

The ampere balance here before you is of the newest type, and is not adapted for the measurement of alternating currents; but a like instrument, slightly less advantageous, can be made for that purpose. Its range is from 5 ampere to 25 amperes. The two movable coils each contain 10 turns of copper ribbon 1·2 centimetres broad and ·11 centimetre thick, and four fixed coils each contain 37 turns of copper ribbon 1·8 centimetres broad and 07 centimetre thick; the total resistance of the instrument being about ·18 ohm. The insulation of these coils is very fine paraffined paper, which takes up exceedingly little space.

In the hekto-ampere balance and in the kilo-ampere balance I have dispensed with one feature of importance with reference to the very highest accuracy, and that is the complete and thorough realisation of the minimum principle. In these we have a single coil—a single rectangular ring rather, not a coil— of copper, movable, and a fixed rectangular coil or ring of copper below it. In the hekto-ampere balance the fixed ring consists of 10 turns of thick copper ribbon insulated from one another, and the range of measurement is from 10 to 500 amperes. In the kilo-ampere balance, whose range is from 50 to (50 × 50 =) 2,500 amperes, the fixed coil is made up exactly as in the hekto-ampere balance, with the difference that instead of insulating paper between the turns they are soldered together and the whole forms a single ring. This method of making the fixed ring was adopted because we found great difficulty in getting forgings or castings of copper of these dimensions. The very first instrument of this kind, which was made to the order of Messrs Paterson and Cooper, had for its fixed rectangle a forging of copper; but we had to wait a long time for it, and, when it came, the forging was not very satisfactory. I have therefore come to the con- clusion that, for the present at all events, it can be made more easily, and with more certainty of a thoroughly satisfactory result, by making it of a copper ribbon and soldering the different turns together. I will just say a word or two regarding the construction and action of the hekto-ampere balance and of the kilo-ampere balance. Imagine a rectangle of copper about 2½ centimetres deep by 1½ centimetres thick, movable about an axis through its centre and parallel to its shorter sides, and imagine a current entering by the middle of one of the longer sides, dividing between the two halves of the rectangle and coming out at the middle of the other longer side: that is the movable part of these instruments. On the other hand, the fixed coil of either instru- ment is so arranged that the whole current passes through it in one direction, once round in the kilo-ampere balance, ten times round in the hekto-ampere balance. Thus it is evident that the current through one side of the fixed coil attracts the current in that side of the movable coil, while the current through the other side, going in the opposite direction, repels the current in the movable coil, so that the movable coil is tilted in the manner of a balance-beam. It has been advisable in this case to give up one part of the principle for highest accuracy, because it would make it so very cumbrous an instrument to carry, and because the suspending ligament is very short and very powerful, so that there is exceedingly little liability to error due to possible change in its length; and therefore I ventured upon the simple form and arrangement which you see.

There are two or three details which I am afraid I must pass over, as the hour is so very far advanced, but just a few words with reference to the ligament, which is, after all, the key of these instruments, as everything depends upon it. There is nothing new in the principle of action of the instruments, but some of the details of construction have involved a good deal of consideration, and have only been arrived at after a vast number of unsatisfactory attempts in various directions. Instead of knife edges, flexible connexions for introducing the current to the movable parts of the circuit are used, and the whole weight of these movable parts is borne by the flexible connections. Weber, I think I am right in saying, made an ordinary balance with flexible springs instead of knife edges. And we all know the admirable use of the elastic suspension in the pendulum of the common clock, and of the astronomical clock—the instrument of the very greatest accuracy that exists in the whole range of practical science: in the astronomical clock the elastic suspension is found practically superior to supporting it by knife edges. I will not say whether ordinary chemical balances may or may not be made more advantageously with suspension ligaments than, as now, with knife edges; but for electric current-measuring instruments of this kind, in which, besides bearing the weight, we have the affair of introducing the current from a fixed coil to a movable coil, the elastic suspension has great advantages. Well, after trying copper ribbons, and finding exceedingly unsatisfactory results through buckling, owing to the impossibility of getting the force uniformly distributed through each broad ribbon, I saw the desirability of trying narrow segments of ribbon, and thus developed the latest ligaments, which consist of very fine copper wires about $\frac{3}{4}$ centimetre long, laid close side by side, and numbering from 10 up to

1,000, according to the different types of the instrument. The number in each ligament of this kilo-ampere balance before you is 800. The ligament is not seen on account of the strengthening guard which is fixed across the middle of the instrument, covering the ligament itself altogether and providing against rough usage and chance of injury. To make this instrument portable a piece of brass is put in beside the ligament, and two screws are screwed down so that the whole weight of the movable coil is taken off the ligament, and the fixed coil is screwed firmly by the divided trunnion against this brass plate.

I have only now to say, in conclusion, that there is on the table before you a set of instruments capable of measuring continuously, and each adjustable by comparison with the one below it, through a range of from one milli-ampere up to 2,500 amperes, or the ratio of 1 to 2,500,000. It is exceedingly convenient to have absolute standard instruments through so wide a range, but most people will be perfectly satisfied with a much less range of *standard*; and accordingly I should just call attention to the range that we get by a deci-ampere balance ranging from 1 to 50 deci-amperes, a milli-ampere magneto-static instrument, and a deka-ampere magneto-static instrument, to extend the range of accurate measurement down to one milli-ampere and up to 50 or 100 or 200 amperes.

I have here a magneto-static instrument which is very conveniently set at one ampere to the division, but it may perfectly readily be set at either two amperes or half an ampere to the division. The index needle is carried by a silk-fibre suspension, and those who know the constancy of silk-fibre suspension in the laboratory would not willingly give it up for anything else. I showed this instrument at the British Association meeting at Manchester, and proved to the meeting the very rough usage it might get, and the tumbling about it might suffer, without breaking the fibre. The glass will break, any part will break, before the suspension gives way; but the fibre will wear. I have learned from experience that a railway journey from Glasgow to London is very apt to *wear* out the fibre, though I believe that fifty years' usage *in situ* would scarcely wear it out. Accordingly there is in this instrument a little trouble in putting on a proper guard for the suspension, to make it perfectly portable and handy;

but that trouble being taken, the fibre shows no tendency to break down, and is perhaps as perennial as any part of the instrument. We can adjust the zero in an instant, and the sensibility can be readily altered by raising or lowering the magnet, which may be securely fixed in any position by its clamping screws. An inspectional instrument readable from one to 100 times one is, I think, practically only obtainable by the magneto-static method, and therefore, though this is not a standard instrument in itself, it is yet an instrument that is a very useful adjunct to absolute standard instruments. For a large variety of laboratory and practical purposes an equipment of useful instruments to give the means of measuring, within a quarter per cent., any current whatever, from a milli-ampere up to 100 amperes, might be as follows:—The deci-ampere standard-balance; for stronger currents, a magneto-static instrument with a coil consisting of a single turn of thick copper; and for smaller currents another magneto-static instrument on exactly the same plan, but with about 300 turns of very fine copper wire. With these three instruments the range from one milli-ampere to 200 amperes is very readily obtained; the magneto-static for stronger currents being standardised by the deci-ampere balance at the top of the deci-ampere meter's range, the magneto-static for smaller currents being also standardised by the deci-ampere meter at the bottom of the range of the deci-ampere meter.

With reference to the vertical scale voltmeter, it is designed for use in an electric light station, or at any place where it is desired to have an inspectional potential meter of the highest accuracy. The instrument before you is of the earlier type. I am sorry that I am not able to show you the most modern type, but one is at present in the Glasgow Exhibition as part of the exhibit of my instrument maker, Mr White, of Glasgow. In the instrument now before you one movable coil is repelled by one fixed coil; but this construction I have modified to one movable coil at one end of the balance repelled down by a fixed coil, and another movable coil repelled upwards by another fixed coil at the other end of the balance. I departed from the simple construction because the instrument threatened to become, not a pure electrical measuring instrument, but the measurer of a compound quality partly depending upon very complex considerations of temperature and of moisture of the air, and only partly upon the things that we wish to measure; it was, in fact, a mixed hygrometer and voltmeter, and the hygrometric part of its indications sometimes disturbed the electrical measurements as much as a half per cent. The best way of eliminating that was to take away the counterpoise weight and substitute a coil, with, of course, a corresponding fixed coil; and that has got over the hygroscopic error. One point remains—the temperature correction. I wish to have no error amounting to as much as one-fifth per cent. in this instrument. We have a platinoid resistance here below, amounting to about 700 ohms, and about 50 ohms or 60 ohms in the instrument itself, and the change of temperature of the platinoid produces just a sensible effect, while the change of resistance of the part of the circuit that is copper is of course very considerable; but as this last is only a small part of the whole, the temperature error is not large. With 20° or 30° C. temperature it might amount to more than one per cent., and accordingly a thermometer is used, and when an absolutely correct reading is required a weight corresponding to the temperature observed may be placed in the pan. But in the new type I have simplified that, and if one wants to make an accurate measurement he looks at the thermometer and sets a balancing flag to the number on the scale corresponding to the reading on the thermometer; so that without touching the weight at all the temperature correction is at once applied.

Mr Granville's question suggests very interesting considerations regarding magnetic induction in iron, which I first learned in working in quite another subject—the Flinders bar for the correction of the mariner's compass, so that the compass, if corrected for one latitude, may be also correct in any other; that a compass corrected for the British .Channel, for instance, could be also correct at Port Elizabeth or the Cape of Good Hope. A large part of the magnetism of the ship is due to the vertical component of the earth's magnetism, and this reverses when the ship goes to the Southern hemisphere: the question, then, is to annul the effect on the compass of this varying part of the ship's magnetism due to the vertical component of the earth's magnetism. Flinders, eighty years ago, placed a vertical bar of soft iron before or behind the compass for that purpose, but it did not then get into general use. Some persons who wished to improve the mariner's compass have tried again and again to effect this annulment, and among others Mr James Napier—one of the sons of Robert Napier, of engineering renown—endeavoured to bring about this result. I questioned him about his experience with the Flinders bar, and said: "If that bar gets a kick from anyone's foot when the ship is heeling over, its magnetism will be reversed, and the compass will show a permanent change when the ship is upright again"; and I might have added, "particularly if she is on an east or west course." He answered, "That is just what we have found"; and a *long thin* bar really is exceedingly subject to the disturbances to which Mr Granville has alluded. If you take a long bar of soft iron in such proportions as this pen—of length more than twenty times its diameter (a common domestic poker answers well)—and hold it vertically, or, better, parallel to the terrestrial magnetic force (the line of "dip"), and place a pocket compass near one end of it, you will see splendidly interesting results. First, you will generally find that the upper end acts as with "blue magnetism" (which is the same as the magnetism of the north pole), and the lower end with "red." If you invert it very gently you will find that the end which was upper still acts as with "blue magnetism" on the compass. If you now give a very gentle tap with the finger to the inverted poker, in an instant the action on the compass is reversed—the "blue magnetism" changes to "red." I do not know whether this result of tapping with the finger is generally known, but I mention it as interesting to the meeting; it cannot be observed with short thick bars. In trying to bring the Flinders bar into practical use I simply tried a bar of a certain diameter, and shorter and shorter lengths, and found whether by inverting it by the side of the compass and striking it with blows from a mallet I could make a difference in its magnetic effect. When the length of the bar was thirty times its diameter, then it was very sensitive to the blows of a mallet; but when its length was not more than eight or ten times the diameter, I found I could invert it in the neighbourhood of the compass and give it a blow without it showing any sensible retentiveness under the small force employed. Under the influence of a large magnetising force the effect would of course be magnified, but under the feeble force of the earth's magnetism a bar whose length is not more than eight or ten times its diameter shows no sensible effect of retentiveness—no effect sensible in comparison with the directive effect of the earth's magnetism on the compass. I simply adopted that principle—shortening the bar—for electric measuring instruments of the marine voltmeter class. The practical result is that until a current of two or three times the strength of the greatest current provided for is passed through the marine voltmeter we find no signs of retentiveness. If I work from zero up to 120 milli-amperes in this instrument, up and down, and with reversed currents, I find no signs of retained magnetism; but a current of two or three hundred milli-amperes leaves a large effect of residual magnetism, which, however, is easily shaken out by turning the little reversing key which forms part of the instrument.

I have to thank you for the patience and kindness which you have shown during the time I have been addressing you.

On an Alternate Current Generator.

[From Discussion of W. M. Mordey's Paper in *Instit. Elec. Engrs. Journ.*
Vol. xviii. [May 30, 1889], pp. 667, 668.]

Before calling on Mr Mordey to reply I may perhaps be
allowed to make a few remarks on his beautiful experiments in
the case in which two alternators, one with a potential of
2,000 volts, and the other 1,000 volts, are coupled in parallel.
The secret of the surprising result of this novel and original combination has not, I think, been quite touched by either Mr Mordey
himself or any of those who have spoken in the discussion, except
Mr Kapp. The 500 volts difference of potential would, notwithstanding ohmic resistance and impedance by self-induction,
produce current in the circuit of the two armatures vastly
greater than any that Mr Mordey has found in his experiments, without some equalising influence which has not been
hitherto suggested except by Mr Kapp. To find what this
influence is, suppose, for simplicity, the ohmic resistances of the
armatures to be zero, and suppose the two armatures to be simply
joined in parallel, ready to do external work but not as yet set
to do it. Thus we have a simple circuit of the two armatures.
The two shafts might be mechanically constrained to run
synchronously in such relative positions that the electromotive
forces of the two armatures conspire in the circuit. The
3,500 volts would produce a prodigious current for all that self-
induction could do to impede it; but this current would
enormously pull down the magnetisation of the iron claws of
the field magnets; and might even annul and reverse that of
the weaker, and thus after the first making of the circuit of the
two armatures the "prodigious" current through it would almost
instantly become much moderated, even supposing the supposed
initial phase-relation to be mechanically maintained. This is
the phase-relation for series co-operation; and, as shown by Dr
Hopkinson in his lecture on electric lighting, delivered before
the Institution of Civil Engineers in 1885, it could only be

maintained by rigid mechanical connection between the two shafts. But, in Mr Mordey's actual experiment, the two shafts are free and independent, and they therefore "fall into step" co-phasally for parallel working. Their electromotive forces become thus exactly opposed in the circuit of the two armatures, and would yield exactly 500 volts as working E.M.F., if the two field magnets remained unchanged. The ohmic resistance being still, for simplicity, supposed zero, the phase of the current kept going in virtue of the E.M.F. would be a quarter-period behind the phase of the E.M.F. Thus the maximum of the current would be at the instant when the nine magnetic fields are in the middle of the apertures of nine alternate armature coils, and the direction of the current would be that tending to demagnetise the iron prongs of the 2,000-volt machine, and to augment the magnetism of those of the 1,000-volt machine. Thus (the period in Mr Mordey's experiment being a 100th of a second) two hundred times per second, the former experiences a demagnetising, and the latter an enhancing, magnetic force. The prongs are of continuous iron, and there must therefore be ample self-induction to prevent any considerable change of magnetisation in the 200th of a second period of this varying magnetic force. Thus the magnetic fields are kept, one of them weaker, and the other stronger, than they were before the armature circuits were connected; and I believe so much weaker and stronger as to reduce the electromotive forces of the two to very near equality. If the iron of the 2,000-volt machine is nearly saturated, its magnetism would be less diminished than the other's is increased, and the resulting E.M.F. might be nearer 2,000 than 1,000; but Mr Mordey has found it to be actually very near to 1,500. A few turns of insulated wire round one prong of either machine, connected to a ballistic galvanometer, would give a ready means of testing the suggested changes of magnetisation.

There is really not much more left for me to say, except that I do wish to be allowed to join in the general and most sincere chorus of admiration at the results put before us by Mr Mordey. We are all admirers of his previously-known alternate-current generator. We have now before us for the first time another form, invented also by Mr Mordey—a form which strikes me as exceedingly beautiful, very admirable in many respects, and, I am pleased to add, quite wonderful.

On the Early History of Submarine Cable Enterprise.

[From Remarks on Retiring from the Presidential Chair, in *Inst. Elect. Engrs. Journ.* Vol. xix. [Jan. 9, 1890], pp. 2—6.]

Gentlemen,

Before retiring from the Presidential chair of the Institution of Electrical Engineers, I should like to say a few words with reference to the origin of our Society. I feel that at this stage it is good for us to remember that this is not a Society merely for electric lighting and for the electrical transmission of power. The magnificent spring of energy which within the last six years has been imminent, and which is now beginning to be active—which has been active during the past year—makes us for the time almost forget everything but electric lighting. In fact, we have almost forgotten the transmission of power in our zeal for electric lighting; and a subject to be seriously considered is how far the great systems of central stations and underground conductors, now being made, can be rendered usefully available for the electric transmission of power as well as for electric lighting. I do not, naturally, in two or three minutes, intend to enter on so large a subject as that is; but I do wish to call the attention of members to the fact that electrical engineering originated with the electric telegraph, and notably in submarine telegraphy. The founders of this Society were all submarine telegraphists, and almost every one of them had had to do with work in deep-sea telegraphy—ocean telegraphy.

I must, before vacating the chair to our new President, congratulate the Institution on our four new Members of Council— Sir James Anderson, Sir Henry Mance, Sir Albert Cappel, and Colonel Jackson. Every one of them is connected with telegraphy, whether under the sea or in the air. I am quite sure the whole Society is pleased that we have men connected with telegraphy, and especially with ocean telegraphy, thus acting along with us, and so valuably in line with us, in the work of the Institution. One might almost have begun to be afraid that our old friends the ocean telegraphers were going to desert us. Those of us who have come more recently into this Society, and who have been connected more with the newer developments to which I have referred, would be exceedingly sorry to lose our association with the founders of the scientific enterprise which has led to such beautiful results.

I am sure you will all be very glad to see with us here this evening Sir James Anderson, who commanded the "Great Eastern" for the tremendous feat of laying and lifting cables in mid-Atlantic in 1865 and 1866.

The Institution of Electrical Engineers has for its province practical applications of electricity and allied branches of science. But science, whether of electric instruments or of nautical operations, or of the heavy engineering work of stress and strength of materials and machines for working with cables in depths of from two to three nautical miles, cannot go on without the sinews of war; and I feel that those who have supplied the sinews of war, and have stood by from the beginning of the submarine cable enterprise up to the present time, ought not to be forgotten by an Institution of Electrical Engineers. I had myself the honour of being a shareholder in the first Atlantic cable of 1858. I had the honour and pleasure of having as a co-director Mr Pender, now Sir John Pender. We did all we could in 1857 and 1858. We laid a cable a fifth of the way across in 1857, then a cable the whole way across in 1858. A great many people do not know that in 1858 there was a cable completed across the Atlantic to Heart's Content, in Newfoundland, and that messages were transmitted through it for three weeks—some exceedingly important public messages that saved many thousands of pounds, as countermanding the moving of two regiments ordered over from Canada to help to quell the Indian Mutiny. The return of those two regiments from Canada was countermanded because the Mutiny was already quelled while they were still under orders to leave for India. I remember the words being spelled out— "To the Military Secretary, Montreal.—The 60th Regiment is not to leave; the 61st Regiment is not to leave"; or words to that effect. I know it was to the Military Secretary; I am

sure the 60th Regiment was one of the two. The return of
the regiments was effectively countermanded through this cable,
the very existence of which seems to be regarded as a myth.
Amongst the directors of that enterprise Mr Pender remained
from 1858 to 1865. I myself, I am sorry to say, was able to do
nothing for cable work from the time of the failure in 1858–59,
except in consulting and advising as to designs, until my
assistance in the nautical and instrumental work was called for in
1865–66. Without the persistent, strenuous, and self-sacrificing
determination of such men as Sir John Pender and some of his
colleagues the enterprise would have died—absolutely died. Cyrus
Field, from the other side of the Atlantic, helped to keep it alive ;
he gave help and impulse where they were required ; worked with
those who did not require revivification ; and he, with his English
colleagues, revived the undertaking in 1865. Then came the sad
and dreadful discouragement in 1865, when half the cable was laid,
and the "Great Eastern" returned, having just succeeded in
proving that Sir Samuel Canning's and Sir James Anderson's
and my own belief was correct—that it was possible to lift it. In
1865 they got the broken end three times from a depth of
2,000 up to 500 fathoms from the surface, and each time it fell
away by reason of breakage of the grapnel-rope. The third time
of grappling—I think it was the third time—the cable came up
with the rope foul of the grapnel. The grapnel was let down a
fourth time, and brought up the cable once more within 500 yards,
when it again broke, and the "Great Eastern" returned. We
were all dispirited in a sense, but not discouraged. Then, when
we got back to England, with a fresh prospectus drawn up on
board the "Great Eastern" for an enterprise to make and lay a
new cable to Trinity Bay in 1865, to come back to mid-ocean, lift
the broken end of the 1865 cable, and complete that cable also to
Trinity Bay, the crisis came, and the crisis was a financial one.
Sir John Pender gave a guarantee of a quarter of a million sterling
for the funds required to complete the enterprise of two cables
across the Atlantic. Without that guarantee many and many a
year would have passed before we should have had the Atlantic
crossed by cables, and before ocean telegraphy would have had the
great advance it has had. I repeat that a Society of Electrical
Engineers has a double duty—a duty to science, to get all the good
it can get out of science ; a duty to satisfy those who pay for the

work that the work is good, a duty to encourage them to bring forward new work and to show that new enterprise has a chance to be successful.

It is interesting in connection with the objects of this Society to consider something about how much money is spent in various forms of electrical enterprise. I have in my hand a table giving particulars of the land lines of Europe, North and South America, Egypt, Asia—including China and Japan—and they amount to 1,714,000 English land miles. The estimated value of all this work amounts to £51,728,500. Land line enterprise, according to this estimate—which is, I believe, substantially correct—runs up to about £52,000,000. Cable enterprise does not reach quite so large a figure, but not far short of it. 107,547 miles have been laid by companies—private companies in the sense of being companies got up by private enterprise, public in the sense of being under the Public Companies Act. The capital laid out is £36,000,000. The capital expended by the Government in submarine cables is £3,700,000. So that we have a total of £39,700,000, or within a few thousands of being £40,000,000. Suppose we speak of submarine enterprise alone, and compare it with electric lighting work. Electric lighting looks very large now—it is large already; it is destined to be larger and larger, and perhaps before five or six years are out a great many more millions may be beneficially spent in electric lighting and other allied enterprises. Still, for the present, it only runs up to £3,000,000 or £4,000,000 of capital spent and subscribed for, in view of the great development of electric lighting, for which we are now promising such delightful results to the public. But we have actually at present from ten to fifteen times as much spent in submarine cable enterprise, and about thirty times as much spent in electro-telegraphic enterprise altogether, as in electric lighting. I think we should consider this, and look forward to electric light work emulating, both in enterprise and the application of science, the great work of submarine telegraphy that preceded it.

I must not speak of introducing our new President. Dr Hopkinson might as well introduce me, as I Dr Hopkinson. I have much regret in leaving the chair, but much pleasure, since I must leave it, in yielding it to Dr Hopkinson.

REMARKS ON THE PRESIDENT'S (DR J. HOPKINSON)
INAUGURAL ADDRESS ON MAGNETISM.

[From *Inst. Elec. Engrs. Journ.* Vol. XIX. [Jan. 9, 1890], p. 36.]

MR PRESIDENT,

I venture to take the initiative in saying a few words,
for this reason, if for no other: I am sorry I shall not be present
at our next meeting, when, no doubt, there will be much occasion
for further discussion on the very important and interesting
matter you have put before us. Dr Hopkinson has done well to
emphasise the marvel in nature that is presented by those three
metals, iron, nickel, and cobalt. Faraday long ago showed that
there is continuity between what he called the ferro-magnetic
quality and the diamagnetic: he showed that all bodies have
something of susceptibility to the magnet which may be not only
compared with this susceptibility of iron, but determined in
absolute measure; so that, for example, we may get the numerical
measure of the absolute susceptibility, negative or positive, of
glass, or of bismuth, or of antimony, or of vegetable substances,
a great variety of which were qualitatively experimented upon
by Faraday. A large field was then opened, which has not yet
been sufficiently cultivated, in determining the absolute values
of magnetic susceptibilities. But Dr Hopkinson has done well
to emphasise the prodigious differences between those three
substances, iron, nickel, and cobalt, and all other known substances.
Faraday's splendid discovery of the universality of magnetic
susceptibility would be delusive if it led us to overlook that
prodigious difference in degree—so great a difference that we
must look upon it almost as a distinct property of matter which
is possessed by the three magnetic bodies; and it is really one

of the marvels of nature that there should be three bodies
differing so enormously from all others in a matter of such great
importance as magnetic susceptibility. Much of the properties
of iron which have been put before us by Dr Hopkinson is new
and valuable produce of his own experimental work; especially
this last admirable discovery regarding the thermal capacity of
iron at temperatures of transition between its magnetic and non-
magnetic conditions; and the manifestly intense interest with
which his Address has been listened to this evening proves how
appreciative the Institution of Electrical Engineers is of genuine
science.

HISTORICAL REMARKS ON SUBMARINE CABLES.

[From *Inst. Elec. Engrs. Journ.* Vol. XVIII. [Feb. 11, 1890], p. 863.]

THE UNIVERSITY, GLASGOW,
February 11th, 1890.

WITH regard to the remarks upon submarine telegraphy by
Mr Preece, at the meeting of the 12th December, 1889*, which
have just come under my notice, the following may be of interest
to members:—

The first Atlantic cable was never coiled in tanks at Greenwich.
It was made, half at Birkenhead, and half at Greenwich. The
two parts were joined in one circuit for the first time on board
the "Niagara" and "Agamemnon," at Queenstown, at the end of
July, 1857, and I then got signals through it at something less
than one word per minute with the special instruments which
had been prepared by Mr Whitehouse. When the cable was laid,
in 1858, we obtained about three words per minute by hand-
signalling, with my mirror galvanometer as receiver, during the
three weeks of successful working of the cable. In this connection
I may quote the following extract from a letter of mine to
the *Athenæum,* dated October 24th, 1856, and published

* *Journal,* p. 844.

November 1st, 1856:—"A mode of operating so as to clear the
wire rapidly of residual electricity, which I have worked out
from theory, and a plan for telegraphic receiving instruments to
take the most full advantage of it, which has recently occurred
to me, allow me now to feel confident of the possibility of
sending a distinct letter every $3\frac{1}{2}$ seconds by such a cable*," or
about $3\frac{1}{2}$ words per minute.

It was the second (1865), and not the first (1857), Atlantic
cable that was coiled in tanks at Greenwich, and it was in
connection with this cable that Mr Varley and Mr Jenkin acted
along with me.

Varley and Jenkin and I were perfectly aware of electro-
magnetic induction†, and we were more pleased than surprised
when we found that, after having promised to the company eight
words per minute, we actually obtained, by hand-signalling, with
mirror-galvanometer as receiver, 15 words per minute through
each of the two cables (the second and third Atlantic cables),
when the laying of both across the Atlantic was completed
in 1866.

* *Collected Mathematical and Physical Papers*, Article LXXVI. Vol. II. p. 101.

† See "Remarks on the Discharge of a Coiled Electric Cable," *British Association Report*, 1859, Part 2; or my *Collected Mathematical and Physical Papers*,
Vol. II. Article LXXX.

ADVANTAGES OF TRANSMISSION BY CONTINUOUS CURRENT.

[From *Inst. Elec. Engrs. Journ.* Vol. xx. [May 21, 1891], pp. 466—469.]

ONE question occurs to me with reference to both Mr Crompton's and Mr Preece's papers. What is the ratio of the number of units of energy supplied to and paid for by the consumer to the total produced at the generating station in each of the two systems? When practical answers to that question are taken into account, I think the result will tell considerably in favour of the low-pressure direct-current system in all cases in which it is applicable, and will lead us rather to stretch the distance by which we should prefer to choose it than give any bias in the direction of choosing the other system.

There is just one other point, and that is, the economy of copper in the conductors. Ideas seem to be ripening towards carrying out low-pressure network, to be fed either directly by low-pressure feeders on the continuous-current system, or by high-pressure feeders with transformers on the alternating high-pressure system. Now it does not seem to have been taken into account that in a low-pressure network we must have very large conductors to feed anything like a dense district, say any part of London. Whenever we go above wire of perhaps 2 or 3 centimetres—whenever we have a wire equivalent to continuous copper of circular section exceeding 2 or 3 centimetres in diameter—we have a very great loss of efficiency indeed over the copper in the alternating-current system. Take, for example, an alternating-current system at 80 periods per second. With a continuous copper wire of 2 centimetres diameter the effect of the conductance of the copper is less by 8 per cent., or the ohmic resistance—the ohmic effect of resistance of the copper— is more by 8 per cent., for the alternating current than for the continuous current. If we take 3 centimetres, the effect of ohmic resistance is 31 per cent. greater with the alternating current than with the direct current. If we take current at a frequency of 100, then 2·7 centimetres diameter of copper would entail a loss of 31 per cent. If we take such a powerful conductor as that suggested in Mr Preece's paper, of a square inch diameter—that is, round wire of 2·9 centimetres diameter—the ohmic resistance of such a wire as that actually suggested by Mr Preece would be nearly 30 per cent. more for the alternating current of the lower frequency—80—than for the continuous current. These things must be taken into account for estimating low-pressure networks. If a conductor for 500 amperes, consisting of a large copper wire cable with a central strand of copper wires surrounded by six similar strands, were altered by taking jute instead of copper for the central strand, it would be found, with alternate currents of 80 or 100 frequency, that the cable with the jute core would conduct practically as well as the cable with the copper core.

I may say I feel, on the whole, that there seems to be what, to my mind, is not quite a practically judicious tendency to rush to alternating current, and to give up the simple and convenient and direct system of steady current and low pressure. I cannot but think that if, irrespectively of division among companies, the whole of London were worked out for electric light, there would not be alternating current anywhere—there would be nothing but the low-pressure direct system. I feel it is due to Edison to name his name in connection with what I believe to be the best system, which nine years ago he proposed for the supply of electric light and power to cities—the system in which consumption districts are connected to one another and fed by one or more central stations at pressures of perhaps 15 per cent. above the pressures that are to be used in the consumption circuit. I am perfectly aware of the reasons that have dictated the choice of the alternate-current system in many cases. I know also how splendidly successful it has been in some cases. Both in London and elsewhere the alternating-current system has done, and is doing, what could not be done by any other system. I do not wish it to be said that I do not appreciate the real merit of, and the great advantages obtained in some cases by, the alternating-current system. But I do think, looking to the future, we must consider that the direct-current system, with its great simplicity, and its thoroughly convenient and economical use for power as well as light, is destined to predominate, and ultimately to supersede alternating current for all densely populated districts.

There is a common idea that it will pay to put the generating system at a distance of several miles away, in order to obtain cheaper coal and possibility of condensing water for the engines. I think, if practical men consider the subject in all its bearings, they will find that the few shillings per ton saved on the coal is far short of compensating the disadvantages of the more distant station; and what has been done hitherto in the use of condensing water, even when freely available, is, I believe, *nil.* I do not know of a single case in which triple or quadruple expansion condensing engines are used for electric lighting, even although the station may be on the side of a canal or river. I do not say it would not be a very great advantage if, by the use of an abundance of condensing water, we could get an economy equal to that of marine engines taking $1\frac{1}{4}$ lbs. of coal per indicated horse-power per hour, instead of the 3 lbs. or 4 lbs. taken by the non-condensing engines at present in use. The varying loads during the 24 hours, and other specialities of the requirements for electric supply, seem, however, to render it impossible, at present, to realise practically in an electric light station anything approaching to the splendid economy of marine engines.

It is satisfactory to know that all these details of engineering are in good hands: and we may look forward to better and better results as the work extends. But I return to this: no great saving has ever yet been realised on the expense for the same power obtained by. putting the generating station at a distant place. On the other hand, it is often difficult and expensive to get a site in the middle of a town, and it may be necessary on this account to put the station a little way off. No other consideration whatever ought to weigh with engineers against putting the generating station as near as possible to where the light is to be used, and doing away with every possible complication between the central station and the consumer.

Electrical Measuring Instruments [Wattmeters for
Alternating Currents].

[From Discussion of J. Swinburne's Paper in *Inst. Civ. Engrs. Proc.*
Vol. cx. [April 26, 1892], pp. 52—54.]

Lord Kelvin said it was, of course, impossible to present a
thorough treatment of the subject in so short a Paper, but as
a record, or an index, of what had been done it was very
comprehensive. The various principles that had been adopted
for electrical measuring instruments had been touched upon
in a fair and appreciative spirit. The Author had remarked,
with reference to his (Lord Kelvin's) multi-cellular electrostatic
voltmeter, "This instrument has a horizontal dial, but it is
provided with a mirror, so that the readings are visible in
front, if it is on the level of the eyes." Although by this
method readings could be taken with fair convenience, he still
felt that, especially for a central station of electric lighting,
it would be an improvement if the mirror could be done away
with, and the instrument made to show at once on a vertical
scale. He had therefore introduced a modification, which had
only been finally tested in his laboratory a few weeks ago. It
gave its readings on a vertical cylindric scale, which could be read
at distances of many yards. There was another point against the
electrostatic voltmeter, which the Author had not mentioned, but
which he thought was of some importance—the provision of some
automatic checking arrangement, so that the instrument would be
practically dead-beat. He had found that that could be done in
the simplest manner by a well-known expedient, corresponding
to the dash-pot in engineering. Without disturbing the general
structure of the instrument, he could readily hang a vane on to
the lower end of the stem that carried the needles; and that vane
working under oil, was so arranged that it produced a practically

dead-beat instrument. The Author's remarks upon the subject of the measurement of electric activity were very important, and ought to be taken to heart. There was a general impression that a wattmeter for alternate currents was impracticable without a certain complication of scientific adjustment which the Author had described. The Author, Professor Ayrton, Mr Sumpner, and Dr Fleming had gone into the subject in a very thorough manner, and had brought out scientific developments of remarkable and interesting character. He nevertheless heartily concurred with the Author's remark—"It is, however, useless to employ complicated methods of measuring power when a simple one is better. The prejudice against the ordinary wattmeter has no real foundation." Let the fine wire coil of the ordinary wattmeter have a sufficient anti-inductive resistance in its circuit, to produce the result that the current through the fine wire coil should be, nearly enough for practical purposes, the same as it would be were there no inductive impedance, and were there no effect of inductive action from the fixed coil upon the movable coil: these conditions being fulfilled, the wattmeter's measurement of activity was correct. The Author had rightly remarked that, "if any one will take a real instrument, designed with a little care, and substitute the approximate values of R and L, he will find that he has been straining at a gnat." He might have added, "and swallowing a camel," taking into account the very difficult methods adopted to avoid the inconveniences arising from the action of that gnat. The fact was that, with a judiciously-arranged wattmeter, there was no difficulty whatever in rendering the error due to the quasi inertia of the electric current, when the wattmeter was applied to an alternate current, practically insensible. They had always the means of testing whether it was or was not insensible and of ascertaining its value in the latter case. There was one difficulty which the Author had not thoroughly dealt with, although he was quite aware of it. This was, however, perhaps the most serious difficulty in connection with wattmeters for alternate currents—namely, the law of distribution of the electric current in the conductor. When conductors were too thick to allow the effect of their current to be calculated as if it were a current in an infinitely thin line, like a wire in an ordinary coil, then an exceedingly difficult subject for mathematical investigation came before them—the law of distribution of the current in the thickness of the conductor. He would not enter into that subject, which would require very elaborate treatment; but would merely say, that the wattmeter ought in every case to have such an arrangement of the thick conductor for carrying the main conductor, that the difference of distribution of the current, for different frequencies of alternation, should affect the result as little as possible. There were ways of doing that which he had discussed with the Author many years ago, and he would not then enter into the question. He had pointed out that any such arrangement was, after all, only an approximate elimination of error; the degree to which it was eliminated could always be ascertained by experiment; and at the worst, a wattmeter might require correction according to the frequency of the current for which it was used.

ELECTRICAL TRACTION ON RAILWAYS.

[From DISCUSSION OF W. M. MORDEY AND B. M. JENKIN'S PAPER in
Inst. Civ. Engrs. Proc. Vol. CXLIX. [Feb. 18, 1902], pp. 87—89.]

LORD KELVIN, G.C.V.O., remarked that he would occupy much time if he were to refer to all he had been led to think of during the reading of the exceedingly interesting Paper, to which every one present had listened with extreme interest and with a great deal of instruction. It was well known how much complexity there seemed to be in electric-traction systems, and yet, how, in some of the arrangements of polyphase, three-phase, two-phase, and single-phase currents, that which at first looked very complicated resolved itself into an exceedingly simple apparatus. The three-phase motor had had full justice done to it by the Authors. In regard to the great variety of appliances necessary to adapt it, on one system or another, to the conditions that had to be observed on railways, the question of the object to be secured was a complicated one. For short railways everything was comparatively simple; but for long railways, whether with few or with many stoppages, for central stations which had also to supply cross lines, it was impossible but that there must be a great complication of means for such a variety of ends. He was glad to see that the Authors boldly attacked the problem of long lines. It was to be looked forward to without any very chimerical optimism as to electrical resources, that in the future long lines would be worked electrically as successfully as short lines were worked already. There were many ways of carrying out the electrical method for long lines, and, as the Authors had justly said, they essentially involved high-tension currents. He had admired the courageous argument and the strong case which the Authors had put forward in favour of the simplest of all alternating-current systems, namely, the single-phase system; and it was almost dramatic to find how, after being led

in Appendix III. of the Paper and the diagrams illustrating it,
through a splendid maze of apparatus, the members were brought
back to the point from which they started. To railways the simplicity with an alternating-current system, if practically achieved,
would certainly be a great triumph. He would only ask for
a little consideration for the continuous-current system. It had
been said that there was but one possibility, viz., constant voltage,
whether alternating or continuous current was used. That idea had
been learned long ago; but while it was being learned there had
been many who had thought of the other system, the constant-
current system. Probably almost every practical electrician would
say that was quite impracticable, and out of the question, in regard
to electric traction on railways. Still, he thought it was worthy
of more consideration than it had obtained; and he suggested
that after all the extreme simplicity of constant continuous current
in one insulated line for the high-pressure circuit might possibly
be found good for traction on long lines. That method had been
advocated long ago by Messrs Siemens, and they had shown how
it could be practised. It had been much better known and
more in favour in some applications 15 years ago than it was
now; but he would suggest that the many acute, intelligent, and
active minds concerned with the subject might be turned back
upon some of those old questions. It must not be imagined,
however, that he was putting it forward as a ripe practical proposal.
He was only suggesting it as a subject for reflection, with the
possible outcome of falling back on the continuous-current system
throughout, instead of merely in the locomotive motors. He had
listened with much interest to the suggestion of a locomotive sub-
station. On any system, even the simplest, there were central
stations and sub-stations, and there was a considerable degree of
complexity. It was an interesting idea for alternating-current
distribution, that the sub-station might be as economically carried
on a locomotive as placed in a fixed position. According to the
Authors, the only real objection to it was the weight carried on
the locomotive, and even that objection was practically disposed
of by the statement that in some instances several tons of useless
material had been added for the purpose of giving stability.
Whether the space occupied, the number of parts and conductors,
and the electrical complication on a locomotive would form serious
objections to the Authors' proposal he did not know; but they did

not seem to him likely to be so, because things became greatly simplified when they were perfectly methodized. The idea of an electric locomotive carrying the sub-station, although perhaps it now looked no more practicable than the steam-engine on wheels had looked to James Watt and his contemporaries, might yet be realized and found good; and he would watch its development with great interest. One thing in the Paper which might surprise many was the voltage lost in the transmission of the return current by iron rails. He had been somewhat startled to hear how much loss there was owing to the fact of the currents not passing equally through the whole material of the rail, but being in some degree "skin" currents. A great deal was known about that effect in relation to copper conductors, and something in regard to iron conductors. A highly complicated physical and mathematical problem was introduced by the magnetization of the iron itself, and the variation of that magnetization which was essential to the carrying of an alternating current by an iron conductor. In conclusion, he considered the Paper was of great practical importance, and he was sure he was expressing the feelings of the members in thanking the Authors for their valuable communication.

Transmission of Energy by Direct Current.

[From Discussion of Mr Highfield's Paper in *Inst. Elect. Engrs. Journ.*
Vol. xxxviii. [March 7, 1907], pp. 502—504.]

I HAVE listened with extreme interest to Mr Highfield's paper; it is really one of the best scientific papers I have ever heard read. It has been so clear on every point that every one present has been thoroughly carried along with the author.

For myself the subject is one of extreme interest. I have never swerved from the opinion that the right system for long-distance transmission of power by electricity is the direct-current system. I do not say a word in depreciation of the beauty of the polyphase idea, and of the development of that idea, especially in the 3-phase working. The great convenience of the 3-phase system is well known to all; the ease with which, by transformers with no moving part, an augmentation can be made from, let us say, 10,000 volts in the generator to 20,000, 30,000, or 40,000 volts in the transmission line is very valuable and important. In fact, it is, I believe, the great convenience of the static transformer that has led to the wonderful and splendid development of the alternating-current system in modern electrical engineering.

We have now come to a position in electrical engineering, looked forward to twenty or thirty years ago, in which we are practically confronted with the problem of the transmission of electric power through very great distances. After what we have heard from Mr Highfield, I have no suggestion to make in respect to the advantages that he has demonstrated of the direct-current system.

Some of the most subtle problems of electricity, problems depending even on resonance, are more or less familiar to all electrical engineers at the present time. The perfect freedom of the direct-current system from complications of that kind is very remarkable.

You all understand something about square root 2 being 1·41. Starting with that scientific fact, we have this practical application that 141 volts of direct current have no more tendency to produce sparks than 100 volts of alternating current. That, I think, is a statement that is generally understood. The 141 to 100 we have all known as the merit in respect to voltage attainable by direct current, as compared with voltage attainable by alternating current; but now we hear from Mr Highfield that the advantage is far more than the ratio 141 to 100. Mr Highfield will correct me, but my recollection is that he said it is more like $2\frac{1}{4}$ to 1 than $1\frac{1}{2}$ to 1. The properties of matter concerned in the breaking down of insulation, and the gradual heating up by the alternating current till the insulation breaks, determine a result which to me was unexpected. Mr Highfield finds, in respect to insulation, a much greater advantage of the direct current over the alternating current than has been hitherto known to us. I think I may safely say, as proved by the experiments which he has put before us, that 80,000 volts direct current are more easily worked than 40,000 volts alternating current.

That is enough to prove the case for direct current, unless there are difficulties in the practical management of direct current as compared with alternating current. But the difficulties are all the other way; the practical difficulties are in the management of alternating current; the ease (except in respect to the brushes) is all for the direct current. So that, not only have we much greater capacities for bearing high pressure, but very much greater ease and convenience in practical working, when we adopt the direct current for long-distance transmission.

I do not go so far as to say that the beautiful systems that are now worked out for distributing power over somewhat long distances would be better carried out by direct current than by alternating current. I do think, however, with regard to the ordinary power-supply stations, which are now with but few exceptions founded on the alternating-current system, it is just possible that, in future, there may be a change to direct current. The penetrating skill and perseverance of Mr Thury having led the way, we may, with the very clear ideas that have been put before us by Mr Highfield, begin to see that, even in the ordinary distribution of power by electricity at 20,000 or 30,000 volts, the advantages of the direct current will bring it to pass sooner or later, that it may supersede the alternating current.

I am reminded of a little prophetic conversational statement made by Lord Rayleigh a good many years ago. He rejoiced to see the use of alternating current coming in, "Because," he said, "the whole world will now learn the subtleties of electrical science, which they had no chance of learning before." That prophecy has been fulfilled. "And," he added, "after that, they will come back to the continuous current." I do not know that you will all agree with that anticipation, but you will, I am sure, all enthusiastically agree with Lord Rayleigh in rejoicing that we have had this twenty years of alternating current, and of electrical science in its most interesting characters, including in fact wireless telegraphy, put before beginners in electrical science, as they now are in practical schools of electricity.

APPENDIX.—ON THE AGE OF THE EARTH.

(Cf. pp. 205—230 *supra*, and previous papers.)

EARLY in 1895 a discussion, including correspondence with various physicists, was published by Prof. J. Perry* in which he enforced the view that the arguments as to the age of the Earth, drawn by Lord Kelvin from the present value of the gradient of surface temperature in the rock strata, were not conclusive. In fact, taking the terrestrial conditions to be represented by a slab of unlimited thickness, initially all at a uniform temperature, say 4000° C., and with one face thereafter kept at zero temperature, the theory of dimensions showed that if conductivity is increased m times and all lengths are increased n times, all temperatures remain the same if the times are increased n^2/m times. It is true that the gradient of temperature at the surface will be altered n^{-1} times; but that can be brought back again to the assigned value, with the same value of the surface temperature, by altering the conductivity and the thickness of an outer surface stratum. With the increased conductivity inside the slab, the loss of heat will in the same time extend deeper into the slab; but provided it does not penetrate sensibly to say the inner half of the radius, the size of the actual sphere which the slab represents need not enter into the dimensional relations, all being nearly the same as if it were infinite. On the other hand, if the conductivity inside is very great, the limitation of size will sooner make itself felt, and the time may be shortened instead of prolonged.

In the special case of a globe of radius $R = 6\cdot38 \times 10^8$ cm., which is that of the Earth, internal conductivity $k = 0\cdot47$, which is 79 times that of surface rock, $k/c = 0\cdot165$, which is 14 times that of surface rock, c being thermal capacity, and with a shell

---

* *Nature*, Jan. 3, 1895.

of rock of thickness 4×10^5 cm., about $2\frac{1}{2}$ miles, superposed, Perry found from Fourier's solution for a sphere that the dimensional conditions above stated were sensibly satisfied, and that the time of cooling to the present terrestrial surface gradient, namely $1°$ C. in 2740 cm., would be as much as 98×10^8 years. With k equal to 195 times and k/c equal to 35 times the value for surface rock, and with a shell of rock of depth $3·272 \times 10^6$ cm., about 20 miles, the time would be 127×10^8 years.

The following letter from Lord Kelvin comes at the end of this discussion.

THE UNIVERSITY, GLASGOW,
December 13, 1894.

DEAR PERRY,

Many thanks for sending me the printed copy of your letter to Larmor and the other papers, which I found waiting my arrival here on Saturday evening. I have been much interested in them and in the whole question that you raise, as to the effect of greater conductivity and greater thermal capacity in the interior. Your $n^2 \div m$ theorem is clearly right, and not limited to the case of the upper stratum being infinitely thin. Twenty or thirty kilometres may be as good as infinitely thin for our purposes. But your solution on the supposition of an upper stratum of *constant* thickness, having smaller conductivity and smaller thermal capacity than the strata below it, is very far from being applicable to the true case in which the qualities depend on the temperature. This is a subject for mathematical investigation which is exceedingly interesting in itself, quite irrespectively of its application to the natural problem of underground heat.

For the natural problem, we must try and find how far Robert Weber's results can be accepted as trustworthy, and I have written to Everett to ask him if he can send me the separate copy of Weber's paper, which it seems was sent to him some time before 1886; but in any case it will be worth while to make farther experiments on the subject, and I see quite a simple way, which I think I must try, to find what deviation from uniformity of conductivity there is in slate, or granite, or marble between ordinary temperatures and a red heat.

38—3

For all we know at present, however, I feel that we cannot assume as in any way probable the enormous differences of conductivity and thermal capacity at different depths which you take for your calculations. If you look at Section 11 of "Secular Cooling" (*Math. and . Phys. Papers*, Vol. III. p. 300), you will see that I refer to the question of thermal conductivities and specific heats at high temperatures. I thought my range from 20 millions to 400 millions was probably wide enough, but it is quite possible that I should have put the superior limit a good deal higher, perhaps 4000 instead of 400.

The subject is intensely interesting; in fact, I would rather know the date of the *Consistentior Status* than of the Norman Conquest; but it can bring no comfort in respect to demand for time in Palæontological Geology. Helmholtz, Newcomb, and another, are inexorable in refusing sunlight for more than a score or a very few scores of million years of past time (see *Popular Lectures and Addresses*, Vol. I. p. 397).

So far as underground heat alone is concerned you are quite right that my estimate was 100 millions, and please remark (*P. L. and A.*, Vol. II. p. 87) that that is all Geikie wants; but I should be exceedingly frightened to meet him now with only 20 million in my mouth.

And, lastly, don't despise secular diminution of the earth's moment of momentum. The thing is too obvious to every one who understands dynamics.

Yours always truly,

KELVIN.

Later* Lord Kelvin returns to the subject in a paper based on fresh data for conductivities (cf. *supra*, p. 215) which had been sent to him from R. Weber, and on other estimates by Barus and Clarence King; and he sums up as follows:

"By the solution of the conductivity problem to which I have referred above, with specific heat increasing up to the melting point, as found by Rücker and Roberts-Austen and by Barus, but with the conductivity assumed constant, and by taking into account the augmentation of melting temperature with pressure

------

* *Nature*, March 7, 1895.

in a somewhat more complete manner than that adopted by Mr Clarence King, I am not led to differ much from his estimate of 24 million years. But, until we know something more than we know at present as to the probable diminution, or still conceivably possible augmentation, of thermal conductivity with increasing temperature, it would be quite uninteresting to publish any closer estimate." Cf. *supra*, p. 216.

Prof. Perry wound up the discussion in a paper* in which he reviewed the whole of the evidence of this and other kinds regarding the age of the Earth, and concluded as follows:

"To sum up, we can find no published record of any lower maximum age of life on the earth as calculated by physicists (I leave out the estimates based upon the assumption of uniform density in the sun, and also that of Mr Clarence King) than 400 million years. From the three physical arguments, Lord Kelvin's higher limits are 1000, 400, and 500 million years. I have shown that we have reasons for believing that the age, from all three, may be very considerably under-estimated. It is to be observed that if we exclude everything but the arguments from mere physics, the *probable* age of life on the earth is much less than any of the above estimates; but if the palæontologists have good reasons for demanding much greater times, I see nothing from the physicist's point of view which denies them four times the greatest of these estimates."

On Prof. Perry's suggestion the problem of the cooling of a uniform sphere covered by a badly conducting stratum was taken up in some detail by Mr O. Heaviside, whose discussion has been reprinted in his *Electromagnetic Theory*, Vol. II. 1899, Ch. I. pp. 12—28, to which reference may conveniently be made. He finds that in the examples selected by Prof. Perry the prolongation of the time happens to be not far from the maximum that is possible.

More recently this subject has entered into another phase, initiated by E. Rutherford and followed up by J. Joly and R. J. Strutt. It appears that the present flow of heat outwards could be entirely accounted for by supply due to the degradation of radium existing in the superficial strata, and that the escape of the original internal heat of consolidation may therefore be far

* *Nature*, April 18, 1895.

more slow than the temperature gradient downwards had previously indicated. And similar considerations would give support to a large prolongation of the estimated life of the Sun; for though his surface temperature, being dominated by radiation, is not greatly above what can be attained by refracting materials in our electric furnaces, yet the rapid increase beneath the surface must carry the conditions beyond our ken, so that copious supplies of heat arising from chemical degradation of unknown types may be continually arriving.

INDEX

CAMBRIDGE: PRINTED BY JOHN CLAY, M.A. AT THE UNIVERSITY PRESS.

Printed in the United States
By Bookmasters